Functional Ceramics Through Mechanochemical Activation

Functional Ceramics Through Mechanochemical Activation

Edited by
Ling Bing Kong
Shenzhen Technology University, Shenzhen, China

IOP Publishing, Bristol, UK

ISBN 978-0-7503-2191-4 (ebook)
ISBN 978-0-7503-2189-1 (print)
ISBN 978-0-7503-2192-1 (myPrint)
ISBN 978-0-7503-2190-7 (mobi)

DOI 10.1088/978-0-7503-2191-4

Version: 20191101

IOP ebooks

British Library Cataloguing-in-Publication Data: A catalogue record for this book is available from the British Library.

Published by IOP Publishing, wholly owned by The Institute of Physics, London

IOP Publishing, Temple Circus, Temple Way, Bristol, BS1 6HG, UK

US Office: IOP Publishing, Inc., 190 North Independence Mall West, Suite 601, Philadelphia, PA 19106, USA

Contents

Preface

Mechanochemical activation, which is also known as mechanical alloying, high-energy ball milling, high-energy activation and so on, has been used to synthesize various materials. It was originally developed to prepare metallic alloys dispersed with oxides in order to obtain composite materials with enhanced mechanical strengths. In the last few decades, the technology has been explored to synthesize oxide-based ceramic nanosized powders. Because it can replace the mixing step in the conventional ceramic process, while the calcination step is skipped, the production cycle of ceramics prepared in this way can be effectively shortened. In addition, the as-synthesized powders have particles at the nanometer scale, thus having higher sinterability. Furthermore, even unreacted powders can be used to fabricate ceramics, so that the fabrication time can be further reduced. In this case, the ceramics are formed through reactive sintering.

The purpose of this book is to offer an overview of the progress in the synthesis of powders and the fabrication of ceramics of selected materials, namely ferroelectrics, ferrite, mullite and so on, using mechanochemical activation. It consists of ten chapters, covering the topics of ferroelectric materials, ferrite ceramics with magneto-dielectric properties, and mullite ceramics with controllable microstructure and anisotropic grain growth behaviors. It is believed that this book can be used as a reference for senior undergraduate students, postgraduate students, researchers and engineers, in materials science and engineering, applied physics, solid-state lasers, solid-state physics, and so on.

Acknowledgments

The financial support from Shenzhen Technology University (SZTU) through a start-up grant (2018) and a grant from the Natural Science Foundation of Top Talent of SZTU (grant no. 2019010801002) is acknowledged.

Contributors

Ling Bing Kong
College of New Materials and New Energies, Shenzhen Technology University, Shenzhen, 518118, Guangdong, China

Zhuohao Xiao
School of Materials Science and Engineering, Jingdezhen Ceramic Institute, Jingdezhen, 333001, Jiangxi, China

Xiuying Li
School of Materials Science and Engineering, Jingdezhen Ceramic Institute, Jingdezhen, 333001, Jiangxi, China

Shijin Yu
School of Mechanical and Electronic Engineering, Jingdezhen Ceramic Institute, Jingdezhen 333001, Jiangxi, China

Wenxiu Que
Electronic Materials Research Laboratory, School of Electronic and Information Engineering, Xi'an Jiaotong University, Xi'an 710049, Shaanxi, China

Yin Liu
School of Science and Engineering, Anhui University of Science & Technology, Huainan 232001, Anhui, China

Tianshu Zhang
School of Science and Engineering, Anhui University of Science & Technology, Huainan 232001, Anhui, China

Kun Zhou
School of Mechanical & Aerospace Engineering, Nanyang Technological University, 50 Nanyang Avenue, Singapore 639798

Hongfang Zhang
Department of Physics, Suzhou University of Science and Technology, Suzhou 215009, China

Functional Ceramics Through Mechanochemical Activation

Ling Bing Kong

Chapter 1

Introduction

Ling Bing Kong, Zhuohao Xiao, Xiuying Li, Shijin Yu, Wenxiu Que, Yin Liu, Tianshu Zhang, Kun Zhou and Hongfang Zhang

1.1 Brief history

Mechanochemical activation is also called mechanical alloying, high-energy mechanical milling, high-energy ball milling, high-energy activation and so on, and these terms are not differentiated throughout this book, unless otherwise indicated. This method was originally developed to obtain oxide-dispersed metallic alloys that have enhanced mechanical performance [1–16]. After that, the strategy was employed to extend metallic solid solubility, synthesize intermetallics, increase disordering of intermetallics, trigger solid-state amorphization, create nanostructured materials, and directly prepare oxide and metal nanosized powders [17–32]. Only oxide-based materials will be covered in this book. The applications of mechanochemical activation can be classified into four groups: (i) size reduction or refinement of oxides [33–41]; (ii) production of the oxide phase through displacement reactions or reactions similar to chemical precipitation [42–57]; (iii) direct phase formation [58–86]; and (iv) enhancement of reactivity [87–107].

The most distinctive characteristic of mechanochemical activation is that the phase formation or reaction is facilitated by mechanical energy, instead of thermal energy which is needed in the conventional solid-state reaction and most of the wet-chemical synthetic methods. In this regard, mechanochemical activation is more advantageous over the conventional approaches because the calcination step is skipped. Therefore, the fabrication process can be shortened compared to conventional ceramic processing. Moreover, cheap precursors, such as oxides and carbides, can be directly used, thus leading to cost-effectiveness. Also, because the reaction occurs at room temperature and the vials are well-sealed, the volatile components are effectively prevented from escaping, such as lead (Pb), bismuth (Bi) and lithium (Li) [108].

doi:10.1088/978-0-7503-2191-4ch1

1.2 Organization of the book

In chapter 2, the working principles and characteristics of various high-energy milling devices will be qualitatively described, while the materials that have been synthesized or processed using high-energy mechanochemical activation will be reported in detail. Specifically, ferroelectric materials will be presented in chapters 3–5, with a focus on phase formation of the powders, as well as the properties of the final ceramics for some samples. In chapters 6 and 7, ferrite ceramics processed using mechanochemical activation will be descibed, with emphasis on their dielectric and magnetic properties. These special materials are known as magneto-dielectric materials, with equal or close values of real permeability and permittivity over a certain frequency range. They also have sufficiently low magnetic and dielectric loss tangents that they can be used for miniaturization of specific antennae. In chapters 8 and 9, mullite based ceramics, with variable microstructures, grain morphologies and densification behaviors, will be described. The effects of milling media and dopants will be discussed. Finally, several other oxides powders and their corresponding ceramics, developed using mechanochemical activation, will be summarized in chapter 10.

Acknowledgments

Shenzhen Technology University (SZTU) is acknowledged for the financial support of a start-up grant (2018) and also the Natural Science Foundation of Top Talent of SZTU (grant no. 2019010801002).

References

[1] Czyrska-Filemonowicz A and Dubiel B 1997 Mechanically alloyed, ferritic oxide dispersion strengthened alloys: structure and properties *J. Mater. Process. Technol.* **64** 53–64

[2] Grahle P and Arzt E 1997 Microstructural development in dispersion strengthened NiAl produced by mechanical alloying and secondary recrystallization *Acta Mater.* **45** 201–11

[3] Welham N J, Willis P E and Kerr T 2000 Mechanochemical formation of metal–ceramic composites *J. Am. Ceram. Soc.* **83** 33–40

[4] Ryu H J and Hong S H 2003 Fabrication and properties of mechanically alloyed oxide-dispersed tungsten heavy alloys *Mater. Sci. Eng.* A **363** 179–84

[5] Radev D D 2010 Mechanical synthesis of nanostructured titanium–nickel alloys *Adv. Powder Technol.* **21** 477–82

[6] Kubaski E T, Cintho O M and Capocchi J D T 2011 Effect of milling variables on the synthesis of NiAl intermetallic compound by mechanical alloying *Powder Technol.* **214** 77–82

[7] Sicre-Artalejo J, Campos M, Molina-Aldareguia J M and Torralba J M 2011 Quantification of hardening in Fe–Mn master alloys prepared by a mechanical alloying process via nanoindentation experiments *J. Mater. Res.* **26** 1726–33

[8] Azabou M, Ibn Gharsallah H, Escoda L, Sunol J J, Kolsi A W and Khitouni M 2012 Mechanochemical reactions in nanocrystalline Cu–Fe system induced by mechanical alloying in air atmosphere *Powder Technol.* **224** 338–44

[9] Yelsukov E P, Ul'yanov A L, Protasov A V and Kolodkin D A 2012 Solid-state reactions upon mechanical alloying of an $Fe_{32}Al_{68}$ binary mixture *Phys. Met. Metall.* **113** 602–11
[10] Zadorozhnyi M Y, Kaloshkin S D, Klyamkin S N, Bermesheva O V and Zadorozhnyi V Y 2013 Mechanochemical synthesis of a TiFe nanocrystalline intermetallic compound and its mechanical alloying with third component *Met. Sci. Heat Treat.* **54** 461–65
[11] Zadorozhnyy V Y, Klyamkin S N, Zadorozhnyy M Y, Bermesheva O V and Kaloshkin S D 2014 Mechanical alloying of nanocrystalline intermetallic compound TiFe doped by aluminum and chromium *J. Alloys Compd.* **586** S56–60
[12] Dobrovolsky V D, Ershova O H and Solonin Y M 2016 Thermal resistance and the kinetics of hydrogen desorption from hydrides of the Mg–Al–Ni–Ti mechanical alloy *Mater. Sci.* **51** 457–64
[13] Zadorozhny V Y *et al* 2017 Preparation and hydrogen storage properties of nanocrystalline TiFe synthesized by mechanical alloying *Prog. Nat. Sci.-Mater. Inter.* **27** 149–55
[14] Zhao Z Q, Xiao Z, Li Z, Zhu M N and Yang Z Q 2017 Characterization of dispersion strengthened copper alloy prepared by internal oxidation combined with mechanical alloying *J. Mater. Eng. Perform.* **26** 5641–47
[15] Ulbrich K F and Campos C E M 2018 Nanosized tetragonal beta-FeSe phase obtained by mechanical alloying: structural, microstructural, magnetic and electrical characterization *RSC Adv.* **8** 8190–98
[16] Takacs L 2013 The historical development of mechanochemistry *Chem. Soc. Rev.* **42** 7649–59
[17] Weeber A W and Bakker H 1988 Amorphization by ball milling—a review *Physica* B **153** 93–135
[18] Hong L B and Fultz B 1996 Two-phase coexistence in Fe–Ni alloys synthesized by ball milling *J. Appl. Phys.* **79** 3946–55
[19] Koch C C and Whittenberger J D 1996 Mechanical milling/alloying of intermetallics *Intermetallics* **4** 339–55
[20] Murty B S and Ranganathan S 1998 Novel materials synthesis by mechanical alloying/milling *Int. Mater. Rev.* **43** 101–41
[21] Suryanarayana C 2001 Mechanical alloying and milling *Prog. Mater Sci.* **46** 1–184
[22] Suryanarayana C, Ivanov E and Boldyrev V V 2001 The science and technology of mechanical alloying *Mater. Sci. Eng.* A **304** 151–58
[23] Zhang D L 2004 Processing of advanced materials using high-energy mechanical milling *Prog. Mater Sci.* **49** 537–60
[24] Li S B and Zhai H X 2005 Synthesis and reaction mechanism of Ti_3SiC_2 by mechanical alloying of elemental Ti, Si, and C powders *J. Am. Ceram. Soc.* **88** 2092–98
[25] Yagodkin Y D, Minakova S M, Ketov S V, Glebov V A, Nefedov V S and Popova O I 2005 Nanocrystalline alloys of the Nd–Fe–B system obtained by mechanical treatment *Met. Sci. Heat Treat.* **47** 467–69
[26] Garroni S, Delogu F, Mulas G and Cocco G 2007 Mechanistic inferences on the synthesis of $Co_{50}Fe_{50}$ solid solution by mechanical alloying *Scr. Mater.* **57** 964–67
[27] Sani R and Beitollahi A 2008 Phase evolution and magnetic properties of Co/α–Fe_2O_3 powder mixtures with different molar ratios treated by mechanical alloying *J. Non-Cryst. Solids* **354** 4635–43
[28] Anvari S Z, Karimzadeh F and Enayati M H 2009 Synthesis and characterization of NiAl–Al_2O_3 nanocomposite powder by mechanical alloying *J. Alloys Compd.* **477** 178–81

[29] Forouzanmehr N, Karimadeh F and Enayati M H 2009 Synthesis and characterization of TiAl/α–Al_2O_3 nanocomposite by mechanical alloying *J. Alloys Compd.* **478** 257–59
[30] Mostaan H, Abbasi M H and Karimzadeh F 2010 Mechanochemical assisted synthesis of Al_2O_3/Nb nanocomposite by mechanical alloying *J. Alloys Compd.* **493** 609–12
[31] Zhu S G, Wu C X and Luo Y L 2010 Effects of stearic acid on synthesis of nanocomposite WC–MgO powders by mechanical alloying *J. Mater. Sci.* **45** 1817–22
[32] Enayati M H and Mohamed F A 2014 Application of mechanical alloying/milling for synthesis of nanocrystalline and amorphous materials *Int. Mater. Rev.* **59** 394–416
[33] Sen S, Ram M L, Roy S and Sarkar B K 1999 The structural transformation of anatase TiO_2 by high-energy vibrational ball milling *J. Mater. Res.* **14** 841–48
[34] Ren R M, Yang Z G and Shaw L L 2000 Polymorphic transformation and powder characteristics of TiO_2 during high energy milling *J. Mater. Sci.* **35** 6015–26
[35] Indris S *et al* 2005 Preparation by high-energy milling, characterization, and catalytic properties of nanocrystalline TiO_2 *J. Phys. Chem.* B **109** 23274–78
[36] Coste S, Bertrand G, Coddet C, Gaffet E, Hahn H and Sieger H 2007 High-energy ball milling of Al_2O_3–TiO_2 powders *J. Alloys Compd.* **434** 489–92
[37] Ali M 2014 Transformation and powder characteristics of TiO_2 during high energy milling *J. Ceram. Proces. Res.* **15** 290–93
[38] Petrovic S *et al* 2019 Effect of high energy ball milling on the physicochemical properties of TiO_2–CeO_2 mixed oxide and its photocatalytic behavior in the oxidation reaction *Reac. Kin. Mechan. Catal.* **127** 175–86
[39] Jiang J Z, Poulsen F W and Morup S 1999 Structure and thermal stability of nano-structured iron-doped zirconia prepared by high-energy ball milling *J. Mater. Res.* **14** 1343–52
[40] Jiang J Z, Wynn P, Morup S, Okada T and Berry F J 1999 Magnetic structure evolution in mechanically milled nanostructured $ZnFe_2O_4$ particles *Nanostruct. Mater.* **12** 737–40
[41] Xiao Z H *et al* 2018 Sintering and electrical properties of commercial PZT powders modified through mechanochemical activation *J. Mater. Sci.* **53** 13769–778
[42] Yang H M, Hu Y H, Tang A D, Jin S M and Qiu G Z 2004 Synthesis of tin oxide nanoparticles by mechanochemical reaction *J. Alloys Compd.* **363** 271–4
[43] McCormick P G, Tsuzuki T, Robinson J S and Ding J 2001 Nanopowders synthesized by mechanochemical processing *Adv. Mater.* **13** 1008
[44] Cai S, Tsuzuki T, Fisher T A, Nener B D, Dell J M and McCormick P G 2002 Mechanochemical synthesis and characterization of GaN nanocrystals *J. Nanopart. Res.* **4** 367–71
[45] Cukrov L M, McCormick P G, Galatsis K and Wlodarski W 2001 Gas sensing properties of nanosized tin oxide synthesised by mechanochemical processing *Sens. Actuat.* B **77** 491–95
[46] Cukrov L M, Tsuzuki T and McCormick P G 2001 SnO_2 nanoparticles prepared by mechanochemical processing *Scr. Mater.* **44** 1787–90
[47] Dodd A C and McCormick P G 2001 Solid-state chemical synthesis of nanoparticulate zirconia *Acta Mater.* **49** 4215–20
[48] Dodd A C and McCormick P G 2001 Synthesis of nanoparticulate zirconia by mechanochemical processing *Scr. Mater.* **44** 1725–29
[49] Dodd A C and McCormick P G 2002 Synthesis of nanocrystalline ZrO_2 powders by mechanochemical reaction of $ZrCl_4$ with LiOH *J. Eur. Ceram. Soc.* **22** 1823–29

[50] Dodd A C, Raviprasad K and McCormick P G 2001 Synthesis of ultrafine zirconia powders by mechanochemical processing *Scr. Mater.* **44** 689–94
[51] Dodd A C, Tsuzuki T and McCormick P G 2001 Nanocrystalline zirconia powders synthesised by mechanochemical processing *Mater. Sci. Eng.* A **301** 54–8
[52] Hos J P and McCormick P G 2003 Mechanochemical synthesis and characterisation of nanoparticulate samarium-doped cerium oxide *Scr. Mater.* **48** 85–90
[53] Muroi M, Street R, McCormick P G and Amighian J 2001 Magnetic properties of ultrafine $MnFe_2O_4$ powders prepared by mechanochemical processing *Phys. Rev.* B **63** 184414
[54] Tsuzuki T and McCormick P G 2000 Synthesis of Cr_2O_3 nanoparticles by mechanochemical processing *Acta Mater.* **48** 2795–801
[55] Tsuzuki T and McCormick P G 2001 Mechanochemical synthesis of niobium pentoxide nanoparticles *Mater. Trans.* **42** 1623–28
[56] Tsuzuki T and McCormick P G 2001 Synthesis of ultrafine ceria powders by mechanochemical processing *J. Am. Ceram. Soc.* **84** 1453–58
[57] Tsuzuki T and McCormick P G 2001 ZnO nanoparticles synthesised by mechanochemical processing *Scr. Mater.* **44** 1731–34
[58] Sepelak V, Duvel A, Wilkening M, Becker K D and Heitjans P 2013 Mechanochemical reactions and syntheses of oxides *Chem. Soc. Rev.* **42** 7507–20
[59] Kong L B, Ma J, Huang H and Zhang R F 2002 $(1-x)$PZN-xBT ceramics derived from mechanochemically synthesized powders *Mater. Res. Bull.* **37** 1085–92
[60] Kong L B, Ma J, Huang H T, Zhu W and Tan O K 2001 Lead zirconate titanate ceramics derived from oxide mixture treated by a high-energy ball milling process *Mater. Lett.* **50** 129–33
[61] Kong L B, Ma J, Zhang T S, Zhu W and Tan O K 2001 $Pb(Zr_xTi_{1-x})O_3$ ceramics via reactive sintering of partially reacted mixture produced by a high-energy ball milling process *J. Mater. Res.* **16** 1636–43
[62] Kong L B, Ma J, Zhang T S, Zhu W and Tan O K 2002 Preparation of antiferroelectric lead zirconate titanate stannate ceramics by high-energy ball milling process *J. Mater. Sci. Mater. Elec.* **13** 89–94
[63] Kong L B, Ma J, Zhu W and Tan O K 2001 Reaction sintering of partially reacted system for PZT ceramics via a high-energy ball milling *Scr. Mater.* **44** 345–50
[64] Kong L B, Ma J, Zhu W and Tan O K 2001 Preparation and characterization of PLZT ceramics using high-energy ball milling *J. Alloys Compd.* **322** 290–97
[65] Kong L B, Ma J, Zhu W and Tan O K 2001 Preparation of PMN powders and ceramics via a high-energy ball milling process *J. Mater. Sci. Lett.* **20** 1241–43
[66] Kong L B, Ma J, Zhu W and Tan O K 2002 Transparent PLZT8/65/35 ceramics from constituent oxides mechanically modified by high-energy ball milling *J. Mater. Sci. Lett.* **21** 197–99
[67] Kong L B, Ma J, Zhu W and Tan O K 2002 Preparation and characterization of PLZT (8/65/35) ceramics via reaction sintering from ball milled powders *Mater. Lett.* **52** 378–87
[68] Kong L B, Ma J, Zhu W and Tan O K 2002 Translucent PMN and PMN–PT ceramics from high-energy ball milling derived powders *Mater. Res. Bull.* **37** 23–32
[69] Kong L B, Ma J, Zhu W and Tan O K 2002 Preparation of PMN–PT ceramics via a high-energy ball milling process *J. Alloys Compd.* **336** 242–46
[70] Kong L B, Zhu W and Tan O K 2000 $PbTiO_3$ ceramics derived from high-energy ball milled nano-sized powders *J. Mater. Sci. Lett.* **19** 1963–66

[71] Ang S K, Wang J, Wang D M, Xue J M and Li L T 2000 Mechanical activation-assisted synthesis of $Pb(Fe_{2/3}W_{1/3})O_3$ *J. Am. Ceram. Soc.* **83** 1575–80
[72] Ang S K, Wang J and Xue J M 2000 Phase stability and dielectric properties of $(1-x)$PFW $+x$PZN derived from mechanical activation *Sol. State Ion* **127** 285–93
[73] Ang S K, Xue J M and Wang J 2000 Mechanical activation and dielectric properties of 0.48PFN–0.36PFW–0.16PZN from mixed oxides *J. Alloys Compd.* **311** 181–87
[74] Khim A S, Wang J and Xue J M 2000 Mechanical activation synthesis and dielectric properties of 0.48PFN–0.36PFW–0.16PZN from mixed oxides *J. Alloys Compd.* **311** 181–87
[75] Lee S E, Xue J M, Wan D M and Wang J 1999 Effects of mechanical activation on the sintering and dielectric properties of oxide-derived PZT *Acta Mater.* **47** 2633–39
[76] Soon H P, Xue J M and Wang J 2004 Dielectric behaviors of $Pb_{1-3x/2}La_xTiO_3$ derived from mechanical activation *J. Appl. Phys.* **95** 4981–88
[77] Tan Y L, Xue J M and Wang J 2000 Stablization of perovskite phase and dielectric properties of 0.95PZN–0.05BT derived from mechanical activation *J. Alloys Compd.* **297** 92–8
[78] Wan D M, Xue J M and Wang J 2000 Nanocrystalline 0.54PZN–0.36PMN–0.1PT of perovskite structure by mechanical activation *Mater. Sci. Eng.* A **286** 96–100
[79] Wan D M, Xue J M and Wang J 2000 Mechanochemical synthesis of $0.9[0.6Pb(Zn_{1/3}Nb_{2/3})O_3{-}0.4Pb(Mg_{1/3}Nb_{2/3})O_3]{-}0.1PbTiO_3$ *J. Am. Ceram. Soc.* **83** 53–9
[80] Wang J, Xue J M and Wan D M 2000 How different is mechanical activation from thermal activation? A case study with PZN and PZN-based relaxors *Sol. State Ion* **127** 169–75
[81] Wang J, Xue J M, Wan D M and Gan B K 2000 Mechanically activating nucleation and growth of complex perovskites *J. Sol. State Chem.* **154** 321–28
[82] Xue J M, Wan D M, Lee S E and Wang J 1999 Mechanochemical synthesis of lead zirconate titanate from mixed oxides *J. Am. Ceram. Soc.* **82** 1687–92
[83] Xue J M, Wan D M and Wang J 1999 Mechanochemical synthesis of nanosized lead titanate powders form mixed oxides *Mater. Lett.* **39** 364–69
[84] Xue J M, Wang J and Rao T M 2001 Synthesis of $Pb(Mg_{1/3}Nb_{2/3})O_3$ in excess lead oxide by mechanical activation *J. Am. Ceram. Soc.* **84** 660–62
[85] Xue J M, Wang J and Wan D M 2000 Nanosized barium titanate powder by mechanical activation *J. Am. Ceram. Soc.* **83** 232–34
[86] Xue J M, Wang J and Weiseng T 2000 Synthesis of lead zirconate titanate from an amorphous precursor by mechanical activation *J. Alloys Compd.* **308** 139–46
[87] Kong L B, Ma J and Huang H 2002 $MgAl_2O_4$ spinel phase derived from oxide mixture activated by a high-energy ball milling process *Mater. Lett.* **56** 238–43
[88] Kong L B, Ma J and Huang H 2002 Low temperature formation of yttrium aluminum garnet from oxides via a high-energy ball milling process *Mater. Lett.* **56** 344–48
[89] Kong L B, Ma J, Huang H and Zhang R F 2002 Crystallization of magnesium niobate from mechanochemically derived amorphous phase *J. Alloys Compd.* **340** L1–4
[90] Kong L B, Ma J, Huang H, Zhang R F and Que W X 2002 Barium titanate derived from mechanochemically activated powders *J. Alloys Compd.* **337** 226–30
[91] Kong L B, Ma J, Huang H, Zhang R F and Zhang T S 2002 Zinc niobate derived from mechanochemically activated oxides *J. Alloys Compd.* **347** 308–13
[92] Kong L B, Li Z W, Lin G Q and Gan Y B 2007 Electrical and magnetic properties of magnesium ferrite ceramics doped with Bi_2O_3 *Acta Mater.* **55** 6561–72

[93] Kong L B, Li Z W, Lin G Q and Gan Y B 2007 Magneto-dielectric properties of Mg–Cu–Co ferrite ceramics: I. Densification behavior and microstructure development *J. Am. Ceram. Soc.* **90** 3106–12
[94] Kong L B, Li Z W, Lin G Q and Gan Y B 2007 Magneto-dielectric properties of Mg–Cu–Co ferrite ceramics: II. Electrical, dielectric, and magnetic properties *J. Am. Ceram. Soc.* **90** 2104–12
[95] Kong L B, Li Z W, Lin G Q and Gan Y B 2008 $Mg_{1-x}Co_xFe_{1.98}O_4$ ceramics with promising magnetodielectric properties for antenna miniaturization *IEEE Trans. Magn.* **44** 559–65
[96] Kong L B, Teo M L S, Li Z W, Lin G Q and Gan Y B 2008 Development of magneto-dielectric materials based on Li-ferrite ceramics—III. Complex relative permeability and magneto-dielectric properties *J. Alloys Compd.* **459** 576–82
[97] Liew X T, Chan K C and Kong L B 2009 Magnetodielectric Ni ferrite ceramics with Bi_2O_3 additive for potential antenna miniaturizations *J. Mater. Res.* **24** 324–32
[98] Teo M L S, Kong L B, Li Z W, Lin G Q and Gan Y B 2008 Development of magneto-dielectric materials based on Li-ferrite ceramics—I. Densification behavior and microstructure development *J. Alloys Compd.* **459** 557–66
[99] Teo M L S, Kong L B, Li Z W, Lin G Q and Gan Y B 2008 Development of magneto-dielectric materials based on Li-ferrite ceramics—II. DC resistivity and complex relative permittivity *J. Alloys Compd.* **459** 567–75
[100] Kong L B, Gan Y B, Ma J, Zhang T S, Boey F and Zhang R F 2003 Mullite phase formation and reaction sequences with the presence of pentoxides *J. Alloys Compd.* **351** 264–72
[101] Kong L B *et al* 2003 Growth of mullite whiskers in mechanochemically activated oxides doped with WO_3 *J. Eur. Ceram. Soc.* **23** 2257–64
[102] Kong L B, Ma J and Huang H 2002 Mullite whiskers derived from an oxide mixture activated by a mechanochemical process *Adv. Eng. Mater.* **4** 490–94
[103] Kong L B, Zhang T S, Chen Y Z, Ma J, Boey F and Huang H 2004 Microstructural composite mullite derived from oxides via a high-energy ball milling process *Ceram. Int.* **30** 1313–17
[104] Kong L B, Zhang T S, Ma J and Boey F 2003 Anisotropic grain growth of mullite in high-energy ball milled powders doped with transition metal oxides *J. Eur. Ceram. Soc.* **23** 2247–56
[105] Kong L B, Zhang T S, Ma J and Boey F Y C 2009 Mullitization behavior and microstructural development of B_2O_3–Al_2O_3–SiO_2 mixtures activated by high-energy ball milling *Sol. State Sci.* **11** 1333–42
[106] Zhang T S, Kong L B, Du Z H, Ma J and Li S 2010 *In situ* interlocking structure in gel-derived mullite matrix induced by mechanoactivated commercial mullite powders *Scr. Mater.* **63** 1132–35
[107] Zhang T S, Kong L B, Du Z H, Ma J and Li S 2010 Tailoring the microstructure of mechanoactivated Al_2O_3 and SiO_2 mixtures with TiO_2 addition *J. Alloys Compd.* **506** 777–83
[108] Harris J R, Wattis J A D and Wood J V 2001 A comparison of different models for mechanical alloying *Acta Mater.* **49** 3991–4003

IOP Publishing

Functional Ceramics Through Mechanochemical Activation

Ling Bing Kong

Chapter 2

Principles of mechanochemical activation

Ling Bing Kong, Zhuohao Xiao, Xiuying Li, Shijin Yu, Wenxiu Que, Yin Liu, Tianshu Zhang, Kun Zhou and Hongfang Zhang

2.1 High-energy mechanochemical activation

Comprehensive review articles on the applications of high-energy activations or ball milling processes for the synthesis of various metallic materials have been available in the literature [1–7]. Although the experimental details in reports are different, it is necessary to briefly describe the types of devices, properties and applications in a more general manner. There are different high-energy milling facilities that have been utilized to synthesize different materials [4]. For different milling machines, both the principles and the production capabilities are different. The production capabilities of high-energy synthetic processes could be from several grams to as high as thousands of kilograms. High-energy milling machines that have been widely adopted for research purposes include mainly vibrational shake mills (SPEX), planetary mills and attrition mills. Other types of devices, like multi-ring-type mills (Model MICROS: MIC-0, Nara Machinery, Tokyo, Japan), are also utilized in the mechanochemical production of several ferroelectric materials [8]. Stainless steel and tungsten carbide milling media have been commonly employed in experiments on high-energy milling processes, with the tungsten carbide media able to provide higher energies due to their higher density.

2.1.1 Vibrational shake mills

The SPEX vibrational shake mill is among of the most widely used devices in the mechanochemical synthesis research community. The common SPEX shaker consists of one vial, containing the sample powders and grinding balls, secured with a clamp. The system experiences a strong swing movement back and forth at speeds of up to thousands of rpm. The back-and-forth shaking movement is accompanied by a lateral motion at the end of the vial. As a result, the vial travels following a figure-of-eight or infinity sign. As the vial swings, the balls impact the powders to be activated and the end of the vial, so that the powder is thoroughly mixed initially and then activated. Due to the large amplitude (~5 cm) and the high speed (~1200 rpm) of the clamp motion, the balls move at very high velocities (e.g. on the order of 5 m s^{-1}). As a consequence, the force applied to the powder by the balls is very large, so that it is called high-energy milling or activation. However, this type of mill can only treat a relatively small quantity in each batch. To address this issue, two-vial devices have been designed [1]. Another brand of shake mill is Fritsch Pulverizette 0, supplied by Fritsch GmbH, Germany. Both types of mills, SPEX 8000 [9–26] and Fritsch Pulverizette 0 [27–36], have been employed to synthesize a wide range of nanosized powders. Photographs of representative SPEX mills and vials/balls are shown in figures 2.1–2.3 (adopted from website). Usually, the vial has a cylindrical shape, with both a diameter and length of 40 mm, while the milling balls are 12.7 mm in diameter. The speed is up to 900 rpm.

2.1.2 Planetary ball mills

Planetary ball milling machines have higher production capacities than the SPEX mills, because larger vials can be used. In these mills, the vials are fixed on a rotating support disk, while they rotate around their own axes following a special drive mechanism. The centrifugal forces generated by the vial that rotates around its own axis and the rotating support disk are applied to the powder in the vials. Therefore, the powder is effectively ground due the moving balls. Because the vial and the supporting disk rotate in opposite directions, centrifugal forces are created alternately in the two directions. In addition, the grinding balls run down the inside wall of the vial, thus leading to friction effects, so that the powder is ground. Also, the grinding balls travel freely inside the vial and collide with the powder, producing impact effects. When milling media with high densities are used, such as stainless steels and tungsten carbides, high-energy ball milling or activation can be achieved using planetary mills. The most popular planetary milling machines that are utilized in the literature are supplied by Retsch and Fritsch in Germany, as illustrated in figures 2.4 and 2.5, respectively. Various nanosized powders have been prepared using different types of planetary mills [37–57].

Figure 2.1. Photograph of an SPEX 8000 type shaker mill (old version).

Figure 2.2. Photographs of an SPEX 8000D type shaker mill (new version).

Figure 2.3. Photographs of the stainless steel vial and balls used with SPEX 8000 type shaker mills.

2.1.3 Attritor mills

Comparatively, the use of attritor mills to synthesize nanosized materials is not very popular. This is simply because they have relatively large production capacities, while a very small quantity is used in research activities. The production capacities

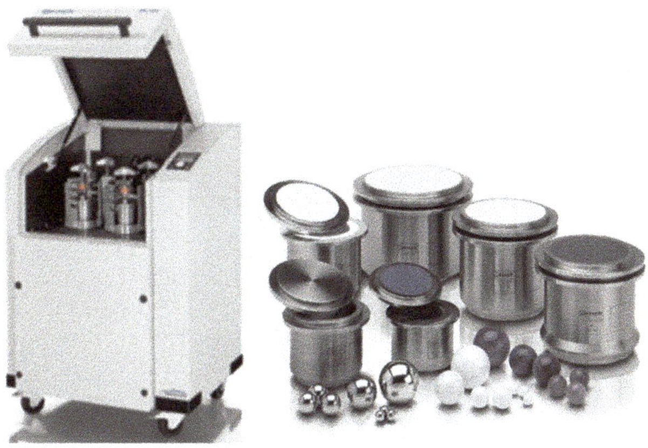

Figure 2.4. Photograph of a Retsch PM 400 planetary mill, with different types of vials and balls. (Adopted from the company website).

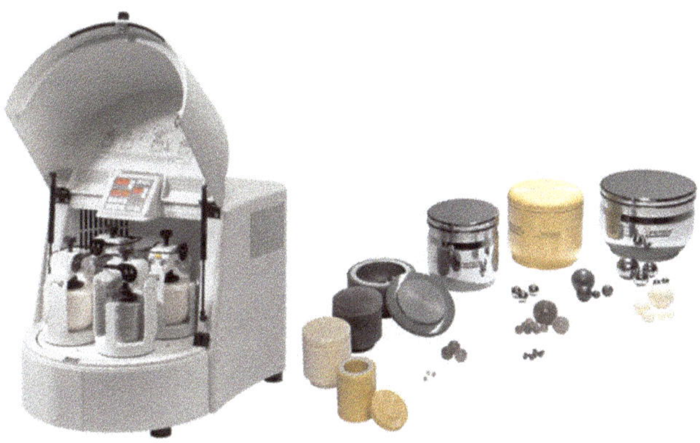

Figure 2.5. Photograph of a Fritsch Pulverisette 5/4 type planetary mill, with different types of vials and balls (adopted from the company website).

of attrition mills could be in the range of 0.5–40 kg. A typical attrition ball mill has a rotating horizontal drum, which is usually filled with small balls to its half volume. As the drum is rotated, the balls will drop repeatedly on the powders to be milled. Therefore, the efficiency of the grinding is increased with increasing rotating speed. However, when the rotating movement is too fast, the centrifugal force applying to the flying balls will be larger than that of gravity, and the balls are pinned to the walls of the drum. As a consequence, the grinding effect vanishes. An attrition mill is composed of a vertical drum with a series of impellers inside. Strong impacts occur between the balls and container walls, and between the balls, agitator shaft and impellers, resulting in refinement and activation of the powder. Figure 2.6 shows a photograph of a typical attrition mill and a schematic diagram of the mill configuration.

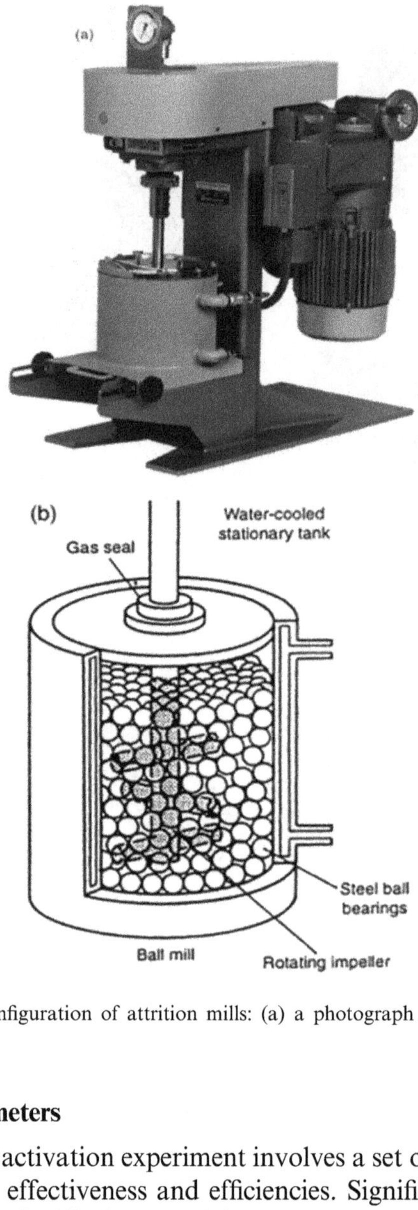

Figure 2.6. Structure and configuration of attrition mills: (a) a photograph of a specific device and (b) a schematic diagram.

2.1.4 Processing parameters

The mechanochemical activation experiment involves a set of parameters that could be used to control the effectiveness and efficiencies. Significant processing parameters include the types of mill, the materials selected for the milling vials and balls, milling speeds, milling time, ball-to-powder weight ratios, milling environmental conditions, process control agents (PCA), temperature control and application of electrical or magnetic fields during the milling. There are actually unlimited combinations of these processing parameters.

Usually, if an SPEX shaker mill is used, 5 g starting powder is loaded. The milling time is set according to the properties of the materials. For planetary mills, as the vial has a volume of 250 ml, the quantity of the powder could be up to 100 g. If 100 tungsten carbide balls with a diameter of 10 mm are used, and 20 g of powder is milled, the ball-to-powder weight ratio is about 40:1. Typically, the milling speed is 200 rpm. The milling is usually stopped for a short period after milling for a certain amount of time, for example 5 min stopping for every 25 min milling, in order to prevent overheating of the system. In most studies, the temperatures are not intentionally controlled. Although it is still difficult to monitor the temperature of the materials during the high-energy activation, the commonly accepted conclusion is that the temperatures must be much lower than the calcination temperatures adopted in the conventional solid-state reaction processes.

2.2 Modeling and simulations

Although there has not been a general model than can be used to predict the milling effect of mechanochemical activation, various works have been reported independently to model or simulate the milling processes of different machines under different conditions [58–60].

In a systematic study, five potential models to describe the particle sizes and the size distributions of the powders during the mechanochemical activation process were analyzed [61]. The models were based on Smoluchowski's coagulation–fragmentation equations, and aggregation and fragmentation rates, which were dependent the particle sizes. In addition, a strategy was proposed to identify the modeling parameters based on the experimental data of average particle size. The size distribution profiles modeled with the selected parameters were in good agreement with the experimental results. In addition, some of the aggregation and fragmentation rates were qualitatively linked to the properties of the starting materials. It was found that the parameter matching was not very effective, which could be improved through a process of trial and error.

The kinetics of batch millings were established to describe the milling process in terms of energy instead of time [62]. The proposed models are effective to various processes in a mill, such as particle refinement, mechanochemical activated reaction, phase transformation and so on. The model was similar to the classical chemical kinetics with time and temperature replaced by energy and milling intensity, respectively. The intensity was defined for mills with loose ball media, including tumbling mills, vibrating mills, planetary mills and stirring mills. The general rate constant was derived as the temperature-dependent conventional rate constant was timed with the intensity-dependent factor of energetic efficiency. The energetic efficiency was determined, which was compared with experimental results. The energy-scaled model of milling kinetics could be used as a guide to generalize experimental data collected from different mills under different milling conditions.

2.3 Concluding remarks

Due to the wide range of materials, diverse equipment/facilities and different processing parameters, it is a huge challenge to develop a general model that can be used to describe all the experimental data. However, it is highly recommended that the experimental parameters should be stated in as much detail as possible, no matter which material is being studied or which type of milling equipment is used. The fields of modeling and simulation are far behind the progress in experimental works, and thus much more attention needs to be paid to the former.

Acknowledgments

Shenzhen Technology University (SZTU) is acknowledged for the financial support of a start-up grant (2018) and also the Natural Science Foundation of Top Talent of SZTU (grant no. 2019010801002).

References

[1] Suryanarayana C 2001 Mechanical alloying and milling *Prog. Mater. Sci.* **46** 1–184

[2] Weeber A W and Bakker H 1988 Amorphization by ball milling—a review *Physica* B **153** 93–135

[3] Zhang D L 2004 Processing of advanced materials using high-energy mechanical milling *Prog. in Mater. Sci.* **49** 537–60

[4] Suryanarayana C, Ivanov E and Boldyrev V V 2001 The science and technology of mechanical alloying *Mater. Sci. Eng.* A **304** 151–8

[5] Jeon J H 2014 Mechanochemical synthesis and mechanochemical activation-assisted synthesis of alkaline niobate-based lead-free piezoceramic powders *Curr. Opin. Chem. Eng.* **3** 30–5

[6] Qu J, Zhang Q W, Li X W, He X M and Song S X 2016 Mechanochemical approaches to synthesize layered double hydroxides: a review *App. Cla. Sci.* **119** 185–92

[7] Kong L B, Zhang T S, Ma J and Boey F 2008 Progress in synthesis of ferroelectric ceramic materials via high-energy mechanochemical technique *Prog. Mater. Sci.* **53** 207–322

[8] Baek J G, Isobe T and Senna M 1997 Synthesis of pyrochlore-free $0.9Pb(Mg_{1/3}Nb_{2/3})O_3$–$0.1PbTiO_3$ ceramics via a soft mechanochemical route *J. Am. Cer. Soc.* **80** 973–81

[9] Xue J M, Wan D M and Wang J 1999 Mechanochemical synthesis of nanosized lead titanate powders form mixed oxides *Mater. Lett.* **39** 364–9

[10] Xue J M, Wan D M, Lee S E and Wang J 1999 Mechanochemical synthesis of lead zirconate titanate from mixed oxides *J. Am. Ceram. Soc.* **82** 1687–92

[11] Xue J M, Tan Y L, Wan D M and Wang J 1999 Synthesizing 0.9PZN–0.1BT by mechanically activating mixed oxides *Sol. Stat. Ion.* **120** 183–8

[12] Xue J M, Wan D M and Wang J 2002 Functional ceramics of nanocrystallinity by mechanical activation *Sol. Stat. Ion.* **151** 403–12

[13] Xue J M, Wang J, Ng W and Wang D 1999 Activation-induced prychlore-to-perovskite conversion for a lead magnesium niobate precursor *J. Am. Ceram. Soc.* **82** 2282–4

[14] Xue J M, Wang J and Rao T M 2001 Synthesis of $Pb(Mg_{1/3}Nb_{2/3})O_3$ in excess lead oxide by mechanical activation *J. Am. Ceram. Soc.* **84** 660–2

[15] Xue J M, Wang J and Wan D M 2000 Nanosized barium titanate powder by mechanical activation *J. Am. Ceram. Soc.* **83** 232–4

[16] Xue J M, Wang J and Weiseng T 2000 Synthesis of lead zirconate titanate from an amorphous precursor by mechanical activation *J. Alloys Comp.* **308** 139–46

[17] Wan D M, Xue J M and Wang J 1999 Synthesis of single phase $0.9Pb[(Zn_{0.6}Mg_{0.4})_{1/3}Nb_{2/3}O_3]$–$0.1PbTiO_3$ by mechanically activating mixed oxides *Act. Materialia* **47** 2283–91

[18] Wan D M, Xue J M and Wang J 2000 Nanocrystalline 0.54PZN–0.36PMN–0.1PT of perovskite structure by mechanical activation *Mater. Sci. Eng.* A **286** 96–100

[19] Wan D M, Xue J M and Wang J 2000 Mechanochemical synthesis of 0.9[0.6Pb($Zn_{1/3}Nb_{2/3}$)O_3–0.4Pb($Mg_{1/3}Nb_{2/3}$)O_3]–0.1$PbTiO_3$ *J. Am. Ceram. Soc.* **83** 53–9

[20] Wang J *et al* 2003 Epitaxial $BiFeO_3$ multiferroic thin film heterostructures *Sci. Adv. Mater.* **299** 1719–22

[21] Wang J, Wan D M, Xue J M and Ng W B 1999 Synthesizing nanocryatalline Pb($Zn_{1/3}Nb_{2/3}$)O_3 powders from mixed oxides *J. Am. Ceram. Soc.* **82** 477–9

[22] Wang J, Wan D M, Xue J M and Ng W B 1999 Mechanochemical synthesis of 0.9Pb($Mg_{1/3}Nb_{2/3}$)O_3–0.1$PbTiO_3$ from mixed oxides *Adv. Mater.* **11** 210–3

[23] Wang J, Xue J M and Wan D M 2000 How different is mechanical activation from thermal activation? A case study with PZN and PZN-based relaxors *Sol. Stat. Ion.* **127** 169–75

[24] Wang J, Xue J M, Wan D M and Gan B K 2000 Mechanically activating nucleation and growth of complex perovskites *J. Sol. Stat. Chem.* **154** 321–8

[25] Wang J, Xue J M, Wan D M and Ng W B 1999 Mechanochemically synthesized lead magnesium niobate *J. Am. Ceram. Soc.* **82** 1358–60

[26] Wang J, Xue J M, Wan D M and Ng W B 1999 Mechanochemical fabrication of single phase PMN of perovskite structure *Sol. State Ion.* **124** 271–9

[27] Brzozowski E and Castro M S 2000 Synthesis of barium titanate improved by modifications in the kinetics of the solid state reaction *J. Eur. Ceram. Soc.* **20** 2347–51

[28] Brzozowski E and Castro M S 2003 Lowering the synthesis temperature of high-purity $BaTiO_3$ powders by modification in the processing conditions *Thermochim. Acta* **389** 123–9

[29] Castro A, Begue P, Jimenez B, Ricote J, Jimenez R and Galy J 2003 New $Bi_2Mo_{1-x}W_xO_6$ solid solution: mechanosynthesis, structural study, and ferroelectric properties of the $x = 0.75$ member *Chem. Mater.* **15** 3395–401

[30] Castro A, Jiménez B, Hungría T, Moure A and Pardo L 2004 Sodium niobate ceramics prepared by mechanical activation assisted methods *J. Eur. Ceram. Soc.* **24** 941–5

[31] Castro A, Millan P, Pardo L and Jimenez B 1999 Synthesis and sintering improvement of Aurivillius type structure ferroelectric ceramics by mechanochemical activation *J. Mater. Chem.* **9** 1313–7

[32] Castro A, Millan P, Ricote J and Pardo L 2000 Room temperature stabilisation of γ-$Bi_2VO_{5.5}$ and synthesis of the new fluorite phase f-Bi_2VO_5 by a mechanochemical activation method *J. Mater. Chem.* **10** 767–71

[33] Cintho O M, Tsai H I, Baer M, de Castro M, Monlevade E F and Capocchi J D T 2012 Variation of tungsten oxide reduction mode using different high energy milling conditions *Mater. Sci. Forum* **727–728** 206–9

[34] Moure A, Pardo L, Alemany C, Millan P and Castro A 2001 Piezoelectric ceramics based on Bi_3TiNbO_9 from mechano-chemically activated precursors *J. Euro. Ceram. Soc.* **21** 1399–402

[35] Pardo L, Castro A, Millan P, Alemany C, Jimenez R and Jimenez B 2000 $(Bi_3TiNbO_9)_x(SrBi_2Nb_2O_9)_{1-x}$ Aurivillius type structure piezoelectric ceramics obtained from mechanochemically activated oxides *Acta Materialia* **48** 2421–8

[36] Ricote J, Pardo L, Castro A and Millan P 2001 Study of the process of mechanochemical activation to obtain Aurivillius oxides with $n = 1$ *J. Sol. State Chem.* **160** 54–61

[37] Kong L B, Ma J and Huang H 2002 $MgAl_2O_4$ spinel phase derived from oxide mixture activated by a high-energy ball milling process *Mater. Lett.* **56** 238–43

[38] Kong L B, Ma J and Huang H 2002 Low temperature formation of yttrium aluminum garnet from oxides via a high-energy ball milling process *Mater. Lett.* **56** 344–8

[39] Kong L B, Ma J, Huang H and Zhang R F 2002 (1–x)PZN-xBT ceramics derived from mechanochemically synthesized powders *Mater. Res. Bull.* **37** 1085–92

[40] Kong L B, Ma J, Huang H and Zhang R F 2002 Lead zinc niobate (PZN)-barium titanate (BT) ceramics from mechanochemically synthesized powders *Mater. Res. Bull.* **37** 2491–8

[41] Kong L B, Ma J, Huang H, Zhang R F and Que W X 2002 Barium titanate derived from mechanochemically activated powders *J. Alloys Comp.* **337** 226–30

[42] Kong L B, Ma J, Huang H T, Zhu W and Tan O K 2001 Lead zirconate titanate ceramics derived from oxide mixture treated by a high-energy ball milling process *Mater. Lett.* **50** 129–33

[43] Kong L B, Ma J, Zhang T S, Zhu W and Tan O K 2001 $Pb(Zr_xTi_{1-x})O_3$ ceramics via reactive sintering of partially reacted mixture produced by a high-energy ball milling process *J. Mater. Res.* **16** 1636–43

[44] Kong L B, Ma J, Zhang T S, Zhu W and Tan O K 2002 Preparation of antiferroelectric lead zirconate titanate stannate ceramics by high-energy ball milling process *J. Mater. Sci.-Mater. Electron.* **13** 89–94

[45] Kong L B, Ma J, Zhu W and Tan O K 2001 Preparation and characterization of translucent PLZT8/65/35 ceramics from nano-sized powders produced by a high-energy ball-milling process *Mater. Res. Bull.* **36** 1675–85

[46] Kong L B, Ma J, Zhu W and Tan O K 2001 Preparation of $Bi_4Ti_3O_{12}$ ceramics via a high-energy ball milling process *Mater. Lett.* **51** 108–14

[47] Kong L B, Ma J, Zhu W and Tan O K 2001 Preparation and characterization of PLZT ceramics using high-energy ball milling *J. Alloys Comp.* **322** 290–7

[48] Kong L B, Ma J, Zhu W and Tan O K 2001 Preparation of PMN powders and ceramics via a high-energy ball milling process *J. Mater. Sci. Lett.* **20** 1241–3

[49] Kong L B, Ma J, Zhu W and Tan O K 2002 Transparent PLZT8/65/35 ceramics from constituent oxides mechanically modified by high-energy ball milling *J. Mater. Sci. Lett.* **21** 197–9

[50] Kong L B, Ma J, Zhu W and Tan O K 2002 Preparation and characterization of PLZT (8/65/35) ceramics via reaction sintering from ball milled powders *Mater. Lett.* **52** 378–87

[51] Kong L B, Ma J, Zhu W and Tan O K 2002 Translucent PMN and PMN-PT ceramics from high-energy ball milling derived powders *Mater. Res. Bull.* **37** 23–32

[52] Kong L B, Ma J, Zhu W and Tan O K 2002 Preparation of PMN–PT ceramics via a high-energy ball milling process *J. Alloys Comp.* **336** 242–6

[53] Kong L B, Zhang T S, Chen Y Z, Ma J, Boey F and Huang H 2004 Microstructural composite mullite derived from oxides via a high-energy ball milling process *Ceram. Inter.* **30** 1313–7

[54] Kong L B, Zhang T S, Ma J and Boey F 2003 Anisotropic grain growth of mullite in high-energy ball milled powders doped with transition metal oxides *J. Eur. Ceram. Soc.* **23** 2247–56

[55] Kong L B, Zhu W and Tan O K 2000 $PbTiO_3$ ceramics derived from high-energy ball milled nano-sized powders *J. Mater. Sci. Lett.* **19** 1963–6

[56] Kong L B, Zhu W G and Tan O K 1999 Direct formation of nano-sized $PbTiO_3$ powders by high energy ball milling *Ferroelectrics* **230** 583–8

[57] Burmeister C F and Kwade A 2013 Process engineering with planetary ball mills *Chem. Soc. Rev.* **42** 7660–7

[58] Urakaev F K and Boldyrev V V 2000 Mechanism and kinetics of mechanochemical processes in comminuting devices: 1 *Theor. Powd. Technol.* **107** 93–107

[59] Urakaev F K and Boldyrev V V 2000 Mechanism and kinetics of mechanochemical processes in comminuting devices: 2. Applications of the theory *Exp. Powd. Technol.* **107** 197–206

[60] Gavrilov D, Vinogradov O and Shaw W J D 1999 Simulation of grinding in a shaker ball mill *Powd. Technol.* **101** 63–72

[61] Harris J R, Wattis J A D and Wood J V 2001 A comparison of different models for mechanical alloying *Acta Mater.* **49** 3991–4003

[62] Kheifets A S and Lin I J 1998 Energetic approach to kinetics of batch ball milling *Inter. J. Miner. Proces.* **54** 81–97

Chapter 3

Ferroelectric ceramics (I)

Ling Bing Kong, Zhuohao Xiao, Xiuying Li, Shijin Yu, Wenxiu Que, Yin Liu, Tianshu Zhang, Kun Zhou and Hongfang Zhang

3.1 Background

There are 32 groups in terms of crystal structure, among which 21 classes are noncentrosymmetric and thus could have a piezoelectric effect, while 20 of them possess piezoelectric characteristics in practice. Half of the 20 crystals exhibit a pyroelectric effect, among which there is a subgroup that has ferroelectric properties. Ferroelectricity is defined as the presence of electrically reversible spontaneous polarization [1–3]. Ferroelectricity is named after ferromagnetism due to their various similarities.

The main crystal structures of ferroelectric ceramics include the perovskite, pyrochlore, tungsten–bronze and bismuth layer structures. Among them, the perovskite group has the strongest ferroelectric effect. Perovskite has the general chemical formula of ABO_3, as shown schematically in figure 3.1, where barium titanate ($BaTiO_3$) is used as the example [2, 3]. The structure is composed of a corner-connected network with oxygen octahedral as the building block, so as to form an octahedral chamber, which is taken by B ions. Therefore, there are B-sites and the interstices of A-sites. In this case, the Ti^{4+} ions take the B-sites, whereas the Ba^{2+} ions are at the A-sites.

Rochelle salt, known as sodium potassium tartrate tetrahydrate, $KNa(C_4H_4O_6)\cdot 4H_2O$, was first reported to show ferroelectricity [4–7]. However, the application of ferroelectricity began with the discovery of $BaTiO_3$ ceramics during World War II. After that, various new ferroelectric materials were developed. For example, lead zirconate titanate ($PbZr_{1-x}Ti_xO_3$ or PZT) ceramics, solid solutions of $PbZrO_3$ and $PbTiO_3$, have demonstrated the highest performances among all the ferroelectric materials. Transparent lead lanthanum zirconate titanate ($Pb_{1-x}La_xZr_{1-y}Ti_yO_3$, or PLZT) exhibited unique electro-optical properties and lead magnesium niobate ($PbMg_{1/3}Nb_{2/3}O_3$ or PMN) showed relaxor ferroelectric characteristics [2].

doi:10.1088/978-0-7503-2191-4ch3

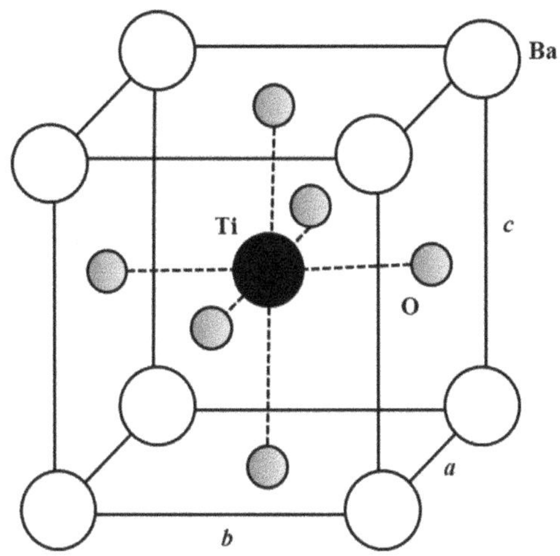

Figure 3.1. Schematic diagram of perovskite $BaTiO_3$.

The ceramic process involves three key steps: powder preparation, compaction and sintering. The quality of the starting powders is critical to the performance of the final ceramics. In other words, the microstructural, electrical and optical properties of ferroelectric ceramics are closely related to the synthetic methods used to prepare the ferroelectric powders [8–10]. The conventional method to synthesize ferroelectric powders is the solid-state reaction process, starting from oxides and/or carbonates as the precursors. However, the commercial oxide/carbonate powders usually have relatively large particle sizes and the homogeneity of the mixed powder cannot be ensured, so that a high calcination temperature is required to facilitate the phase formation of the designed compounds [8–10]. This could be an issue for lead-containing ferroelectric materials, because of the high volatility of PbO. If too much Pb is lost, the properties of the final products will be compromised.

To address this problem, various wet-chemical synthetic approaches have been developed to obtain submicron or nanosized ferroelectric powders, such as chemical coprecipitation [11–23], the sol–gel process [24–30], hydrothermal synthesis [31–46], combustion [47–54], thermal pyrolysis spray [55–57], microemulsion [58, 59], molten salt [60–63], etc. However, these chemical methods have shortcomings. For example, chemical coprecipitation processes use a large amount of water and require washing and calcination steps. The sol–gel process involves expensive metal alkoxides, while it is also very sensitive to environmental conditions, including moisture, light and heat. The hydrothermal process is not suitable for large-scale production. As a result, it is important to find new methods to prepare ferroelectric powders, in particular for those containing lead (Pb) [64–84], bismuth (Bi) [85–97] or lithium (Li) [98] as the major component.

In this regard, the mechanochemical process has various advantages. First, the mechanochemical method is not only completely compatible with the conventional ceramic process, but is even simpler due to the omission of the calcination step. Second, powders synthesized in this way are at nanometer scales, thus they have high sintering activity. Third, the starting materials are either oxides or carbonates, which are relatively cheap and widely available. More importantly, in some cases, for example PZT or PLZT, it is unnecessary to obtain the compounds. That is to say, the unreacted precursor powders can be used to fabricate ceramics directly. Therefore, the fabrication process is very time-saving. Although some ferroelectric materials cannot be directly synthesized using mechanochemical methods, such as $BaTiO_3$ [99–102] and most Aurivillius family ferroelectrics [103–109], the phase formation temperatures of these materials could be greatly reduced. In this chapter, lead-containing unary ferroelectric and antiferroelectric ceramics are discussed, while other ferroelectric materials will be discussed in chapters 4 and 5.

3.2 Lead-containing unary ferroelectric ceramics

3.2.1 Lead titanate and lead lanthanum titanate

Lead titanate, $PbTiO_3$ or PT in short, is a typical ferroelectric material, with a Curie temperature of $T_C \approx 490$ °C [110]. PT ceramics have various special properties, such as a Curie transition temperature, a low ratio for the planar-to-thickness coupling factor, a low aging rate of the dielectric constant and a relatively low dielectric constant, thus being suitable for various potential applications. However, it is well known that pure PT ceramics cannot be prepared via the conventional ceramic process, due to their large tetragonality, i.e. a *c/a* ratio of 1.065. During the cooling process, an anisotropic thermal expansion is generated as the phase transition takes place from the cubic paraelectric phase to the tetragonal ferroelectric phase. As a result, large internal stresses are present, which make the ceramics fragile and lead to numerous microcracks. Therefore, PT ceramics can only be obtained with the presence of various additives when using the conventional ceramic process, where the additives either mitigate the internal stresses or reduce the tetragonality of the perovskite phase [111–115].

It is well known that the internal stress in ceramics is caused by the incompatible strains when there is thermal expansion anisotropy during the cooling process. The formation of microcracks is determined by a critical grain size. The stress per unit grain boundary area is proportional to the fractional volume of the grains, because the smaller the grains the more grain boundaries there are. It was confirmed that the critical grain size of pure PT ceramics was about 3 μm, beyond which spontaneous cracking occurred. As a consequence, the samples disintegrated once cracking was initiated. In other words, to obtain crack-free PT ceramics, the average grain size should be below 3 μm [24, 25, 116].

PT powders have been synthesized using mechanochemical milling, with different milling facilities [65–67, 84, 117–120]. For example, the effect of mechanical milling factors on the phase formation of PT from oxide precursors was studied [64]. It was observed that the kinetic energy for the formation of PT compound from a mixture of PbO and TiO_2 could be reduced. After milling for 0.5 h, the kinetic energy was about 110 kJ mol^{-1}, smaller compared to that (152 kJ mol^{-1}) of the unmilled sample. In addition, the reaction kinetics were also dependent on the characteristics of the starting materials.

By using an SPEX shaker mill, Xue *et al* studied the phase formation of PT from oxide precursors [84]. The commercial PbO and TiO_2 powders had particles of 3–5 μm, and were pre-mixed using the conventional ball mill process. Mechanochemical milling for 5 h resulted in refinement of the precursor powders, evidenced by the broadening of the XRD peaks. In addition, the PT phase started to be present in the mixture. Nearly single-phase PT was achieved after milling for 20 h. The as-synthesized PT powder had particle sizes in the range of 20–30 nm. The process was also applied to promote the crystallization of PT from an amorphous precursor obtained using a chemical coprecipitation method [65].

Kong *et al* used a Fritsch Pulverisette 5 type planetary mill, with tungsten carbide (WC) balls and vials, to synthesize PT powders from oxide mixtures [66]. They found that the phase formation process was independent of the crystal structure of the TiO_2 powder, but dependent on the particle size. After milling for a sufficiently long time, the PT phase was formed, no matter whether anatase or rutile TiO_2 was used. More significantly, crack-free PT ceramics could be derived from the PT powders, due to their small particle size and thus high sintering activity [67].

The effect of milling duration on the phase formation of PT nanosized powders was studied by Wongmaneerung *et al* using a vibro-mill, with a polypropylene jar and polycrystalline Al_2O_3 ceramic bars as milling media [121]. Because of the low density of the milling media, the energy that was applied to the milled powder was not sufficiently high to trigger the phase formation of PT directly from the oxide mixtures. However, the mixture was highly activated by the low energy milling, so that the PT phase could be obtained at 600 °C for 1 h, which was much less than required by the conventional solid-state reaction process.

Starting with commercial oxides of PbO and anatase TiO_2, the phase formation of PT was studied using XRD and Raman technologies [122]. The mechanochemical activation was conducted under ambient conditions with an SPEX 8000 Mixer Mill; the type of milling media were not stated in the reference. The oxide mixture was milled for times of up to 50 h. No attempt was made to fabricate PT ceramics using the synthesized powders.

Figure 3.2 shows the XRD patterns of the oxide mixtures before and after milling for various amounts of time [122]. After milling for 10 h, the XRD peaks of the oxide precursors were absent, indicating that the PT phase formation process was almost complete. After milling for 15 h, the PT phase with a single-phase tetragonal perovskite structure was obtained, confirming that PT was formed due to the mechanochemical activation at near room temperature, without the requirement for thermal treatment. In addition, long duration milling (>30 h) led to broadening of the diffraction peaks, suggesting the refinement of the PT powders. Figure 3.3 shows TEM images of the sample milled for 50 h. The powder consisted of irregular grains with sizes in the range 70–150 nm.

Lanthanum (La) is an element that is commonly used to modify PT. La^{3+} has a radius of 1.032 Å, which could substitute for Pb^{2+} with $r = 1.19$ Å at the A-site. It is not possible for it to sit at the B-site, because Ti^{4+} has a much smaller radius with

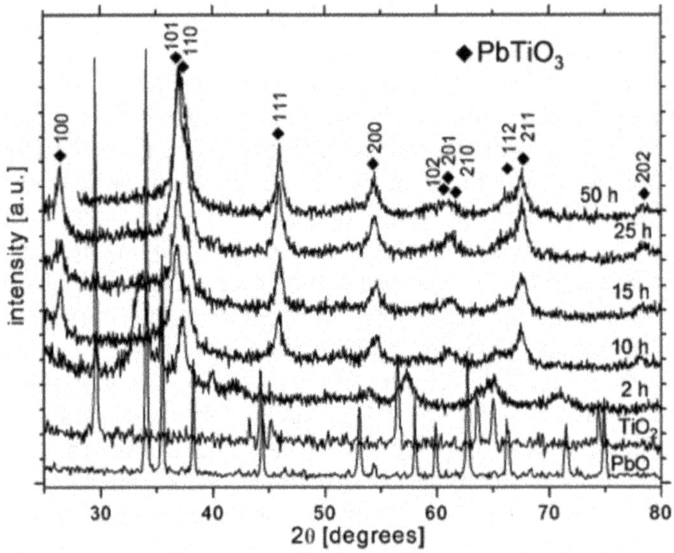

Figure 3.2. XRD patterns of the PbO and TiO_2 mixtures before and after milling for different amounts of time. Reproduced with permission from [122]. Copyright 2006 Wiley.

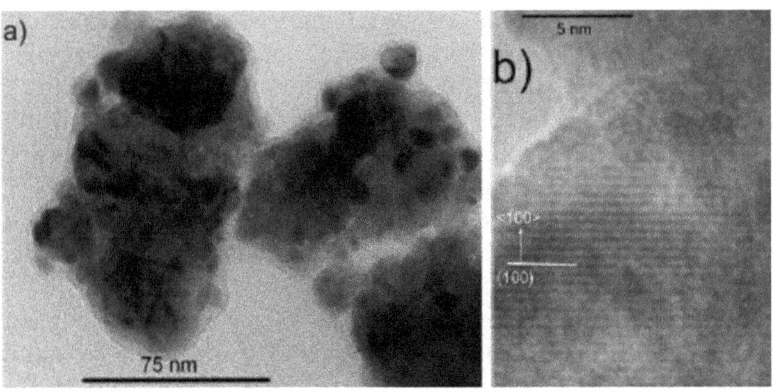

Figure 3.3. TEM images of the sample milled for 50 h: (a) low magnification and (b) high magnification. Reproduced with permission from [122]. Copyright 2006, Wiley.

$r = 0.605$ Å. The balance related to the La^{3+} substitution is maintained through the formation of vacancies at either the A-site or B-site, i.e. $Pb_{1-3x/2}La_xTiO_3$ (PLT-A) or $Pb_{1-x}La_xTi_{1-x/4}O_3$ (PLT-B), respectively. PLT-A had an extremely high dielectric constant, whereas PLT-B exhibited relaxor ferroelectric behavior. Since PbO is highly volatile at high temperatures, the compositions of the PLT could easily deviate from the designed ones, which is readily achieved using high-energy mechanochemical activation [123].

Commercial powders of PbO, TiO_2 and La_2O_3 were thoroughly mixed using the conventional lab ball milling process. The mixtures were subjected to mechanochemical activation with a batch quantity of 5 g. A cylindrical vial of 40 mm in diameter together with a ball of 20 mm in diameter were used in the experiment. Mechanical activation was conducted with an SPEX 8000 high-energy shaker mill operated at 900 rpm for different amounts of time. After mechanochemical activation for 20 h, the powders were cold pressed into pellets of 10 mm in diameter and 1.5 mm in thickness at a uniaxial pressure of 40 MPa. The green bodies were sintered at 1200 °C for 2 h, with heating and cooling rates of 3 °C min^{-1} and 10 °C min^{-1}, respectively.

For comparison, lead titanate compositions with 20 mol% La were annealed under four different conditions: (a) thermal annealing in O_2 for different times at 800 °C; (b) thermal annealing in N_2 for different times at 800 °C; (c) thermal annealing in O_2 for 4 h at 400 °C and (d) thermal annealing in N_2 for 12 h and then in O_2 for 12 h at 800 °C. The heating and cooling rates of the thermal annealing experiments were both 5 °C min^{-1}.

Figure 3.4 shows XRD patterns of the mixtures of PbO, TiO_2 and La_2O_3 for PLT-A with $x = 0.15$, i.e. PLT-A15, before and after mechanochemical activation at room temperature for 5, 10, 15 and 20 h. The conventionally mixed powders had no

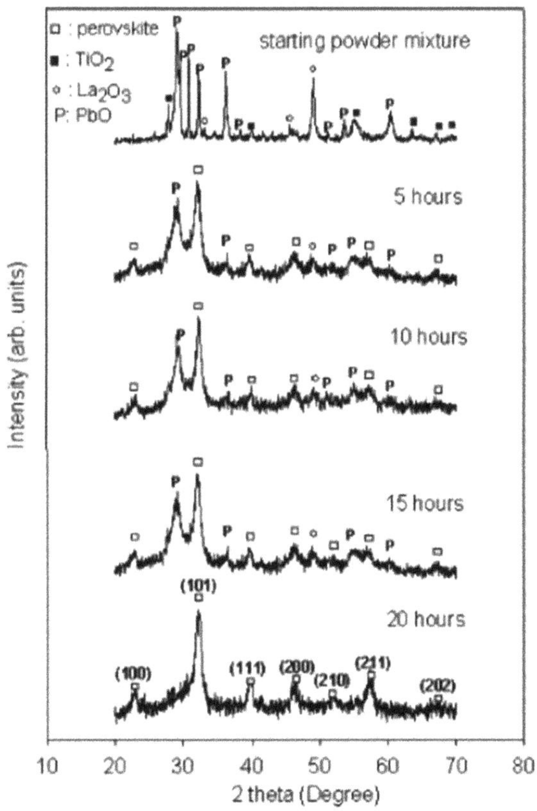

Figure 3.4. XRD patterns of the mixtures of PbO, TiO_2 and La_2O_3 with the composition of $Pb_{0.775}La_{0.15}TiO_3$ mechanically activated for different amounts of time. Reproduced with permission from [123]. Copyright 2004 American Institute of Physics.

reaction. After milling for 5 h, all the peaks were broadened and some peaks disappeared. Also, the peaks of perovskite $Pb_{0.775}La_{0.15}TiO_3$ or PLT-A15 were observable. In this case, the starting oxides were significantly refined, while they reacted to form the PLT phase. As the milling time was increased from 5 h to 15 h, the diffraction peaks of PLT became more and more pronounced. After milling for 20 h, single-phase PLT-A15 was achieved. Similar results were observed for the compositions with different contents of La. In all cases, a milling time of 20 h was sufficient for the mechanochemical synthesis of the PLT samples. The as-obtained PLT powders had particle sizes in the range 10–20 nm.

It was found that the thermal annealing caused almost no phase variation of the PLT samples. At the same time, the sintered density of the PLT-A ceramics was independent of the annealing time and the annealing atmosphere. Figure 3.5 shows representative SEM images of the as-sintered PLTA20 ceramics and those that were annealed in O_2 and N_2, demonstrating nearly no difference in microstructure and average grain size among them. In other words, the dielectric properties of the PLTA ceramics were not influenced by the thermal annealing conditions, instead they were merely dependent on their compositions.

3.2.2 Lead zirconate titanate

Lead zirconate titanate, $Pb(Zr_xTi_{1-x})O_3$ or PZT, is a solid solution of PT and PZ. PZT ceramics have the highest ferroelectric, piezoelectric, pyroelectric and optoelectronic properties, and have been extensively and intensively studied for several decades [2, 3]. There is an almost temperature-independent phase boundary at $x = 0.52$–0.53, separating the rhombohedral Zr-rich phase and the tetragonal Ti-rich phase, which is known as the morphotropic phase boundary (MPB) [124]. PZT powders have been prepared using the mechanochemical activation process, from oxide and coprecipitated precursors, with either an SPEX shaker mill [68, 69, 80, 125] or a planetary ball mill [70–75, 126].

It was found that the mixtures milled for a relatively short time, in which the phase formation is not started or incomplete, can also be used to fabricate PZT ceramics without compromising the microstructure and electrical properties [71]. According to the densification behaviors of the mixtures with different degrees of reaction due to the different mechanochemical activation times, an obvious volumetric expansion was present at certain temperatures. The volumetric expansion is the indicator of the phase formation of the perovskite PZT from the unreacted oxide mixtures. Figure 3.6 shows the XRD patterns of the mixtures milled for different times, which were then sintered at 900 °C for 4 h. Obviously, all the samples are of single-phase perovskite structure, confirming that the unreacted mixture could be employed to obtain PZT ceramics.

Representative SEM images of the PZT ceramics derived from the mixture mechanochemically milled for 4 h are shown in figure 3.7. It was found that the unreacted powder had a very high sinterability. After sintering at 900 °C for 4 h, a fully dense sample was achieved. A nicely intergrained fracture surface could be observed, with a quite uniform grain size distribution and an average grain size of

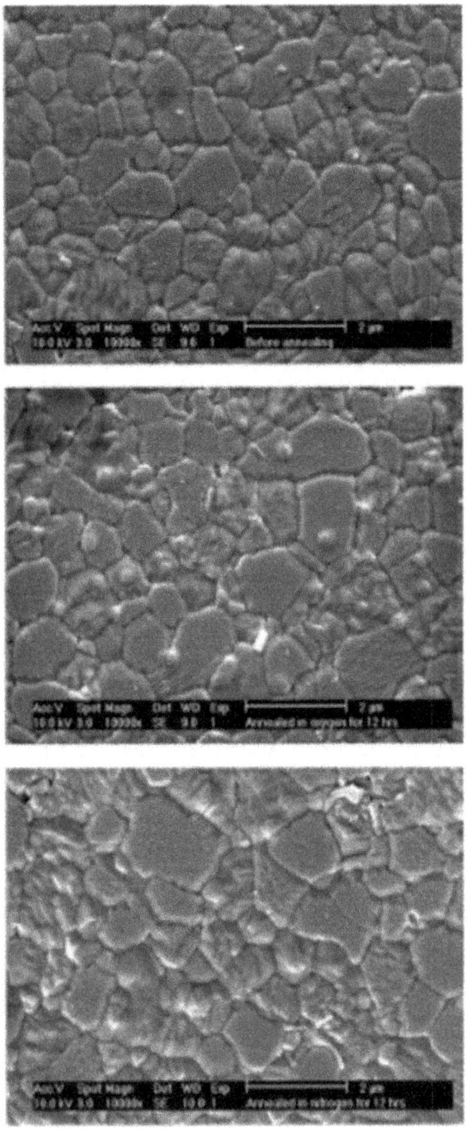

Figure 3.5. SEM images of the polished and etched surfaces of the PLT-A20 samples: (a) before annealing, (b) annealed in O_2 for 12 h at 800 °C and (c) annealed in N_2 for 12 h at 800 °C. Reproduced with permission from [123]. Copyright 2004 American Institute of Physics.

about 2.3 μm. As the sintering temperature was increased, the grain size was gradually increased following an almost linear function, while the microstructure was always homogeneous. The average grain size versus sintering temperature is shown in figure 3.8.

Figure 3.9 shows *P–E* hysteresis loops of the samples derived from the 4 h milled mixture, which were sintered at different temperatures for 4 h. The samples sintering

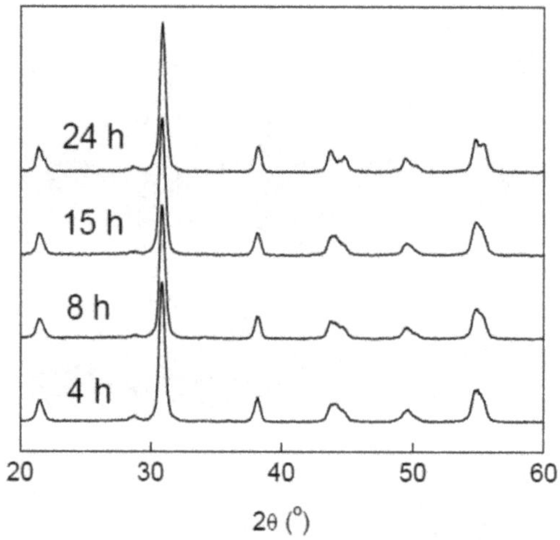

Figure 3.6. XRD patterns of the mixtures of PbO, ZrO_2 and TiO_2 milled for different times, followed by sintering at 900 °C for 4 h.

at 900 °C and 925 °C exhibited well-shaped *P–E* loops, yet they were not saturated. Also, the remnant polarization was at a relatively low level. However, after sintering at 950 °C, a saturated *P–E* loop was obtained. Any further increase in sintering temperature thereafter resulted in no significant enhancement in ferroelectric performances. The results suggested that the 4 h milled mixture could be sintered at such a low temperature, but with electrical properties close to the theoretical values of PZT with that composition. This sintering temperature is much lower than those required by the conventional ceramic process and most of the wet-chemical synthetic routes. Figure 3.10 shows remnant polarization versus sintering temperature for the samples derived from the 4 h milled oxide mixture.

Here, it is worth mentioning that the fabrication of PZT ceramics could be completed in as little as two days. In the conventional ceramic process, various time-consuming steps are involved. For example, the initial oxide powders should be thoroughly mixed by using the conventional ball milling process. The milling is usually conducted in the presence of water or ethanol. The mixed slurries should be dried to remove the liquid phase. After that, the samples need to be crushed and sieved. In comparison, the mechanochemical activation process is conducted without the use of any liquid. The as-milled powders can be directly compacted without any additional treatment. As a result, the whole process can be significantly shortened.

The mechanochemical activation process has also been applied to prepare rare-earth element-doped PZT ($Pb_{0.92}Gd_{0.08}(Zr_{0.53}Ti_{0.47})_{0.98}O_3$ (PGZT) and $Pb_{0.92}Nd_{0.08}(Zr_{0.53}Ti_{0.47})_{0.98}O_3$ (PNZT)) nanoceramics [75, 127]. It was found that the PGZT and PNZT ceramics possessed interesting dielectric and ferroelectric properties, which were not observed in the samples prepared using the conventional ceramic process. For example, the PGZT ceramics exhibited a higher dielectric constant compared to those made of the unmilled powders. The high dielectric constants of the mechanochemically derived ceramics were closely related to the

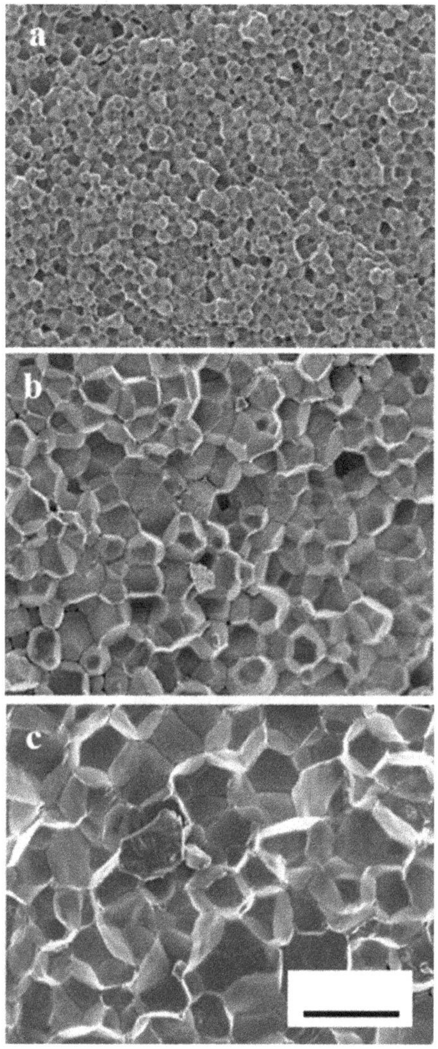

Figure 3.7. SEM images of mixtures of PbO, ZrO_2 and TiO_2 milled for 4 h after sintering at different temperatures for 4 h (scalebar = 5 μm): (a) 900 °C, (b) 950 °C and (c) 1000 °C.

nanocrystalline characteristics of the activated powders, leading to enhancement in microstructure and thus electrical properties. On the other hand, the PNZT ceramics displayed a diffused ferroelectric–paraelectric phase transition behavior, thus resulting in a high dielectric constant over a wide range of temperatures.

Nanosized PZT ($Pb(Zr_{0.7}Ti_{0.3})O_3$) powders were obtained by mechanochemically activating the hydrous oxide precursors of lead–zirconium–titanium (Pb–Zr–Ti) in the presence of NaCl [126]. After the synthetic process, the NaCl was washed out with deionized water. The content of NaCl had a significant effect on the particle size of the PZT powders, where the optimal weight ratio was 4:1, which led to PZT powders with an average grain size of 110 nm. The PZT had an optimal sintering

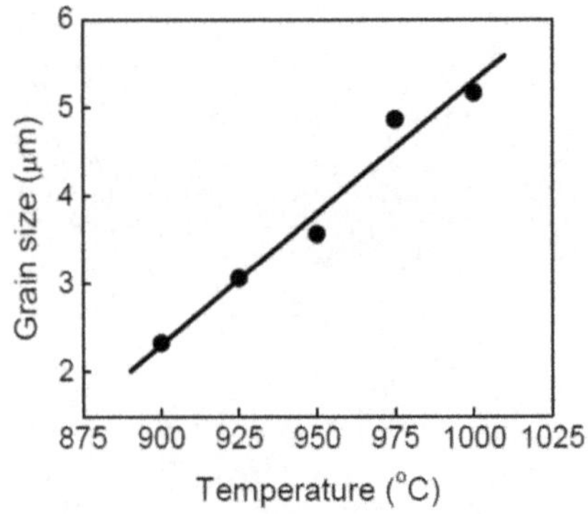

Figure 3.8. Grain size versus sintering temperature for the 4 h milled samples.

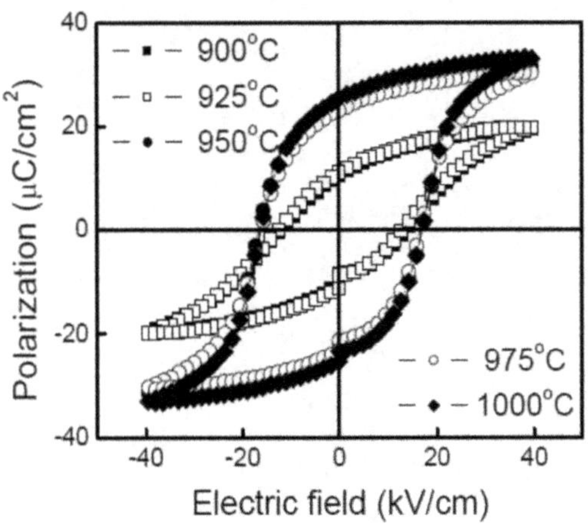

Figure 3.9. *P–E* hysteresis curves of the mixtures of PbO, ZrO_2 and TiO_2 milled for different times, followed by sintering at different temperatures for 4 h.

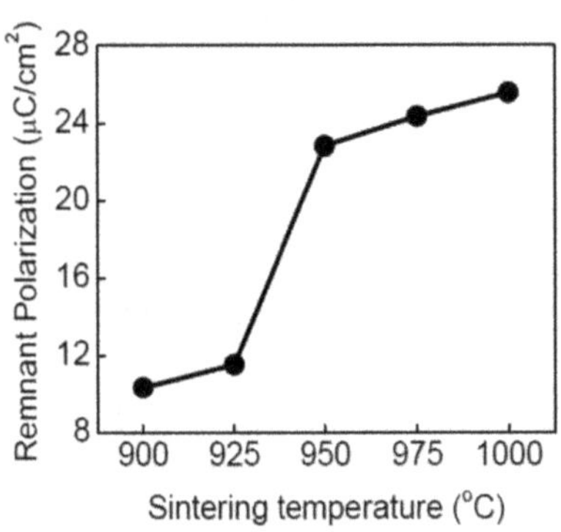

Figure 3.10. Remnant polarization versus sintering temperature for the 4 h milled samples.

temperature of 1150 °C, which was lower than those used for the conventional ceramic process by 50 °C–150 °C. However, the electrical properties of the PZT ceramics have been well retained.

If intensive milling, i.e. a high milling speed and large ball-to-powder weight ratio, is adopted, the PZT phase can be synthesized after milling for only 1 h [74]. In other words, the fabrication process of PZT ceramics could be further accelerated, thus leading to enhanced productivity and reduced cost of the products. The milling experiment was carried out with stainless steel vials (V = 500 ml) and balls (d = 13.4 mm), while the ball-to-powder weight ratio was 40:1 and the powder weight was 20 g. The milling was conducted in air atmosphere, at a basic disk rotation speed of 317 min^{-1}, with a rotation speed of disks with jars of 396 min^{-1}, and the milling times were in the range of 5–480 min.

Figure 3.11 shows the XRD patterns of the powder mixtures milled for various times [74]. After milling for just 5 min, perovskite structure was present as a minor phase. Various intermediate phases were also detected, such as Ti_9O_{17}, $Ti_{10}O_{19}$, Ti_4O_7, $Pb_{12}O_{19}$, etc. Also, the diffraction peaks of the starting oxides were all reduced and broadened, indicating the refinement of their particles. After milling for 15 min, both TiO_2 and PbO were completely amorphized, while crystalline ZrO_2 was still observed. At the same time, the peak intensity of the perovskite phase gradually increased with increasing milling time. After milling for 1 h, there were only peaks corresponding to perovskite and ZrO_2, while no other phases were detectable. The degree of crystallization of the perovskite phase was enhanced with prolonged milling. Two hours of milling led to near complete phase formation, but further

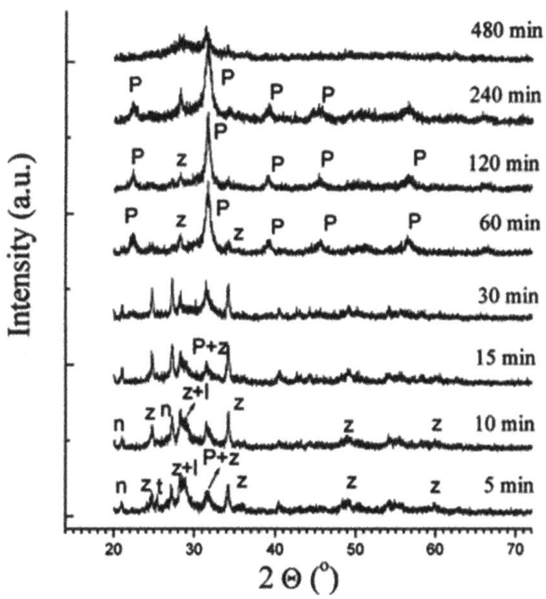

Figure 3.11. XRD patterns of the powder mixtures milled for different times (n = nonstoichiometric compounds of Zr and Ti; z = ZrO_2; t = TiO_2; l = PbO; P = perovskite). Reproduced with permission from [74]. Copyright 2003 Elsevier.

milling resulted in a total amorphization. This amorphization has been reported by other researchers. Furthermore, it is reasonably expected that if tungsten carbide milling media are used, the outcome could be more complicated.

3.2.3 Lead lanthanum zirconate titanate

Lead lanthanum zirconate titanate has a general formula of $(Pb_{1-y}La_y)(Zr_{1-x}Ti_x)_{1-y/4}O_3$, which is simplified as PLZT100y/1 – x/x generally. PLZT ceramics could be manipulated to have a wide range of ferroic phases, including ferroelectric (FE), antiferroelectric (AFE) and paraelectric (PE) phases [128–132]. According to the room-temperature phase diagram of PLZT solid solutions, the important phases include antiferroelectric orthorhombic (AFE_O), ferroelectric rhombohedral (FE_{Rh}), ferroelectric tetragonal (FE_{Tet}), relaxor ferroelectric (RFE) and paraelectric cubic (PE_{Cubic}). For example, PLZT8/65/36 is a typical relaxor ferroelectric phase, with a high dielectric constant and low coercive field. Transparent PLZT8/65/36 ceramics have strong electro-optic effects, with potential applications as electro-optic switches [133–146]. PLZT7/60/40 ceramics exhibit the highest piezoelectric longitudinal coefficients and electromechanical planar coupling factor, which are d_{33} = 710 pC/N and k_p = 0.72, respectively. PLZT powders, including PLZT8/65/35, PLZT15/65/35, PLZT7/60/40 and PLZT2/95/5, have been synthesized using the mechanochemical activation process [147–150].

Commercial oxide powders of PbO, La_2O_3, ZrO_2 and TiO_2 were selected as the starting materials, according to composition of $(Pb_{1-x}La_x)(Zr_{1-y}Ti_y)_{1-(x/4)}O_3$. Three compositions were selected to demonstrate the feasibility of the mechanochemical activation process, which are ferroelectric PLZT8/65/35, non-ferroelectric PLZT15/65/35 and antiferroelectric PLZT2/95/5. A 10 wt% excessive PbO was utilized to compensate for the evaporation of lead during the sintering to form PLZT ceramics at elevated temperatures. The milling experiment was conducted with a Fritsch Pulverisette 5 planetary high-energy ball mill. The operation was carried out in air at room temperature for different times up to 36 h. A 250 ml tungsten carbide vial and 100 tungsten carbide balls with a diameter of 10 mm were employed as the milling media. The milling speed was set at 200 rpm. The milling was stopped for 5 min for every 25 min of milling to cool down the system.

Figure 3.12 shows the XRD patterns of the mixtures for PLZT8/65/35, PLZT15/65/35 and PLZT2/95/5 after milling for 36 h. It can be seen that the perovskite PLZT is the major in all three samples. Chemical reaction as a consequence of high-energy activation is a complex process. The mechanism of the phase formation at such low temperatures has not been clarified. According to the XRD patterns of the three samples, the reaction to the ferroelectric compounds was almost finished in the milling time. The retention of the reactants in the powders would be entirely converted to the desired compounds in the PLZT ceramics after sintering at high temperatures. Figure 3.13 shows SEM images of the three mixtures, demonstrating the nanometer scale of the as-milled powders. Due to the fine particle size, they have high sinterability.

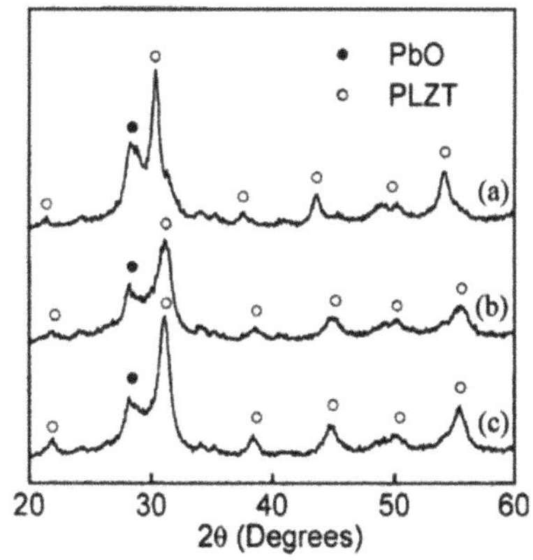

Figure 3.12. XRD patterns for the milled oxide mixtures: (a) PLZT8/65/35, (b) PLZT15/65/35 and (c) PLZT2/95/5. Reproduced with permission from [147]. Copyright 2001 Elsevier.

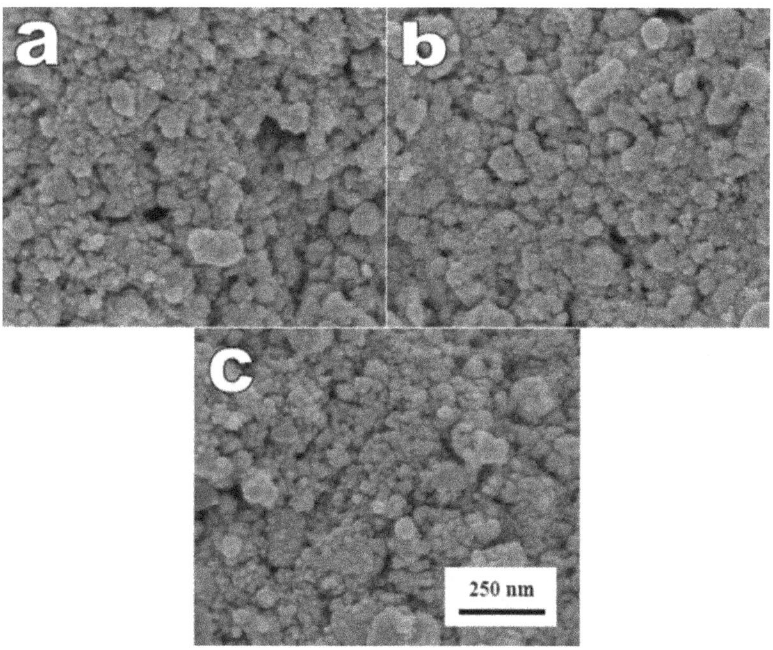

Figure 3.13. SEM images showing particle profiles of the milled oxide mixtures: (a) PLZT8/65/35, (b) PLZT15/65/35 and (c) PLZT2/95/5. Reproduced with permission from [147]. Copyright 2001 Elsevier.

Figure 3.14 shows sintering profiles of the three powder samples. PLZT8/65/35 and PLZT15/65/35 have a similar densification behavior. There is one expansion peak in their sintering rate curves, at 780 °C for PLZT8/65/35 and 790 °C for PLZT15/65/35, respectively. This expansion is attributed to the reaction of the oxide precursors that are not consumed during the high-energy milling process. The expansion is ascribed to the fact that the PLZT phases have a larger molar volume than the summation of the reactants. The maximum shrinkage peaks of these two

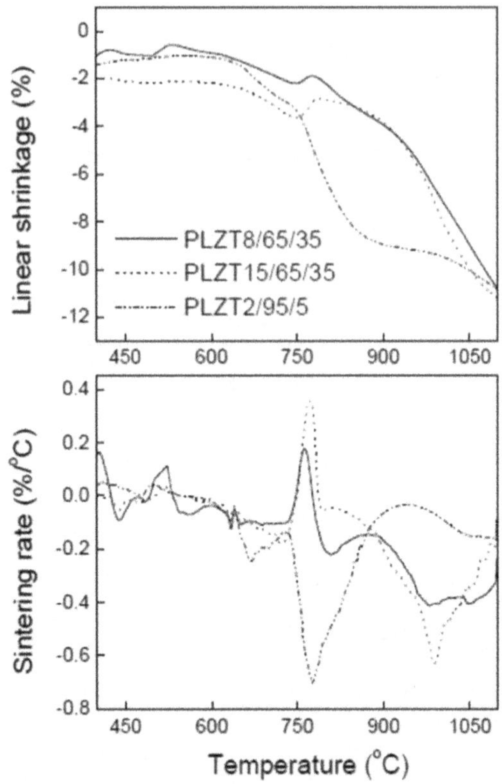

Figure 3.14. Sintering behaviors of the as-milled oxide mixtures. Reproduced with permission from [147]. Copyright 2001 Elsevier.

samples are both near 950 °C, which is similar to the literature data [151]. In comparison PLZT2/95/5 exhibits a lower maximum sintering temperature. It is therefore suggested that the higher the content of zirconium and the lower the content of lanthanum, the lower the densification temperature would be.

Cross-sectional SEM images of the three PLZT ceramics sintered at different temperatures for 1 h are shown in figures 3.15–3.17. The grain size of the PLZT ceramics as a function of sintering temperature is depicted in figure 3.18. The PLZT8/65/35 sample after sintering at 900 °C has an almost fully densified microstructure, with well-established grains and an average grain size of 0.2 μm, as demonstrated in figure 3.15(a). When the sintering temperature is raised to 1000 °C, an increase is observed in the density and average grain size. After sintering at 1100 °C, the average grain size is further increased to 0.8 μm, while the density is also increased. As illustrated in figure 3.15(d), the PLZT8/65/35 ceramic sintered at 1200 °C shows a highly densified fractural surface, while the average grain size was about 2 μm. The variations of average grain size and microstructure as a function of increasing sintering temperature are similar for the other two samples, with their grain sizes being in the order PLZT8/65/35 < PLZT15/65/35 < PLZT2/95/5. Therefore, the grain size of the PLZT ceramics is dependent on their chemical composition.

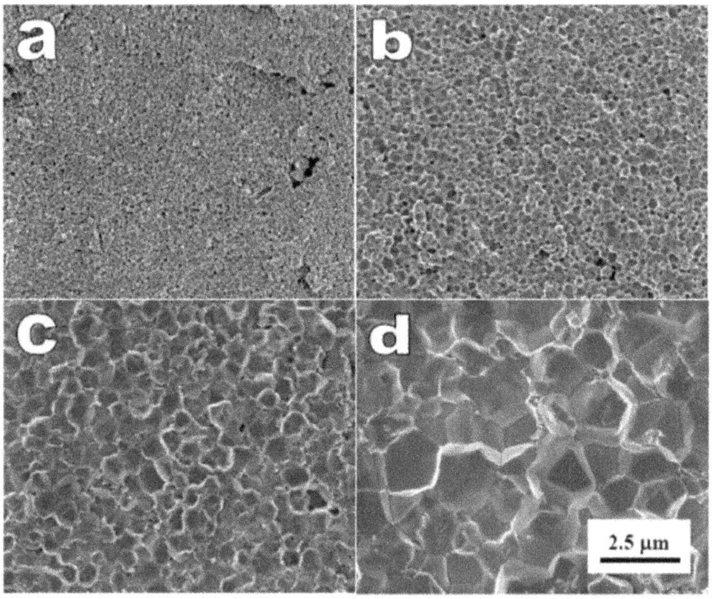

Figure 3.15. Cross-sectional SEM images of the PLZT8/65/35 ceramics sintered at different temperatures: (a) 900 °C, (b) 1000 °C, (c) 1100 °C and (d) 1200 °C. Reproduced with permission from [147]. Copyright 2001 Elsevier.

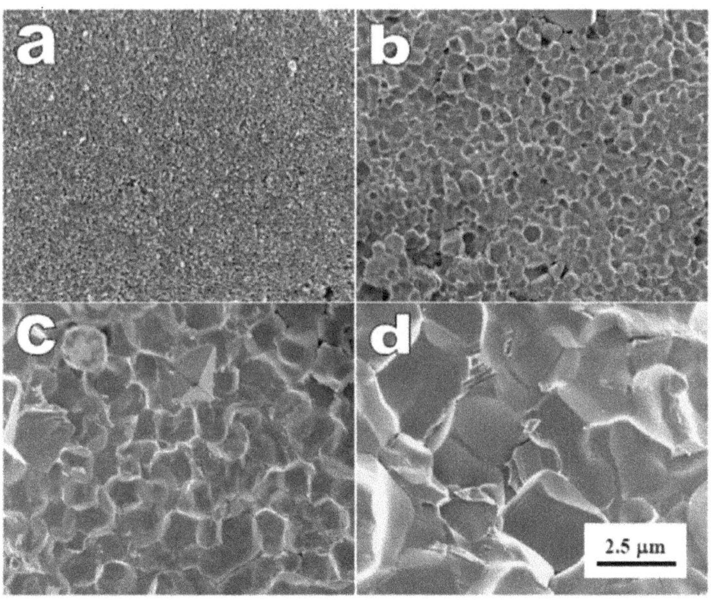

Figure 3.16. Cross-sectional SEM images of the PLZT15/65/35 ceramics sintered at different temperatures: (a) 900 °C, (b) 1000 °C, (c) 1100 °C and (d) 1200 °C. Reproduced with permission from [147]. Copyright 2001 Elsevier.

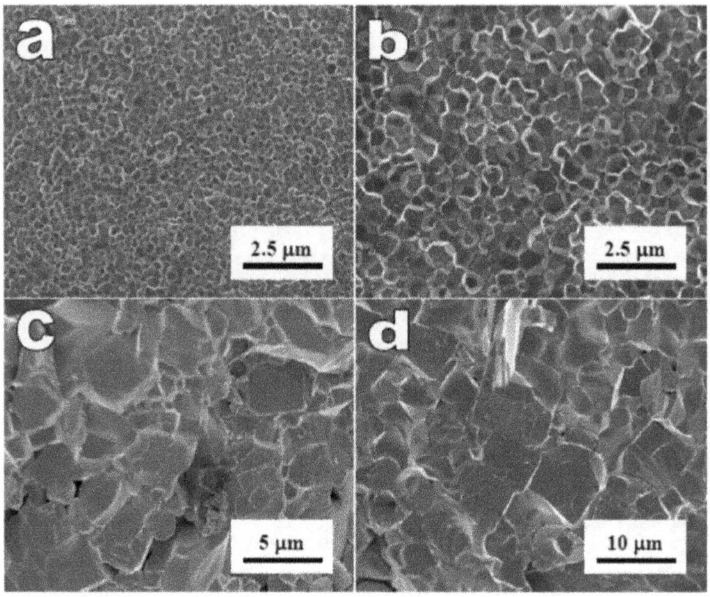

Figure 3.17. Cross-sectional SEM images of the PLZT2/95/5 ceramics sintered at different temperatures: (a) 900 °C, (b) 1000 °C, (c) 1100 °C and (d) 1200 °C. Reproduced with permission from [147]. Copyright 2001 Elsevier.

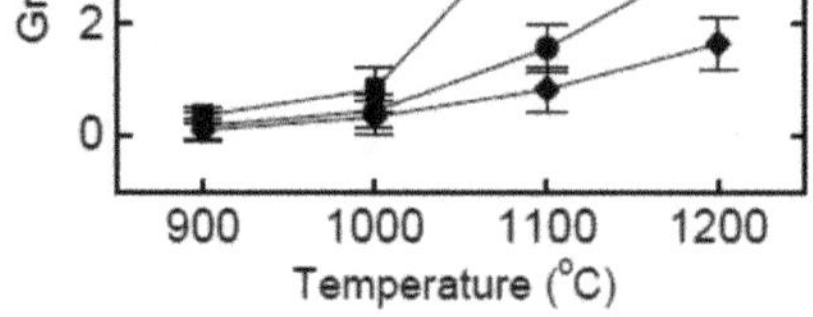

Figure 3.18. Grain size variations of the PLZT ceramics with sintering temperature. Reproduced with permission from [147]. Copyright 2001 Elsevier.

Figure 3.19 shows dielectric constants (at 1 kHz) of the three PLZT ceramic samples as a function of sintering temperature. The PLZT ceramics derived from the mechanochemically activated powders have dielectric constants that are comparable with the reported values of their counterparts processed with the conventional methods [2, 132, 137]. The dielectric constant is monotonically increased, as the sintering temperature is raised from 900 °C to 1200 °C. The variation in dielectric constant of the fully sintered PLZT ceramics with increasing sintering temperature

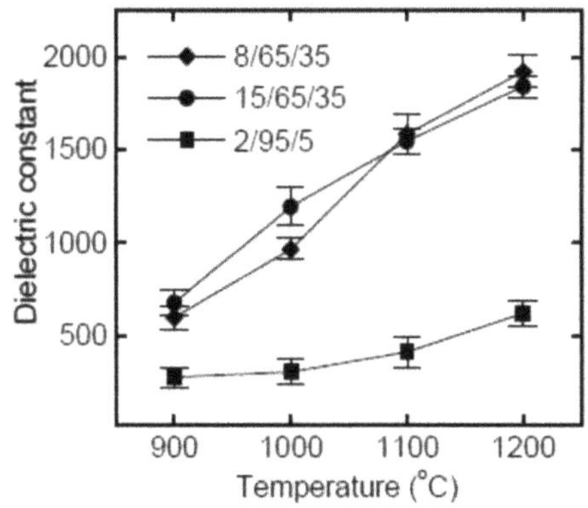

Figure 3.19. Dielectric constant of the PLZT ceramics versus sintering temperature. Reproduced with permission from [147]. Copyright 2001 Elsevier.

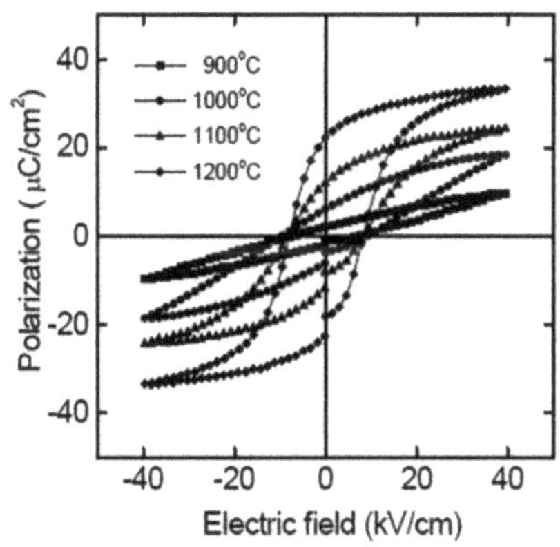

Figure 3.20. *P–E* hysteresis loops of the PLZT8/65/35 ceramics measured at room temperature. Reproduced with permission from [147]. Copyright 2001 Elsevier.

can be directly linked to their microstructure and grain size, as demonstrated in figures 3.15–3.18.

Figure 3.20 illustrates room-temperature *P–E* hysteresis loops of the PLZT8/65/35 ceramics. PLZT8/65/35 samples exhibit typical slim hysteresis loops, which is in good agreement with the literature data. Ferroelectric characteristics, such as saturation polarization, remnant polarization and coercive field of the samples are summarized in table 3.1. The remnant polarization and saturation polarization of the PLZT8/65/35 samples exhibit a monotonous increase with increasing sintering temperature. The coercive field of the samples is maximized at 1000 °C. A constant decrease in coercive field is observed as the sintering temperature is increased from 1000 °C to 1200 °C. The variation in saturation and remnant polarization of the PLZT8/65/35 samples with sintering temperature is also directly related to the grain size of the ceramics, because the achievable domain alignment usually increases with increasing grain size.

Table 3.1. Ferroelectric parameters of the PLZT8/65/35 ceramics sintered at different temperatures for 1 h. Reproduced with permission from [147]. Copyright 2001 Elsevier.

	900 °C	1000 °C	1100 °C	1200 °C
P_S (μC cm^{-2})	9.7	18.8	24.3	32.8
P_r (μC cm^{-2})	2.1	6.3	11.8	22.1
E_C (kV cm^{-1})	6.9	9.6	8.7	8.2

Electrical properties of ferroelectric materials, including dielectric constant, dielectric loss tangent, saturation polarization, remnant polarization and coercive field, all depend on the grain size and microstructure the ceramic materials in one way or another. According to the ferroelectric properties of $BaTiO_3$ and $Pb(Zr_{1-x}Ti_x)O_3$ ceramics, it is commonly recognized that the larger the grain size, the more the domain variants that are available [124, 152, 153]. This is because the volume fraction of the grain boundary is inversely proportional to the grain size. Therefore, if the grain size is reduced, the coupling effect between the grain boundaries and the domain wall, which makes domain reorientation more difficult and severely constrains the domain wall motion, will be reduced. As a result, the domain wall mobility is enhanced so that the dielectric constant would be increased.

The dependence of coercive field on sintering temperature can also be explained by taking into account the change in grain size. As mentioned above, for materials with certain compositions, the smaller the grain size the more the grain boundaries. The equivalent electrical circuit of a PLZT ceramic capacitor can be assumed to consist of a grain component and a grain boundary component connected in series. The coercive voltage (E_C) is a sum of the voltages from the grain part (E_G) and grain boundary part (E_{GB}). E_G is usually taken as a constant, so that E_C will be decreased with increasing grain size, due to the reduction in E_{GB}. The observation that the coercive field of the PLZT8/65/35 samples increases from 900 °C to 1000 °C (table 3.1) should be explained in another way. It is mainly because the grains of the PLZT ceramics sintered at 900 °C are lower than the critical value, so that the ferroelectric domains of the materials are not matured, thus leading to almost linear *P–E* characteristics. In this case, a relatively low coercive field is observed.

P–E hysteresis loops of the PLZT15/65/35 ceramics are depicted in figure 3.21. It is not surprising that the PLZT15/65/35 ceramics are non-ferroelectric without loop characteristics. The increase in slope of the *P–E* curves corresponds to an increase in capacitance, which is also linked to the increase in grain size with increasing sintering temperature.

Typical double-loop *P–E* hysteresis curves are present in the antiferroelectric PLZT2/95/5 ceramics after sintering at temperatures of >1000 °C, as depicted in figure 3.22. Parameters, including maximum polarization, forward-switching electric field $E_{AFE\text{-}FE}$ (antiferroelectric to ferroelectric phase transformation) and backward field $E_{FE\text{-}AFE}$ (ferroelectric to antiferroelectric) are listed in table 3.2. The polarization of the antiferroelectric PLZT samples is increased, while the values of

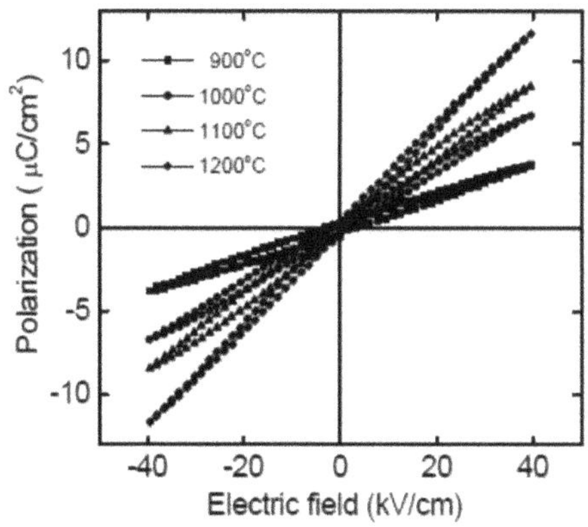

Figure 3.21. *P–E* hysteresis loops of the PLZT15/65/35 ceramics measured at room temperature. Reproduced with permission from [147]. Copyright 2001 Elsevier.

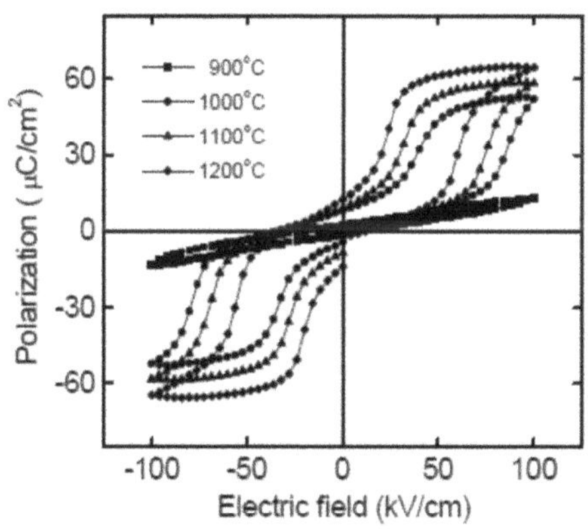

Figure 3.22. *P–E* hysteresis loops of the PLZT2/95/5 ceramics measured at room temperature. Reproduced with permission from [147]. Copyright 2001 Elsevier.

Table 3.2. Antiferroelectric parameters of the PLZT2/95/5 ceramics sintered at different temperatures. Reproduced with permission from [147]. Copyright 2001 Elsevier.

	900 °C	1000 °C	1100 °C	1200 °C
P_S (μC cm^{-2})	12	51	58	65
$E_{AFE\text{-}FE}$ (kV cm^{-1})	/	86	79	63
$E_{FE\text{-}AFE}$ (kV cm^{-1})	/	45	33	24

$E_{AFE\text{-}FE}$ and $E_{FE\text{-}AFE}$ are reduced, as the sintering temperature is raised, which could be similarly attributed to the change of grain size of the samples. The weakly visible remnant polarization of the PLZT2/95/5 ceramics can be assigned to thermal effect in the samples during the measurement and the presence of a small amount of the ferroelectric state [154–156].

PLZT7/60/40 ceramics were prepared from a mixture of PbO, La_2O_3, TiO_2 and ZrO_2, which was milled using a mechanochemical activation process [157]. The powders were milled using a 250 ml agate bowl with high wear resistant zirconia grinding media, with a Fritsch Pulverisette 6 mill. The balls had a diameter of 3 mm, while the ball-to-powder ratio was 10:1 and the speed of milling was 150 rpm. The milling was conducted for 5 h. The milling was stopped for 5 min after every 30 min of milling in order to prevent over-heating of the system. The milled powders were dried for 2 h at 500 °C and then calcined at 800 °C for 4 h. The PLZT were compacted using a CIP-sintering technique, at a pressure of 300 MPa. The samples were sintered in air at 1150 °C for 4 h in a double sealed covered crucible.

In this case, only partial phase formation of PLZT was observed, due to the low energy of the milling process, as shown in figure 3.23. According to DTA-TGA results, the reaction kinetics of PLZT phase formation was greatly reduced. Fully dense PLZT ceramics were obtained at relatively low sintering temperatures, compared to those in the conventional ceramic process. Figure 3.24 shows a representative TEM image of the as-milled powder, indicating that the average particle sizes were in the range 20–30 nm. The bulk density of sintered PLZT ceramics, measured using the Archimedes method, was >98% of the theoretical density. Figure 3.25 shows a representative SEM image of the PLZT ceramics. Dense PLZT ceramic pellets had a remnant polarization of Pr = 38 μC cm^{-2}, together with a coercive field of E_C = 5.83 kV cm^{-1}.

The same authors fabricated PLZT8/60/40 ceramics from mechanochemically treated oxide mixture, focusing on the microstructure and electrical properties of the ceramics [158]. After activation for 5 h, the mixed powder had spherical particles, with a relatively broad distribution of particle size. Importantly, the particle sizes after the high-energy ball milling are at the nanometric scale, as shown in figure 3.26.

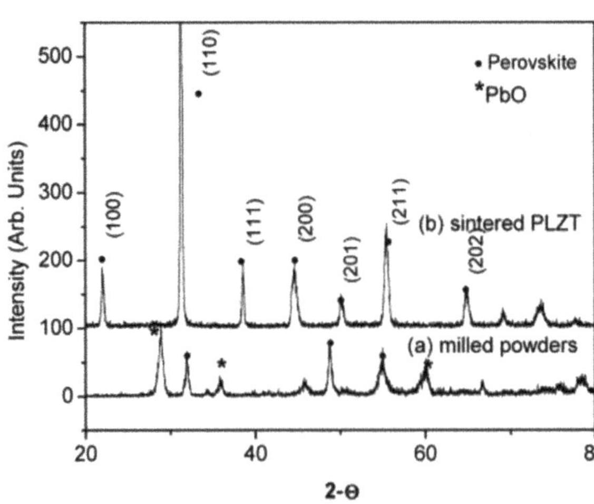

Figure 3.23. XRD patterns of the mixture of PbO, TiO_2 and La_2O_3 with the composition of PLZT7/60/40 mechanically activated for 5 h (a) and the sintered sample (b). Reproduced with permission from [157]. Copyright 2006 Springer.

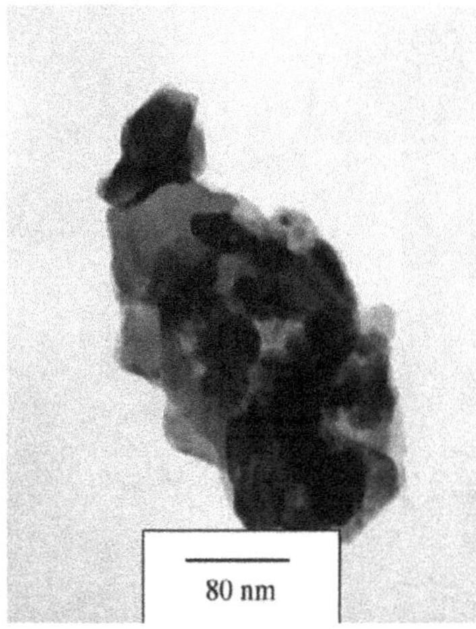

Figure 3.24. Representative TEM image of the mixture of PbO, TiO_2 and La_2O_3 with the composition of PLZT7/60/40 mechanically activated for 5 h. Reproduced with permission from [157]. Copyright 2006 Springer.

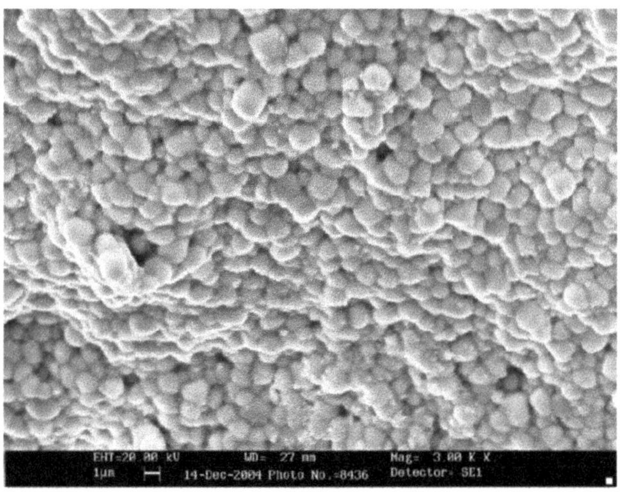

Figure 3.25. Representative SEM image of the PLZT7/60/40 ceramics. Reproduced with permission from [157]. Copyright 2006 Springer.

The decrease in particle size led to an increment in the surface-to-volume ratio, making the particles highly reactive and having reduced sintering temperatures. The final PLZT ceramics had an average grain size of 1.5 μm, as demonstrated in figure 3.27. The piezoelectric charge coefficient (d_{33}) was about 561 pC N^{-1}, while the remanent polarization (P_r) was 33.29 μC cm^{-2} and the coercive field (E_c) was 10.57 kV cm^{-1}. The electric field induced strain (S–E loop) reached a level of ~0.27% with minimum loss.

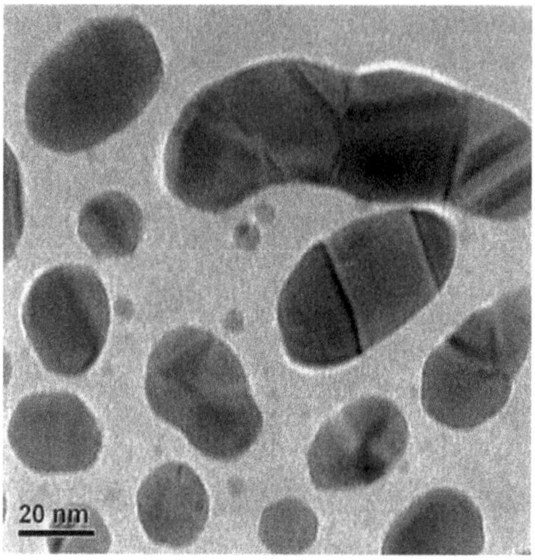

Figure 3.26. Representative TEM image of the oxide mixture after mechanochemical treatment. Reproduced with permission from [158]. Copyright 2014 Elsevier.

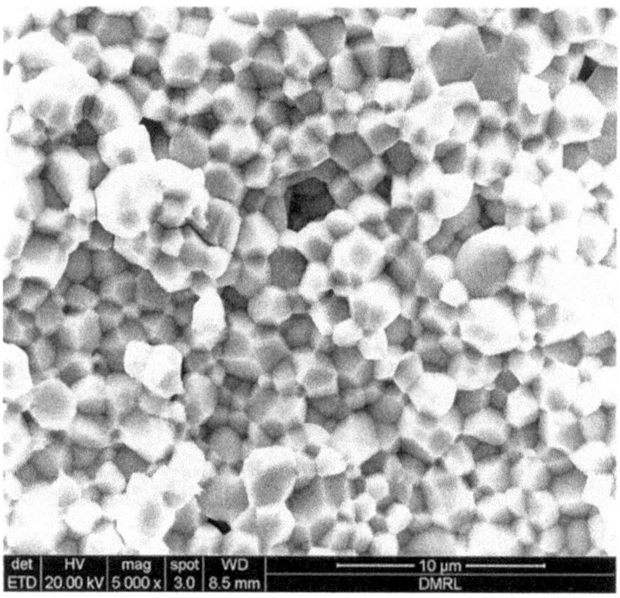

Figure 3.27. Representative SEM image of the PLZT8/60/40 ceramics. Reproduced with permission from [158]. Copyright 2014 Elsevier.

PLZT8/60/40 ceramics were also obtained from the powders made with high-energy mechanical ball milling using microwave sintering, in order to study their sintering behavior, microstructure, dielectric properties and piezoelectric performance [159]. The optimal sintering temperature was 1150 °C, resulting in highest density and more uniform grain size distribution. The average grain size was 1.2 μm, while the dielectric constant was about 2100. The PLZT ceramics exhibited a piezoelectric charge (d_{33}) and electromechanical coupling coefficient (k_p) of ~574 pC N^{-1} and ~67%, respectively.

Figure 3.28 shows XRD patterns of the PLZT8/60/40 ceramics after microwave sintering. The milled oxide mixture was significantly refined to be more reactive. During the sintering process PLZT phase formation was completed. In addition, no secondary phase was observed in the sintered samples, suggesting that the contamination was negligible during the milling process. However, after sintering at 1200 °C, additional peaks at 28° and 34° were present, indicating the formation of a non-ferroelectric pyrochlore phase. The presence of the pyrochlore phase was attributed to the volatilization of PbO at this temperature.

As mentioned earlier, the grain size has a large effect on tetragonality (c/a ratio), phase transition temperature (T_C), dielectric properties, ferroelectric properties, and the piezoelectric and pyroelectric performance of ferroelectric ceramics. Figure 3.29 shows SEM images of the microwave PLZT ceramics sintered at various temperatures. The samples sintered at 900 °C and 1000 °C were actually not fully densified, although the grain sizes were already at the submicron scale. After sintering at temperatures ⩾1050 °C, the densification was greatly promoted, while the grain size

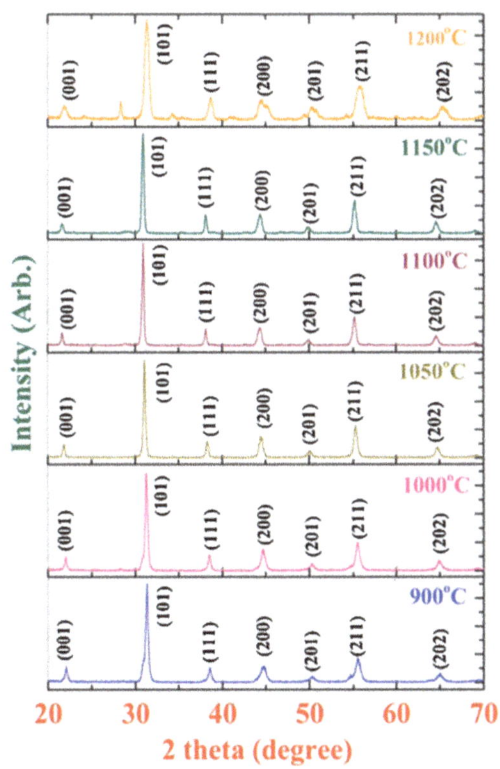

Figure 3.28. XRD patterns of the PLZT8/60/40 ceramics from the mechanochemically milled oxide mixture sintered at different temperatures. Reproduced with permission from [159]. Copyright 2016 Elsevier.

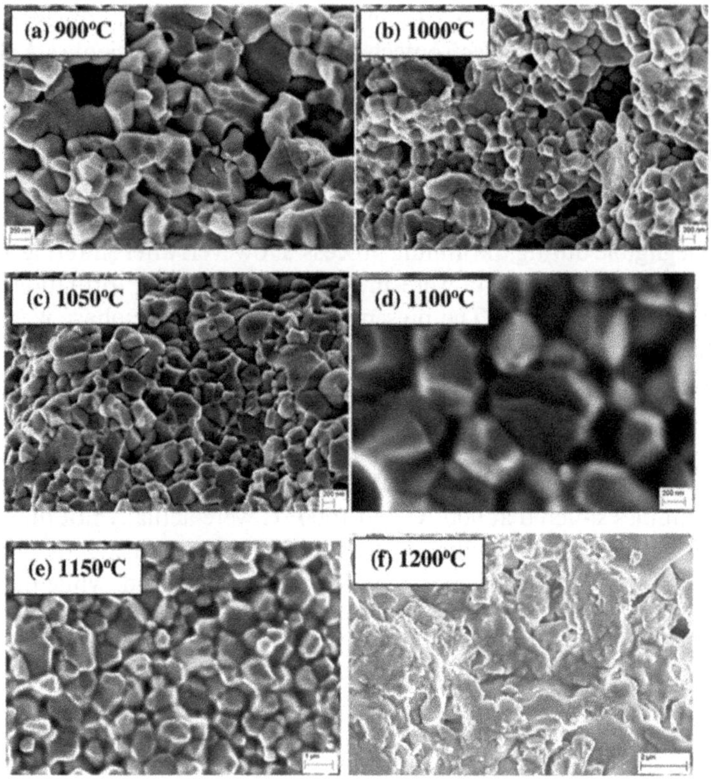

Figure 3.29. SEM images of the PLZT8/60/40 ceramics sintered at different temperatures for 20 min: (a) 900 °C, (b) 1000 °C, (c) 1050 °C, (d) 1100 °C, (e) 1150 °C and (f) 1200 °C. Reproduced with permission from [159]. Copyright 2016 Elsevier.

was constantly increased. The PLZT8/60/40 ceramics sintered at 1150 °C had a dense microstructure, with relatively uniform size distribution and an average grain size of about 1.2 μm, as shown in figure 3.29(e). Figure 3.29(f) indicates that the sample seemed to be over-sintered at 1200 °C.

3.3 Antiferroelectric ceramics

In antiferroelectric materials, spontaneous polarization dipoles are arranged antiparallel. At sufficiently high external electric fields, the antiferroelectric (AFE) could be transferred to the ferroelectric (FE) phase [160–164]. During the phase transition process, there is a large change in volume, because the FE phase has a relatively larger unit cell than the AFE phase, thus leading to a longitudinal strain. Also, the AFE–FE phase transition could be used for energy storage [165–169]. Additionally, antiferroelectric ceramics have strong pyroelectric, electro-optical and other effects. Due to their multiple component compositions, the synthesis of antiferroelectric materials using wet-chemical routes is still a challenge [170–172]. In this case, mechanochemical process has been a promising alternative to prepare antiferroelectric powders and ceramics.

PZ is a typical antiferroelectric material and pure PZ has a Curie temperature of $T_C = 230$ °C [173]. Fabrication of PZ ceramics from mechanochemically activated mixtures of PbO and ZrO_2 has been demonstrated [174]. It was found that the PZ phase formation from the mixture of PbO and ZrO_2 after milling for 24 h was still not complete. Compared to PT or PZT, this was mainly because the reactivity of ZrO_2 was less than that of TiO_2. Figure 3.30 shows the XRD patterns of the PZ samples from the milled mixture sintered at different temperatures. It is illustrated that single-phase PZ was formed after sintering at 900 °C, indicating the complete reaction of the oxide component remaining in the milled mixture.

After sintering at 900 °C, the sample was almost fully densified, which had an average grain size of about 1 μm, with a relative density of about 96%. As the sintering temperature increased from 900 °C to 1100 °C, both the average grain size and relative density of the PZ ceramics increased monotonically with increasing sintering temperature. Figure 3.31 shows SEM images of the PZ ceramics sintered at different temperatures. After that, the average grain size slowly increased from about 6.3 μm to about 8.2 μm, while the relative density significantly decreased from about 98% to about 93%, as the sintering temperature was increased from 1100 °C to 1300 °C, mainly due to the roughening of the grains and the volatilization of PbO at higher temperatures.

Figure 3.32 shows the dielectric constant of the PZ samples as a function of sintering temperature. It was found that the dielectric constant was monotonically increased as the sintering temperature was raised from 900 °C to 1200 °C. The increase in dielectric constant was ascribed to the change in grain size with sintering temperature. With increasing grain size, the fraction of grain boundaries decreases, thus leading to an increase in dielectric constant. The decline in the dielectric constant of the PZ ceramics after sintering at 1300 °C could be attributed to the

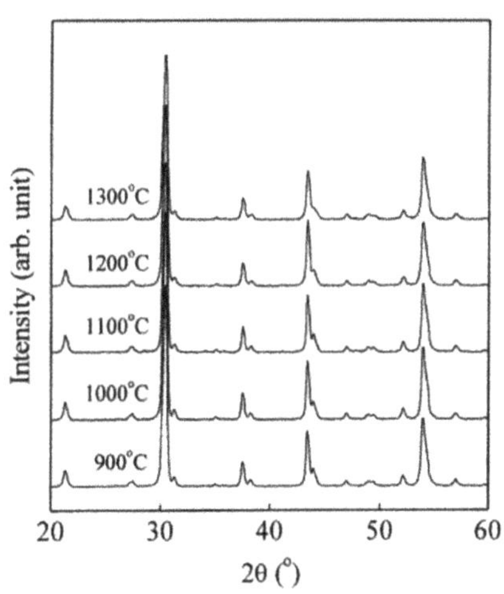

Figure 3.30. XRD patterns of the samples derived from the oxide mixture milled for 24 h after sintering at different temperatures. Reproduced with permission from [174]. Copyright 2001 Elsevier.

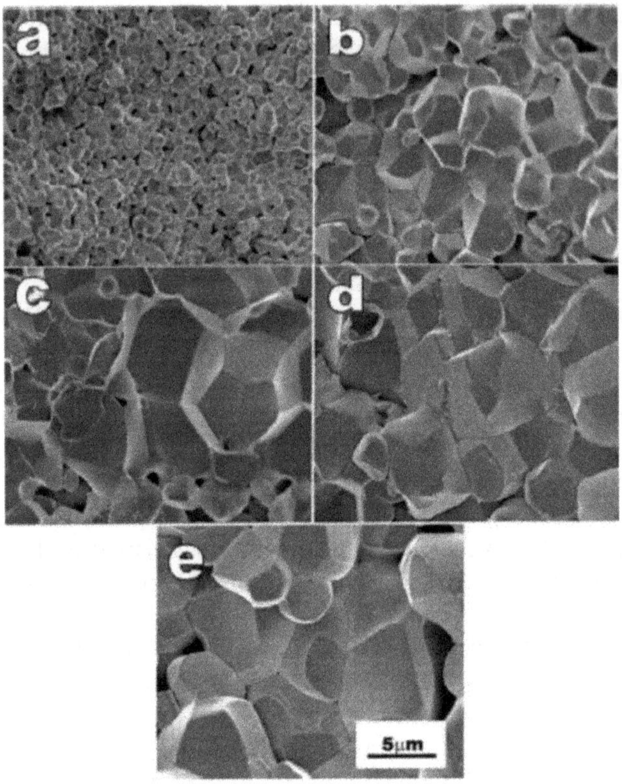

Figure 3.31. SEM images of the samples derived from the oxide mixture milled for 24 h after sintering at different temperatures. Reproduced with permission from [174]. Copyright 2001 Elsevier.

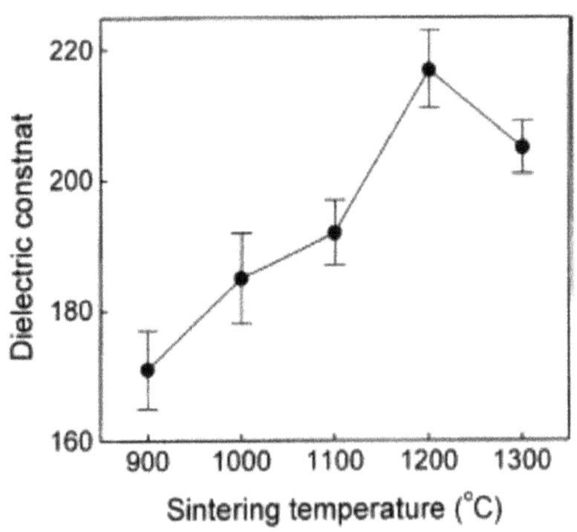

Figure 3.32. Dielectric constant of the PZ ceramics from the mechanochemically activated oxide mixture as a function of sintering temperature. Reproduced with permission from [174]. Copyright 2001 Elsevier.

volatilization of PbO during the sintering process, which was in agreement with the relative density results.

$Pb_{1-x}Ca_xZrO_3$ (with $x = 0.02$, 0.04, 0.06, 0.08, 0.10 and 0.15), $Pb_{1-x}Sr_xZrO_3$ (with $x = 0.02$, 0.04, 0.06, 0.08, 0.10 and 0.15) and $Pb_{1-x}Bi_xZrO_3$ (with $x = 0.01$, 0.02, 0.04, 0.06, 0.08, 0.10, 0.15 and 0.20) ceramics have been obtained directly from mixtures of their respective oxide/carbonate components, which were subjected to mechanochemical activation, without involving the calcination step. The starting powders were all commercially available and include PbO, ZrO_2, $CaCO_3$, $SrCO_3$ and Bi_2O_3. The milling was conducted by using the above-mentioned planetary mill, with similar milling conditions. The XRD patterns of the samples after sintering are shown in figures 3.33–3.35. The three groups of ceramics all had a single perovskite phase, with an orthorhombic crystal structure.

Lattice parameters of the three groups of samples are illustrated in figures 3.36–3.38. According to JCPDS cards No. 20–608 and No. 35–739, the lattice parameters of pure PZ are $a = 5.886 \pm 0.003$, $b = 11.749 \pm 0.008$ and $c = 8.248 \pm 0.006$ [175]. In all cases, the three lattice constants were shortened after the incorporation of the three elements, mainly because the ionic radii of Ca^{2+}, Sr^{2+} and Bi^{3+} are smaller than that of Pb^{2+}. In other words, over the composition range studied, solid solutions of $Pb_{1-x}Ca_xZrO_3$, $Pb_{1-x}Sr_xZrO_3$ and $Pb_{1-x}Bi_xZrO_3$ were formed, with the presence of obvious secondary phases, indicating the effectiveness and efficiency

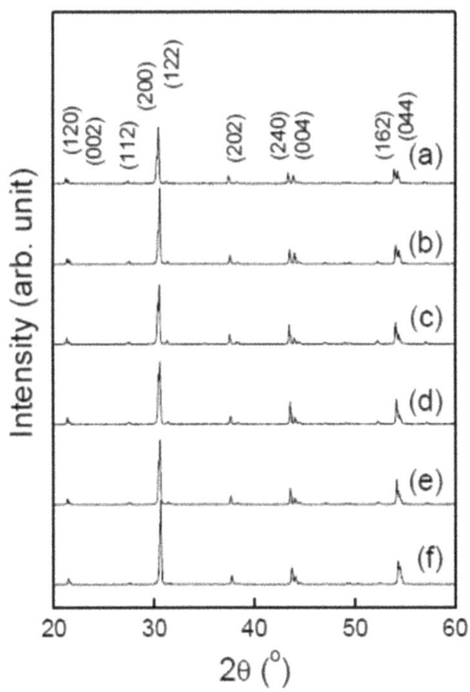

Figure 3.33. XRD patterns of the $Pb_{1-x}Ca_xZrO_3$ samples with different compositions (x): (a) 0.02, (b) 0.04, (c) 0.06, (d) 0.08, (e) 0.10 and (f) 0.15.

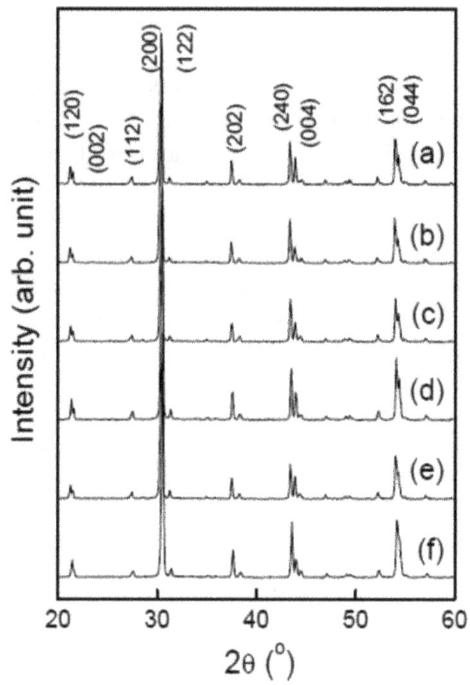

Figure 3.34. XRD patterns of the $Pb_{1-x}Sr_xZrO_3$ samples with different compositions (x): (a) 0.02, (b) 0.04, (c) 0.06, (d) 0.08, (e) 0.10 and (f) 0.15.

of the high-energy mechanochemical activation process in the synthesis and fabrication of Pb-containing ferroelectric/antiferroelectric materials.

SEM images of the PZ-based solid solution ceramics are presented in figures 3.39–3.41. It was found that the addition of the three elements into PZ led to a decrease in grain size of the final ceramics. Specifically, figure 3.42 shows the grain size of the $Pb_{1-x}Bi_xZrO_3$ samples as a function of the composition. As the content of Bi was increased from $x = 0.01$ to $x = 0.10$, the grain size was almost linearly decreased from about 10.9 μm to about 8.2 μm. After that, the rate in grain size reduction was slowed down, but the linearity was retained, with the sample of $x = 0.20$ having an average grain size of about 7.3 μm.

For $Pb_{1-x}Ca_xZrO_3$, the samples with $x = 0.02$ and $x = 0.04$ exhibited an intragrain fracture behavior, while the samples with $x = 0.06$ and $x = 0.08$ displayed an intergrain fracture, which was evidenced by the presence of whole grains, in particular for the sample with $x = 0.08$, as seen in figure 3.39(d). However, as the content of Ca was further increased to $x = 0.10$ and $x = 0.15$, intragrain fracture was present again, which could not be fully understood. In comparison, a monotonic variation in the grain fracture behavior from intragrain to intergrain was observed in the $Pb_{1-x}Sr_xZrO_3$ group.

In contrast to the above two groups of samples, an opposite trend in grain morphology variation was observed in the $Pb_{1-x}Bi_xZrO_3$ samples. At low concentrations of Bi, the samples experienced intragrain fracture, in particular the sample

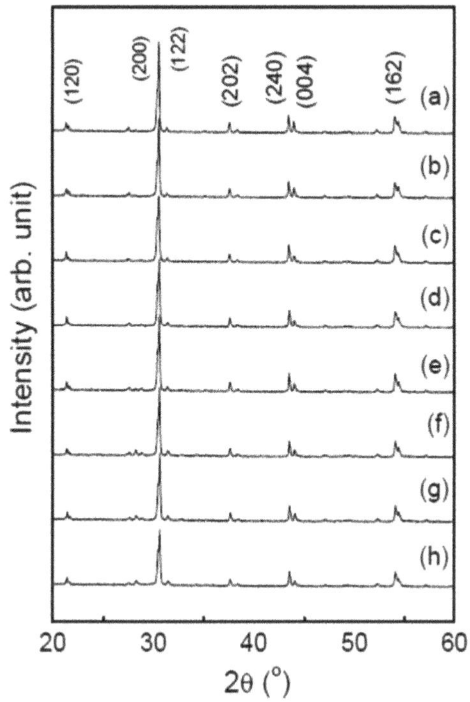

Figure 3.35. XRD patterns of the $Pb_{1-x}Bi_xZrO_3$ samples with different compositions (x): (a) 0.01, (b) 0.02, (c) 0.04, (d) 0.06, (e) 0.08, (f) 0.10, (g) 0.15 and (h) 0.20.

with $x = 0.01$. With increasing content of Bi, the intergrain fracture behavior became more and more pronounced. The phenomenon could be explained in terms of the difference in sintering mechanisms of the samples with different doping elements. For the $Pb_{1-x}Bi_xZrO_3$ group, liquid phase sintering could be dominant during the densification of the samples. In the presence of the liquid phase, the mass transport was enhanced, so that the samples of the $Pb_{1-x}Bi_xZrO_3$ group had larger grain sizes than those of the $Pb_{1-x}Ca_xZrO_3$ and $Pb_{1-x}Sr_xZrO_3$ groups.

Dielectric constants of the three groups of ceramic samples, as a function of composition, are demonstrated in figures 3.43–3.45. For the two groups of $Pb_{1-x}Ca_xZrO_3$ and $Pb_{1-x}Sr_xZrO_3$, the dielectric constant monotonically decreased with increasing content of Ca/Sr. As the doping levels of Ca and Sr were increased from $x = 0.02$ to $x = 0.15$, the dielectric constants decreased from about 99 to 79 and 103 to 81, respectively. For the $Pb_{1-x}Bi_xZrO_3$ group, the dielectric constant increased from about 110 to 430 as the concentration of Bi was raised from $x = 0.01$ to $x = 0.10$. However, as the concentration of Bi was further increased from $x = 0.10$ to $x = 0.20$, the dielectric constant was 610.

The decrease in dielectric constant of the $Pb_{1-x}Ca_xZrO_3$ and $Pb_{1-x}Sr_xZrO_3$ ceramics could be attributed to the decrease in their grain size with increasing content of the dopants and smaller polarizability of Ca^{2+}/Sr^{2+} than Pb^{2+}. For the $Pb_{1-x}Bi_xZrO_3$ group, the variation in dielectric constant was a concurrent effect of

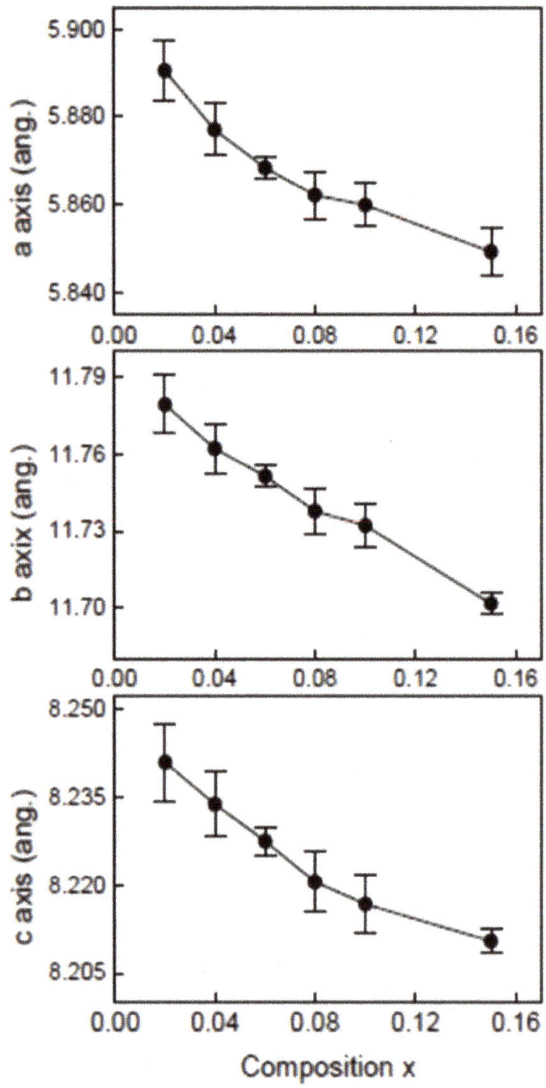

Figure 3.36. Lattice parameters of the $Pb_{1-x}Ca_xZrO_3$ samples as a function of the composition.

two factors. On one hand, the decrease in grain size would result in a decrease in dielectric constant. On the other hand, Bi^{3+} has a larger polarizability than Pb^{2+}, thus leading to an increase in dielectric constant with increasing concentration of Bi. Because the effect of polarizability was stronger than that of grain size, the dielectric constant of the $Pb_{1-x}Bi_xZrO_3$ ceramics was increased as the content of Bi was increased.

Antiferroelectric PZST ceramics, with typical compositions of $Pb_{0.99}Nb_{0.02}(Zr_{0.85}Sn_{0.13}Ti_{0.02})_{0.98}O_3$ (PNZST) and $Pb_{0.97}La_{0.02}(Zr_{0.65}Sn_{0.31}Ti_{0.04})O_3$ (PLZST), have been prepared using the oxide precursors activated with high-energy

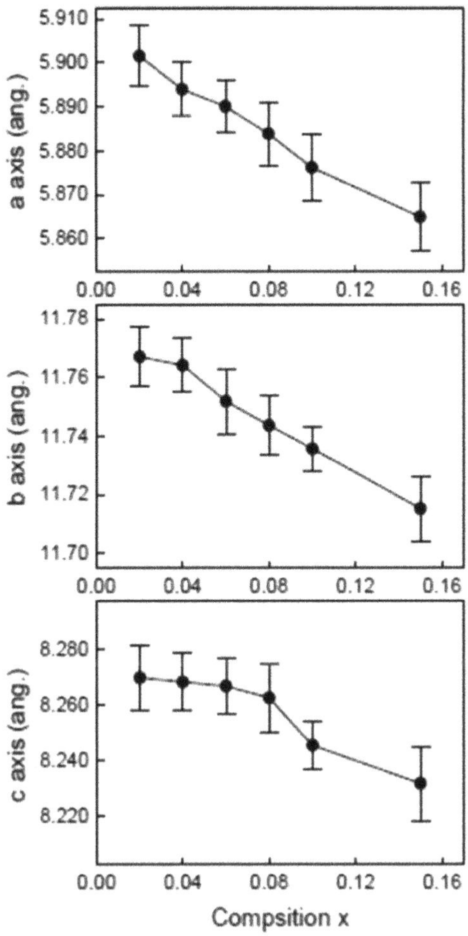

Figure 3.37. Lattice parameters of the $Pb_{1-x}Sr_xZrO_3$ samples as a function of the composition.

mechanochemical milling, with a planetary mill and tungsten carbide milling media [176]. Commercial powders of PbO, La_2O_3, Nb_2O_5, ZrO_2, SnO_2 and TiO_2 were used as the starting materials. Figure 3.46 shows XRD patterns of the oxide mixtures with the compositions of PNZST and PLZST. In both groups, the perovskite phase was the major phase, indicating the reaction of the oxides triggered by the high-energy activation. No contamination of the milling media materials was observed from the XRD patterns. SEM examination demonstrated that the as-milled powders consisted of particles with an average size at the scale of tens of nanometers.

According to the sintering curves of the mixtures, the shrinkage peaks were 805 °C and 795 °C for PLZST and PZSZT, respectively, confirming the incompleteness of the reaction of the constituent oxides during the mechanochemical activation process. This also supported the earlier statement that the incomplete reaction during the

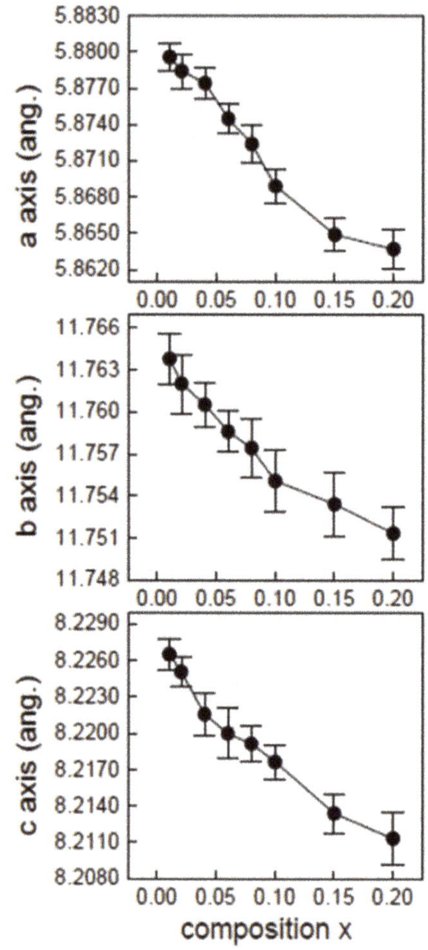

Figure 3.38. Lattice parameters of the $Pb_{1-x}Bi_xZrO_3$ samples as a function of the composition.

milling process had no effect on the fabrication of the final ceramics. Figure 3.47 shows SEM images of the PNSZT ceramics sintered at different temperatures. The two groups of samples exhibited a similar evolution profile in both microstructure and grain morphology with sintering temperature. There was an abrupt increase in grain size as the sintering temperature was increased from 1100 °C to 1200 °C.

Their density monotonically increased with increasing sintering temperature from 900 °C to 1100 °C. The density declined after sintering at 1200 °C, which was believed to be related to the volatilization of the PbO at high temperatures. The dielectric constant of PNZST and PLZST was increased with increasing sintering temperature, with a sharp increase as sintering temperature was increased from 1100 °C to 1200 °C. This observation suggested that the positive effect of grain size on dielectric constant was stronger than the negative effect of PbO loss at high temperatures.

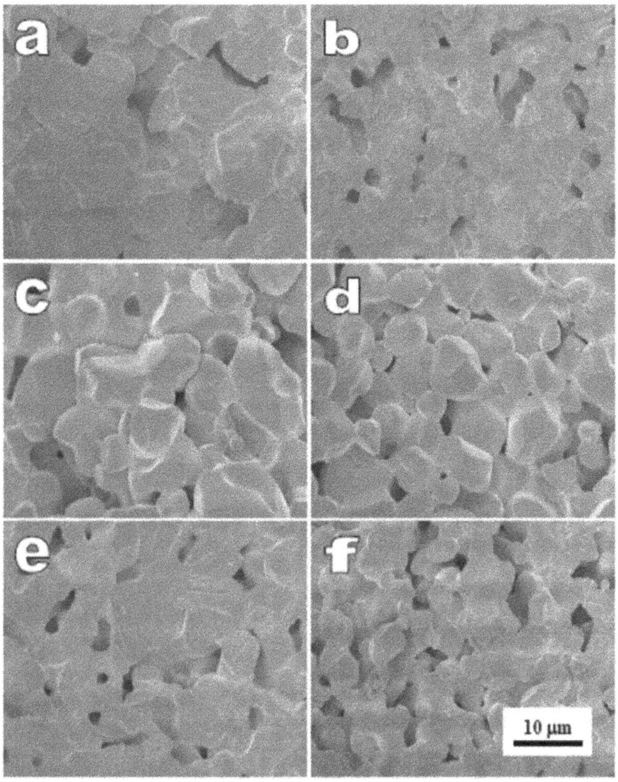

Figure 3.39. XRD patterns of the $Pb_{1-x}Ca_xZrO_3$ samples with different compositions (x): (a) 0.02, (b) 0.04, (c) 0.06, (d) 0.08, (e) 0.10 and (f) 0.15.

Figure 3.48 shows room-temperature P–E hysteresis loops of the PNZST and PLZST ceramics sintered at 1200 °C for 1 h, indicating the presence of typical double-loop characteristics. The maximum polarization values were 29 and 53 μC cm^{-2}, for PNZST and PLZST, respectively. Their corresponding forward-switching fields (antiferroelectric to ferroelectric or $E_{\text{AFE-FE}}$) were 76 and 75 kV cm^{-1}, while the backward-switching fields (ferroelectric to antiferroelectric or $E_{\text{FE-AFE}}$) were 33 and 24 kV cm^{-1}, respectively. The remnant polarization was caused by the heat dissipation in the samples during the measurement.

3.4 Discussion and conclusions

Various ferroelectric materials have been prepared using high-energy mechanochemical activation technology. In most cases, Pb-based ferroelectric phases in the form of nanosized powders can be obtained directly from oxide mixtures through mechanochemical activation, without the requirement for high temperature calcination or annealing. The as-synthesized powders exhibited high sinterability, thus leading to relatively low sintering temperatures to fabricate their corresponding ceramics, while the properties and performance of the ceramics are not compromised. Moreover, some relaxor ferroelectric phases, such as PZN, PMN and PFW,

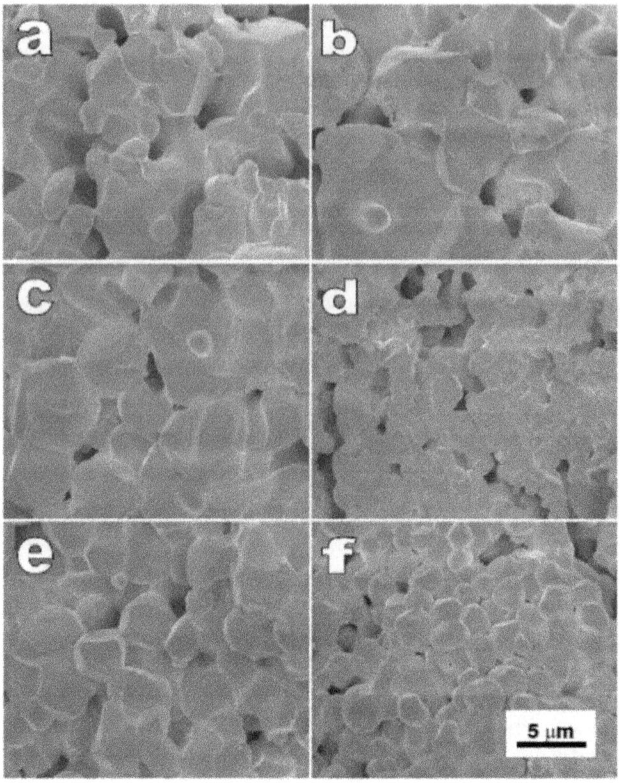

Figure 3.40. XRD patterns of the $Pb_{1-x}Sr_xZrO_3$ samples with different compositions (x): (a) 0.02, (b) 0.04, (c) 0.06, (d) 0.08, (e) 0.10 and (f) 0.15.

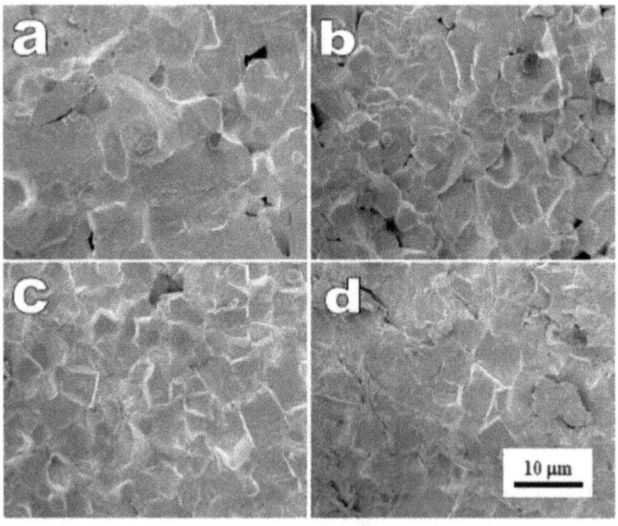

Figure 3.41. Representative SEM images of the $Pb_{1-x}Bi_xZrO_3$ samples with different compositions (x): (a) 0.02, (b) 0.06, (c) 0.08 and (d) 0.15.

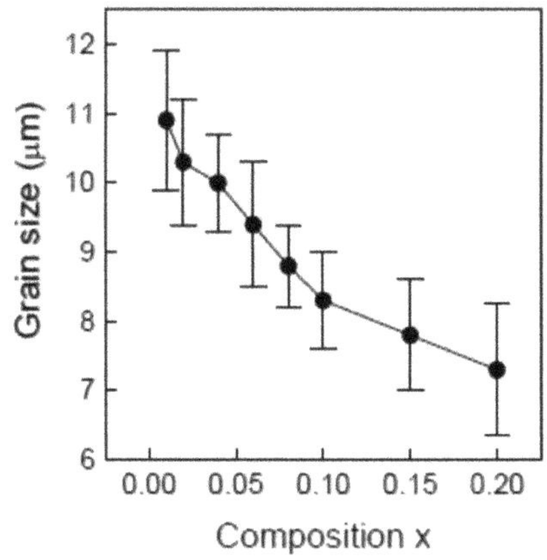

Figure 3.42. Average grain size of the $Pb_{1-x}Bi_xZrO_3$ samples as a function of the composition.

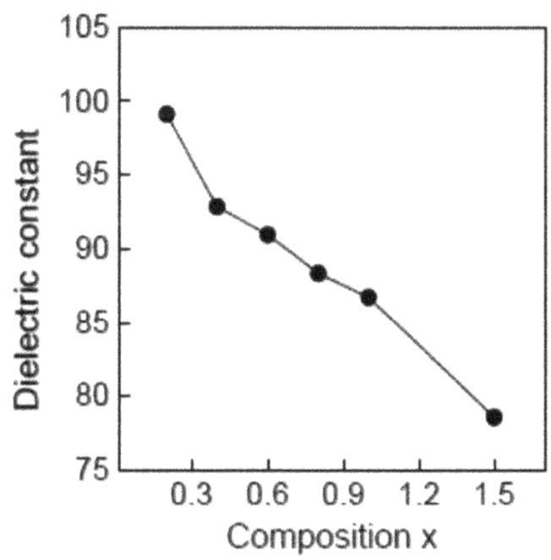

Figure 3.43. Dielectric constant of the $Pb_{1-x}Ca_xZrO_3$ samples as a function of composition.

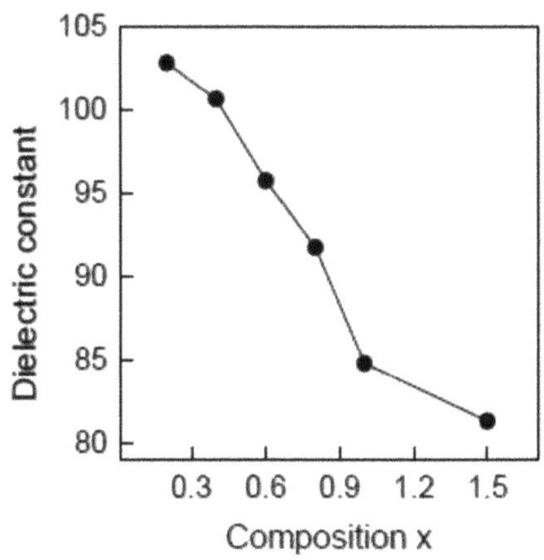

Figure 3.44. Dielectric constant of the $Pb_{1-x}Sr_xZrO_3$ samples as a function of composition.

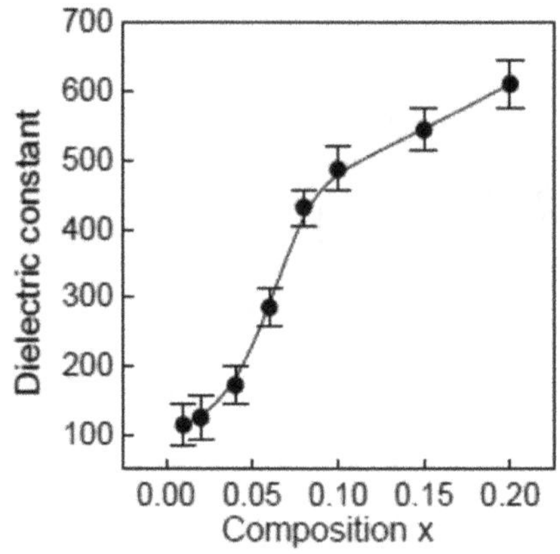

Figure 3.45. Dielectric constant of the $Pb_{1-x}Bi_xZrO_3$ samples as a function of composition.

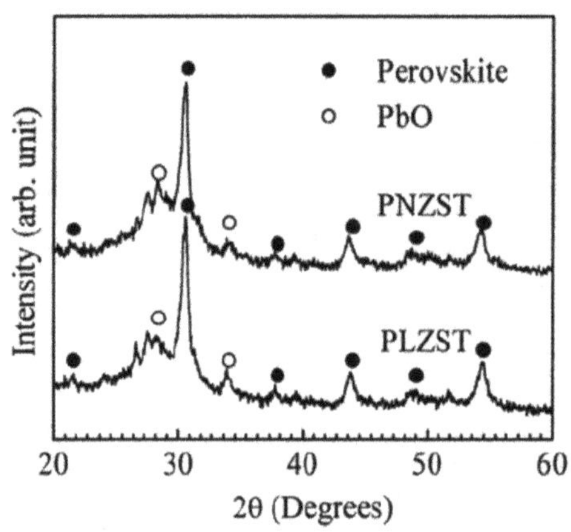

Figure 3.46. XRD patterns of the two oxide mixtures after high-energy mechanochemical activation. Reproduced with permission from [176]. Copyright 2002 Springer.

which are difficult to create using the conventional solid-state reaction process, can be synthesized directly. In addition, the process can easily be incorporated into the conventional ceramic process, while not involving the wet-mixing step, thus significantly shortening the whole fabrication process. In fact, it has been observed that the mixtures activated for a certain time can be processed to final ceramics, even when the designed phases have not been formed. Such precursors would allow reaction sintering to occur, which in turn additionally enhances the densification effect of the ceramic materials. In other words, the process can be further shortened.

Bismuth-based ferroelectric materials, well known as the Aurivillius family, have also been synthesized using the mechanochemical activation process. Some of them can be directly derived from oxide or carbonate precursors, while most of them are not obtained in the form of crystalline phases, but in an amorphous state instead. This is highly dependent on the milling conditions, including the type of milling machine, milling media and milling parameters. The amorphized precursors, with a

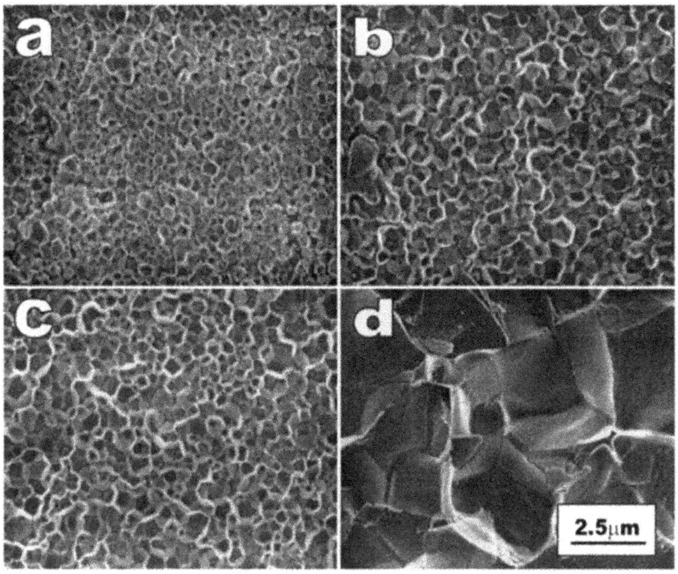

Figure 3.47. SEM images of the PNZST ceramics sintered at different temperatures for 1 h: (a) 900 °C, (b) 1000 °C, (c) 1100 °C and (d) 1200 °C. Reproduced with permission from [176]. Copyright 2002 Springer.

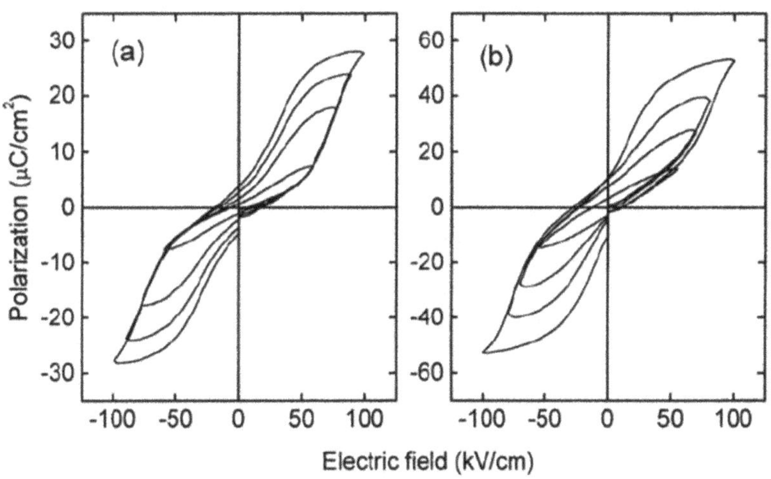

Figure 3.48. *P–E* hysteresis loops of the two ceramics sintered at 1200 °C for 1 h: (a) PNZST and (b) PLZST. Reproduced with permission from [176]. Copyright 2002 Springer.

high degree of particle refinement and compositional homogeneity, can easily be crystallized into target compounds through calcination at much lower temperatures, compared to those in other synthetic processes. Ceramics with similar or higher performance can be developed with the amorphous powders.

The mechanisms behind the phase formation or amorphization as a result of the high-energy mechanochemical activation for different materials at different milling conditions have not been clarified. However, it has been widely accepted that the milling is started as the milled precursors are highly refined and homogenized. During this process, a large number of defects, fresh surfaces or interfaces are generated in the starting materials, so that they become highly reactive. In addition, the particle refinement results in shortened diffusion paths, thus promoting mass transport. Further activation would facilitate nucleation and further trigger the crystallization of the desired compounds. For some materials, amorphization instead of crystallization occurs, due to the increased degree of structural and chemical disordering of the precursors.

However, it is necessary to pay more attention to the mechanisms during the mechanochemical activation process, although tremendous progress has already been achieved, as reported in the available literature. In future research, the focus should be on the following: (i) systematic study of the effects of processing parameters on the phase formation of the material to be synthesized; (ii) identifying or monitoring the microstructural and compositional evolution during the activation process and (iii) strengthening theoretical simulation and modeling of the reaction process. It is strongly believed that this special method could be applied to other materials.

Acknowledgments

Shenzhen Technology University (SZTU) is acknowledged for the financial support of a start-up grant (2018) and also the Natural Science Foundation of Top Talent of SZTU (grant no. 2019010801002).

References

[1] Kong L B, Zhang T S, Ma J and Boey F 2008 Progress in synthesis of ferroelectric ceramic materials via high-energy mechanochemical technique *Prog. Mater Sci.* **53** 207–322

[2] Haertling G H 1999 Ferroelectric ceramics: history and technology *J. Am. Ceram. Soc.* **82** 797–818

[3] Damjanovic D 1998 Ferroelectric, dielectric and piezoelectric properties of ferroelectric thin films and ceramics *Rep. Prog. Phys.* **61** 1267–324

[4] Valasek J 1921 Piezo-electric and allied phenomena in Rochelle salt *Phys. Rev.* **17** 475–81

[5] Valasek J 1922 Piezo-electric activity of Rochelle salt under various conditions *Phys. Rev.* **19** 478–91

[6] Valasek J 1922 Properties of Rochelle salt related to the piezo-electric effect *Phys. Rev.* **20** 639–63

[7] Valasek J 1924 Dielectric anomalies in Rochelle salt crystals *Phys. Rev.* **24** 560–8

[8] Keizer K, Janssen E H J, Vries K J D and Burggraaf A J 1973 Influence of particle-size and structure of ZrO_2 on microstructure development and dielectric constant of $PbZr_{0.5}Ti_{0.5}O_3$ *Mater. Res. Bull.* **8** 533–44

[9] Yamamoto T 1992 Optimum preparation methods for piezoelectric ceramics and their evaluation *Am. Ceram. Soc. Bull.* **71** 978–85

[10] Arlt G 1990 The influence of microstructure on the properties of ferroelectric ceramics *Ferroelectrics* **104** 217–27

[11] Camargo E R, Frantti J and Kakihana M 2001 Low-temperature chemical synthesis of lead zirconate titanate (PZT) powders free from halides and organics *J. Mater. Chem.* **11** 1875–9
[12] Yoshikawa Y and Tsuzuki K 1992 Fabrication of transparent lead lanthanum zirconate titanate ceramics from fine powders by two-stage sintering *J. Am. Ceram. Soc.* **75** 2520–8
[13] Cerqueira M, Nasar R S, Leite E R, Longo E and Varela J A 1998 Synthesis and characterization of PLZT (9/65/35) by the Pechini method and partial oxalate *Mater. Lett.* **35** 166–71
[14] Tas A C 1999 Preparation of lead zirconate titanate ($Pb(Zr_{0.52}Ti_{0.48})O_3$) by homogeneous precipitation and calcination *J. Am. Ceram. Soc.* **82** 1582–4
[15] Oren E E, Taspinar E and Tas A C 1997 Preparation of lead zirconate by homogeneous precipitation and calcination *J. Am. Ceram. Soc.* **80** 2714–6
[16] Bochenek D 2010 Magnetic and ferroelectric properties of $PbFe_{1/2}Nb_{1/2}O_3$ synthesized by a solution precipitation method *J. Alloys Compd.* **504** 508–13
[17] Dubey S and Kurchania R 2017 Effect of sintering temperature on morphology, dielectric and ferroelectric properties of five layer Aurivillius oxides: $A_2Bi_4Ti_5O_{18}$ (A = Ba, Pb and Sr) synthesized by co-precipitation route *J. Mater. Sci.-Mater. Electron.* **28** 8266–77
[18] Li G, Yang T Q, Wang J F, Chen S C and Yao X 2013 Preparation of PLZST antiferroelectric ceramics by hydroxide coprecipitation method *Ceram. Int.* **39** S345–8
[19] Simon-Seveyrat L, Hajjaji A, Emiane Y, Guiffard B and Guyomar D 2007 Re-investigation of synthesis of $BaTiO_3$ by conventional solid-state reaction and oxalate coprecipitation route for piezoelectric applications *Ceram. Int.* **33** 35–40
[20] Tang J L, Zhu M K, Zhong T, Hou Y D, Wang H and Yan H 2007 Synthesis of fine Pb ($Fe0_{.5}Nb_{0.5})O_3$ perovskite powders by coprecipitation method *Mater. Chem. Phys.* **101** 475–9
[21] Tarale A N, Premkumar S, Reddy V R and Mathe V L 2017 Microstructural evolution of 0.75PMN–0.25PT ferroelectrics synthesized by hydroxide co-precipitation method and their dielectric properties *J. Mater. Sci.-Mater. Electron.* **28** 5485–97
[22] Xiang P H, Kinemuchi Y, Nagaoka T and Watari K 2005 Sintering behaviors of bismuth titanate synthesized by a coprecipitation method *Mater. Lett.* **59** 3590–4
[23] Xue L H, Li Q, Guo Q Y, Liu R and Zhang Y L 2005 Synthesis and characterization of PLZST prepared by coprecipitation *J. Mater. Sci.* **40** 2697–9
[24] Kim S, Jun M C and Hwang S C 1999 Preparation of undoped lead titanate ceramics via sol–gel processing *J. Am. Ceram. Soc.* **82** 289–96
[25] Tartaj J, Moure C, Lascano L and Duran P 2001 Sintering of dense ceramics bodies of pure lead titanate obtained by seeding-assisted chemical sol–gel *Mater. Res. Bull.* **36** 2301–10
[26] Chauhan A K S, Gupta V and Sreenivas K 2006 Dielectric and piezoelectric properties of sol–gel derived Ca doped $PbTiO_3$ *Mater. Sci. Eng.* B **130** 81–8
[27] Lanki M, Nourmohammadi A and Feiz M H 2013 A precise investigation of lead partitioning in sol–gel derived $PbTiO_3$ nanopowders *Ferroelectrics* **448** 123–33
[28] Selbach S M, Wang G Z, Einarsrud M A and Grande T 2007 Decomposition and crystallization of a sol–gel-derived $PbTiO_3$ precursor *J. Am. Ceram. Soc.* **90** 2649–52
[29] Singh A, Gupta V, Sreenivas K and Katiyar R S 2007 Influence of Ca additives on the optical and dielectric studies of sol–gel derived $PbTiO_3$ ceramics *J. Phys. Chem. Solids* **68** 119–23
[30] Zhao C W, Luo B C, Guo S J and Chen C L 2017 Enhanced electrical and photocurrent characteristics of sol–gel derived Ni-doped $PbTiO_3$ thin films *Ceram. Int.* **43** 7861–5

[31] Sato S, Murakata T, Yanagi H, Miyasaka F and Iwaya S 1994 Hydrothermal synthesis of fine perovskite $PbTiO_3$ powders with a simple mode of size distribution *J. Mater. Sci.* **29** 5657–63

[32] Moon J, Li T, Randall C A and Adair J H 1997 Low temperature synthesis of lead titanate by a hydrothermal method *J. Mater. Res.* **12** 189–97

[33] Peterson C R and Slamovich E B 1999 Effect of processing parameters on the morphology of hydrothermally derived $PbTiO_3$ powders *J. Am. Ceram. Soc.* **82** 1702–10

[34] Zimmermann-Chopin R and Auer S 1994 Spray drying of sol–gel precursors for the manufacturing of PZT powders *J. Sol–Gel Sci. Technol.* **3** 101–7

[35] Muralidharan B G, Sengupta A, Rao G S and Agrawal D C 1995 Powders of $Pb(Zr_xTi_{1-x})O_3$ by sol–gel coating of PbO *J. Mater. Sci.* **30** 3231–7

[36] Wu A Y, Vilarinho P M, Salvado I M M and Baptista J L 2000 Sol–gel preparation of lead zirconate titanate powders and ceramics: effect of alkoxide stabilizers and lead precursors *J. Am. Ceram. Soc.* **83** 1379–85

[37] Chao C Y *et al* 2013 Hydrothermal synthesis of ferroelectric $PbTiO_3$ nanoparticles with dominant (001) facets by titanate nanostructure *Cryst. Eng. Comm.* **15** 8036–40

[38] Delahaye T, Al-Zein A, Berger M H, Bril X and Hochepied J F 2014 Hydrothermal synthesis of ferroelectric mixed potassium niobate-lead titanate nanoparticles *J. Am. Ceram. Soc.* **97** 1456–64

[39] Golic D L *et al* 2016 Structural, ferroelectric and magnetic properties of $BiFeO_3$ synthesized by sonochemically assisted hydrothermal and hydro-evaporation chemical methods *J. Eur. Ceram. Soc.* **36** 1623–31

[40] Lim J B, Suvorov D, Kim M H and Jeon J H 2012 Hydrothermal synthesis and characterization of $(Bi,K)TiO_3$ ferroelectrics *Mater. Lett.* **67** 286–8

[41] Liu Y W, Pu Y P and Sun Z X 2014 Enhanced relaxor ferroelectric behavior of BCZT lead-free ceramics prepared by hydrothermal method *Mater. Lett.* **137** 128–31

[42] Miao H Y, Zhou Y H, Tan G Q and Dong M 2008 Microstructure and dielectric properties of ferroelectric barium strontium titanate ceramics prepared by hydrothermal method *J. Electroceram.* **21** 553–6

[43] Qi L *et al* 2005 Low-temperature paraelectric–ferroelectric phase transformation in hydrothermal $BaTiO_3$ particles *Mater. Lett.* **59** 2794–8

[44] Um M H and Kumazawa H 2000 Hydrothermal synthesis of ferroelectric barium and strontium titanate extremely fine particles *J. Mater. Sci.* **35** 1295–300

[45] Wang F *et al* 2013 Shape-controlled hydrothermal synthesis of ferroelectric $Bi4Ti_3O_{12}$ nanostructures *Cryst. Eng. Comm.* **15** 1397–403

[46] Wei N, Zhang D M, Han X Y, Yang F X, Zhong Z C and Zheng K Y 2007 Synthesis and mechanism of ferroelectric potassium tantalate niobate nanoparticles by the solvothermal and hydrothermal processes *J. Am. Ceram. Soc.* **90** 1434–7

[47] Aghayan M, Zak A K, Behdani M and Hashim A M 2014 Sol–gel combustion synthesis of Zr-doped $BaTiO_3$ nanopowders and ceramics: dielectric and ferroelectric studies *Ceram. Int.* **40** 16141–6

[48] Bongkarn T, Chootin S, Pinitsoontorn S and Maensiri S 2016 Excellent piezoelectric and ferroelectric properties of KNLNTS ceramics with Fe_2O_3 doping synthesized by the solid state combustion technique *J. Alloys Compd.* **682** 14–21

[49] Dubey S, Subohi O and Kurchania R 2017 Solution combustion synthesis: effect of calcination and sintering temperature on structural, dielectric and ferroelectric properties of five layer Aurivillius oxides *Physica* B **521** 73–83

[50] Kornphom C, Rittisak J, Laowanidwatana A and Bongkarn T 2018 Enhanced dielectric and ferroelectric behavior in 0.94BNT–0.06BCTS lead free piezoelectric ceramics synthesized by the solid state combustion technique *Integr. Ferroelectr.* **187** 20–32

[51] Selvamurugan V *et al* 2017 Ferroelectric and dielectric behavior of samarium-substituted $Bi_4Ti_3O_{12}$ nanomaterials synthesized by gel combustion method *Trans. Indian Inst. Met.* **70** 903–8

[52] Subohi O, Kumar G S, Malik M M and Kurchania R 2015 Study of influence of fuel on dielectric and ferroelectric properties of bismuth titanate ceramics synthesized using solution based combustion technique *Mater. Res. Exp.* **2** 036302

[53] Wang J L, Tang J, Lei Z W, Liu M, Knize R J and Lu Y L 2014 Pyrochlore-free ferroelectric $0.64Pb(Ni_{1/3}Nb_{2/3})O_3$–$0.36PbTiO_3$ ceramics synthesized by the combustion method *J. Am. Ceram. Soc.* **97** 2130–4

[54] Narendar Y and Messing G L 1997 Kinetic analysis of combustion synthesis of lead magnesium niobate from metal carboxylate gels *J. Am. Ceram. Soc.* **80** 915–24

[55] Ghasemifard M 2012 Optical properties of BMN–BT ferroelectric thin films prepared by spray pyrolysis deposition *Mod. Phys. Lett.* B **26** 1150009

[56] Nautiyal A, Sekhar K C, Pathak N P and Nath R 2010 Study of ferroelectric properties of spray pyrolysis deposited cesium nitrate films *Thin Solid Films* **518** E143–5

[57] Tomashpol'skii Y Y, Rybakova L F, Lunina T V, Fedoseeva O F, Prutchenko S G and Men'shikh S A 2001 Ferroelectric lead zirconate titanate films prepared by spray pyrolysis of carboxylate solutions *Inorg. Mater.* **37** 500–7

[58] Calzada M L, Torres M, Fuentes-Cobas L E, Mehta A, Ricote J and Pardo L 2007 Ferroelectric self-assembled $PbTiO_3$ perovskite nanostructures onto (100) $SrTiO_3$ substrates from a novel microemulsion aided sol–gel preparation method *Nanotechnology* **18** 375603

[59] Torres M, Alonso M, Lourdes Calzada M and Pardo L 2009 Influence of the substrate surface on the self-assembly of ferroelectric $PbTiO_3$ nanostructures obtained by microemulsion assisted chemical solution deposition *Ferroelectrics* **390** 122–9

[60] Arendt R H, Rosolowski J H and Szymaszek J W 1979 Lead zirconate titanate ceramics from molten-salt solvent synthesized powders *Mater. Res. Bull.* **14** 703–9

[61] Chiu C C, Li C C and Desu S B 1991 Molten-salt synthesis of a complex perovskite $Pb(Fe_{0.5}Nb_{0.5})O_3$ *J. Am. Ceram. Soc.* **74** 38–41

[62] Kannan B R and Venkataraman B H 2016 Dielectric and electrical conductivity characteristics of undoped and samarium doped ferroelectric $SrBi_2Ta_2O_9$ ceramics derived from molten salt synthesis route *Ferroelectrics* **493** 110–9

[63] Porob D G and Maggard P A 2006 Synthesis of textured $Bi_5Ti_3FeO_{15}$ and $LaBi_4Ti_3FeO_{15}$ ferroelectric layered Aurivillius phases by molten-salt flux methods *Mater. Res. Bull.* **41** 1513–9

[64] Aning A O, Hong C and Desu S B 1995 Novel synthesis of lead titanate by mechanical alloying *Mater. Sci. Forum* **179–181** 207–13

[65] Yu T, Shen Z X, Xue J M and Wang J 2002 Nanocrystalline $PbTiO_3$ powders from an amorphous Pb–Ti–O precursor by mechanical activation *Mater. Chem. Phys.* **75** 216–9

[66] Kong L B, Zhu W G and Tan O K 1999 Direct formation of nano-sized $PbTiO_3$ powders by high energy ball milling *Ferroelectrics* **230** 583–8

[67] Kong L B, Zhu W and Tan O K 2000 $PbTiO_3$ ceramics derived from high-energy ball milled nano-sized powders *J. Mater. Sci. Lett.* **19** 1963–6
[68] Xue J M, Wan D M, Lee S E and Wang J 1999 Mechanochemical synthesis of lead zirconate titanate from mixed oxides *J. Am. Ceram. Soc.* **82** 1687–92
[69] Lee S E, Xue J M, Wan D M and Wang J 1999 Effects of mechanical activation on the sintering and dielectric properties of oxide-derived PZT *Acta Mater.* **47** 2633–9
[70] Kong L B, Zhu W and Tan O K 2000 Preparation and characterization of $Pb(Zr_{0.52}Ti_{0.48})O_3$ ceramics from high-energy ball milling powders *Mater. Lett.* **42** 232–9
[71] Kong L B, Ma J, Zhang T S, Zhu W and Tan O K 2001 $Pb(Zr_xTi_{1-x})O_3$ ceramics via reactive sintering of partially reacted mixture produced by a high-energy ball milling process *J. Mater. Res.* **16** 1636–43
[72] Kong L B, Ma J, Huang H T, Zhu W and Tan O K 2001 Lead zirconate titanate ceramics derived from oxide mixture treated by a high-energy ball milling process *Mater. Lett.* **50** 129–33
[73] Kong L B, Ma J, Zhu W and Tan O K 2001 Reaction sintering of partially reacted system for PZT ceramics via a high-energy ball milling *Scr. Mater.* **44** 345–50
[74] Brankovic Z, Brankovic G, Jovalekic C, Maniette Y, Cilense M and Varela J A 2003 Mechanochemical synthesis of PZT powders *Mater. Sci. Eng.* A **345** 243–8
[75] Parashar S K S, Choudhary R N P and Murty B S 2003 Ferroelectric phase transition in $Pb_{0.92}Gd_{0.08}(Zr_{0.53}Ti_{0.47})_{0.98}O_3$ nanoceramic synthesized by high-energy ball milling *J. Appl. Phys.* **94** 6091–6
[76] Wang J, Xue J M, Wan D M and Ng W B 1999 Mechanochemically synthesized lead magnesium niobate *J. Am. Ceram. Soc.* **82** 1358–60
[77] Wang J, Xue J M, Wan D M and Ng W B 1999 Mechanochemical fabrication of single phase PMN of perovskite structure *Solid State Ionics* **124** 271–9
[78] Xue J M, Wang J and Rao T M 2001 Synthesis of $Pb(Mg_{1/3}Nb_{2/3})O_3$ in excess lead oxide by mechanical activation *J. Am. Ceram. Soc.* **84** 660–2
[79] Wang J, Xue J M, Wan D M and Gan B K 2000 Mechanically activating nucleation and growth of complex perovskites *J. Solid State Chem.* **154** 321–8
[80] Xue J M, Wan D M and Wang J 2002 Functional ceramics of nanocrystallinity by mechanical activation *Sol. State Ion* **151** 403–12
[81] Kong L B, Ma J, Zhu W and Tan O K 2001 Preparation of PMN powders and ceramics via a high-energy ball milling process *J. Mater. Sci. Lett.* **20** 1241–3
[82] Kong L B, Ma J, Zhu W and Tan O K 2002 Translucent PMN and PMN–PT ceramics from high-energy ball milling derived powders *Mater. Res. Bull.* **37** 23–32
[83] Wang J, Xue J M and Wan D M 2000 How different is mechanical activation from thermal activation? A case study with PZN and PZN-based relaxors *Sol. State Ion* **127** 169–75
[84] Xue J M, Wan D M and Wang J 1999 Mechanochemical synthesis of nanosized lead titanate powders form mixed oxides *Mater. Lett.* **39** 364–9
[85] Kong L B, Ma J, Zhu W and Tan O K 2001 Preparation of $Bi_4Ti_3O_{12}$ ceramics via a high-energy ball milling process *Mater. Lett.* **51** 108–14
[86] Ng S H, Xue J M and Wang J 2002 Bismuth titanate from mechanical activation of a chemically coprecipitated precursor *J. Am. Ceram. Soc.* **85** 2660–5
[87] Shantha K and Varma K B R 1999 Preparation and characterization of nanocrystalline powders of bismuth vanadate *Mater. Sci. Eng.* B **60** 66–75

[88] Shantha K, Subbanna G N and Varma K B R 1999 Mechanically activated synthesis of nanocrystalline powders of ferroelectric bismuth vanadate *J. Solid State Chem.* **142** 41–7

[89] Shantha K and Varma K B R 2000 Characterization of fine-grained bismuth vanadate ceramics obtained using nanosized powders *J. Am. Ceram. Soc.* **83** 1122–8

[90] Ricote J, Pardo L, Castro A and Millan P 2001 Study of the process of mechanochemical activation to obtain Aurivillius oxides with $n = 1$ *J. Solid State Chem.* **160** 54–61

[91] Castro A, Millan P, Ricote J and Pardo L 2000 Room temperature stabilisation of γ-$Bi_2VO_{5.5}$ and synthesis of the new fluorite phase f-Bi_2VO_5 by a mechanochemical activation method *J. Mater. Chem.* **10** 767–71

[92] Castro A, Begue P, Jimenez B, Ricote J, Jimenez R and Galy J 2003 New $Bi_2Mo_{1-x}W_xO_6$ solid solution: mechanosynthesis, structural study, and ferroelectric properties of the $x = 0.75$ member *Chem. Mater.* **15** 3395–401

[93] Zhao M L, Wang C L, Zhong W L, Zhang P L, Wang J F and Chen H C 2003 Dielectric and pyroelectric properties of $SrBi_4Ti_4O_{15}$-based ceramics for high-temperature applications *Mater. Sci. Eng.* B **99** 143–6

[94] Zhu J, Mao X Y and Chen X B 2004 Properties of vanadium-doped $SrBi_4Ti_4O_{15}$ ferroelectric ceramics *Solid State Commun.* **129** 707–10

[95] Zheng L Y, Li G R, Zhang W Z, Chen D R and Yin Q R 2003 The structure and piezoelectric properties of $(Ca_{1-x}Sr_x)Bi_4Ti_4O_{15}$ ceramics *Mater. Sci. Eng.* B **99** 363–5

[96] Lu C H and Wu C H 2002 Preparation, sintering, and ferroelectric properties of layer-structured strontium bismuth titanium oxide ceramics *J. Eur. Ceram. Soc.* **22** 707–14

[97] Gelfuso M V, Thomazini D and Eiras J A 1999 Synthesis and structural, ferroelectric, and piezoelectric properties of $SrBi_4Ti_4O_{15}$ ceramics *J. Am. Ceram. Soc.* **82** 2368–72

[98] de Figueiredo R S, Messai A, Hernandes A C and Sombra A S B 1998 Piezoelectric lithium niobate obtained by mechanical alloying *J. Mater. Sci. Lett.* **17** 449–51

[99] Xue J M, Wang J and Wan D M 2000 Nanosized barium titanate powder by mechanical activation *J. Am. Ceram. Soc.* **83** 232–4

[100] Abe O and Suzuki Y 1996 Mechanochemically assisted preparation of $BaTiO_3$ powder *Mater. Sci. Forum* **225** 563–8

[101] Kong L B, Ma J, Huang H, Zhang R F and Que W X 2002 Barium titanate derived from mechanochemically activated powders *J. Alloys Compd.* **337** 226–30

[102] Berbenni V, Marini A and Bruni G 2001 Effect of mechanical milling on solid state formation of $BaTiO_3$ from $BaCO_3$–TiO_2 (rutile) mixtures *Thermochim. Acta* **374** 151–8

[103] Castro A, Millan P, Pardo L and Jimenez B 1999 Synthesis and sintering improvement of Aurivillius type structure ferroelectric ceramics by mechanochemical activation *J. Mater. Chem.* **9** 1313–7

[104] Ricote J, Pardo L, Moure A, Castro A, Millan P and Chateigner D 2001 Microcharacterisation of grain-oriented ceramics based on Bi_3TiNbO_9 obtained from mechanochemically activated precursors *J. Eur. Ceram. Soc.* **21** 1403–7

[105] Moure A, Pardo L, Alemany C, Millan P and Castro A 2001 Piezoelectric ceramics based on Bi_3TiNbO_9 from mechano-chemically activated precursors *J. Eur. Ceram. Soc.* **21** 1399–402

[106] Jimenez B, Castro A, Pardo L, Millan P and Jimenez R 2001 Electric and ferro-piezoelectric properties of $(SBN)_{1-x}(BTN)_x$, ceramics obtained from amorphous precursors *J. Phys. Chem. Solids* **62** 951–8

[107] Pardo L, Castro A, Millan P, Alemany C, Jimenez R and Jimenez B 2000 $(Bi_3TiNbO_9)_x(SrBi_2Nb_2O_9)_{1-x}$ Aurivillius type structure piezoelectric ceramics obtained from mechanochemically activated oxides *Acta Mater.* **48** 2421–8

[108] Moure A, Castro A and Pardo L 2004 Improvement by-recrystallisation of Aurivillius-type structure piezoceramics from mechanically activated precursors *Acta Mater.* **52** 945–57

[109] Moure A, Alemany C and Pardo L 2004 Electromechanical properties of SBN/BTN Aurivillius-type ceramics up to the transition temperature *J. Eur. Ceram. Soc.* **24** 1687–91

[110] Shirane G and Hoshino S 1951 On the phase transition in lead titanate *J. Phys. Soc. Jap.* **6** 265–70

[111] Takeuchi H and Yamauchi H 1981 Strain effects on surface acoustic wave velocities in modified $PbTiO_3$ ceramics *J. Appl. Phys.* **52** 6147–50

[112] Chu S Y and Chen T Y 2004 The influence of Cd doping on the surface acoustic wave properties of Sm-modified $PbTiO_3$ ceramics *J. Eur. Ceram. Soc.* **24** 1993–8

[113] Yoo J H, Hong J I and Suh S 1999 Effect of MnO_2 impurity on the modified $PbTiO_3$ system ceramics for power supply *Sens. Actuat.* **78** 168–71

[114] Zeng Y, Xue W, Benedetti A and Fagherazzi G 1994 Microstructure study of Sm, Mn-modified $PbTiO_3$ piezoelectric ceramics by XRD profile-fitting technique *J. Mater. Sci.* **29** 1045–50

[115] Ikegami S, Ueda I and Nagata T 1971 Electromechanical properties of $PbTiO_3$ ceramics containing La and Mn *J.Acous. Soc. Am.* **50** 1060–6

[116] Udomporn A, Pengpat K and Ananta S 2004 Highly dense lead titanate ceramics form refined processing *J. Eur. Ceram. Soc.* **24** 185–8

[117] Komatsubara S, Isobe T and Senna M 1994 Effect of preliminary mechanical treatment on the microhomogenization during heating of hydrous gels as precursors for lead titanate *J. Am. Ceram. Soc.* **77** 278–82

[118] Hamada K and Senna M 1996 Mechanochemical effects on the properties of starting mixtures for $PbTiO_3$ ceramics by using a novel grinding equipment *J. Mater. Sci.* **31** 1725–8

[119] Durović D, Dostić E, Kiss S J and Zec S 1998 Mechanochemical synthesis of $PbTiO_3$ from PbO and TiO_2 *J. Alloys Compd.* **279** L1–3

[120] Leit E R *et al* 2001 Phololuminescence of nanostructured $PbTiO_3$ processed by high-energy mechanical milling *Appl. Phys. Lett.* **78** 2148–50

[121] Wongmaneerung R, Yimnirun R and Ananta S 2006 Effect of vibro-milling time on phase formation and particle size of lead titanate nanopowders *Mater. Lett.* **60** 1447–52

[122] Szafraniak I, Połomska M and Hilczer B 2006 XRD, TEM and Raman scattering studies of $PbTiO_3$ nanopowders *Cryst. Res. Technol.* **41** 576–9

[123] Soon H P, Xue J M and Wang J 2004 Dielectric behaviors of $Pb_{1-3x/2}La_xTiO_3$ derived from mechanical activation *J. Appl. Phys.* **95** 4981–8

[124] Randall C A, Kim N, Kucera J P, Cao W W and Shrout T R 1998 Intrinsic and extrinsic size effects in fine-grained morphotropic-phase-boundary lead zirconate titanate ceramics *J. Am. Ceram. Soc.* **81** 677–88

[125] Xue J M, Wang J and Weiseng T 2000 Synthesis of lead zirconate titanate from an amorphous precursor by mechanical activation *J. Alloys Compd.* **308** 139–46

[126] Liu X Y, Akdogan E K, Safari A and Riman R E 2005 Mechanically activated synthesis of PZT and its electromechanical properties *Appl. Phys.* A **81** 531–7

[127] Parashar S K S, Choudhary R N P and Murty B S 2005 Size effect of $Pb_{0.92}Nd_{0.08}(Zr_{0.53}Ti_{0.47})_{0.98}O_3$ nanoceramic synthesized by high-energy ball milling *J. Appl. Phys.* **98** 104305

[128] Brown L M and Mazdiyasni K S 1972 Cold-pressing and low-temperature sintering of alkoxy-derived PLZT *J. Am. Ceram. Soc.* **55** 541–4

[129] Yao X, Chen Z and Cross L E 1983 Polarization and depolarization behavior of hot pressed lead lanthanum zirconate titanate ceramics *J. Appl. Phys.* **54** 3399–403

[130] Viehland D, Jang S J, Cross L E and Wuttig M 1991 Internal strain relaxation and the glassy behavior of La-modified lead zirconate titanate relaxors *J. Appl. Phys.* **69** 6595–602

[131] Jiang Q Y, Subbarao E C and Cross L E 1994 Effect of composition and temperature on electric fatigue of La-doped lead zirconate titanate ceramics *J. Appl. Phys.* **75** 7433–43

[132] Viehland D, Dai X H, Li J F and Xu Z 1998 Effects of quenched disorder on La-modified lead zirconate titanate: long-and short-range ordered structurally incommensurate phases, and glassy polar clusters *J. Appl. Phys.* **84** 458–71

[133] Haertling G H and Land C E 1972 Recent improvements in the optical and electrooptic properties of PLZT ceramics *Ferroelectrics* **3** 269–80

[134] James A D and Messer R M 1972 The preparation of transparent PLZT ceramics from oxide powders by liquid-phase sintering *Trans. British Ceram. Soc.* **77** 152–8

[135] Snow G S 1973 Improvements in atmosphere sintering of transparent PLZT ceramics *J. Am. Ceram. Soc.* **56** 479–80

[136] Yoshikawa Y and Tsuzuki K 1992 Fabrication of transparent lead lanthanum zirconate titanate ceramics from fine powders by two-stage sintering *J. Am. Ceram. Soc.* **75** 2520–8

[137] Brodeur R P, Gachigi K W, Pruna P M and Shrout T R 1994 Ultra-high strain ceramics with multiple field-induced phase transitions *J. Am. Ceram. Soc.* **77** 3042–4

[138] Akbas M A and Lee W E 1995 Synthesis and sintering of PLZT powder made by freeze/alcohol drying or gelation of citrate solutions *J. Eur. Ceram. Soc.* **15** 57–63

[139] Cerqueira M, Nasar R S, Leite E R, Longo E and Varela J A 1998 Synthesis and characterization of PLZT(9/65/35) by the Pechini method and partial oxalate *Mater. Lett.* **35** 166–71

[140] Choi J J, Ryu J and Kim H E 2001 Microstructural evolution of transparent PLZT ceramics sintered in air and oxygen atmospheres *J. Am. Ceram. Soc.* **84** 1465–9

[141] Abe Y, Kakegawa K, Ushijima H, Watanabe Y and Sasaki Y 2002 Fabrication of optically transparent lead lanthanum zirconate titanate ($(Pb,La)(Zr,Ti)O_3$) ceramics by a three-stage-atmoshere-sintering technique *J. Am. Ceram. Soc.* **85** 473–5

[142] Zhang J H, Mo S J, Wang H Y, Zhen S X, Chen H W and Yang C R 2011 Microstructure and electrical properties of PLZT ceramics from Pb_3O_4 as the lead source *J. Alloys Compd.* **509** 2838–41

[143] Cui Z H, Gregori G, Ding A L, Guo X X and Maier J 2012 Electrical transport properties of transparent PLZT ceramics: bulk and grain boundaries *Sol. State Ionics* **208** 4–7

[144] Liu Z W, Shi D L, Zhou H, Deng L H and Li K 2015 Fabrication and evaluation of Pb $(W_{0.5}Cu_{0.5})O_3$ modified PLZT piezoelectric ceramics *Ceram. Int.* **41** 941–6

[145] Somwan S, Ngamjarurojana A and Limpichaipanit A 2016 Dielectric, ferroelectric and induced strain behavior of PLZT 9/65/35 ceramics modified by Bi_2O_3 and CuO co-doping *Ceram. Int.* **42** 10690–6

[146] Dimza V *et al* 2017 Effects of Mn doping on dielectric properties of ferroelectric relaxor PLZT ceramics *Curr. Appl Phys.* **17** 169–73

[147] Kong L B, Ma J, Zhu W and Tan O K 2001 Preparation and characterization of PLZT ceramics using high-energy ball milling *J. Alloys Compd.* **322** 290–7

[148] Kong L B, Ma J, Zhu W and Tan O K 2002 Preparation and characterization of PLZT (8/65/35) ceramics via reaction sintering from ball milled powders *Mater. Lett.* **52** 378–87

[149] Kong L B, Ma J, Zhu W and Tan O K 2002 Transparent PLZT8/65/35 ceramics from constituent oxides mechanically modified by high-energy ball milling *J. Mater. Sci. Lett.* **21** 197–9

[150] Kong L B, Ma J, Huang H and Zhang R F 2002 Effect of excess PbO on microstructure and electrical properties of PLZT7/60/40 ceramics derived from a high-energy ball milling process *J. Alloys Compd.* **345** 238–45

[151] Hammer M and Hoffmann M J 1998 Sintering model for mixed-oxide-derived lead zirconate titanate ceramics *J. Am. Ceram. Soc.* **81** 3277–84

[152] Damjanovic D and Demartin M 1996 Dependence of the direct piezoelectric effect in coarse and fine grain barium titanate ceramics on dynamic static pressure *Appl. Phys. Lett.* **68** 3046–8

[153] Mishra S K and Pandey D 1995 Effect of particle size on the ferroelectric behaviour of tetragonal and rhombohedral $Pb(Zr_xTi_{1-x})O_3$ ceramics and powders *J. Phys. Condens. Matter* **7** 9287–303

[154] Sengupta S S, Roberts D, Li J F, Kim M C and Payne D A 1995 Field-induced phase switching and electrically driven strain in sol–gel derived antiferroelectric (Pb,Nb)(Zr,Sn, Ti)O_3 thin layers *J. Appl. Phys.* **78** 1171–7

[155] Xu B, Ye Y, Wang Q M and Cross L E 1999 Dependence of electrical properties on film thickness in lanthanum-doped lead zirconate titanate stannate antiferroelectric thin films *J. Appl. Phys.* **85** 3753–8

[156] Bharadwaja S S N and Krupanidhi S B 1999 Growth and study of antiferroelectric lead zirconate thin films by pulsed laser ablation *J. Appl. Phys.* **86** 5862–9

[157] James A R and Subrahmanyam J 2006 Processing and structure-property relation of fine-grained PLZT ceramics derived from mechanochemical synthesis *J. Mater. Sci. Mater. Electron.* **17** 529–35

[158] Kumar A, Prasad V V B, Raju K C J and James A R 2014 Ultra high strain properties of lanthanum substituted PZT electro-ceramics prepared via mechanical activation *J. Alloys Compd.* **599** 53–9

[159] Kumar A, Emanic S R, Bhanu Prasada V V, James Raju K C and James A R 2016 Microwave sintering of fine grained PLZT 8/60/40 ceramics prepared via high energy mechanical milling *J. Eur. Ceram. Soc.* **36** 2505–11

[160] Li J F, Viehland D, Tani T, Lakeman C D E and Payne D A 1994 Piezoelectric properties of sol–gel-derived ferroelectric and antiferroelectric thin layers *J. Appl. Phys.* **75** 442–8

[161] Tani T, Li J F, Viehland D and Payne D A 1994 Antiferroelectric–ferroelectric switching and induced strains for sol–gel derived lead zirconate thin layers *J. Appl. Phys.* **75** 3017–23

[162] Xu C H *et al* 2016 High charge–discharge performance of $Pb_{0.98}La_{0.02}(Zr_{0.35}Sn_{0.55}Ti_{0.10})_{0.995}O_3$ antiferroelectric ceramics *J. Appl. Phys.* **120** 074107

[163] Xu R, Tian J J, Zhu Q S, Feng Y J, Wei X Y and Xu Z 2017 Effect of temperature-driven phase transition on energy storage and release properties of $Pb_{0.97}La_{0.02}[Zr_{0.55}Sn_{0.30}Ti_{0.15}]O_3$ ceramics *J. Appl. Phys.* **122** 024104

[164] Xu R, Golinveaux F S, Sheng M, Xu Z, Feng Y J and Lynch S S 2019 Effects of compressive stress on electric-field-induced phase transition of antiferroelectric ceramics *J. Appl. Phys.* **125** 204104
[165] Zhang Q F *et al* 2017 Effects of composition and temperature on energy storage properties of (Pb,La)(Zr,Sn,Ti)O_3 antiferroelectric ceramics *Ceram. Int.* **43** 11428–32
[166] Dan Y *et al* 2019 Superior energy-storage properties in (Pb,La)(Zr,Sn,Ti)O_3 antiferroelectric ceramics with appropriate La content *Ceram. Int.* **45** 11375–81
[167] QP X, Zhang Y F, Chen X F, Zhou M M, Wang G S and Dong X L 2019 Effect of Mn-doping on dielectric and energy storage properties of $(Pb_{0.91}La_{0.06})(Zr_{0.96}Ti_{0.04})O_3$ antiferroelectric ceramics *J. Alloys Compd.* **780** 581–7
[168] Hao X H, Li W, Zhai J W and Chen H 2019 Progress in high-strain perovskite piezoelectric ceramics *Mater. Sci. Eng.* R **135** 1–57
[169] Hao X H, Zhai J W, Kong L B and Xu Z K 2014 A comprehensive review on the progress of lead zirconate-based antiferroelectric materials *Prog. Mater Sci.* **63** 1–57
[170] Ibrahim D M and Hennicke H W 1981 Preparation of lead zirconate by a sol–gel method *Trans. J. Br. Ceram. Soc.* **80** 18–22
[171] Rao Y S and Sunandana C S 1992 Low-temperature synthesis of lead zirconate *J. Mater. Sci. Lett.* **11** 595–7
[172] Oren E E, Taspinar E and Tas A C 1997 Preparation of lead zirconate by homogeneous precipitation and calcinations *J. Am. Ceram. Soc.* **80** 2714–6
[173] Hao X H *et al* 2011 Structure and electrical properties of $PbZrO_3$ antiferroelectric thin films doped with barium and strontium *J. Alloys Compd.* **509** 271–5
[174] Kong L B, Ma J, Zhu W and Tan O K 2001 Preparation and characterization of lead zirconate ceramics from high-energy ball milled powder *Mater. Lett.* **49** 96–101
[175] Chotsawat M, Sarasamak K, Thanomngam P, Limpijumnong S and Thienprasert J 2016 First-principles study of Bi and Al in orthorhombic $PbZrO_3$ *Comput. Mater. Sci.* **115** 99–103
[176] Kong L B, Ma J, Zhang T S, Zhu W and Tan O K 2002 Preparation of antiferroelectric lead zirconate titanate stannate ceramics by high-energy ball milling process *J. Mater. Sci. Mater. Electron.* **13** 89–94

IOP Publishing

Functional Ceramics Through Mechanochemical Activation

Ling Bing Kong

Chapter 4

Ferroelectric ceramics (II)

Ling Bing Kong, Zhuohao Xiao, Xiuying Li, Shijin Yu, Wenxiu Que, Yin Liu, Tianshu Zhang, Kun Zhou and Hongfang Zhang

4.1 Brief introduction

As observed in chapter 3, a large number of the Pb-containing ferroelectric and antiferroelectric powders could be synthesized directly from an oxide mixture through mechanochemical activation. In this chapter, the processing and characterization of relaxor ferroelectric powders, including unary, binary and ternary compounds, will be presented and discussed.

Perovskite relaxor ferroelectric compounds have a general formula of Pb(B'B')O$_3$, where B' site usually hosts low-valence cations, such as Mg^{2+}, Zn^{2+}, Fe^{3+} or Sc^{3+}, while B' is occupied by high-valence cations, such as Nb^{5+}, Ta^{5+} or W^{6+} [1–8]. The presence of the pyrochlore phase is a serious issue for the synthesis of relaxor ferroelectric materials. Usually, the columbite and Wolframite routes are employed to prepare perovskite relaxor phases, using the conventional ceramic process [1, 2]. Relaxor ferroelectric examples include lead magnesium niobate ($Pb(Mg_{1/3}Nb_{2/3})O_3$ or PMN), lead zinc niobate ($Pb(Zn_{1/3}Nb_{2/3})O_3$ or PZN), lead iron niobate ($Pb(Fe_{1/2}Nb_{1/2})O_3$ or PFN) and lead scandium tantalate ($Pb(Sc_{1/2}Ta_{1/2})O_3$ or PST). In addition, the typical relaxors have been stabilized by incorporating them with other perovskites, such as PT, PZT, BT ($BaTiO_3$) and ST ($SrTiO_3$), to form binary or ternary compositions.

4.2 Unary phase

4.2.1 Lead magnesium niobate

PMN nanosized powders have been directly derived from mixtures of PbO, MgO and Nb_2O_5 powders using high-energy mechanochemical activation methods, with either an SPEX shaker mills (stainless steels) [9–14] or a planetary ball mills (tungsten carbides) [15, 16]. It has been demonstrated that the phase formation of PMN from the oxide mixtures is possible without involving any pyrochlore phase,

doi:10.1088/978-0-7503-2191-4ch4

which is frequently observed once the conventional solid-state reaction processes, with the calcination step at high temperatures, are used. The as-milled powders can be used to make PMN ceramics without further treatment, which is similarly to that for PZT, as discussed previously.

PMN nanosized powder was synthesized directly from the mixed oxide powders of PbO, MgO and Nb_2O_5 using mechanochemical activation with an SPEX shake mill [9] The oxide was mixed according to the composition of PMN using the conventional ball milling process with the addition of ethanol. The mixed powder was subjected to high-energy mechanochemical milling of the powder mixture, carried out in an Al_2O_3 cylindrical vial with a diameter of 40 mm and a length of 40 mm, together with a stainless steel ball 12.7 mm in diameter. The milling was conducted for 20 h at a speed of 900 rpm.

Figure 4.1 shows the XRD patterns of the oxide powder before and after mechanochemical activation for 20 h. Before mechanochemical activation, the mixture had sharp peaks of PbO, MgO and Nb_2O_5 in the XRD pattern, suggesting that the conventional milling could not trigger the reaction of the starting oxides. However, after high-energy milling for 20 h, perovskite PMN was formed as

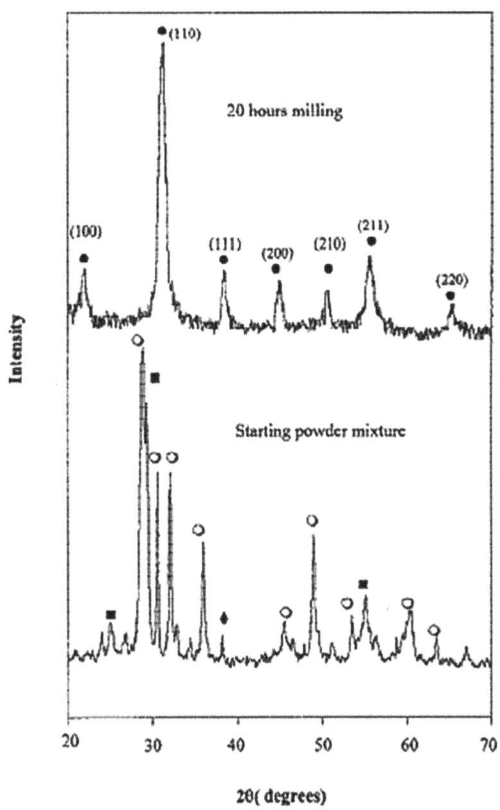

Figure 4.1. XRD patterns of the powder mixture and before and after mechanochemical treatment for 20 h: (○) PbO, (■) Nb_2O_5, (♦) MgO and (●) PMN. Reproduced with permission from [9]. Copyright 1999 Wiley.

single-phase in the XRD pattern. According to the Scherrer's equation, the average crystallite sizes of the PMN phase were in the range 10–15 nm, based on the half-width of the (110) peak of PMN. More significantly, no pyrochlore phase was present in the as-activated powder. This suggested that the reaction mechanism of PMN becuase of the mechanochemical activation was different from that in the conventional solid-state reaction process, in which a calcination/presintering step was used. Figure 4.2 depicts a typical TEM image of the PMN powder, with particle sizes being 20–30 nm, which was in good agreement with the values estimated based on the XRD pattern.

The PMN powder was utilized to fabricate PMN ceramics. As the sintering temperature was increased from 900 °C to 1050 °C, the relative density of the samples gradually increased and was maximized at 1050 °C, with an optimal density of about 99%. If the sintering temperature was further increased, the density slightly decreased, mainly due to the loss of PbO and over-sintering. In addition, the average grain size of the ceramics was increased monotonically with sintering temperature up to 1100 °C. After sintering at 1100 °C, about 0.5 wt% PbO was lost and the grains were exaggerated, which was responsible for the decrease in density. Figure 4.3 shows a representative SEM image of the PMN ceramics sintered at 1050 °C for 1 h, showing grain sizes in the range 8–10 μm. Specifically, the 1050 °C sintered PMN ceramic sample has a maximum dielectric constant of about 18 000 at 1 kHz.

The same authors further examined the phase formation of PMN using mechanochemical activation in the presence of 50 wt% excess PbO [10]. They found that the PMN phase with finer particles was obtained after mechanochemical milling for 20 h in 50 wt% excess PbO. The degree of crystallization of the PMN phase was enhanced, after the as-activated PMN–PbO mixture was calcined at 800 °C. After the excess PbO was removed by washing with acetic acid solution at room temperature, the powder was used to fabricate PMN ceramics. The sample sintered at 1200 °C for 1 h had a relative density of about 99% and a dielectric constant of about 14 000 at 100 Hz.

Figure 4.4 shows the XRD patterns of the powder mixtures of PbO, MgO and Nb_2O_5 for PMN with 50 wt% excess PbO, after mechanical activation for different

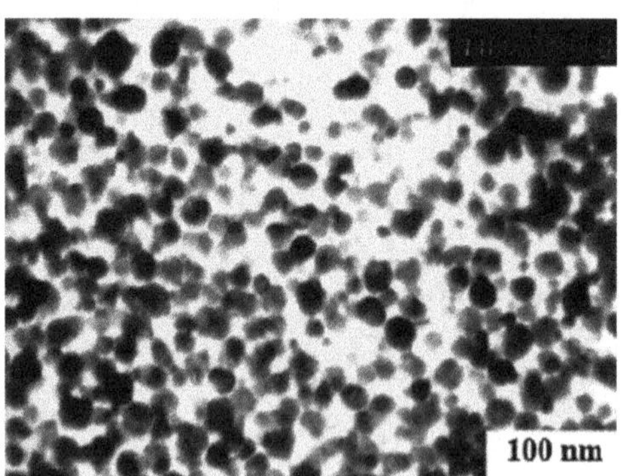

Figure 4.2. TEM image of the mechanochemically synthesized PMN nanosized powder after milling for 20 h. Reproduced with permission from [9]. Copyright 1999 Wiley.

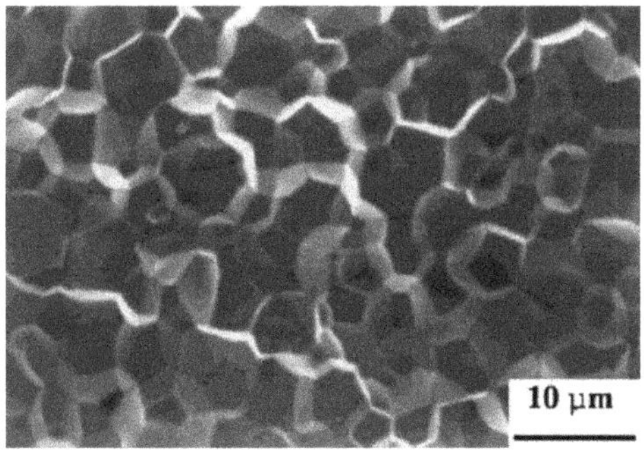

Figure 4.3. Cross-sectional SEM image of the PMN derived from mechanochemically activated powder after sintering at 1050 °C for 1 h. Reproduced with permission from [9]. Copyright 1999 Wiley.

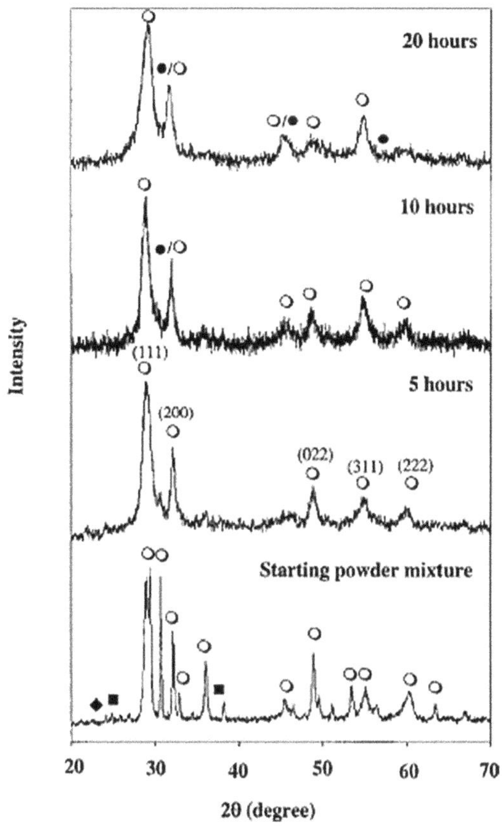

Figure 4.4. XRD patterns of the oxide mixture of PbO, MgO and Nb_2O_5 for PMN with 50 wt% excess PbO mechanically activated for different times: (●) PMN, (○) PbO, (■) Nb_2O_5 and (♦) MgO. Reproduced with permission from [10]. Copyright 2001 Wiley.

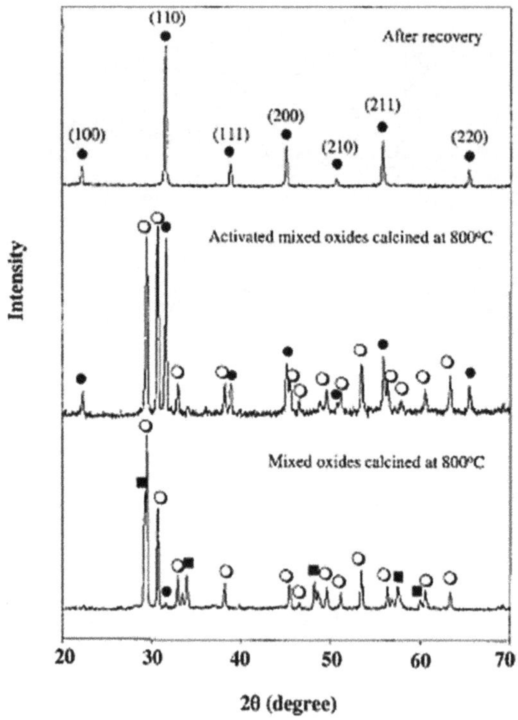

Figure 4.5. XRD patterns of the oxide mixture of PbO, MgO and Nb_2O_5 for PMN with 50 wt% excess PbO mechanically activated for 0 and 20 h, together with the sample calcined at 800 °C for 2 h, followed by removal of the excess PbO through washing with acetic acid: (●) PMN, (○) PbO and (■) $Pb_3Nb_2O_8$. Reproduced with permission from [10]. Copyright 2001 Wiley.

times. It was found that no reaction took place in the unactivated mixture after the general ball milling process. Mechanical activation for 5 and 10 h gradually broadened and reduced the diffraction peaks of the oxides. This implied that the particles were refined and the crystallite sizes were decreased due to the mechanical activation. Mechanical activation for 20 h led to the formation of PMN. A HRTEM image revealed that crystallized perovskite PMN particles were distributed in the matrix of PbO. No pyrochlore phase was detected, suggesting that the perovskite phase was directly derived from the oxides, and was not affected by the excess PbO.

Figure 4.5 shows the XRD patterns of the sample with 50 wt% excess PbO before and after mechanochemical milling for 20 h, which was calcined at 800 °C for 2 h, as well the PMN powder after removing the excess PbO. Without the mechanochemical activation, calcination at 800 °C for 2 h resulted in almost no phase formation of PMN, whereas pyrochlore $Pb_3Nb_2O_8$ was the major phase. However, after mechanical activation for 20 h, the PMN phase was predominant upon calcination at 800 °C. More importantly, no pyrochlore phase was detected.

Figure 4.6 shows SEM images of the PMN powders with and without 50 wt% PbO, which were mechanochemically milled for 20 h and subsequently calcined at 800 °C for 2 h. As observed in figure 4.6(A), for the sample without the presence of

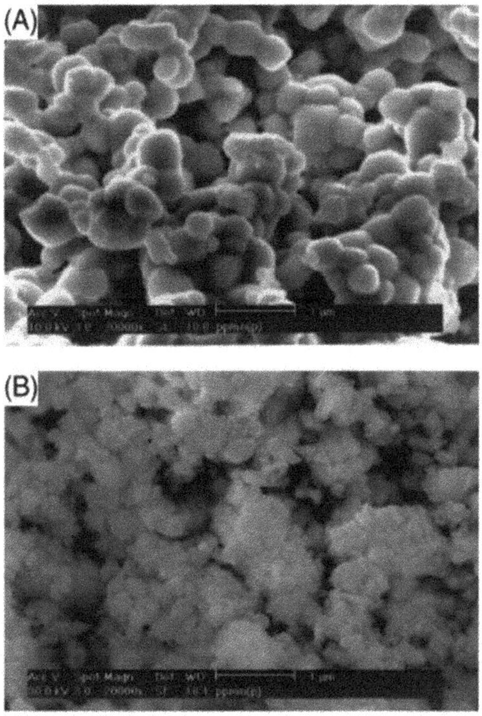

Figure 4.6. SEM images of the mixed oxide powders for PMN after mechanochemical activation without (A) and with (B) 50 wt% excess PbO, with a milling time of 20 h and calcination at 800 °C for 2 h. Reproduced with permission from [10]. Copyright 2001 Wiley.

excess PbO, the particles had relatively large sizes and they were agglomerated. The particle sizes were in the range 0.7–1.0 μm, with obvious necks between every two neighboring particles. In comparison, in the presence of 50 wt% PbO, the PMN powder possessed much smaller particles, with sizes in the range 0.2–0.3 μm, as demonstrated in figure 4.6(B).

It was interesting to demonstrate that the PMN phase could be derived from the pyrochlore phase through mechanical activation [11]. The PMN powder, with particle sizes of 30–50 nm, was well crystallized with stable perovskite structure. Commercial $Pb(NO_3)_2$, $Mg(NO_3)_2 \cdot 6H_2O$ and $NbCl_5$ were used as the precursors. To synthesize the PMN precursor, 2 M ammonia solution was used to precipitate hydrated niobate ($Nb_2O_5 \cdot xH_2O$) from $NbCl_5$ aqueous solution. The precipitate was thoroughly washed and then dried. Then, aqueous solutions of $Pb(NO_3)_2$ and Mg $(NO_3)_2 \cdot 6H_2O$ were prepared, which were mixed with the ratio of Pb:Mg being 3:1. After that, the dried hydrated niobate powder was introduced to the mixed solution. The mixture was controlled to have a pH value of about 7.0 with ammonia solution, followed by heating to form a viscous slurry. The powder sample was obtained after freeze-drying. The pyrochlore phase was formed from the freeze-dried precursor powder through calcination at 400 °C for 5 h. The pyrochlore precursor powder was subjected to high-energy mechanochemical activation as described previously.

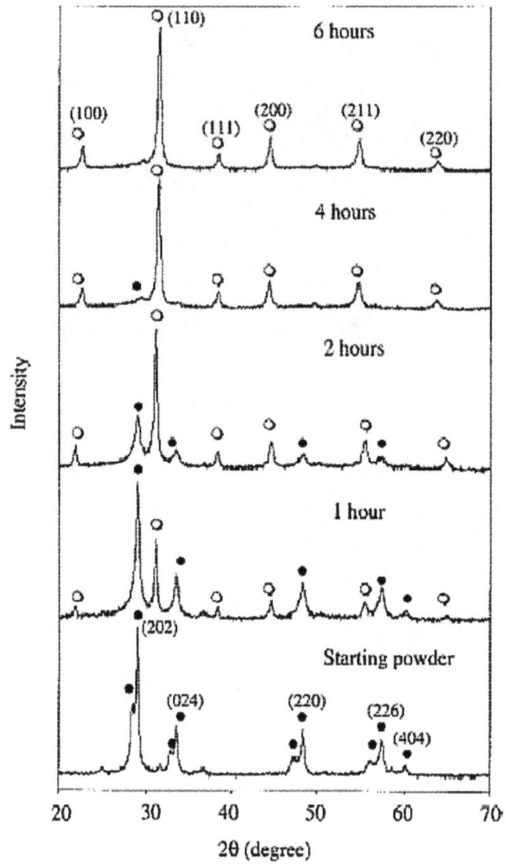

Figure 4.7. XRD patterns of the freeze-dried precursor PMN powders mechanically activated for different time durations: (○) perovskite and (●) pyrochlore. Reproduced with permission from [11]. Copyright 1999 Wiley.

Figure 4.7 shows the XRD patterns of the precursor powders before and after mechanical milling for different times. Obviously, the unmilled sample was crystalline pyrochlore, $Pb_2Nb_2O_7$. After mechanochemical activation for 1 h, a perovskite PMN phase was detected, indicating the phase transformation from pyrochlore to perovskite, as a result of the activation. As the activation time was increased, the content of PMN perovskite gradually increased, while that of the pyrochlore phase deceased, evidenced by the XRD patterns of the samples milled for 2 and 4 h. After milling for 6 h, perovskite PMN was present as the major phase in the sample.

A systematic study was conducted to understand the phase formation mechanism of PMN from oxide powders using the high-energy mechanochemical activation process [17]. The evolution in the compositions of the crystalline and amorphous phases as a function of milling time was examined. At the early stage of activation, particle size was reduced, while amorphization occurred, during which both the perovskite and pyrochlore phases were formed.

The planetary mill was combined with a 250 ml tungsten carbide milling jar and 15 WC balls 20 mm in diameter. The total quantity of the powder was 200 g. The rotational speed of the supporting disk was 300 rpm. It is worth mentioning that this configuration was slightly different from that reported by Kong *et al*, as stated earlier. PbO, MgO and Nb_2O_5 for PMN were milled for times of up to 60 h. Samples after milling for 8, 16, 32, 48 and 60 h were collected and analyzed using XRD. The powders milled for 8 and 60 h were calcined at 800 °C for 1 h. In addition, pure PbO, MgO and Nb_2O_5 were similarly milled for times of up to 48 h for comparison. The Rietveld refinement technique was employed to conduct quantitative analysis according to the XRD patterns.

PbO was composed of two phases, orthorhombic massicot and tetragonal litharge. The massicot–litharge transition was triggered after a short time of activation. In addition, the background was increased and the diffraction peaks were broadened. It was found that, after milling for 48 h, the amorphous phase and the crystalline phase had the same weight percentage, with the crystalline phase being litharge with a trace of massicot. During the early state of milling a small quantity of hydrocerussite, $PbCO_3{\cdot}PbO{\cdot}(H_2O)_2$, was detected, mainly due to the adsorption of water and CO_2 molecules. No phase transition was observed for orthorhombic Nb_2O_5 during high-energy mechanochemical activation for up to 48 h, but the particle size was significantly reduced after milling for just 1 h. Finally, the content of the amorphous phase was about 28 wt%. MgO had no phase transition during the high-energy activation, while the amorphization degree was just about 5 wt%, but the particles were refined.

After mechanochemical activation for 8 h, the mixture consisted of Nb_2O_5, MgO, litharge PbO and massicot PbO. At the same time, the cubic pyrochlore phase was detected, whereas the perovskite PMN was not clearly detected in the XRD pattern. After milling for 16 h, the cubic pyrochlore became the major crystalline phase, with a formula of $Pb_{1.86}Mg_{0.24}Nb_{1.76}O_{6.5}$. In this sample, perovskite PMN was detected together with small contents of MgO, Nb_2O_5 and PbO. For the sample milled for 32 h, the diffraction peak of the pyrochlore phase was still stronger than that of perovskite. A significant increase in the content of perovskite was observed after the sample was milled for 48 h. As the milling time was increased to 60 h, the phase formation of PMN was nearly completed, but the pyrochlore phase was still visible in the XRD pattern.

The weight percentages of the crystalline phases, including the products of PMN, $Pb_{1.86}Mg_{0.24}Nb_{1.76}O_{6.5}$, and the constituent oxides of MgO, Nb_2O_5 and PbO, as well as the amorphous phase, as a function of milling time, are illustrated in figure 4.8. The content of PMN was constantly increased with increasing milling time, whereas the amount of pyrochlore phase was maximized at 16 h, corresponding to the minimum levels of crystalline Nb_2O_5 and PbO. The contents of these two oxides were almost not decreased as the milling time was further increased. In contrast, the content of MgO was decreased gradually with increasing milling time, due to the formation of PMN.

The lattice constant of PMN was estimated to be 0.4056 nm, which was almost unchanged over the milling time from 32 h to 60 h, implying that the composition of

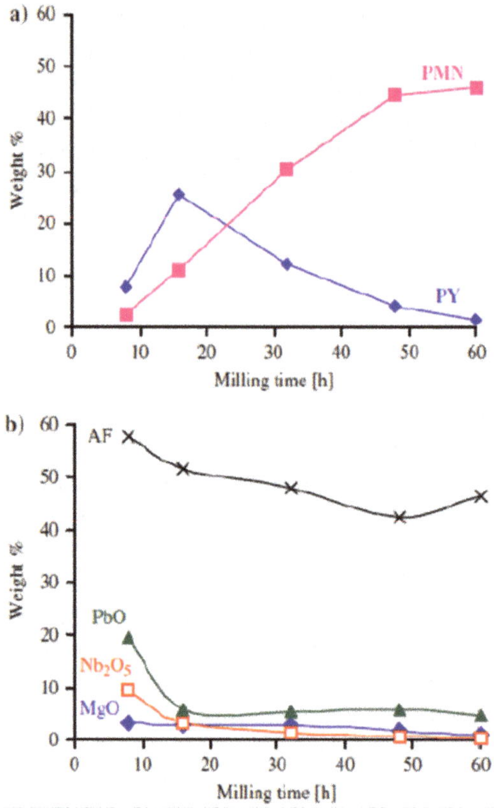

Figure 4.8. (a) Weight percentages of perovskite PMN and pyrochlore $Pb_{1.86}Mg_{0.24}Nb_{1.76}O_{6.5}$ in the crystalline state as a function of milling time. (b) Weight percentages of MgO, Nb_2O_5 and PbO in the crystalline state and the weight percentage of the amorphous phase as a function of milling time. Reproduced with permission from [17]. Copyright 2006 Wiley.

PMN was varied. However, the lattice parameter of the PMN in the sample milled for 16 h was 0.4061 nm, which was smaller than the literature values, suggesting its deviation in composition.

The phase diagram indicated that no reaction occurs between MgO and PbO [18]. For MgO and Nb_2O_5, two phases, $MgNb_2O_6$ and $Mg_4Nb_2O_9$, could be formed. For PbO and Nb_2O_5 there could be five compounds, including $Pb_3Nb_2O_8$, $Pb_5Nb_4O_{15}$, $Pb_2Nb_2O_7$, $Pb_3Nb_4O_{13}$ and $PbNb_2O_6$. However, it was found that only the pyrochlore phase with a slight deviation from the stoichiometric composition was observed in the mixture triggered with mechanochemical activation. In summary, this study presented a different phase formation route of PMN from the mixture of constituent oxide precursors compared to those reported by other researchers, as discussed above. To date this difference has not been fully understood. However, it is strongly believed that the differences in milling conditions, such as type of milling media, milling ball diameter, ball-to-powder weight ratio, milling speed, etc, are responsible for the differences in phase formation of PMN.

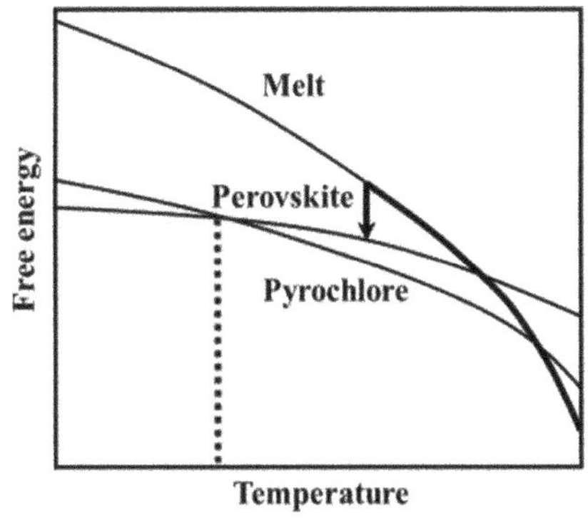

Figure 4.9. Free energies of $PbZn_{1/3}Nb_{2/3}O_3$ (PZN) as a function of temperature with different phases. PZN single crystal is precipitated from supercooled melt at the nonequilibrium state. Reproduced with permission from [20] with modification. Copyright 1995 Elsevier.

4.2.2 Lead zinc niobate

It is interesting that PZN single crystals can be obtained from the flux with excess PbO at high temperatures, but single-phase PZN ceramics have never been formed using conventional ceramic processing [19]. This can be understood according Ostward's step rule, with figure 4.9 showing a schematic diagram demonstrating the free energies of the different phases for PZN [20]. PZN single crystals could be grown from supercooled melt in PbO flux in the unequilibrium state. The free energy of the molten system was changed along the thick curve from high temperatures. Once the supercooled melt was supersaturated, crystallization would occur.

In this case, PZN perovskite, instead of pyrochlore, was formed, because the difference in free energy between the perovskite PZN and the melt was smaller than that between the pyrochlore phase and the melt. However, the free energy of pyrochlore was lower than that of the perovskite, and the PZN single crystal was metastable. Figure 4.9 also reveals that the free energy–temperature curves of perovskite and pyrochlore cross over each other. Below this point, the free energy of perovskite was lower than that of pyrochlore. In other words, perovskite was stable at low temperatures. It has been reported that the cross-over temperature is about 600 °C. Generally, solid-state reaction requires temperatures >800 °C for the pyrochlore phase to be stable. However, the mechanochemical activation process is conducted at room temperature. Even though there could be a localized instantly high temperature, the overall temperature should be never higher than 600 °C. As a result, it is highly possible to obtain perovskite PZN using high-energy mechanochemical milling.

The formation of single-phase perovskite PZN nanosized powder has been demonstrated using a high-energy mechanochemical activation process [21]. Two systems were used to check the phase formation process. One was from the mixture of PbO, ZnO and Nb_2O_5, while the other consisted of PbO and $ZnNb_2O_6$ derived from ZnO and Nb_2O_5 through the conventional solid-state reaction process. The mechanochemical milling process was conducted using an SPEX shake mill, with milling conditions similar to those for PZT and PMN by the authors discussed above. The as-synthesized PZN powders were phases of pure of perovskite structure, with crystallite sizes of 10–15 nm.

Figure 4.10 shows the XRD patterns of the oxide mixture of PbO, ZnO and Nb_2O_5 after mechanochemical activation for different times [21]. After milling for 5 h, perovskite PZN was present, while the peaks of PbO were largely reduced and broadened. It suggested that the phase formation of PZN was accompanied by the amorphization of the precursor oxides. As the milling time was increased, the peak intensity of perovskite PZN was gradually increased. After milling for 20 h, the phase formation of perovskite PZN was almost complete. Similarly, no pyrochlore phases were detected in the samples.

Figure 4.11 shows the XRD patterns of the mixture of PbO and $ZnNb_2O_6$ powders, which were mechanochemically activated for different times [21]. After

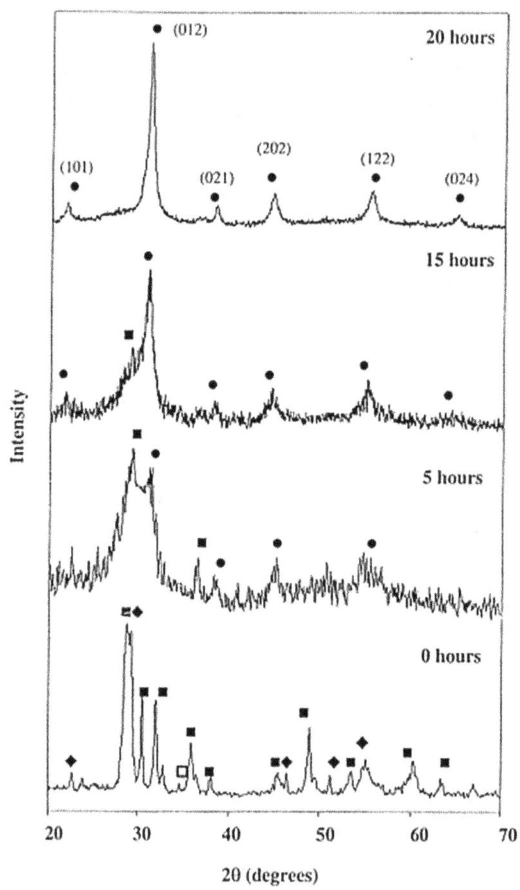

Figure 4.10. XRD patterns of the powder mixture of PbO, ZnO and Nb_2O_5 milled for various times: (■) Nb_2O_5, (◆) ZnO and (●) PZN. Reproduced with permission from [21]. Copyright 1999 Wiley.

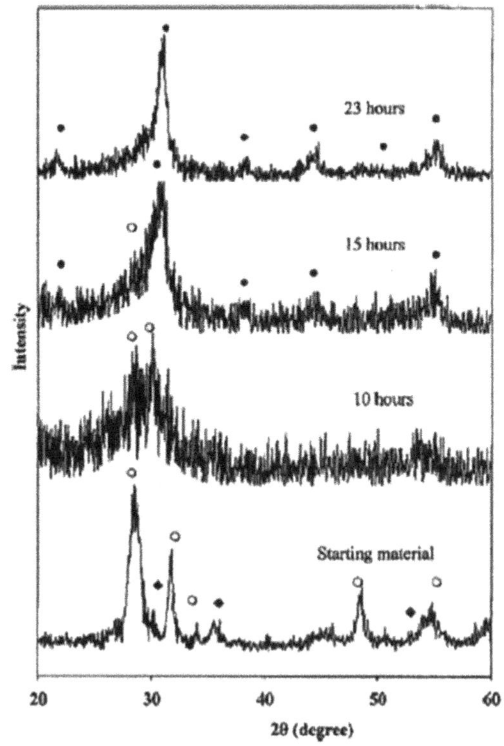

Figure 4.11. XRD patterns of the mixture of PbO and $ZnNb_2O_6$ milled for different times: (○) PbO, (◆) ZN and (●) PZN. Reproduced with permission from [21]. Copyright 1999 Wiley.

milling for 10 h, no obvious diffraction peaks of PZN were observed, suggesting that there was no reaction between PbO and $ZnNb_2O_6$, and only particle sizes were reduced. However, as the milling time was increased to 15 h, the perovskite PZN phase became the major phase in the sample. The average crystallite size of the sample was about 10 nm. After milling for 23 h, almost single-phase perovskite PZN was achieved. In addition, the (110) peak of PZN was increased, implying that the degree of crystallization of the PZN phase was increased due to the prolonged milling time. Although perovskite PZN was obtained using the mechanochemical activation process, the powder cannot be used to fabricate PZN ceramics. This is simply because high temperature sintering will induce the phase transition from perovskite to pyrochlore, since pyrochlore is more stable than perovskite at high temperatures, as mentioned above.

4.2.3 Lead iron niobate and lead iron tungstate

$Pb(Fe_{1/2}Nb_{1/2})O_3$ (PFN) and $Pb(Fe_{2/3}W_{1/3})O_3$ (PFW) possess strong relaxor ferroelectric effects [22–28]. They essentially have a high dielectric constant and a wide temperature range of ferroelectric–paraelectric phase transition. In addition, they have relatively low sintering temperatures and are suitable as candidates for multilayer ceramic capacitors (MLCs) with low-cost metals as the internal electrodes. Moreover, they exhibit weak magnetic properties, thus making them promising materials for multiferroic applications.

The mechanochemical activation process can be used to synthesize PFN either directly from an oxide mixture and or indirectly from the Columbite precursors of $FeNbO_4$ and PbO using an SPEX shake mill [29]. In both cases, single-phase PFN was obtained after milling for 30 h. The as-synthesized PFN nanosized powders had particle sizes in the range 5–15 nm. It was interesting to observe that the PFN phase obtained directly from the oxide mixture of PbO, Fe_2O_3 and Nb_2O_5 was less stable than that obtained indirectly from the Columbite precursor. After calcination at 500 °C–900 °C, the oxide derived PFN would partially decompose to pyrochlore, but the pyrochlore phase was not present in the Columbite sample. Furthermore, two PFNs displayed different sintering behaviors and the final ceramics exhibited different dielectric properties. According to the Raman spectroscopic results, it was found that the difference in the properties of the PFN nanosized powders was attributed to the difference in their compositional inhomogeneity. In this regard, mechanochemical activation has similar characteristics to the traditional thermal reaction, at least specifically for PFN. This observation is slightly different from those discussed above for PZT, PLZT and PMN.

For the oxide mixture of PbO, Fe_2O_3 and Nb_2O_5, perovskite phase PFN was detected after milling for only 5 h. The content of PFN was gradually increased with increasing milling time. Twenty hours of mechanochemical milling resulted in complete phase formation of PFN. As the milling time was increased to 30 h, only a slight increase in crystallinity of PFN occurred. The final PFN powder had an average crystalline size of about 12 nm based on the broadening of the XRD peak. The TEM image revealed that the particle sizes were in the range 15–35 nm. Similarly, the PFN phase was also formed in the sample of Columbite precursor milled for 5 h. In addition, the precursors were refined due to the high-energy activation. The perovskite phase formation was complete after milling for 20 h, while further milling for 30 h had no effect on the composition of the PFN phase, only a slight enhancement in the crystallinity. The particle sizes estimated from TEM image were 20–35 nm.

The DTA curve of the unmilled sample had two strong exotherms and three endotherms. The phase transition of PbO occurred at about 294 °C, while the formation of the pyrochlore phase was at 618 °C. A pyrochlore-to-pyrochlore phase transition was observed at 743°, while perovskite PFN was formed at 885 °C. After milling for 30 h, the phase transition of PbO and the phase formation of pyrochlore and perovskite were absent, due to the complete formation of PFN. For the unmilled mixture of PbO and $FeNbO_4$, the phase transition temperature of PbO was 304 °C, while the perovskite PFN was obtained at about 712 °C. After activation for 30 h, no thermal events were observed, confirming the complete phase formation of PFN.

The XRD patterns of the two samples calcinated at different temperatures are shown in figure 4.12. As seen in figure 4.12(a), the pyrochlore phase was detected in the oxide mixture group calcined at temperatures >500 °C. The content of the

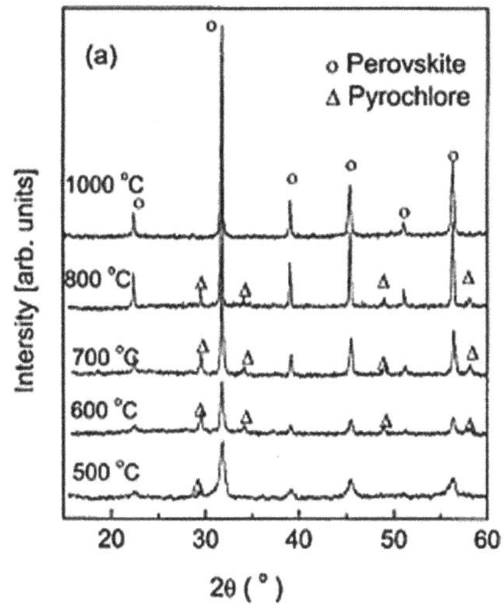

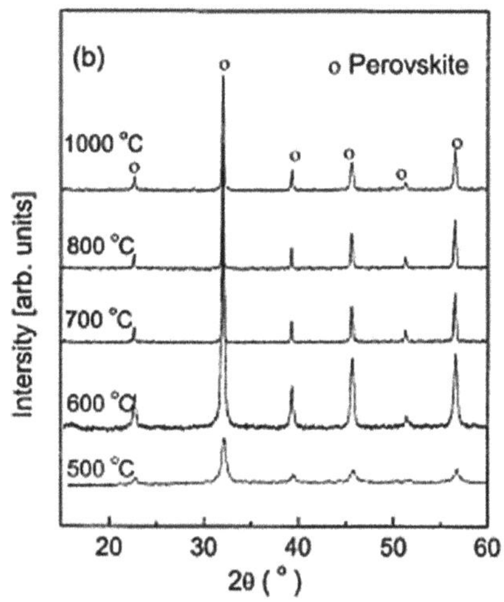

Figure 4.12. XRD patterns of the two PFN powders calcined at different temperatures, from different precursors after mechanochemical activation for 30 h: (a) an oxide mixture of PbO, Fe_2O_3 and Nb_2O_5, and (b) a mixture of PbO and $FeNbO_4$. Reproduced with permission from [29]. Copyright 2002 Wiley.

pyrochlore phase was maximized at 600 °C and declined with further increase in the calcination temperature. In contrast, the perovskite PFN phase in the Columbite precursor group had no pyrochlore phase, over the calcination temperature range of 400 °C–900 °C, as observed in figure 4.12(b).

According to the Raman spectroscopy results, a signal from Fe_2O_3 was present in the perovskite PFN powder that was formed by mechanochemically activating the oxide mixture. In comparison, no oxide was detected in the sample from the columbite precursor. This clearly suggested that the reaction to form the perovskite PFN was not complete after mechanochemical activation for 30 h. However, due to the small quantities, the constituent oxides could not be detected using XRD. It was the unreacted oxides in the sample that were responsible for the formation of the pyrochlore phase from the oxide mixture. It was strongly expected that the constituent oxides could be isolated from one another, so that they could not be gathered to realize the reaction, because three components must be involved at the same time. In contrast, there were only two components for the phase formation of PFN in the columbite precursor group. It was further illustrated that the PFN ceramics from the columbite precursor had a higher density than those from the oxide mixture.

Figure 4.13 shows cross-sectional SEM images of the PFN samples made from the oxide mixture after sintering at different temperatures. After sintering at 950 °C, the ceramics were highly porous, with an average grain size of 1.5 μm. With increasing sintering temperature, the average grain size increased gradually to 4 μm and 7 μm as the sintering temperature was increased to 1000 °C and 1100 °C, respectively. In addition, the porosity of the ceramics was reduced. Unfortunately, the pores could not be completely eliminated even after sintering at 1150 °C.

Cross-sectional SEM images of the PFN ceramics, which were made with the powder from the columbite precursor, after sintering at different temperatures, are shown in figure 4.14. Comparatively, these PFN ceramics possessed a denser and more homogeneous microstructure. At the same time, they had smaller grain sizes than the samples from the oxide mixture, and the grains were also smaller than those of the material derived from mixed oxides at each sintering temperature. The average grain sizes of the samples sintered at 1000 °C and 1100 °C were 0.8 μm and 4 μm, respectively. Although the sample sintered at 1150 °C experienced grain coarsening, the porosity was comparatively much lower.

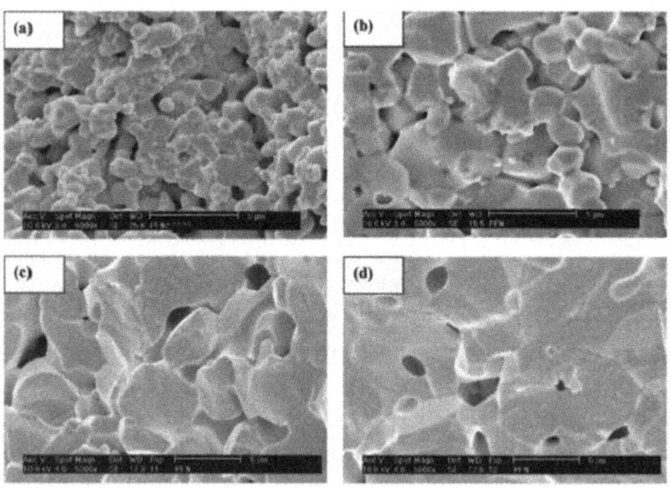

Figure 4.13. Cross-sectional SEM images of the PFN ceramics from the oxide mixture (PbO, Fe_2O_3 and Nb_2O_5) sintered at different temperatures: (a) 950 °C, (b) 1000 °C, (c) 1100 °C and (d) 1150 °C. Reproduced with permission from [29]. Copyright 2002 Wiley.

Figure 4.14. Cross-sectional SEM images of the PFN ceramics from the columbite precursor (PbO and $FeNbO_4$) sintered at different temperatures: (a) 950 °C, (b) 1000 °C, (c) 1100 °C and (d) 1150 °C. Reproduced with permission from [29]. Copyright 2002 Wiley.

The difference in densification process and microstructural development between the two groups of PFN ceramics could be readily attributed to the difference in properties between the two PFN nanosized powders. For the oxide mixture group, due to the presence of the unreacted oxides, the pyrochlore phase was formed during the sintering process at certain low temperatures. Because the pyrochlore phase has a larger molar volume than the perovskite PFN, the phase transition from pyrochlore to perovskite will leave pores behind. In addition, the first groups of powders had compositional fluctuation, which led to Kirkendall swelling, due to the difference in the diffusion rate of the phases. In contrast, the second group of PFN powders had no pyrochlore phase, thus resulting in much better sintering behavior. The columbite precursor derived PFN ceramics also exhibited higher dielectric properties.

Although the perovskite PFW phase could not be directly derived from the oxide mixture of PbO, Fe_2O_3 and WO_3 through the mechanochemical activation route, the milling enhanced the phase formation during the post calcination step [30]. Figure 4.15 shows the XRD patterns of the oxide mixtures mechanochemically milled for different times. After milling for 5 h, most of the diffraction peaks of the oxides were largely broadened, while some of them were absent. In addition, several new peaks were present, indicating the formation of new phases. The new phases included $PbWO_4$ and pyrochlore $Pb_2FeWO_{6.5}$. However, it was found that phase composition of the system was not obviously changed as the milling time was increased up to 25 h.

Figure 4.16 shows the XRD patterns of the powders with and without 20 h mechanochemical activation calcined at 500 °C for 2 h. The milled powder consisted of $PbWO_4$ and $Pb_2FeWO_{6.5}$ as the major phases, while only $Pb_2FeWO_{6.5}$ was present in the unmilled sample. The unmilled and milled powders were calcined at different temperatures to observe their phase evolution behaviors. For the milled

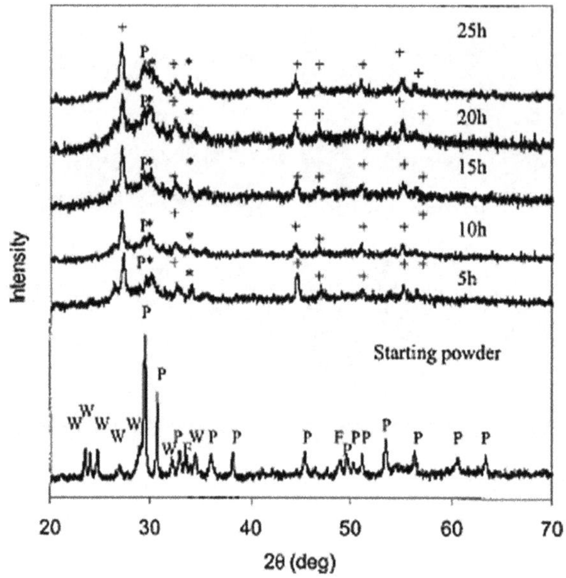

Figure 4.15. XRD patterns of the oxide mixture mechanochemically activated for different times: (+) $PbWO_4$, (*) $Pb_2FeWO_{6.5}$, (P) PbO, (F) Fe_2O_3 and (W) WO_3. Reproduced with permission from [30]. Copyright 2000 Wiley.

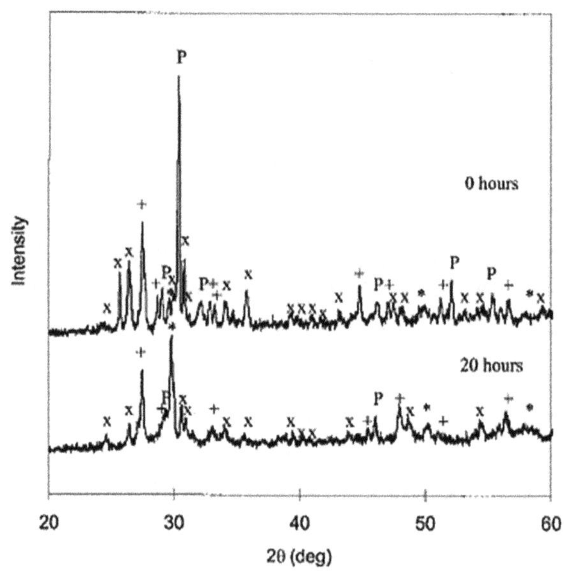

Figure 4.16. XRD patterns of powders before and after mechanochemical activation for 20 h after calcination at 500 °C for 2 h: (P) PbO, (+) $PbWO_4$, (*) $Pb_2FeWO_{6.5}$ and (X) Pb_2WO_5. Reproduced with permission from [30]. Copyright 2000 Wiley.

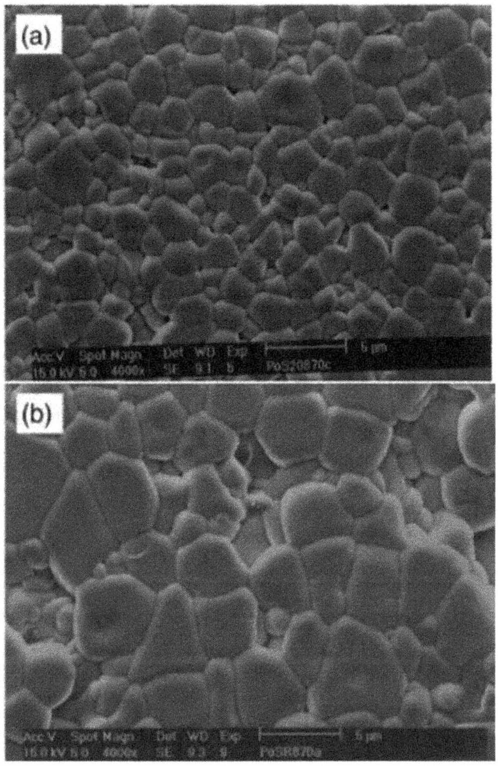

Figure 4.17. Surface SEM images of the PFW ceramics from different powders after sintering at 870 °C for 2 h and thermal etching at 770 °C for 10 min: (a) mechanochemical activation for 20 h and (b) without activation. Reproduced with permission from [30]. Copyright 2000 Wiley.

sample, $PbWO_4$ was absent after calcination at 650 °C, whereas pyrochlore $Pb_2FeWO_{6.5}$ was the only major phase. However, after calcination at 700 °C, an almost complete conversion of pyrochlore to perovskite phase was observed. The perovskite was stable up to 870 °C. For the unmilled mixture, pyrochlore was the major phase, while the constituent oxides were also detected after calcination at 650 °C. As the calcination temperature was increased to 700 °C, pyrochlore was still dominant, but the perovskite PFW phase was observed. Nearly single-phase perovskite was obtained at 750 °C. Therefore, it was confirmed that the phase formation of perovskite PFW was greatly promoted by the mechanochemical activation.

Figure 4.17 shows representative SEM images of the PFW ceramics from the milled and unmilled powders, after sintering at 870 °C for 2 h. The PFW ceramics derived from the activated powder had an average grain size of 1.8 μm, while the average grain size of the sample from the unmilled powder was 3.7 μm. The milled powder exhibited a higher densification rate and the resultant PFW ceramics possessed a more homogeneous microstructure. As a result, they also displayed higher dielectric properties. For example, the PFW ceramics from the mechanochemically activated powder had a peak dielectric constant of 9800 at 10 kHz, while the value of the sample from the unmilled powder was only about 7500.

The same authors came up with the idea to use perovskite PFW powder as a seed to promote the phase formation of PFW from an oxide mixture with mechanochemical activation [31]. It was found that the addition of 0.4 mole PFW seeding powder could enhance the phase formation of PFW from the oxide mixture. Figure 4.18 shows the XRD patterns of oxide mixture that was mechanochemically milled for 20 h, followed by calcination at different temperatures. After calcination at 400 °C, the diffraction peaks of $PbWO_4$ and $Pb_2FeWO_{6.5}$ were increased, suggesting an increase in their crystallinity. The calcination at 500 °C brought out a new phase of Pb_2WO_5. The sample calcinated at 650 °C contained $Pb_2FeWO_{6.5}$ as the dominant phase, but the perovskite PFW phase was present at the same time. Single-phase PFW was achieved after the sample was calcined at 700 °C. This implied that the PFW was formed through the $PbWO_4$ and $Pb_2FeWO_{6.5}$ phases.

Figure 4.19 shows the XRD patterns of the oxide mixture that was mechanochemically milled for 20 h, with the addition of 0.4 mole PFW as the seed, followed by calcination at different temperatures. The calcination at temperatures of up to 650 °C led to almost no variation in the phase composition of the samples. After calcination at 700 °C, the perovskite PFW phase was formed. This demonstrated that the phase evolution behaviors of the powders with and without the seed were

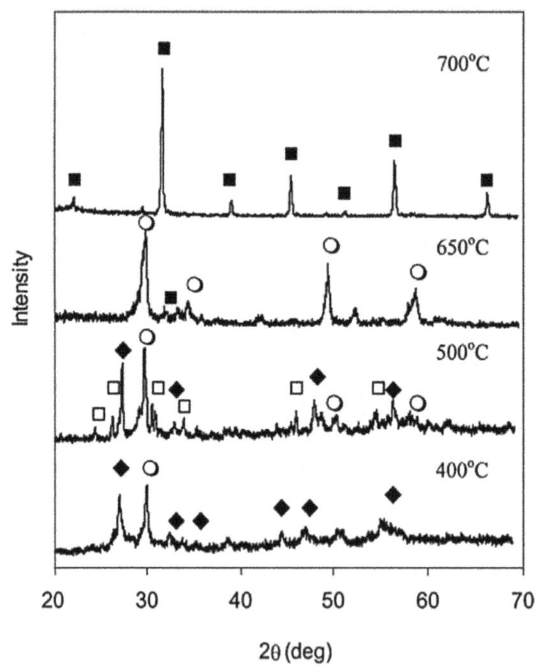

Figure 4.18. XRD patterns of the oxide mixture of PbO, Fe_2O_3 and WO_3 after mechanochemical activation for 20 h followed by calcination at different temperatures: (P) PbO, (♦) $PbWO_4$, (□) Pb_2WO_5, (○) $PbFe_2WO_{6.5}$ and (■) $Pb(Fe_{2/3}W_{1/3})O_3$. Reproduced with permission from [31]. Copyright 2000 Elsevier.

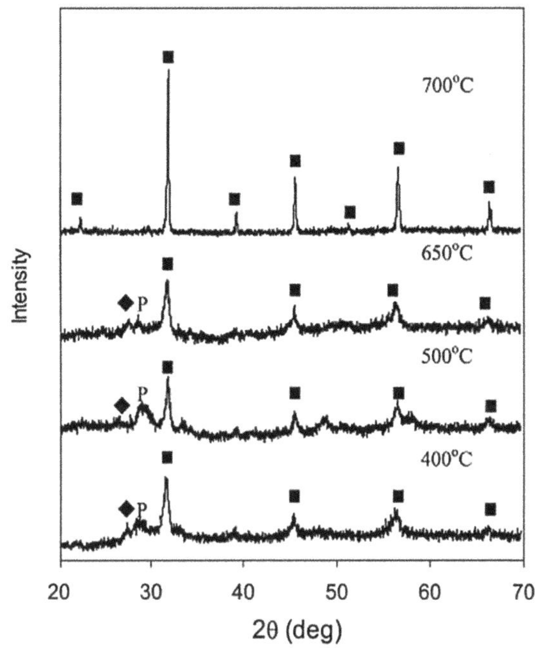

Figure 4.19. XRD patterns of the oxide mixture of PbO, Fe_2O_3 and WO_3 with 0.4 mole perovskite PFW powder after mechanochemical activation for 20 h followed by calcination at different temperatures: (P) PbO, (♦) $PbWO_4$ and (■) $Pb(Fe_{2/3}W_{1/3})O_3$. Reproduced with permission from [31]. Copyright 2000 Elsevier.

essentially different. With the presence of 0.4 mole PFW seed, the formation of single-phase PFW occurred without involving the pyrochlore phase and other intermediate phases.

Single-phase perovskite PFW has also been obtained from a mixture of $Pb_3Fe_2O_6$ and WO_3 through mechanochemical activation [32]. The phase formation of PFW from the mixture occurs after milling for only 5 h. The resultant powder had particle sizes in the range 20–30 nm, with near-spherical particle morphology. The PFW phase was stable up to 870 °C during the sintering process. The PFW ceramics could reach a maximum relative density of about 99% at the optimal sintering temperature of 870 °C. A maximum dielectric constant of 9000 was observed at 10 kHz.

$Pb_3Fe_2O_5$ powder was synthesized using a chemical coprecipitation method from $Pb(NO_3)_2$ and $Fe(NO_3)_3{\cdot}9H_2O$. The two salts were dissolved in distilled water to form aqueous solutions. Ammonia solution was used to precipitate the Pb^{2+} and Fe^{3+} ions. The precipitate was thoroughly washed, followed by drying at 80 °C and calcination at 600 °C, in order to form $Pb_3Fe_2O_5$ powder. The $Pb_3Fe_2O_5$ powder was mixed with WO_3, followed by the mechanochemical activation process. The formation of PFW was realized without involving any intermediate phases. The final PFW powder had particle sizes in the range 5–10 nm. PFW ceramics could be developed from the nanosized powder at relatively low sintering temperatures.

The XRD patterns of the mixture of $Pb_3Fe_2O_5$ and WO_3 after mechanochemical activation for different times, followed by calcining at 600 °C and 870 °C, are shown

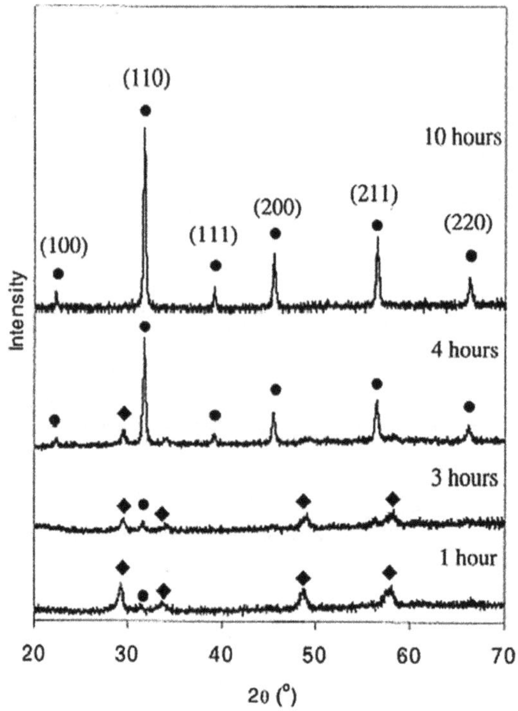

Figure 4.20. XRD patterns of the mixture of $Pb_3Fe_2O_6$ and WO_3 with different activation times calcined at 600 °C for 2 h: (●) PFW and (♦) $Pb_3Fe_2O_6$. Reproduced with permission from [32]. Copyright 2002 Elsevier.

in figures 4.20 and 4.21, respectively. For the samples milled for <3 h, $Pb_2Fe_2O_6$ was still the main phase, after calcination at 600 °C. In addition, perovskite was present as a minor phase. This was simply because the phase formation of PFW was not completed due to the insufficient milling time. Single-phase perovskite was achieved for the powder milled for 10 h at this calcination temperature. As the calcination temperature was increased to 870 °C, all the samples exhibited single-phase perovskite PFW, regardless of the activation time.

The densities of the PFW ceramics derived from the powders milled for different times and sintered at different temperatures were measured. It was found that, at a given sintering temperature, the density of the PFW ceramics rapidly increased as the activation time was increased from 1 h to 4 h. If the activation time was longer than 5 h, the samples would have similar values of density. For the samples with the same activation time, the density was increased with increasing sintering temperature. The low densification behavior of the powders activated for a relatively short time was ascribed to the incomplete phase formation of PFW. The presence of the pyrochlore phase and the reaction induced volume effect were responsible for the poor sintering process. In addition, to ensure high quality PFW ceramics with high dielectric properties, the mechanochemical activation time should be sufficiently long.

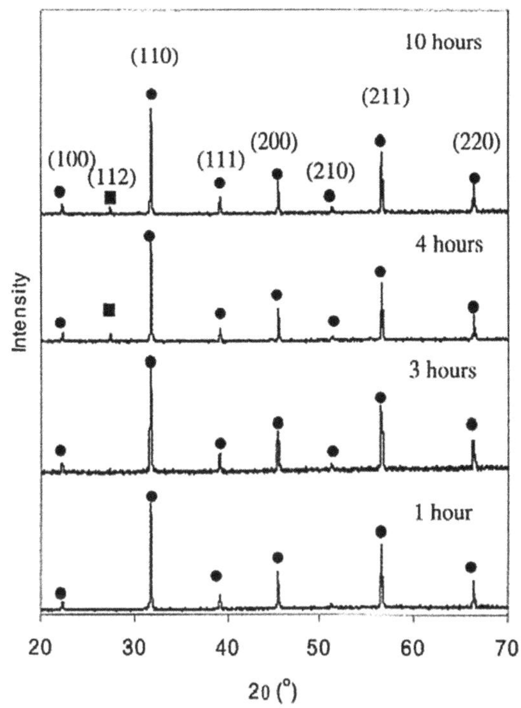

Figure 4.21. XRD patterns of the mixture of $Pb_3Fe_2O_6$ and WO_3 with different activation times calcined at 870 °C for 2 h: (●) PFW and (■) $PbWO_4$. Reproduced with permission from [32]. Copyright 2002 Elsevier.

4.2.4 Lead scandium tantalate

$Pb(Sc_{1/2}Ta_{1/2})O_3$ (PST) is an interesting ferroelectric material [33–35]. Ordered PST has a normal ferroelectric effect, while disordered PST exhibits relaxor ferroelectricity. It can be tailored in the degree of B-site ordering through thermal treatment, thus leading to electrical properties that can be adjustable. Perovskite PST was derived from an oxide mixture activated with high-energy mechanochemical milling [36]. Twenty hours of milling led to the formation of single-phase PST. The starting powders were PbO, Sc_2O_3 and Ta_2O_5. The mechanochemical activation was conducted using an SPEX 8000 shake mill. PST was also made using the conventional solid-state reaction method for comparison. Wolframite precursor was synthesized with Sc_2O_3 and Ta_2O_5 at a calcination temperature of 1350 °C for 4 h, which was then reacted with PbO at 1000 °C for 3 h to form PST.

Figure 4.22 shows the XRD patterns of the PST samples directly synthesized from the oxide mixture through the mechanochemical activation and through the reaction from the Wolframite precursor. The two samples had the same perovskite structure. However, the mechanochemically activated powder possessed broad diffraction peaks, while the peaks of the solid-state reaction product were very sharp. In fact, the (111) superlattice reflection was absent in the PST powder formed using the mechanochemical activation process, which indicated that the mechanochemically synthesized perovskite phase lost the long-range structural ordering.

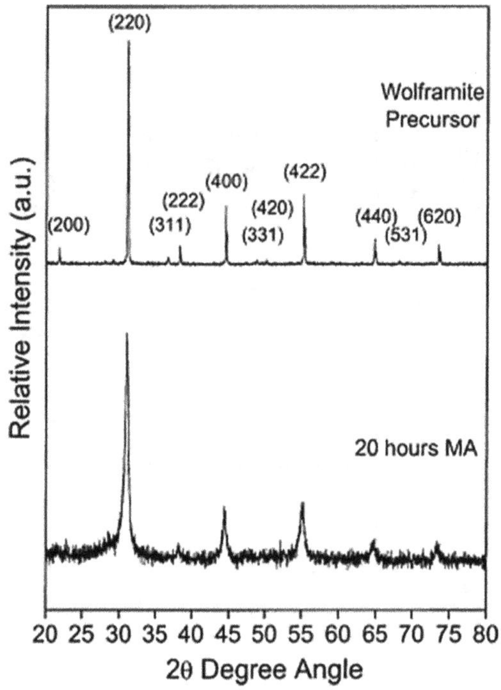

Figure 4.22. XRD patterns of the PST samples formed through 20 h mechanical activation and the reaction with the Wolframite precursor. Reproduced with permission from [36]. Copyright 2002 Elsevier.

It was found that the structural ordering of the PST from the mechanically activated oxide mixture was not recovered even after sintering at 1256 °C. However, the order parameters of the mechanochemical activation sample and the solid-state reaction sample were increased from about zero to 0.23 and 0.729 to 0.764, respectively, after they were sintered at 1250 °C for 2 h. The mechanochemically derived PST ceramic reached a relative density of 97.1%, while that of the solid-state reaction sample was only 92.6%. This could be readily ascribed to the more reactive characteristics of the mechanochemically obtained powder, due to its smaller particle sizes.

The PST ceramics from the mechanochemical powder exhibited the highest dielectric constant after sintering at 1200 °C, because the sample sintered at 1250 °C contained a small quantity of pyrochlore phase, although the latter had the highest density. The peak dielectric constant had a high value of about 13 850 at 100 Hz. In comparison, the PST ceramics from the conventional solid-state reaction powder had a peak dielectric constant of 4530 at 100 Hz.

4.3 Binary solid solutions

Various binary phases have been generated through the combination of relaxor ferroelectrics with ABO_3 compounds, such as BT, ST and PT. Combination of a relaxor and a relaxor is also an effective approach to synthesize materials with

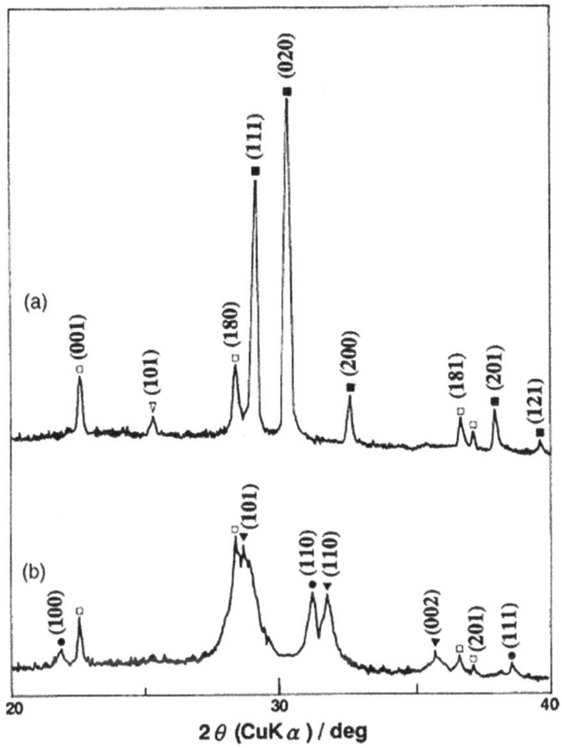

Figure 4.23. XRD patterns of the mixture before (a) and after (b) mechanochemical activation: (■) PbO (massicot), (▼) PbO (litharge), (□) Nb_2O_5, (▽) TiO_2 (anatase) and (●) perovskite. Reproduced with permission from [43]. Copyright 1997 Wiley.

specific properties for specific applications. Typical binary systems such as PMN–PT [37–39], PZN–BT and PZN–PT [40–42], will be described to demonstrate the strong capabilities of high-energy mechanical activation in preparing nanosized ferroelectric powders.

4.3.1 PMN based solid solutions

PMN–PT binary systems could be tailored with different compositions to exhibit a wide range of properties and performances. A low-energy soft-mechanochemical route was used to prepare single-phase 0.9PMN–0.1PT from a mixture of PbO, TiO_2, $Mg(OH)_2$ and Nb_2O_5 [43]. After milling for 1 h, the powder was used to fabricate 0.9PMN–0.1PT ceramics. It was found that both the perovskite phase formation and densification of precursors were enhanced due to the activation. As a result, the final 0.9PMN–0.1PT ceramics exhibited higher dielectric properties compared to those from the unmilled powder. Figure 4.23 shows XRD patterns of the mixture before and after milling for 1 h. The diffraction peaks of the components were significantly broadened, indicating a reduction in particle size. In addition, perovskite was formed, although the phase formation was not finished. The PMN–PT ceramics from the mechanochemically treated powder after sintering at 1250 °C for 2 h had a peak dielectric constant of 2900 at 1 kHz, with an average grain size of about 9 μm. In comparison, the samples from the unmilled powder had a peak dielectric constant of 19 000, with an average grain size of about 20 μm.

Direct synthesis of PMN–0.1PT nanosized powder was realized by high-energy mechanochemically milling the oxide mixture of PbO, MgO, Nb_2O_5 and TiO_2, with an SPEX 8000 shake mill [44]. The milling media were stainless steel. Figure 4.24 shows XRD patterns of the mixtures mechanochemically activated for different times. After treatment for 5 h, the diffraction peaks of PbO and other components were tremendously reduced and broadened, suggesting the refinement of their particles as a result of the mechanochemical action. As the mixture was milled for 10 h, the peak of perovskite PMN–PT was present, implying that the reaction of the component was triggered by the mechanochemical treatment. The reaction was still not finished, because the peaks of PbO were present. An obvious increase in the peak intensity of the perovskite phase was observed in the sample after milling for 15 h. Complete phase formation was achieved after milling for 20 h. The resulting PMN–PT powder had an average particle size of 20–30 nm, which showed very

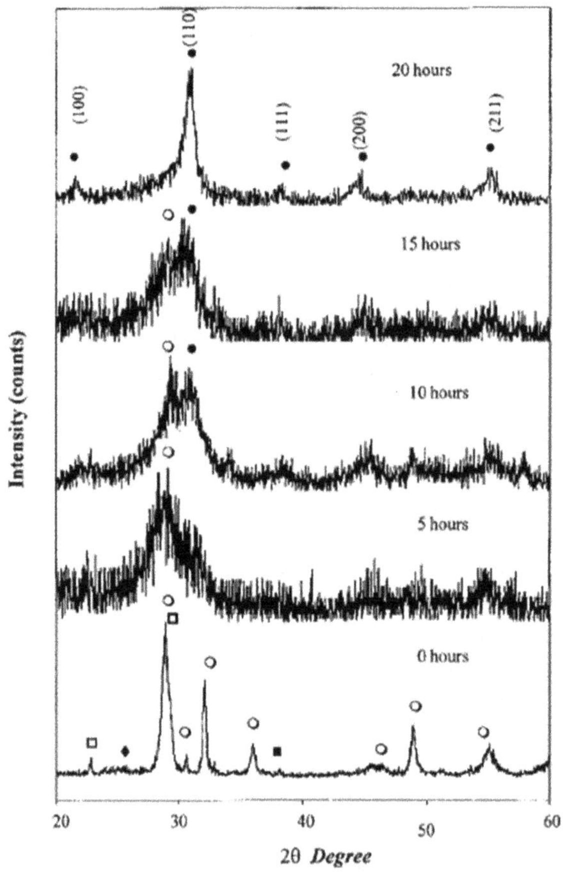

Figure 4.24. XRD patterns of the mixtures mechanochemically activated for different times: (○) PbO, (■) MgO, (□) Nb_2O_5, (♦) TiO_2 and (●) perovskite. Reproduced with permission from [44]. Copyright 2000 Wiley.

high sinterability. After sintering at 1050 °C for 1, the PMN–PT ceramics could reach a relative density of 99.3%. The PMN–PT sample sintered at 1150 °C for 1 h possessed a peak dielectric constant of 26 500 at 100 Hz.

PMN–PT powders with different compositions have been synthesized from oxide mixtures by using mechanochemical activation with a planetary mill [16, 45, 46]. Figure 4.25 shows XRD patterns of the oxide powder mixtures for PMN and PMN–PT milled for 20 h [46]. The three samples all exhibited single-phase perovskite structure, with the presence of any secondary phases, including the component oxides or pyrochlore phase. The crystal sizes of the as-obtained powders were 10–30 nm according the peak broadening of the XRD patterns. Therefore, the phase formation of PMN based powder was different from that observed in the conventional solid-state reaction method. Usually, PbO and Nb_2O_5 first reacted at about 500 °C to form cubic pyrochlore ($Pb_3Nb_4O_{13}$), which further reacted with PbO to result in rhombohedral pyrochlore ($Pb_2Nb_2O_7$) at about 600 °C. $Pb_2Nb_2O_7$ then reacted with MgO to obtain the final perovskite PMN, while $Pb_3Nb_4O_{13}$ was preset at 800 °C. The content of $Pb_3Nb_4O_{13}$ could be minimized by introducing excessive PbO or MgO, but it was hard to eliminate it from the final product.

Figure 4.26 shows cross-sectional SEM images of PMN and PMN–PT ceramics after sintering at 1100 °C, demonstrating a dense and homogeneous structure. Their relative densities were in the range 97%–99%, over a sintering temperature range of 950 °C–1150 °C. This implied that that mechanochemically activated PMN and PMN–PT powders had high sinterability. Figure 4.27 shows the grain sizes of the PMN and PMN–PT ceramics as a function of sintering temperature. After sintering at 950 °C, PMN, PMN–0.1PT and PMN–0.35PT ceramics possessed average grain sizes of 1.2, 1.1 and 0.8 μm, respectively. The average grain size increased constantly with increasing sintering temperature.

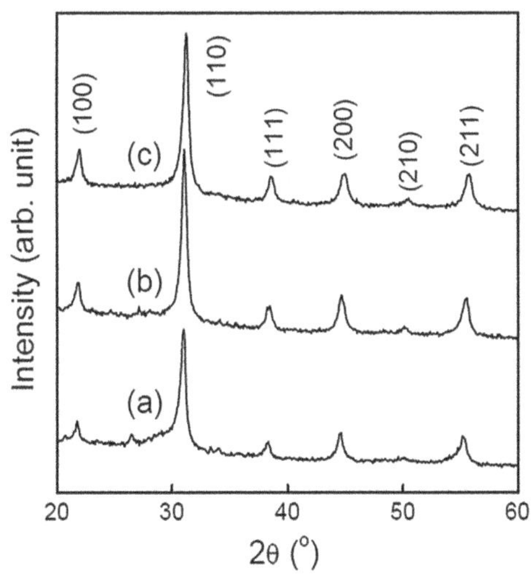

Figure 4.25. XRD patterns of the oxide powder mixtures for PMN and PMN–PT milled for 20 h: (a) PMN, (b) PMN–0.10PT and (c) PMN–0.35PT. Reproduced with permission from [46]. Copyright 2002 Elsevier.

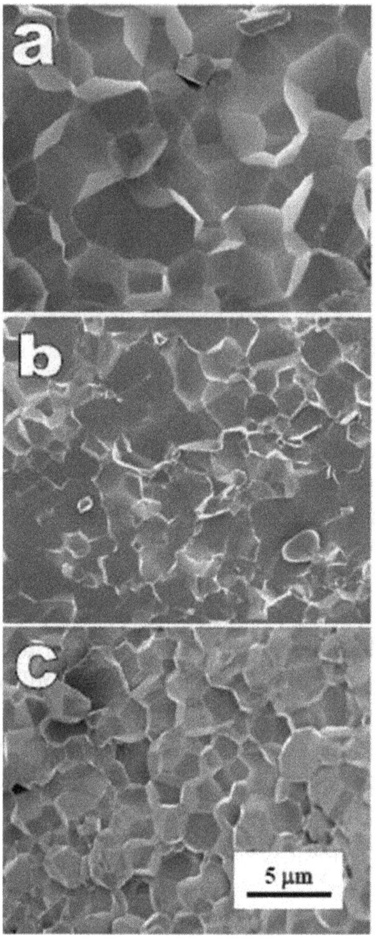

Figure 4.26. SEM images of the PMN and PMN–PT ceramics sintered at 1100 °C for 1 h: (a) PMN, (b) PMN–0.10PT and (c) PMN–0.35PT. Reproduced with permission from [46]. Copyright 2002 Elsevier.

Figure 4.28 shows the dielectric constants, recorded at 1 kHz at room temperature, of the PMN and PMN−PT ceramics as a function of sintering temperature. At a given sintering temperature, the average grain size was decreased with increasing content of PT in the samples. The PMN–0.1PT ceramics sintered at 1100 °C for 1 h had a room temperature dielectric constant of 22 237 at 1 kHz, which was close to the data for the samples with the same compositions derived from other synthetic methods with a sintering temperature of 1200 °C. The PMN–0.35PT sample sintered at 1100 °C had a dielectric constant of 8050 at 1 kHz, which was even higher than that of the ceramics prepared through the conventional solid-state reaction method.

PMN–0.20PT and PMN–0.35PT ceramics were fabricated from powders obtained using mechanochemical activation process ceramics [47]. The PMN−PT ceramics were developed through pressureless sintering in air and the hot-pressing

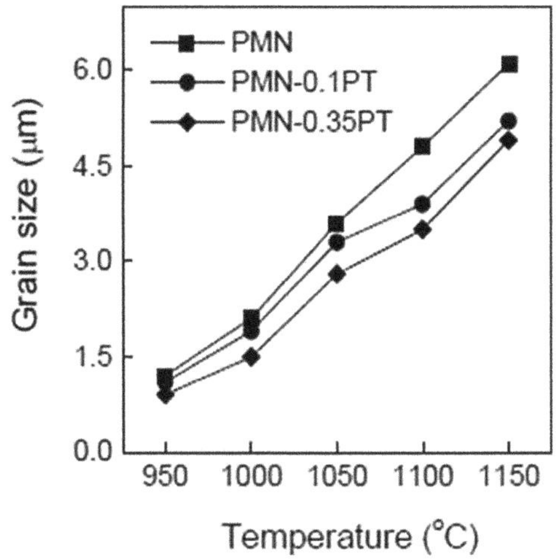

Figure 4.27. Average grain sizes of the PMN and PMN–PT ceramics as a function of sintering temperature. Reproduced with permission from [46]. Copyright 2002 Elsevier.

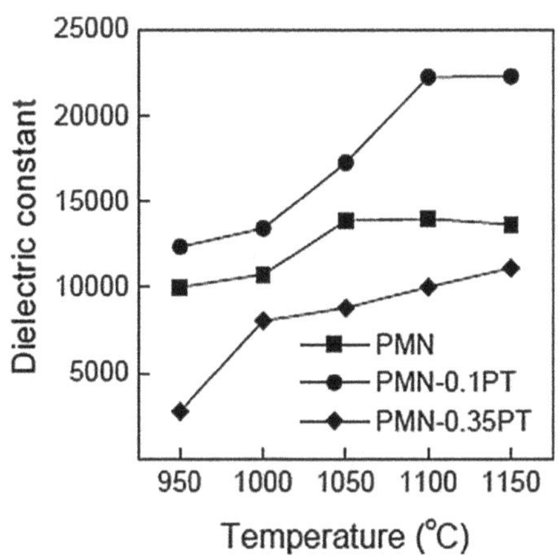

Figure 4.28. Dielectric constants (at 1 kHz) of the PMN and PMN–PT ceramics as a function of sintering temperature. Reproduced with permission from [46]. Copyright 2002 Elsevier.

method. $PbZrO_3$ powder was used to mitigate the loss of PbO during the sintering process as the sintering temperature was >1000 °C. The samples sintered at 1200 °C had a relative density of about 90% and an average grain size of 4 μm. In comparison, the ceramics fabricated using hot-pressing were almost fully densified, with grain sizes in the range 0.1–0.5 μm. The PMN–PT ceramics exhibited electrical properties similar to those of the same materials prepared using other methods.

The effect of PT seeding on the phase formation of PMN–0.10PT was evaluated when using mechanochemical activation from the oxide precursor [48]. It was observed that the perovskite formation could be complete after milling for 64 h when starting with PT as the seeds, while a time of 143 h was required to complete the perovskite phase formation from the mixture of the oxide constituents without PT seeds. The involvement of the PT seeds promoted the kinetics of PMN–0.10PT perovskite phase formation. In this case, the transformation of the amorphous to pyrochlore phase and that from the pyrochlore to perovskite phase were found with the presence of the PT seeds. Because the PT crystals could offer nucleation sites because of the presence of low-energy interfaces, the PT seeds would also help create an amorphous matrix, within which the perovskite phase was crystallized. The matrix would shorten the diffusion paths, thus boosting the kinetics of the perovskite phase formation.

4.3.2 PZN based solid solutions

Because of the instability at high temperatures, single-phase PZN ceramics cannot be obtained using the conventional fabrication process, although the synthesis of PZN perovskite nanosized powder has been realized using the mechanochemical activation process, as presented earlier. In practice, PZN based ceramics are prepared using other stable perovskite phases as a stabilizer, such as $BaTiO_3$ (BT), $SrTiO_3$ (ST) and $PbTiO_3$ (PT). Various PZN−BT, PZN−ST or PZN−PT powders and ceramics have been developed using the mechanochemical process.

Single-phase 0.95PZN–0.05BT and 0.90PZN–0.10BT nanosized powders and ceramics were prepared using the mechanochemical process with an SPEX 8000 shaker mill [49]. Figure 4.29 shows the XRD patterns of the oxide mixture of PbO, ZnO, Nb_2O_5, BaO and TiO_2 for the composition of 0.95PZN–0.05BT after mechanochemical activation for different times [49]. Diffraction peaks of the sample milled for 5 h were all reduced and broadened, due to the refinement of the oxide particles. No reaction was observed in this sample. After milling for 10 h, the perovskite phase started to appear, while the peaks of the precursor oxides were almost all absent. Phase formation was nearly complete after milling for 15 h. Prolonged milling had no effect on the phase composition of the samples.

Phase stability of the sample milled for 20 h was examined by calcining at different temperatures for 1 h. Decomposition was observed after calcination at 450 °C. The content of the perovskite phase was continuously reduced with increasing calcination temperature, reaching the lowest level of only about 20% at 700 °C. Above 700 °C, the content started to increase until 1100 °C. Too high a temperature led to a decrease in the content of perovskite again. When this powder was used to prepare PZN–0.05BT ceramics, the optimal sintering temperature was 1100 °C, which ensured a relative density of 96%. A peak dielectric constant of 14 000 was observed at 100 Hz.

For PZN–0.10BT, the phase formation time for perovskite was relatively short, compared to that for PZN–0.05BT [50]. This could be attributed to the higher content of BT, further confirming the stabilization effect of BT in the solid solution. Milling for 5 h resulted in perovskite phase formation, while it was completed after activation for 10 h. The powder could be sintered at 1100 °C to obtain PZN–0.10BT ceramics, which exhibited a relative density of about 96%. The ceramics displayed a peak dielectric constant of 8800 at 100 Hz.

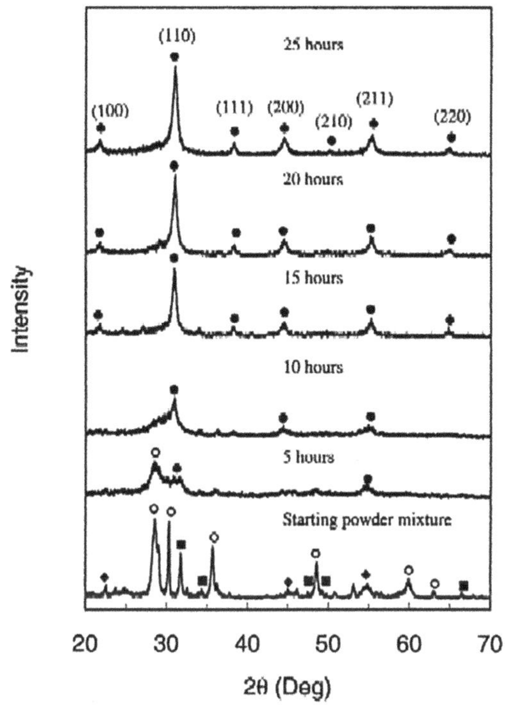

Figure 4.29. XRD patterns of the oxide mixture for 0.95PZN–0.05BT after mechanochemical activation for different times: (○) PbO, (■) ZnO, (♦) Nb_2O_5 and (●) perovskite. Reproduced with permission from [49]. Copyright 2000 Elsevier.

The synthesis of nanosized powders and fabrication of ceramics for the solid solutions of $(1-x)$PZN–xBT, with x being up to 0.30, by using the mechanochemical activation process have been demonstrated [51]. After the mixtures with different compositions were milled for 12 h, the perovskite phase was dominant in all samples. There was still a trace of pyrochlore detected in the XRD patterns, but the content was gradually decreased with increasing content of BT in the solid solutions. The sample with $x = 0.30$ had no pyrochlore phase to be detected, according to the XRD pattern.

After sintering at 1050 °C for 1 h, the samples with $x = 0.10$–0.30 all exhibited a single phase of perovskite, while the 0.95PZN–0.05BT ceramics contained only about 5% pyrochlore phase. Assuming that the PZN–BT solid solutions had a pseudocubic structure, the lattice constant showed a linear decrease with increasing content of BT, confirming the solid solutions to be formed between PZN and BT. The phase pure characteristics of the solid solution ceramics with $x \geqslant 0.10$ were also evidenced by their microstructures, with nearly spherical grains and well intergrain fractured surfaces.

The effect of the milling time and properties of the PZN–0.3BT were specifically studied [52]. Figure 4.30 shows XRD patterns of the oxide mixtures for PZN–0.30BT milled for different times. The perovskite phase was formed as a major phase

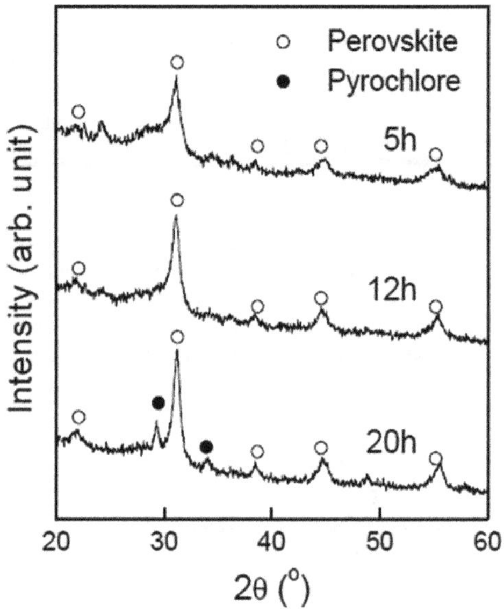

Figure 4.30. XRD patterns of the mixture for PZN–0.30BT activated for different times. Reproduced with permission from [52]. Copyright 2002 Elsevier.

in the sample milled for 5 h. Single-phase perovskite PZN–BT was achieved as the sample was activated for 12 h. However, the pyrochlore phase was present again if the milling time was increased to 20 h. Therefore, there was an optimal milling time in order to maintain phase pure powders. PZN–BT ceramics were fabricated using the powders milled for 12 h. The XRD patterns indicated that the PZN–BT ceramics sintered at 1000 °C, 1050 °C and 1100 °C all had a single phase of perovskite structure, suggesting that the mechanochemically synthesized PZN–BT powders had high stability.

Figure 4.31 shows cross-sectional SEM images of the PZN–BT ceramics, which were made from the powders milled for 12 h, sintered at different temperatures. All the PZN–BT ceramics were nearly fully densified, with well-developed grains and homogeneous microstructure. After sintering at 1000 °C for 1 h, the average grain size was about 0.8 μm. As the sintering temperature was raised to 1050 °C and 1100 °C, the grains grew to 1.9 μm and 3.9 μm in average size, while the relative densities were 97% and 99%, respectively. In this case, the average grain size increased almost linearly as the sintering temperature was increased. In addition, the samples derived from the powders milled for 5 h and 20 h exhibited similar characteristics in microstructure, grain morphology and grain size. This means that the incomplete reaction of the constituent oxides in the powder milled for 5 h was finished, while the pyrochlore phase in the sample milled for 20 h was eliminated, when the powders were sintered at high temperatures.

The synthesis and phase formation process of PZN–0.08PT were studied using DTA/TGA, XRD and TEM technologies [53]. The mixtures of PbO, ZnO, Nb_2O_5 and TiO_2 were treated with a Pulverisette 6 model Fritsch planetary mill operating at

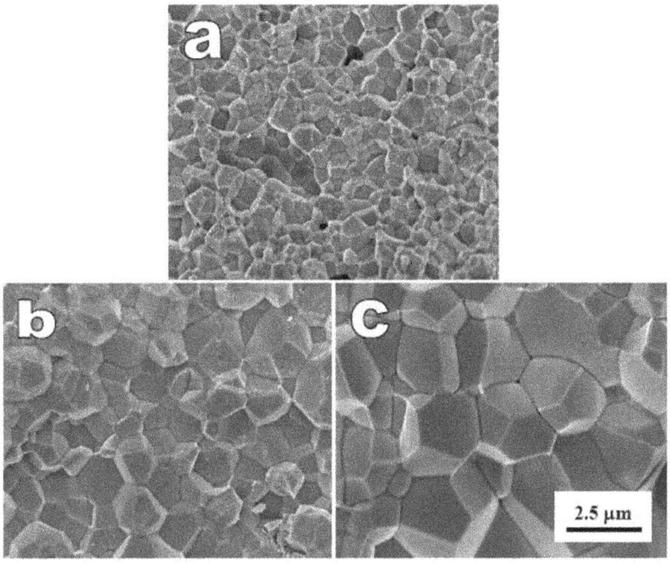

Figure 4.31. Cross-sectional SEM images of the PZN–0.30BT ceramics derived from the powder activated for 12 h after sintering at different temperatures for 1 h: (a) 1000 °C, (b) 1050 °C and (c) 1100 °C. Reproduced with permission from [52]. Copyright 2002 Elsevier.

a speed of 300 rpm. The mechanochemical activation was conducted in air for times of up to 140 h. Perovskite PZN–PT became a major phase after 20 h and single-phase perovskite was achieved after milling for 30 h. Prolonging the milling time from 30 to 70 h led to a slight increase in the intensity of the perovskite peaks, which could be attributed to the enhanced crystallization of the perovskite phase. Further milling for up to 140 h did not result in further change in the XRD pattern of the powders. A careful examination indicated that the diffraction peak (110) shifted to a high angle with increasing milling time (from 30 h to 140 h), which implies that the perovskite cell was shrunk. There was no significant difference in grain size among the samples milled for different times.

Figure 4.32 shows the XRD patterns of the oxide powders after mechanochemical activation for different times. Similar to the previous observations, PbO experienced a phase transition from massicot to litharge, while the diffraction peaks of all the constituent oxides were broadened after milling for just 1 h. PbO was retained after milling for times of up to 20 h. Perovskite PZN–PT phase was present in the sample milled for 10 h. After treating for 20 h, the perovskite phase was dominant, while phase pure perovskite was achieved after activation for 30 h. The sample milled for 30–70 h exhibited an increase in the intensity of the diffraction peaks. A further increase in milling time to 140 h had no effect on phase composition but resulted in a slight reduction in the crystallinity of the solid solution phase. Detailed analysis indicated that the crystallite sizes of the powders after milling for 30 h, 70 h and 140 h were 15.6, 18.1 and 16.7 nm, respectively, together with average strains of 1.19×10^{-2}, 1.05×10^{-2} and 1.11×10^{-2}.

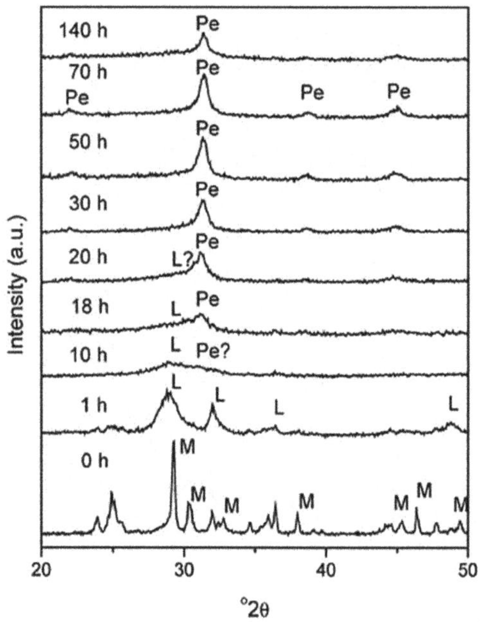

Figure 4.32. XRD patterns of the oxide mixed powders after mechanochemical activation for different times: (M) massicot, (L) litharge and (Pe) perovskite. Reproduced with permission from [53]. Copyright 2004 Wiley.

The authors stated that the mechanochemical milling for up to 20 h resulted in chemical homogenization of the constituent oxides at the nanometer scale, facilitating the phase formation of PZN–0.08PT. During the activation process, agglomerates were developed, within which the perovskite phase was formed and the crystallinity was gradually enhanced until 70 h. Because the crystallite size of the PZN–0.08PT powders was almost unchanged with increasing milling time, it could be concluded that the milled system experienced a continuous nucleation process, instead of growth from initial nuclei. A further increase in milling time could result in a high degree of agglomeration and bring in Fe contamination due to the abrasion of the milling media.

Figure 4.33 shows XRD patterns of the powder milled for 30 h, followed by calcination at different temperatures. Below 400 °C, the samples were all phase pure with perovskite structure. After calcining at 500 °C, the pyrochlore phase began to be present. However, the sample calcined at 550 °C contained the pyrochlore phase as the major phase, while the content of perovskite was significantly reduced. A complete phase transformation from perovskite to pyrochlore was observed in the sample that was calcined at 600 °C. The pyrochlore phase, a composition of $Pb_{1.88}Zn_{0.3}Nb_{1.25}O_{5.305}$, was stable up to 700 °C. The powders milled for longer times had similar variation in phase composition as a result of the thermal calcination.

Figure 4.34 shows the XRD patterns of the samples with different times of mechanochemical activation and different calcining conditions. There were only

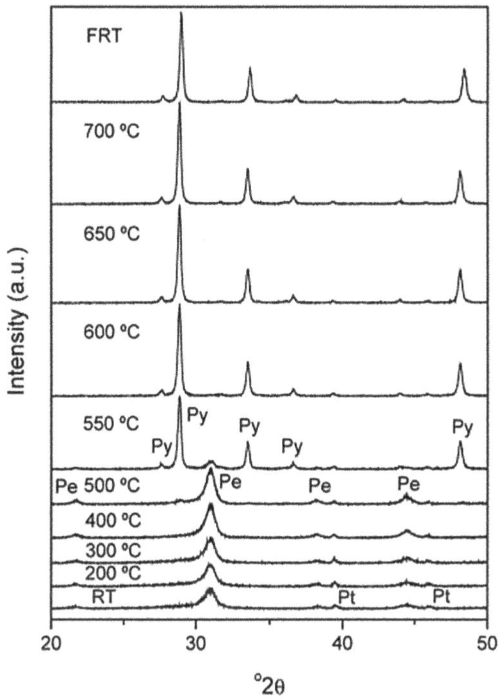

Figure 4.33. XRD patterns of the powders mechanochemically milled for 30 h, followed by calcination at different temperatures: (Pe) perovskite, (Py) pyrochlore and (Pt) platinum (from the sample holder). Reproduced with permission from [53]. Copyright 2004 Wiley.

slight differences in the temperatures for the presence of the pyrochlore phase and the absence of the perovskite phase. The effect of milling time and the subsequent hot-pressing conditions on the phase transformation of the perovskite to pyrochlore phase were studied. By comparing the XRD patterns of the powder milled for 30 h that was hot pressed at 500 °C and 40 MPa and the sample milled for 70 h that was hot pressed at 550 °C and 50 MPa with the patterns of the same powders which were calcined at ambient pressure, it was found that hot-pressing could retain the perovskite phase. In other words, the perovskite phase became more stable at a certain level of pressure. Therefore, the level of 0.08PT was not sufficient to stabilize perovskite PZN at high temperatures, although it could be synthesized using the mechanochemical activation process at low temperatures (most likely ambient temperature).

$0.3Pb(Zn_{1/3}Nb_{2/3})O_3$–$0.7Pb(Zr_{0.51}Ti_{0.49})O_3$ (PZN–PZT) ceramics doped with La at different concentrations were derived from the corresponding powders that were obtained using the mechanochemical activation method [54]. All the ceramics had a single phase of pervoskite structure and dense microstructure with uniform grain size distribution. The doping of La increased the bulk density of the final PZN–PT ceramics. With increasing content of La, the Curie temperature (T_C) of the ceramics was decreased and the degree of diffusion behavior in phase transition was

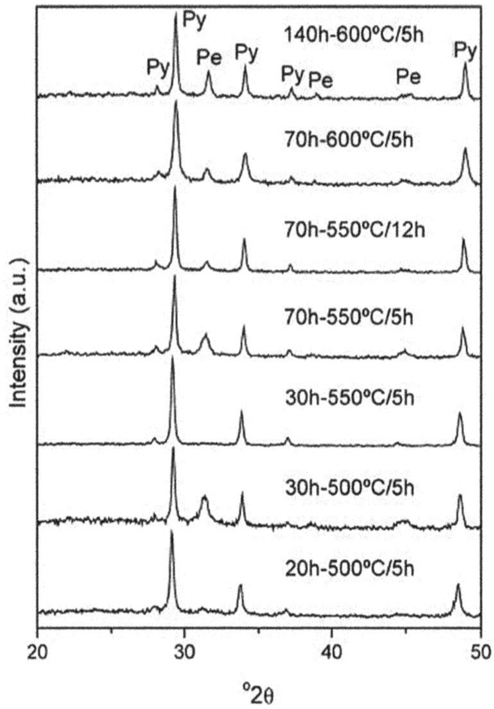

Figure 4.34. XRD patterns of the powders mechanochemically activated for different times followed by calcination under different conditions: (Pe) perovskite and (Py) pyrochlore. Reproduced with permission from [53]. Copyright 2004 Wiley.

enhanced. Both the piezoelectric coefficient and electromechanical coupling constant of the ceramics were increased as a result of the addition of La.

In a separate study, xPb($Zn_{1/3}Nb_{2/3}$)O_3–(1 − x)Pb($Zr_{0.47}Ti_{0.53}$)O_3 binary solid solution powders, with fine grains, were derived from precursor powders prepared using the high-energy mechanochemical activation process [55]. The incorporation of PZN with PZT induced a phase transition from tetragonal to morphotropic and then to rhombohedral. Compared to results of the coarse-grained ceramics made with conventional ceramic processing, the MPB was shifted from 50% to 30% in the content of PZN, which was attributed to the increased internal stress related to the fined grains in the final ceramics. The enhancement in electrical and mechanical performances is closely related to the phase composition and microstructure. Specifically, the 0.3PZN–0.7PZT ceramics had an average grain size of 0.65 μm, with promising piezoelectric properties of d_{33} = 380 pC/N and k_p = 0.49. They also displayed high mechanical performances of H_v = 5.0 GPa and K_{IC} = 1.33 MPa$\cdot$ m$^{1/2}$.

4.3.3 Other relaxor related solid solutions

Relaxor–relaxor solid solutions have been explored to develop ferroelectric ceramics with interesting properties. The phase formation and electrical properties of the PMN–PZN binary relaxor ceramics, resulted from precursors of PbO, $Mg(OH)_2$,

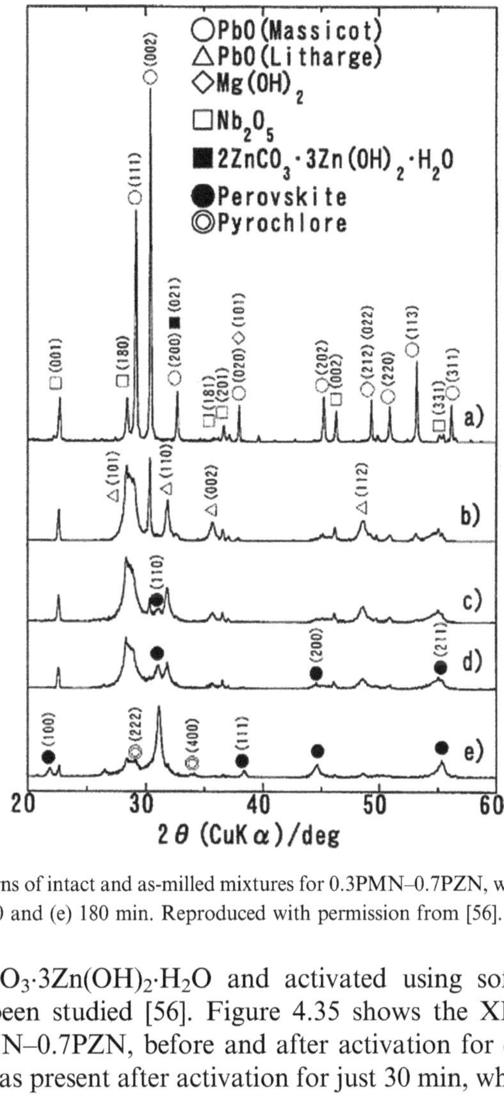

Figure 4.35. XRD patterns of intact and as-milled mixtures for 0.3PMN–0.7PZN, with different milling times: (a) 0, (b) 5, (c) 30, (d) 60 and (e) 180 min. Reproduced with permission from [56]. Copyright 2000 Wiley.

Nb_2O_5 and $2ZnCO_3{\cdot}3Zn(OH)_2{\cdot}H_2O$ and activated using soft-mechanochemical processing, have been studied [56]. Figure 4.35 shows the XRD patterns of the mixture for 0.3PMN–0.7PZN, before and after activation for different times. The perovskite phase was present after activation for just 30 min, whereas it was already dominant after milling for 180 min. For the $(1 - x)$PMN–xPZN samples with $x \leqslant 0.9$, the content of perovskite phase was close to 80%. For pure PZN, the perovskite phase took up about 60%.

The thermal stability of the as-activated PMN–PZN powders decreased with increased concentration of PZN. When the sample with $x = 0.9$ was calcined at 700 °C for 1 h, the content of perovskite was largely reduced from 80% to about 10%. Under the thermal treatment conditions, PZN was completely converted to the pyrochlore phase. For samples with $x \geqslant 0.8$, the content of perovskite phase was decreased first and then increased, with the lowest levels at 700 °C–800 °C, as the calcination temperature was increased. The exact temperature of the minimum content of perovskite phase was closely related to the composition of the materials.

(1 − x)PFW−xPZN (x = 0–0.5) solid solutions were prepared using the mechanochemical activation process, with their thermal stability and electrical properties to be studied [57]. The starting oxide powders were PbO, Fe_2O_3, WO_3, ZnO and Nb_2O_5, which were mixed and mechanochemically activated for 20 h. The XRD analysis results indicated that the perovskite phase was not formed if no PZN component was present. As the content of PZN was increased, the quantity of the perovskite phase was increased continuously.

The sample with x = 0.5 was nearly completely single-phase perovskite. The powders with different compositions had different thermal stabilities, highly dependent on the content of PZN. The sample with x = 0.2 contained only pyrochlore after calcining at 500 °C, while perovskite was only visible after treatment at 700 °C. Calcination at 800 °C led to the formation of single-phase perovskite, whereas pyrochlore was present after calcining at 900 °C. For the sample with x = 0.4, the content of perovskite phase was much higher at the same calcination temperatures. Over the temperature range 500 °C–800 °C, the content of perovskite phase increased with increasing calcination temperature, accompanied by a decrease in the quantity of pyrochlore phase which could not be completely absent.

Ferroelectric $0.57Pb(Sc_{1/2}Nb_{1/2})O_3$–$0.43PbTiO_3$ (0.57PSN–0.43PT) ceramics were made of powders that were synthesized using the mechanochemical activation method, with outstanding electrical properties [58]. The ceramics reached a relative density of 97% after sintering at 1000 °C, which is lower than the temperature required by the conventional solid-state reaction process by 200 °C–300 °C. The starting oxide powders were PbO, Sc_2O_3, TiO_2 and Nb_2O_5, which were thoroughly mixed according to the composition of 0.57PSN–0.43PT. The mixture was mechanochemically treated with a Retsch Model PM 400 planetary mill. The activation was conducted for 24 h at 300 rpm, with 15 vidia balls and 250 ml vidia vessels. The powder was then calcined at 900 °C for 1 h. In this case, some milling parameters, such as the diameter of the vidia balls and powder-to-ball weight ratio, were not mentioned. The authors further optimized the sintering conditions of the ceramics to achieve high electrical performance [59]. They found that the optimal sintering parameters were 1000 °C for 8 h, with the largest polarization, peak-to-peak strain, piezoelectric coefficient and coupling coefficients (k_p and k_t) being 40 μC cm^{-2}, 0.41%, 570 pC/N, 0.71 and 0.56, respectively.

4.4 Ternary solid solutions

Ternary solid solutions, $0.54Pb(Zn_{1/3}Nb_{2/3})O_3$–$0.36Pb(Mg_{1/3}Nb_{2/3})O_3$–$0.1PbTiO_3$ (PZN–PMN–PT) powders, were synthesized from oxide precursors using the mechanochemical activation process [60–62]. Commercial oxide powders, including PbO, ZnO, MgO, Nb_2O_5 and TiO_2, were mixed and activated through mechanochemical milling with an SPEX 8000 shaker mill. The perovskite phase was formed after milling for 10 h. The PZN–PMN–PT powder crystal sizes were in the range 10–15 nm. A relative density of 97% could be achieved after sintering at 950 °C for 1 h. The PZN–PMN–PT ceramics sintered at 1100 °C for 1 h had a peak dielectric constant of 18 986 at 100 Hz.

The color of the mixed powder varied from yellow brown to light yellow gradually as the milling time was increased. During the mechanochemical activation process, PbO had a phase change before the reaction to form perovskite PZN–PMN–PT. Figure 4.36 shows the XRD patterns of the mixed powders mechanochemically milled for different times, in which $Mg_{0.4}Zn_{0.6}Nb_2O_6$ (MZN) was prepared through the conventional solid-state reaction process. After milling for 3 h, diffraction peaks of the starting components were reduced and widened, suggesting refinement of the particles. The average particle size of the starting powder was about 1.23 μm, which was reduced to 0.54 μm after activation for 3 h. In addition, the BET specific surface area was enlarged from 1.51 to 12.45 $m^2\ g^{-1}$. After milling for 4 h, the perovskite phase was present, implying that the reaction of the components was triggered by the mechanochemical activation. Perovskite phase formation was nearly complete as the milling time was increased to 5 h. Crystallinity of the perovskite phase was further increased in the sample milled for 10 h, while no significant variation was apparent in XRD patterns when the milling time was increased to 20 h.

DTA results revealed that the starting powder mixture without mechanochemical activation had two exothermic peaks at 350 °C and 700 °C. The first peak

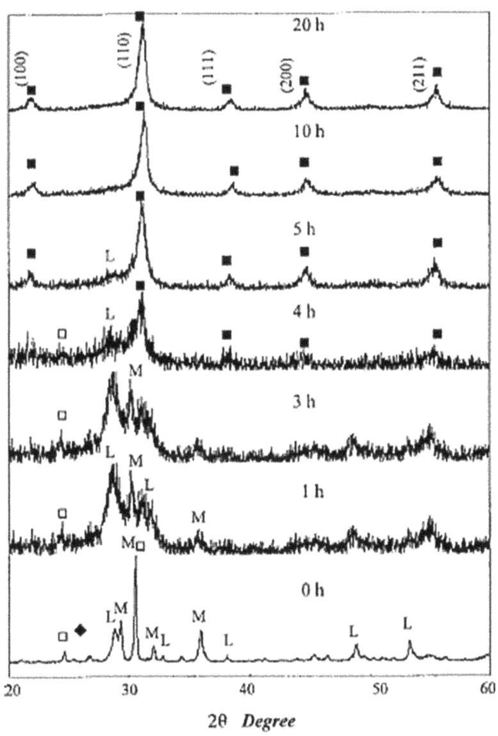

Figure 4.36. XRD patterns of the oxide mixture for PZN–PMN–PT milled for different times: (L) litharge, (M) massicot, (□) $Mg_{0.4}Zn_{0.6}Nb_2O_6$ (MZN), (♦) TiO_2 and (■) perovskite. Reproduced with permission from [62]. Copyright 2000 Wiley.

corresponded to the phase transition of PbO into Pb_3O_4. The second exotherm was an indicator of the reaction to form the PZN–PMN–PT phase. In contrast, no exothermic peak was visible for the sample mechanochemically activated for 10 h. However, a weak hump was observed in the curve, which was mostly attributed to the enhancement in crystallinity of the perovskite in the temperature range 350 °C–380 °C. Therefore, the phase formation of perovskite PZN–PMN–PT was finished in the sample milled for 10 h. In other words, the mechanochemical activation was equivalent to the thermally triggered chemical reaction to form the perovskite phase.

The thermal stability of the powder mechanochemically activated for 10 h was examined by heating at temperatures in the range 500 °C–850 °C. Figure 4.37 shows the XRD patterns of the samples calcined at different temperatures. It was found that the perovskite phase was stable up to 800 °C, without the presence of the pyrochlore phase. After calcining at 850 °C for 4 h, about 2% pyrochlore phase was included in the sample. With increasing calcination temperature, the diffraction peaks of perovskite were intensified, because the crystallinity of the perovskite was increased. In addition, the particle size of the PZN–PMN–PT powder was increased. For example, the sample calcined at 500 °C for 4 h had an average particle size of 1.23 μm, which was larger than that of the as-synthesized sample by more than two times.

The ternary system 0.48PFN–0.36PFW–0.16PZN, composed of three relaxor components, is another example that has been prepared from the oxide mixture of

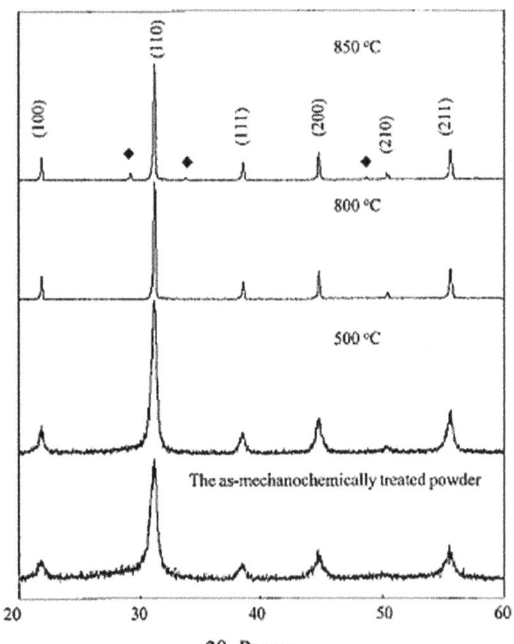

Figure 4.37. XRD patterns of the PZN–PMN–PT powder milled for 10 h followed by calcination at different temperatures: (♦) pyrochlore. Reproduced with permission from [62]. Copyright 2000 Wiley.

PbO, Fe_2O_3, WO_3, Nb_2O_5 and ZnO, using the mechanical activation technique [63]. Figure 4.38 shows the XRD patterns of the mixtures for 0.48PFN–0.36PFW–0.16PZN before and after being mechanochemically activated for different times. After activation for 5 h, the diffraction peaks of all the constituent oxides were absent. Instead, the pyrochlore and perovskite phases were present. As stated earlier, the phase formation mechanisms due to the mechanochemical activation from oxide mixtures were not the same as those in the conventional solid-state reaction process. The pyrochlore was most likely $Pb_2FeWO_{6.5}$, as observed in the mixture of PbO, Fe_2O_3 and WO_3, as discussed previously.

After milling for 10 h, no significant change in the XRD pattern was observed, whereas the relative density of the pyrochlore and perovskite phases varied. Phase conversion to perovskite was completed after milling for 15 h. A further increase in milling time to 25 h had almost no effect on the XRD patterns, suggesting that the perovskite was stable against mechanochemical activation. The refinement of the particles of the precursor oxides during the early stage of activation was confirmed by the increase in the specific surface area of the powder. For example, the oxide powder mixture without activation had a specific surface area of 1.1 $m^2\ g^{-1}$, while it was increased up to 4.5 $m^2\ g^{-1}$ for the milled samples.

For the unmilled oxide powder mixture, the perovskite phase was obtained after calcination at 750 °C, but the pyrochlore phase was also present. As the calcination

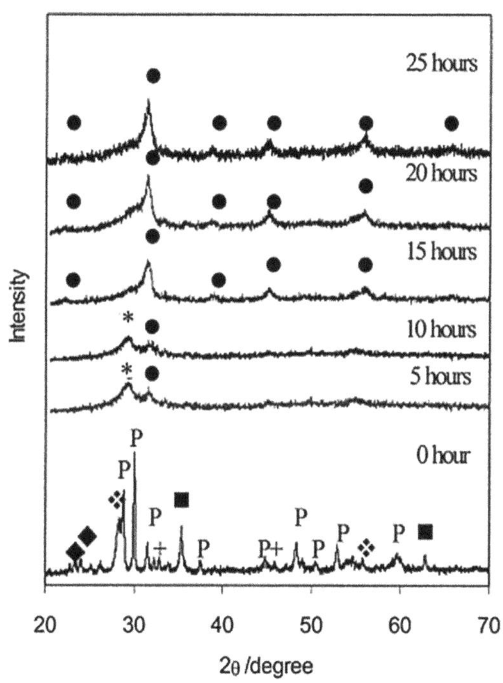

Figure 4.38. XRD patterns of the oxide mixture to form 0.48PFN–0.36PFW–0.16PZN before and after mechanochemical activation for different times: (P) Pb, (■) Fe_2O_3, (♦) WO_3, (+) ZnO and (*) pyrochlore. Reproduced with permission from [63]. Copyright 2000 Elsevier.

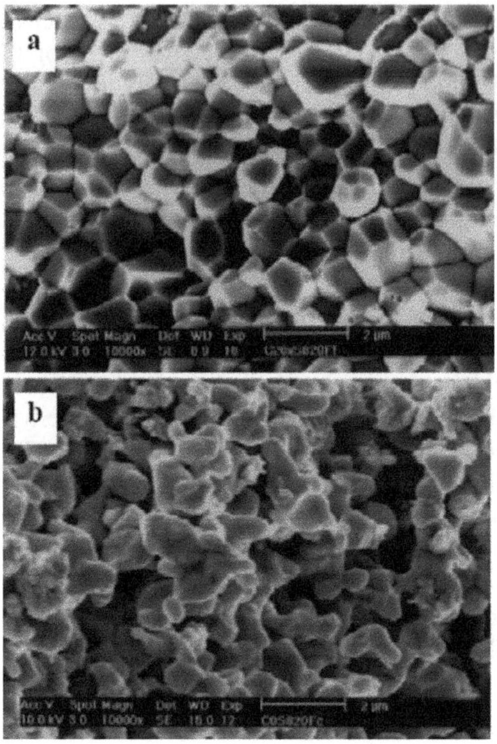

Figure 4.39. Cross-sectional SEM images of the 0.48PFN–0.36PFW–0.16PZN ceramics sintered at 820 °C for 45 min for different powders: (a) oxide mixture mechanochemically activated for 20 h and (b) oxide mixture without activation. Reproduced with permission from [63]. Copyright 2000 Elsevier.

temperature was increased to 820 °C, the peak intensity of the pyrochlore phase was gradually reduced, suggesting a decrease in the content of pyrochlore. If the calcination temperature was further increased, the content of pyrochlore started to increase. A similar trend was observed for the sample activated for 20 h. After calcining at 750 °C, a trace of pyrochlore phase was present. The peak intensity of the pyrochlore phase was decreased with increasing calcination temperature. Single-phase perovskite was achieved after calcining at 820 °C. A further increase in calcination temperature resulted in the presence of the pyrochlore phase. Therefore, the optimal sintering temperature of the ternary perovskite was selected to be 820 °C.

PFN–PFW–PZN ceramics were developed with the mixed oxide powder milled for 20 h by sintering at 820 °C for 45 min, Figure 4.39 shows cross-sectional SEM images of the ceramics derived from the unmilled and 20 h milled oxide mixtures. The sample from the milled powder had an average grain size of 1.3 μm, while that of the sample from the unmilled powder was 1.0 μm. Also, the ceramics from the milled powder exhibited a much denser microstructure than the ceramics from the unmilled powder. The highest relative density was 98%. The PFN–PFW–PZN ceramics displayed a peak dielectric constant of 9357 at 100 Hz. In comparison, the sample from the unmilled powder was less than 500.

Acknowledgments

Shenzhen Technology University (SZTU) is acknowledged for the financial support of a start-up grant (2018) and also the Natural Science Foundation of Top Talent of SZTU (grant no. 2019 010 801 002).

References

[1] Cross L E 1987 Relaxor ferroelectrics *Ferroelectrics* **76** 241–67

[2] Cross L E, Jang S J and Newnham R E 1980 Large electrostrictive effects in relaxor ferroelectrics *Ferroelectrics* **23** 187–92

[3] Shrout T R and Halliyal A 1987 Preparation of lead-based ferroelectric relaxors for capacitors *Am. Ceram. Soc. Bull.* **66** 704–11

[4] Choy J H, Yoo J S, Kang S G, Hong S T and Kim D G 1990 Ultrafine $Pb(Mg_{1/3}Nb_{2/3})O_3$ (PMN) powder synthesized from metal-citrate gel by thermal shock method *Mater. Res. Bull.* **25** 283–91

[5] Chiu C C, Li C C and Desu S B 1991 Molten salt synthesis of a complex perovskite $Pb(Fe_{1/2}Nb_{1/2})O_3$ *J. Am. Ceram. Soc.* **74** 38–41

[6] Jenhi M, Elghadraoui E H, Bali I, Elaatmani M and Rafiq M 1998 Reaction mechanism in the formation of perovskite $Pb(Fe_{1/2}Nb_{1/2})O_3$ by calcining of mixed oxide (CMO) *Eur. J. Sol. State Inorg. Chem.* **35** 221–30

[7] Park Y, Knowles K M and Cho K 1998 Particle-size effect on the ferroelectric phase transition in $Pb(Sc_{1/2}Nb_{1/2})O_3$ ceramics *J. Appl. Phys.* **83** 5702–08

[8] Uršičab H, Malič̌a B, Cilenšek J, Rojac T, Kmet B and Kosec M 2013 Linear thermal expansion coefficients of relaxor–ferroelectric $0.57Pb(Sc_{1/2}Nb_{1/2})O_3$–$0.43PbTiO_3$ ceramics in a wide temperature range *J. Eur. Ceram. Soc.* **33** 2167–71

[9] Wang J, Xue J M, Wan D M and Ng W B 1999 Mechanochemically synthesized lead magnesium niobate *J. Am. Ceram. Soc.* **82** 1358–60

[10] Xue J M, Wang J and Rao T M 2001 Synthesis of $Pb(Mg_{1/3}Nb_{2/3})O_3$ in excess lead oxide by mechanical activation *J. Am. Ceram. Soc.* **84** 660–62

[11] Xue J M, Wang J, Ng W and Wang D 1999 Activation-induced prychlore-to-perovskite conversion for a lead magnesium niobate precursor *J. Am. Ceram. Soc.* **82** 2282–84

[12] Wang J, Xue J M, Wan D M and Ng W B 1999 Mechanochemical fabrication of single phase PMN of perovskite structure *Sol. St. Ionics* **124** 271–79

[13] Wang J, Xue J M, Wan D M and Gan B K 2000 Mechanically activating nucleation and growth of complex perovskites *J. Solid State Chem.* **154** 321–28

[14] Xue J M, Wan D M and Wang J 2002 Functional ceramics of nanocrystallinity by mechanical activation *Sol. State Ion.* **151** 403–12

[15] Kong L B, Ma J, Zhu W and Tan O K 2001 Preparation of PMN powders and ceramics via a high-energy ball milling process *J. Mater. Sci. Lett.* **20** 1241–43

[16] Kong L B, Ma J, Zhu W and Tan O K 2002 Translucent PMN and PMN–PT ceramics from high-energy ball milling derived powders *Mater. Res. Bull.* **37** 23–32

[17] Kuscer D, Holc J, Kosec M and Meden A 2006 Mechano-synthesis of lead–magnesium–niobate ceramics *J. Am. Ceram. Soc.* **89** 3081–88

[18] Guha J P 2001 Reaction chemistry and subsolidus phase equilibria in lead-based relaxor systems, part II *J. Mater. Sci.* **36** 5219–26

[19] Jang H M, Oh S H and Moon J H 1992 Thermodynamic stability and mechanisms of Pb $(Zn_{1/3}Nb_{2/3})O_3$ formation and decomposition of perovskite prepared by the PbO flux method *J. Am. Ceram. Soc.* **75** 82–8

[20] Wakiya N, Ishizawa N, Shinozake K and Mizutani N 1995 Thermal stability of $Pb(Zn_{1/3}Nb_{2/3})O_3$ (PZN) and consideration of stabilization conditions of perovskite type compounds *Mater. Res. Bull.* **30** 1121–31

[21] Wang J, Wan D M, Xue J M and Ng W B 1999 Synthesizing nanocryatalline $Pb(Zn_{1/3}Nb_{2/3})O_3$ powders from mixed oxides *J. Am. Ceram. Soc.* **82** 477–79

[22] Liou Y C 2004 Effect of heating rate on properties of $Pb(Fe_{1/2}Nb_{1/2})O_3$ ceramics produced by simplified Wolframite route *Ceram. Int.* **30** 567–69

[23] Hu X, Chen X M and Wu S Y 2003 Preparation, properties and characterization of $CaTiO_3$-modified $Pb(Fe_{1/2}Nb_{1/2})O_3$ dielectrics *J. Eur. Ceram. Soc.* **23** 1919–24

[24] Fraygola B, Frizon N, Lente M H, Coelho A A, Garcia D and Eiras J A 2013 Phase transitions and magnetoelectroelastic instabilities in $Pb(Fe_{1/2}Nb_{1/2})O_3$ multiferroic ceramics *Acta Mater.* **61** 1518–24

[25] Matteppanavar S, Rayaprol S, Angadi B and Sahoo B 2016 Composition dependent room temperature structure, electric and magnetic properties in magnetoelectric $Pb(Fe_{1/2}Nb_{1/2})O_3$–$Pb(Fe_{2/3}W_{1/3})O_3$ solid-solutions *J. Alloys Compd.* **677** 27–37

[26] Ye Z G and Schmid H 1996 Growth from high temperature solution and characterization of $Pb(Fe_{2/3}W_{1/3})O_3$ single crystals *J. Cryst. Growth* **167** 628–37

[27] Zhou L Q, Vilarinho P M, Mantas P Q, Baptista J L and Fortunato E 2000 The effects of La on the dielectric properties of lead iron tungstate $Pb(Fe_{2/3}W_{1/3})O_3$ relaxor ceramics *J. Eur. Ceram. Soc.* **20** 1035–41

[28] Shivaraja I, Matteppanavar S, Rayaprol S and Angadi B 2019 Evidence of weak ferromagnetic and antiferromagnetic interaction at low temperature in $Pb(Fe_{2/3}W_{1/3})O_3$ multiferroic *Physica* B **561** 114–20

[29] Gao X S, Xue J M, Wang J, Yu T and Shen Z X 2002 Sequential combination of constituent oxides in the synthesis of $Pb(Fe_{1/2}Nb_{1/2})O_3$ by mechanical activation *J. Am. Ceram. Soc.* **85** 565–72

[30] Ang S K, Wang J, Wang D M, Xue J M and Li L T 2000 Mechanical activation-assisted synthesis of $Pb(Fe_{2/3}W_{1/3})O_3$ *J. Am. Ceram. Soc.* **83** 1575–80

[31] Ang S K, Wang J and Xue J M 2000 Seeding effect in the formation of $Pb(Fe_{2/3}W_{1/3})O_3$ via mechanical activation of mixed oxides *Solid State Ionics* **132** 55–61

[32] Ang S K, Xue J M and Wang J 2002 $Pb(Fe_{2/3}W_{1/3})O_3$ by mechanical activation of coprecipitated $Pb_3Fe_2O_6$ and WO_3 *J. Alloys Compd.* **343** 156–63

[33] Chu F, Setter N, Elissalde C and Ravez J 1996 High frequency dielectric relaxation in Pb $(Sc_{1/2}Ta_{1/2})O_3$ ceramics *Mater. Sci. Eng.* B **38** 171–6

[34] Petzelt J, Buixaderas E and Pronin A V 1998 Infrared dielectric response of ordered and disordered ferroelectric $Pb(Sc_{1/2}Ta_{1/2})O_3$ ceramics *Mater. Sci. Eng.* B **55** 86–94

[35] Yang L H, Li Z R, Wang T T and Song K X 2018 Structure and dielectric behavior of $(1-x)$ $Pb(Sc_{1/2}Ta_{1/2})O_{3-x}Pb(In_{1/2}Nb_{1/2})O_3$ solid solution *Ceram. Int.* **44** 8005–11

[36] Lim J, Xue J M and Wang J 2002 Ferroelectric lead scandium tantalate from mechanical activation of mixed oxides *Mater. Chem. Phys.* **75** 157–60

[37] Kelly J, Leonard M, Tantigate C and Safari A 1997 Effect of composition on the electromechanical properties of (1–*x*)Pb($Mg_{1/3}Nb_{2/3}$)$O_{3–x}$PbTiO_3 ceramics *J. Am. Ceram. Soc.* **80** 957–64

[38] Guha J P 2003 Effect of compositional modifications on microstructure development and dielectric properties of Pb($Mg_{1/3}Nb_{2/3}$)O_3–PbTiO_3 ceramics *J. Eur. Ceram. Soc.* **23** 133–39

[39] Suh D H, Lee K H and Kim N K 2002 Phase development and dielectric/ferroelectric responses in the PMN–PT system *J. Eur. Ceram. Soc.* **22** 219–23

[40] Ozgul M, Tekemura K, Trolier-Mckinstry S and Randall C A 2001 Polarization fatigue in Pb($Zn_{1/3}Nb_{2/3}$)O_3–PbTiO_3 ferroelectric single crystal *J. Appl. Phys.* **89** 5100–06

[41] Kuwata J, Uchino K and Nomura S 1981 Phase relations in the Pb($Zn_{1/3}Nb_{2/3}$)O_3–PbTiO_3 system *Ferroelectrics* **37** 579–82

[42] Halliyal A, Kumar U, Newnham R E and Cross L E 1987 Stabilizaion of the perovskite phase and dielectric properties of ceramics in the Pb($Zn_{1/3}Nb_{2/3}$)O_3–BaTiO_3 system *Am. Ceram. Soc. Bull.* **66** 671–76

[43] Baek J G, Isobe T and Senna M 1997 Synthesis of pyrochlore-free 0.9Pb($Mg_{1/3}Nb_{2/3}$) O_3−0.1PbTiO_3 ceramics via a soft mechanochemical route *J. Am. Ceram. Soc.* **80** 973–81

[44] Wang J, Wan D M, Xue J M and Ng W B 1999 Mechanochemical synthesis of 0.9Pb($Mg_{1/3}$ $Nb_{2/3}$)O_3−0.1PbTiO_3 from mixed oxides *Adv. Mater.* **11** 210–13

[45] Kong L B, Ma J, Zhu W and Tan O K 2002 Rapid formation of lead magnesium niobate-based ferroelectric ceramics via a high-energy ball milling process *Mater. Res. Bull.* **37** 459–65

[46] Kong L B, Ma J, Zhu W and Tan O K 2002 Preparation of PMN–PT ceramics via a high-energy ball milling process *J. Alloys Compd.* **336** 242–46

[47] Algueró M, Alemany C, Jiménez B, Holc J, Kosec M and Pardo L 2004 Piezoelectric PMN–PT ceramics from mechanochemically activated precursors *J. Eur. Ceram. Soc.* **24** 937–40

[48] Dragomir M *et al* 2019 Seeding effects on the mechanochemical synthesis of 0.9Pb($Mg_{1/3}$ $Nb_{2/3}$)O_3–0.1PbTiO_3 *J. Eur. Ceram. Soc.* **39** 1837–45

[49] Tan Y L, Xue J M and Wang J 2000 Stablization of perovskite phase and dielectric properties of 0.95PZN–0.05BT derived from mechanical activation *J. Alloys Compd.* **297** 92–8

[50] Xue J M, Tan Y L, Wan D M and Wang J 1999 Synthesizing 0.9PZN–0.1BT by mechanically activating mixed oxides *Solid State Ionics* **120** 183–88

[51] Kong L B, Ma J, Huang H and Zhang R F 2002 (1–*x*)PZN–*x*BT ceramics derived from mechanochemically synthesized powders *Mater. Res. Bull.* **37** 1085–92

[52] Kong L B, Ma J, Huang H and Zhang R F 2002 Lead zinc niobate (PZN)–barium titanate (BT) ceramics from mechanochemically synthesized powders *Mater. Res. Bull.* **37** 2491–98

[53] Alguero M, Ricote J and Castro A 2004 Mechanosynthesis and thermal stability of piezoelectric perovskite 0.92Pb($Zn_{1/3}Nb_{2/3}$)O_3–0.08PbTiO_3 powders *J. Am. Ceram. Soc.* **87** 772–78

[54] Laishram R, Thakur O P, Bhattacharya D K and Harsh 2010 Dielectric and piezoelectric properties of La doped lead zinc niobate–lead zirconium titanate ceramics prepared from mechano-chemically activated powders *Mater. Sci. Eng.* B **172** 172–76

[55] Zheng M P, Hou Y D, Zhu M K, Zhang M and Yan H 2014 Shift of morphotropic phase boundary in high-performance fine-grained PZN–PZT ceramics *J. Eur. Ceram. Soc.* **34** 2275–83

[56] Shinohara S, Baek J G, Isobe T and Mamoru S 2000 Synthesis of phase-pure $Pb(Zn_xMg_{1-x})_{1/3}Nb_{2/3}O_3$ up to $x = 0.7$ from a single mixture via a soft-mechanochemical route *J. Am. Ceram. Soc.* **83** 3208–10
[57] Ang S K, Wang J and Xue J M 2000 Phase stability and dielectric properties of (1–x)PFW +xPZN derived from mechanical activation *Sol. State Ion* **127** 285–93
[58] Uršič H, Tellier J, Holc J, Drnovšek S and Kosec M 2012 Structural and electrical properties of 0.57PSN–0.43PT ceramics prepared by mechanochemical synthesis and sintered at low temperature *J. Eur. Ceram. Soc.* **32** 449–56
[59] Ursic H, Holc J and Kosec M 2013 Influence of the sintering conditions on the properties of 0.57PSN–0.43PT ceramics prepared from mechanochemically activated powder *J. Eur. Ceram. Soc.* **33** 795–803
[60] Wan D M, Xue J M and Wang J 1999 Synthesis of single phase $0.9Pb[(Zn_{0.6}Mg_{0.4})_{1/3}Nb_{2/3}O_3]-0.1PbTiO_3$ by mechanically activating mixed oxides *Acta Mater.* **47** 2283–91
[61] Wan D M, Xue J M and Wang J 2000 Nanocrystalline 0.54PZN−0.36PMN−0.1PT of perovskite structure by mechanical activation *Mater. Sci. Eng.* A **286** 96–100
[62] Wan D M, Xue J M and Wang J 2000 Mechanochemical synthesis of $0.9[0.6Pb(Zn_{1/3}Nb_{2/3})O_3-0.4Pb(Mg_{1/3}Nb_{2/3})O_3]-0.1PbTiO_3$ *J. Am. Ceram. Soc.* **83** 53–9
[63] Ang S K, Xue J M and Wang J 2000 Mechanical activation and dielectric properties of 0.48PFN−0.36PFW−0.16PZN from mixed oxides *J. Alloys Compd.* **311** 181–87

Chapter 5

Ferroelectric ceramics (III)

Ling Bing Kong, Zhuohao Xiao, Xiuying Li, Shijin Yu, Wenxiu Que, Yin Liu, Tianshu Zhang, Kun Zhou and Hongfang Zhang

5.1 Brief introduction

As mentioned before, nonlead-containing ferroelectric compounds, such as $BaTiO_3$ [1–4] and most Aurivillius family ferroelectrics [5–11], cannot be directly formed from their oxide/carbonate mixtures, while the phase formation temperatures can be accelerated, due to the refinement of the precursor powders. More recently, multiferroic materials, in particular those based on bismuth ferrite, have attracted much more attention in the research community. The fabrication of single-phase bismuth ferrite ceramics with high dielectric and electrical performance is still a challenge.

5.2 $BaTiO_3$ based materials

Barium titanate ($BaTiO_3$ or BT) ceramics have a wide range of applications, such as piezoelectric actuators, multilayer ceramic capacitors (MLC) and positive temperature coefficient resistors (PTCR), because of their outstanding electrical properties. Therefore, various strategies have been explored to synthesize BT or BT-based powders, such as the chemical coprecipitation process, sol–gel process, hydrothermal reaction, molten salt synthesis, microemulsion and auto-combustion methods [12–18]. The mechanochemical activation method offers an alternative way to synthesize BT or promote its phase formation.

Phase pure BT nanosized powder was synthesized using the mechanochemical activation process, with oxide mixture consisting of BaO and TiO_2 [1]. The activation process was conducted in N_2 protection at room temperature, with an SPEX 8000 shaker mill. Phase formation of perovskite BT was observed after milling for 5 h, while it was complete after activation for 15 h. The BT powder had particle sizes of 20–30 nm. No attempt was made in that study to fabricate BT ceramics using the nanosized powder.

Single-phase BT, together with MT (M = Mg, Ca, Sr), was obtained using mechanochemical activation in a vacuum [19]. The oxides used were calcined at

doi:10.1088/978-0-7503-2191-4ch5

1300 °C for 4 h in Ar flow to decompose the hydroxides and/or carbonates formed during storage. The calcined oxide powders were immediately used for experiments to minimize potential rehydration. Samples with a total quantity of 6.00 g were mechanochemically treated in a vacuum for different times of up to 100 h at room temperature, with a vertical 316 s stainless steel mill and five 420C stainless steel balls 25.4 mm in diameter. Phase formation of BT was completed after milling for 25 h.

Because it is expected that BaO easily adsorbs CO_2 in air to form $BaCO_3$, which is more stable than BT at low temperatures, the mechanochemical activation must be carried out in the protection of N_2 gas or in a vacuum. However, there was also a report demonstrating the direct synthesis of BT in air [20]. The experiment was conducted with ZrO_2 milling media and a Fritsch Pulverisette 5 planetary ball mill. The ZrO_2 vial had a volume of 500 ml, filled with ZrO_2 balls having a diameter of 10 mm. Rutile TiO_2 and BaO were used as the starting materials. The ball-to-powder weight ratio was 20:1. The angular velocities of the supporting disc and vials were 33.2 rad s^{-1} (317 rpm) and 41.5 rad s^{-1} (396 rpm), respectively. The mixtures were mechanochemically activated for times of up to 4 h.

BT was detected after milling for 2 h, while single-phase perovskite BT powder with crystal sizes of 20–50 nm was achieved after milling for 4 h. Fully densified BT ceramics with grain sizes of 0.5–1.0 μm were derived from the BT nanosized powder after sintering at 1330 °C. The BT ceramics had a room temperature dielectric constant of 2500 at 100 kHz. The Curie temperature (T_C) of the BT ceramics was 135 °C, which was higher than that of the conventionally obtained BT ceramics (120 °C–130 °C). The slightly higher Curie temperature of the mechanochemically derived BT ceramics was closely related to the effect of the high-energy activation, but it has still not been fully understood.

High-energy mechanochemical milling has also been utilized to decrease the phase formation temperature of BT [2–4]. DTA/TGA and the XRD characterization data indicated that the reaction consequences of the phase formation in the activated mixtures was changed during the post-thermal calcination process [4]. $BaCO_3$ and TiO_2 powders were manually mixed with a molar ratio of 1:1. The mixture was then mechanochemically milled for different times of up to 159 h with a Fritsch Pulverisette 7 type planetary mill at a rotation speed of 400 rpm. Two agate balls with diameters of 12 mm were used, while the volume of the vial was not mentioned. The ball-to-powder weight ratio was 10:1. Single-phase BT was achieved by calcining the milled mixture at 750 °C, while the powder had an average particle size of about 40 nm. This temperature is lower than that required when using the conventional solid-state reaction method without the application of mechanochemical activation.

In addition, the sintering behavior and electrical properties of BT ceramics could be modified by mechanochemically activating BT powders [21–26]. For example, commercial BT powder was mechanochemically activated with a planetary ball mill for 60 and 120 min [26]. Densification behaviors of the as-received and milled BT powder were studied by monitoring their shrinkage curves up to 1380 °C at a heating rate of 10 °C/min. The commercial BT powder consisted of spherical particles with an average particle size of 1.0 μm without agglomeration. After activation, the morphology of the powder became irregular with soft agglomeration. The average particle sizes of the samples milled for 60 min and 120 min were 0.7 μm and 0.5 μm, respectively. The size of the agglomerated particles was about 10 μm.

The densification behaviors of the BT powders before and after mechanochemical activation were studied. The activated powders exhibited different shrinkage behaviors compared to the as-received commercial powder. However, they were similar in the trend of densification process, starting with a bulk shrinkage in the temperature range 900 °C–1000 °C and fully dense at >1300 °C. The non-activated powder experienced a one-step shrinkage, because the densification peak was only slightly split. In contrast, the powders with mechanochemical activation exhibited a multiple-step densification process. The sample activated for 60 min displayed three densification peaks at about 1022 °C, 1141 °C and 1208 °C. The densification peaks of the sample activated for 120 min were at 1058 °C, 1154 °C and 1242 °C. All these densification peaks were below the liquid eutectic temperature in the TiO_2–BaO binary system, which was about 1320 °C, during the densification process when using the conventional solid-state reaction technique.

The sintering kinetics of the powders were examined by monitoring the densification rate as a function of both the relative density and sintering temperature. A maximum densification rate was observed for all three samples. The maximum densification rates took place at the relative densities of 66%, 83% and 86%, for the as-received, 60 min milled and 120 min milled samples, respectively. At the low temperature side, the densification rate curve of the non-activated sample was nearly linear and steady with increasing density. However, after mechanochemical activation, the powders had a different profile. There were three relative density ranges, corresponding to three temperature ranges. The slope of the densification rate curve of the non-activated sample was smaller than those of the activated samples.

Variation in the density of the samples was characterized as a function of sintering temperature in the form of non-isothermal sintering. For the milled samples, there was a shoulder in the density curve, indicating the change in the densification mechanism at temperatures of about 1230 °C and 1248 °C for the 60 min and 120 min activation, respectively. They were just above the temperatures of the maximum densification rate. It was accepted that agglomeration in a powder could influence its sintering behavior, where the agglomerates would experience compressive stresses and the matrix would have tensile stresses during the sintering process. The tensile stresses retarded the densification of the sample.

The densification rate curves of the activated samples could be divided into three stages. The first densification stage was related to the action of the intra-agglomerated particles, resulting in localized densification and thus strengthening the bonding of the agglomerates. Moreover, densification through the intra-agglomerate particles involved local rearrangement of the particles near the agglomerates, which generated stresses that could influence the sintering at later stages. As a result, the densification rate was at a relatively low level at the early sintering stage, which occurred at temperatures of <1050 °C. As the sintering temperature was increased, non-agglomerated regions started to densify, leading to redistribution of pores, which corresponds a high densification rate. The final sintering stage was closely related to the agglomerates.

$SrTiO_3$ (ST) and BT can be combined to form solid solutions, $Ba_{1-x}Sr_xTiO_3$ or BST, with perovskite structure in the whole composition range, with $x = 0$ to $x = 1$ [27–31]. The Curie temperature of BST ($Ba_{1-x}Sr_xTiO_3$) is almost linearly decreased with increasing content of Sr, making it possible to develop materials with the desired electrical properties. For example, the compositions with T_C near room temperature are x = 0.25–0.30, which can be used as dielectric materials for multilayered ceramic capacitors (MLCCs). BST possesses a diffused phase transition from the ferroelectric to paraelectric state, which is useful to design materials with a wide range of stability of electrical properties against temperature.

$Ba_{1-x}Sr_xTiO_3$ powders, with $x = 0$, 0.25, 0.50, 0.75 and 1, have been synthesized using the mechanochemical activation process [32]. The powders were used to fabricate fully dense BST ceramics with fine grains using a spark plasma sintering (SPS) process. The precursors included BaO_2, SrO and TiO_2 (anatase). The high-energy milling experiment was performed with a Fritsch Pulverisette 6 type planetary mill, with stainless steel vials and balls as the milling media. Single-phase BST nanosized powders were achieved in the sample activated for 72 h at a rotation speed of 200 rpm, with five stainless steel balls having a diameter of 2 cm. The lowest temperature required to sinter the BST ceramics was as low as 985 °C. This is lower than the sintering temperatures of SBT powders synthesized using the solid-state reaction or chemical synthetic routes by 300 °C–400 °C. The grain sizes of the BST ceramics could be less than 200 nm by adjusting the SPS sintering parameters.

The XRD patterns of the mixtures with compositions of $x = 0 - 1$ after milling for 72 h are shown in figure 5.1(a). Phase pure perovskite was achieved for all the samples. With increasing content of Sr, the diffraction peaks shifted to higher angles, indicating that the unit cell was shrunk gradually, simply because the radius of Sr^{2+} is smaller than that of Ba^{2+}. After annealing at 1200 °C for 12 h, the diffraction peaks were significantly sharpened, suggesting that the samples had enhanced crystallization, as seen in figure 5.1(b). In addition, the relative position of the diffraction peaks was not changed, implying the high stability of the perovskite solid solutions. A representative HRTEM image of the as-synthesized BT powder is shown in figure 5.2(a). DDP indicated that the powder consisted of both cubic and tetragonal structures. The particle size distribution of the BT powder is depicted in figure 5.2(b). The average particle size was 8.7 nm, but the size distribution was relatively wide.

An attrition milling process, combined with hot uniaxial pressing (HUP), was used to develop $Ba_{0.6}Sr_{0.4}TiO_3$ ceramics with submicron-sized grains and temperature stable dielectric properties [33]. The BST powder was attrition-milled with three types of balls for the purpose of comparison. Ce and Y doped zirconia balls displayed higher efficiency in refining the BST powder, due to their relative densities. The milling efficiency of the ZrO_2–Ca balls was lower by about 25%. The Curie temperatures of the BST ceramics derived from the powders milled with ZrO_2–Ce and ZrO_2–Ca balls were decreased compared to that of the ceramics from the unmilled powder, while the dielectric constant was decreased, owing to the doping

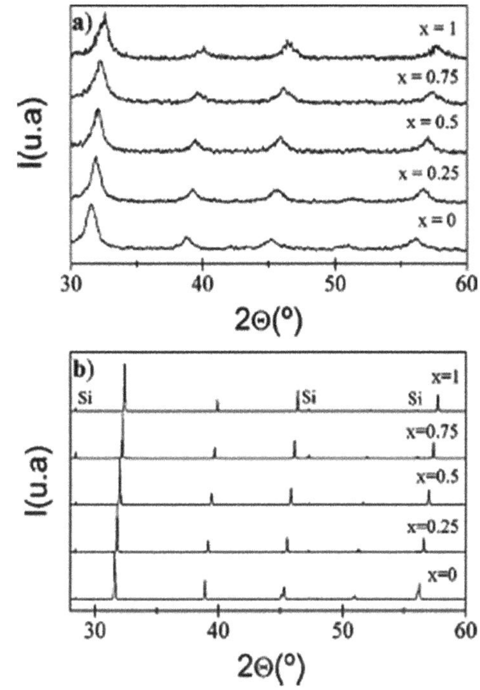

Figure 5.1. XRD patterns of the samples with compositions of $(1-x)BaO_2$–$xSrO$–TiO_2 after milling for 72 h: (a) as-synthesized powders and (b) after annealing at 1200 °C for 12 h (Si = standard). Reproduced with permission from [32]. Copyright 2005 American Chemical Society.

effects of Zr, Ca and Ce in the crystal structure of BST. In comparison, if ZrO_2–Y balls were used, the Curie temperature was slightly increased. Therefore, it is possible to modify the microstructure and electrical properties of the BST ceramics by selecting the materials of the milling media.

$Ba_{1-x}Sr_xTiO_3$ powders, with $x = 0.0$, 0.05, 0.075, 0.10 and 0.15, were prepared using the mechanochemical activation method in order to develop BST ceramics with strong electrical performance for applications such as multilayered ceramic capacitors (MLCCs) and microwave tunable devices [34]. The structural, dielectric and ferroelectric properties of the materials were systematically studied. The XRD patterns of the activated powders were analyzed using Rietveld refinement, confirming that tetragonal phase with *P4mm* space group was formed. The structural variation with the content of Sr was examined using Raman spectra. SEM results indicated that the microstructure and grain size of the final BST ceramics were independent of the concentration of Sr. As expected, the Curie temperature of the materials almost linearly decreased with increasing levels of Sr.

The effects of mechanochemical activation on the microstructure and electrical properties of Sr/Zr co-doped $BaTiO_3$ ceramics, with a composition of $Ba_{0.9}Sr_{0.1}Zr_{0.04}Ti_{0.96}O_3$, were systematically studied [35]. The activation was conducted for times of up to 45 h. According to the Rietveld refinement analysis results of XRD patterns, it was confirmed that all the samples had a tetragonal phase with

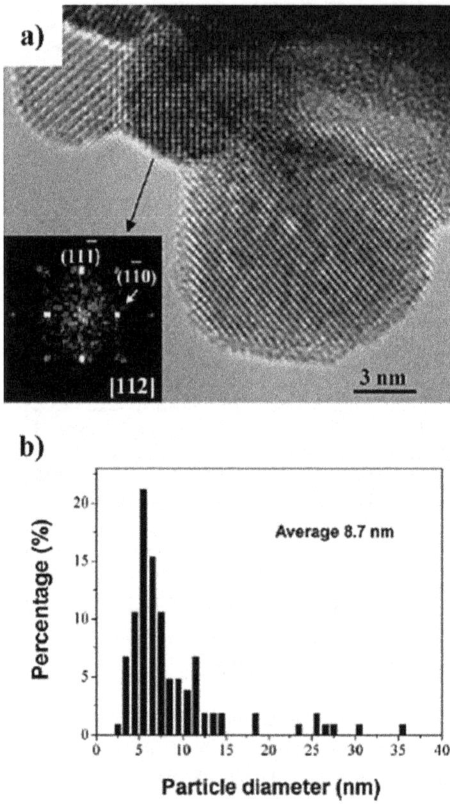

Figure 5.2. (a) HRTEM micrograph of the $BaTiO_3$ powder with the inset being the DDP of a selected particle. (b) Particle size distribution from the HRTEM images of the $BaTiO_3$ powder. Reproduced with permission from [32]. Copyright 2005 American Chemical Society.

P4mm space group, as shown in figure 5.3. The samples exhibited an average particle size of about 40 nm, without significant differences from one another. In addition, with increasing milling time, the lattice parameters and the unit volume of the perovskite phase were slightly decreased.

The $Ba_{0.9}Sr_{0.1}Zr_{0.04}Ti_{0.96}O_3$ powders were prepared with $BaCO_3$, $SrCO_3$, TiO_2 and ZrO_2 commercial powders as the starting materials. The mechanochemical milling was carried out in air at a rotation speed of 450 rpm for 25, 30, 35, 40 and 45 h, with the corresponding samples denoted as BSZT25, BSZT30, BSZT35, BSZT40 and BSZT45, respectively. The facility was a Retsch PM100 planetary ball mill, with the ZrO_2 vial having a diameter of 45 mm and 30 ZrO_2 balls having a diameter of 10 mm as the milling media. The activated powders were calcined at 1100 °C for 4 h. The calcined powders were made into pellets and then sintered at 1200 °C for 4 h. Figure 5.4 shows representative SEM images of the BSZT ceramics from the powders activated for different times. The average grain size of the ceramics was decreased from 0.93 to 0.39 μm, as the milling duration was prolonged from 25 to 45 h. However, the optimal milling condition was 35 h in terms of achieving the highest ferroelectric properties and piezoelectric performance of the final BSZT ceramics.

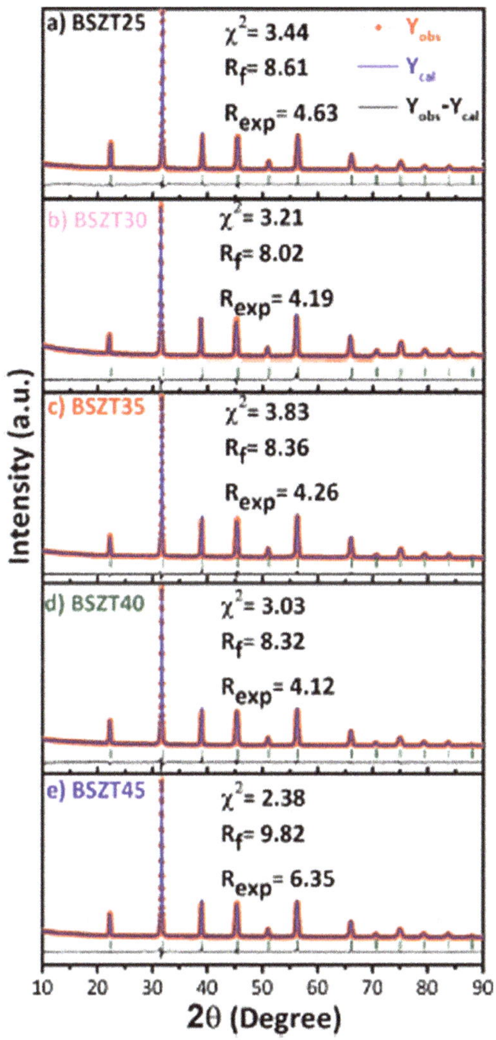

Figure 5.3. Rietveld refined XRD patterns of the BSZT samples milled for different durations: (a) 25 h, (b) 30 h, (c) 35 h, (d) 40 h and (e) 45 h. Reproduced with permission from [35]. Copyright 2016 Elsevier.

5.3 Bismuth containing ferroelectric materials

The Aurivillius structured compounds have a general formula $[Bi_2O_2][A_{n-1}B_nO_{3n+1}]$, consisting of n pseudo-perovskite $[A_{n-1}B_nO_{3n+1}]^{2-}$ layers with alternation of $[Bi_2O_2]^{2+}$ layers [36–41]. They have been extensively explored because of their high Curie temperatures and outstanding piezoelectric properties. In addition, the Aurivillius family has offered various combination and incorporation possibilities to design materials with specific electrical properties. It has been demonstrated that most of the Aurvillius ferroelectric materials could be synthesized using the mechanochemical activation process, as discussed in the following sections.

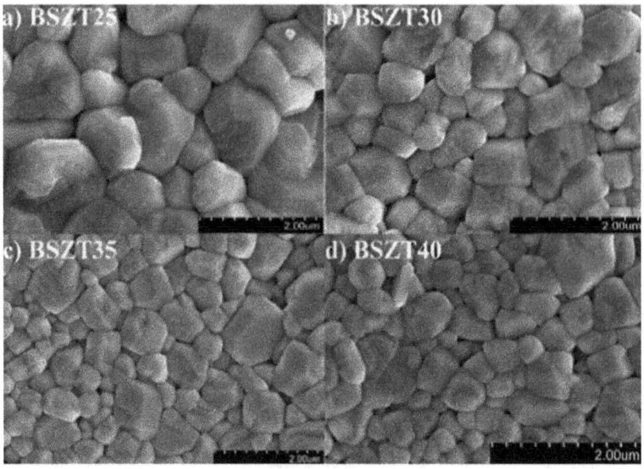

Figure 5.4. SEM images of the sintered BSZT pellets derived from powders milled for different durations: (a) 25 h, (b) 30 h, (c) 35 h and (d) 40 h. Reproduced with permission from [35]. Copyright 2016 Elsevier.

5.3.1 $Bi_4Ti_3O_{12}$

Bismuth titanate ($Bi_4Ti_3O_{12}$ or BiT) with $n = 3$ is a typical member of the Aurivillius family of ferroelectric materials [42]. It has advantages, such as a high Curie temperature (T_C = 675 °C), a strong electro-optical switching effect, special piezoelectric properties and a relatively low sintering temperature, thus finding a wide range of applications [43–47]. BiT nanosized powders have been prepared from oxide mixture using the mechanochemical activation method, which could be used to fabricate BiT ceramics with promising electrical properties.

According to the XRD patterns, the reaction between Bi_2O_3 and TiO_2 was not triggered after activation for 3 h when using planetary milling with tungsten carbide milling media, but the particle size of the starting oxides was largely reduced [48]. The BiT phase was detected when the milling lasted for 9 h, while single-phase BiT was achieved after activation for 15 h. BiT ceramics were prepared, after the powder had been milled for 15 h, at 750 °C, 850 °C and 950 °C for 1 h. The BiT ceramics exhibited typical plate-like grains, whose grain size was increased with increasing sintering temperature. The optimal sintering temperature was 850 °C, corresponding to a relatively density of 98.3%. Too high a temperature led to decreased densification, due to the overgrowth of the plate-like grains with low packing density.

In a separate study, it was found that the product of mechanochemical activation was amorphous instead of crystalline BiT, as shown in figure 5.5 [49]. However, the amorphous precursor could be readily crystallized at much lower temperatures than those required in the conventional solid-state reaction and chemical process. In addition, the authors also found that a planetary mill was more efficient compared to a vibrating mill. For example, when using a vibrating mill, the precursor oxides were still dominant after milling for the long time of 72 h. Complete amorphization

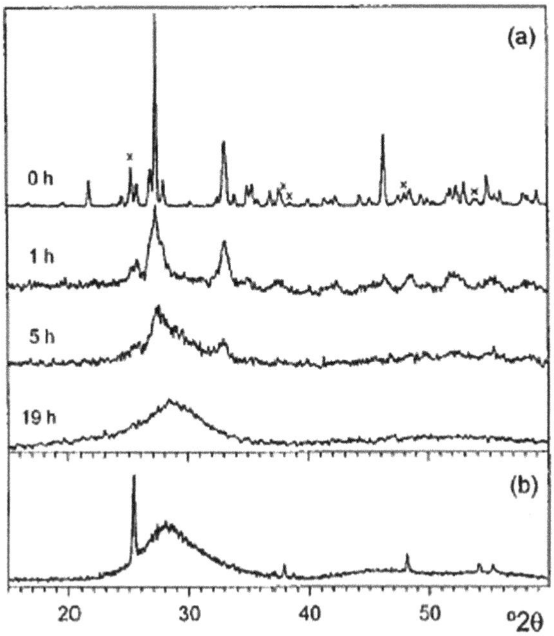

Figure 5.5. (a) XRD patterns of a mixture of Bi_2O_3 and TiO_2 after milling for different times with a planetary mill. (b) XRD pattern of the precursor obtained using the *n*-butylamine method (× = anatase TiO_2). Reproduced with permission from [49]. Copyright 2001 American Chemical Society.

was only observed after activation for 168 h. However, if the planetary mill was employed, an amorphous state was achieved after milling for only 19 h, as demonstrated in figure 5.5(a). This difference in milling efficiency between the vibrating and the planetary mills was closely related to the difference in the amount of energy produced by the two mills. Therefore, selection of the milling facility is an important factor in determining the efficiency of the activation process.

Nanosized BiT powder was also derived from an amorphous precursor using mechanochemical activation [50]. The amorphous precursor was prepared through chemical coprecipitation from a solution of $TiCl_4$ and $Bi(NO_3)_3{\cdot}5H_2O$, with ammonia water as the precipitant. Figure 5.6 shows the XRD patterns of the amorphous precursor before and after mechanochemical activation for different times. The BiT crystalline phase was obtained from the amorphous precursor after activation for 20 h. Similarly, the BiT nanosized powders had an average particle size of 15 nm with high sinterability, compared to the powders synthesized using other methods. After sintering at 875 °C for 2 h, the BiT ceramics made from the powder activated for 20 h approached a relative density of >95%, with a dielectric constant of ~1260 at 1 MHz and a Curie temperature of 646 °C.

It was found that the BiT crystalline phase could be obtained after milling for just 2 h in the presence of excess Bi_2O_3 [51]. The crystal size was about 20 nm, while the particle size was about 30 nm. In this case, Bi_2O_3 played a dominant role in the

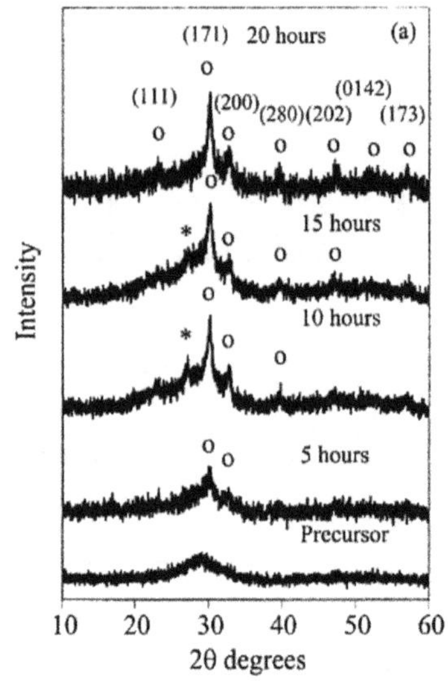

Figure 5.6. XRD patterns of the amorphous precursor mechanochemically treated for different times: (*) Bi_2O_3 and (○) BiT. Reproduced with permission from [50]. Copyright 2002 Wiley.

development of the BiT crystalline phase during the mechanochemical synthesis from oxide precursors. Commercial powders of Bi_2O_3 and anatase TiO_2, with particle sizes of 2–4 μm and 1–5 μm, respectively, were mixed with 3 wt% excess Bi_2O_3. ZrO_2 vials and ZrO_2 balls with a diameter of 10 mm were used as the milling media. Mechanochemical activation was conducted in air with a Fritch Pulverisette 5 planetary ball mill. The ball-to-powder weight ratio was 20:1. The basic disc rotation speed was 317 rpm, while the rotation speed of the disc with vials was 396 rpm. The milling was carried out for times of up to 6 h.

After activation for 1 h, the phase evolutions of the sample with and without excess Bi_2O_3 were different. For the oxide mixture without excess Bi_2O_3, after milling for 2 h amorphization and peak broadening of the precursor oxides were observed. As the milling time was prolonged to 3 h and 4 h, the BiT crystalline phase was still invisible. Only after activation for 6 h did the BiT phase begin to appear. Figure 5.7 shows the XRD patterns of the mixed oxide powder with excess Bi_2O_3 after milling for 2 h and 6 h. Obviously, a BiT phase with a high degree of crystallization occurred in these two samples. According to the diffraction peaks, the BiT phase had an orthorhombic structure. The BiT crystalline phase was further confirmed by the Rietveld analysis results. In addition, the sample milled for 2 h contained less of the amorphous phase than the sample milled for 6 h, which suggested that prolonged milling was not necessary to obtain high crystallinity of the BiT powder. The BiT powder from the oxide mixture with 3 wt% excess Bi_2O_3 activated for 2 h was used to fabricate BiT ceramics. After sintering at 875 °C for 4 h, dense BiT ceramics were obtained, which had plate-like grains and a relative density of 95%. The average grain size was about 3.5 μm.

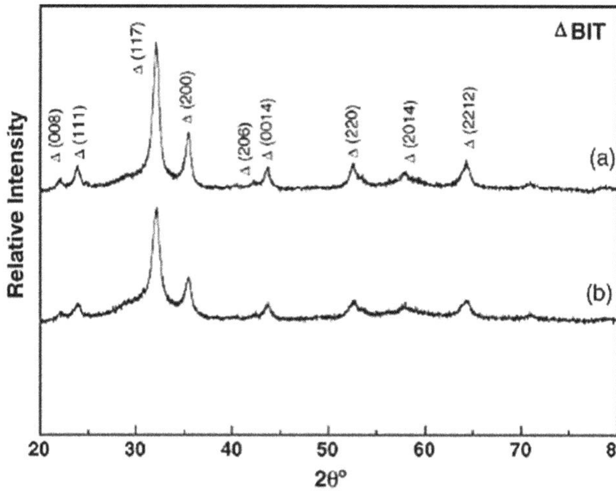

Figure 5.7. XRD patterns of the BiT powders from the oxide mixtures with 3 wt% excess Bi_2O_3 after mechanochemical activation for different times: (a) 2 h and (b) 6 h. Reproduced with permission from [51]. Copyright 2006 Elsevier.

The structural variation and phase evolution in the system of $2Bi_2O_3{\cdot}3TiO_2$ as a result of high-energy mechanochemical activation have been systematically studied [52]. The effect of activation on the structural evolution of BiT powder was also examined for the purpose of comparison. Both the oxide mixture of Bi_2O_3 and TiO_2 and the BiT powder were activated with a Fritsch Pulverisette 5 type planetary ball mill. Stainless steel vials of 500 ml in volume and hardened steel balls of 13.4 mm in diameter were employed as the milling media, while the ball-to-powder mass ratio was set to be 20:1.

Two sets of milling parameters were applied to reveal the effect of the milling power. In the first set, the powders were milled up to 15 h, where the angular velocities of the basic disk and the vials were 180 and 225 rpm, respectively. Then, their angular velocities were raised to 317 and 396 rpm, respectively. The samples were activated for additional times of up to 5 h (denoted as 15 + 5). After that, the angular velocities of the basic disk and the vials were set back to 180 rpm and 225 rpm, at which the samples were treated for 10 h (denoted as 15 + 5 +10). In the second set, the milling was carried out at the higher speeds of 317 rpm and 396 rpm for 20 h. Then, the samples were further milled for 10 h at low velocities of 180 rpm and 225 rpm.

The XRD patterns of the $Bi_4Ti_3O_{12}$ powder and the mixture of $2Bi_2O_3{\cdot}3TiO_2$ mechanochemically treated for different times are illustrated in figures 5.8 and 5.9, respectively. For the $Bi_4Ti_3O_{12}$ powder, the milling led to gradual reduction and broadening of the diffraction peaks, while no new phase was detected. This suggested that amorphization of the $Bi_4Ti_3O_{12}$ phase was the only consequence of the high-energy mechanochemical activation. For the oxide mixture, in addition to the reduction of the diffraction peaks of the components, a new phase corresponding to $Bi_2(CO_3)O_2$ was present before the formation of $Bi_4Ti_3O_{12}$. It was suggested that

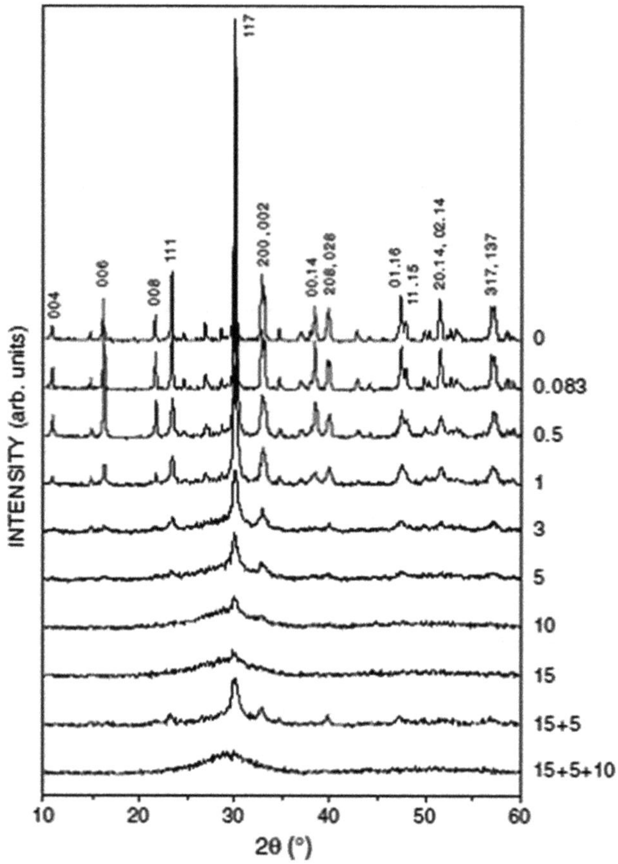

Figure 5.8. XRD patterns of the $Bi_4Ti_3O_{12}$ powder milled for different times with different intensities, P = 0.49 W g^{-1} and P = 2.68 W g^{-1} (milling time 15 + 5 h). Reproduced with permission from [52]. Copyright 2006 Elsevier.

TiO_2 was dispersed into the matrix of α-Bi_2O_3 after activation for 1 h. In addition, α-Bi_2O_3 reacted with CO_2 to form $Bi_2(CO_3)O_2$. Further activation resulted in the generation of amorphous $Bi_4Ti_3O_{12}$. The XRD patterns of the two samples milled for 15 h were obviously different. The $Bi_4Ti_3O_{12}$ phase had been nearly completely amorphized, while the crystalline phase was still visible in the oxide mixture sample.

Interestingly, when the two samples were subjected to an additional stronger activation for 10 h, they both displayed partial crystallization, i.e. the formation of crystalline $Bi_4Ti_3O_{12}$. The crystalline $Bi_4Ti_3O_{12}$ was amorphized again after milling for another 10 h at the initial power intensities (angular velocities). A similar trend in phase evolution was observed for the two groups of samples when they were subjected to the second set of milling parameters. Due to the stronger action, the amorphization progress was comparatively much more rapid. In summary, mechanochemical activation induced phase evolution of the amorphization/crystallization

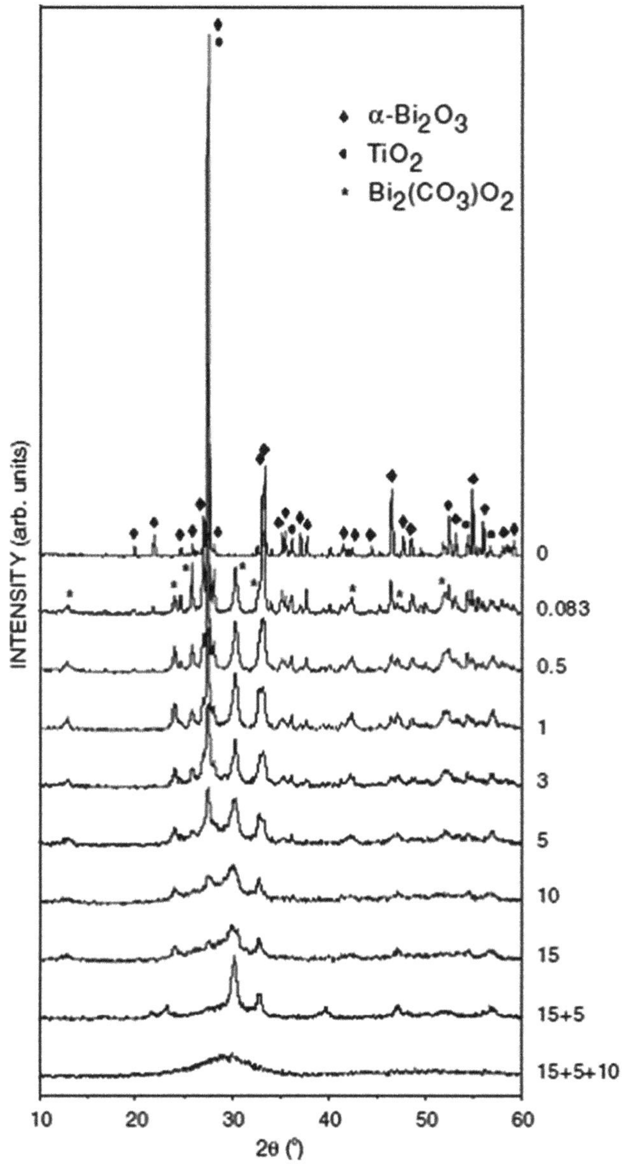

Figure 5.9. XRD patterns of the powder mixture of $2Bi_2O_3{\cdot}3TiO_2$ milled for different times with different intensities, $P = 0.49$ W g^{-1} and $P = 2.68$ W g^{-1} (milling time 15 + 5 h). Reproduced with permission from [52]. Copyright 2006 Elsevier.

of the Bi_2O_3–TiO_2 binary system, no matter whether it was an oxide mixture of Bi_2O_3 and TiO_2 powders or $Bi_4Ti_3O_{12}$ powder with a high degree of crystallinity.

As the crystal size was reduced to <40 nm, the $Bi_4Ti_3O_{12}$ phase began to be amorphized. During the amorphization process, the crystal size was continuously decreased to 12–18 nm. Amorphization started due to the collapse of the crystalline structure, as the critical size was approached. After that, the remaining nanocrystalline phase was gradually dissolved into the amorphous matrix. For the mixture of Bi_2O_3 and TiO_2, the $Bi_4Ti_3O_{12}$ phase was formed due to the reaction of the precursor oxides as a result of the high-energy mechanochemical activation, after the creation of the intermediate phase of $Bi_2(CO_3)O_2$. After activation for 15 h with lower milling intensity, the contents of amorphous phase in the oxide mixture of Bi_2O_3–TiO_2 and $Bi_4Ti_3O_{12}$ powder were 40 wt% and 80 wt%. The lower content of the amorphous phase in the oxide mixture was ascribed to the kinetic retardance caused by the remaining components of Bi_2O_3 and TiO_2 and the presence of $Bi_2(CO_3)O_2$.

It was also demonstrated that high intensity activation tended to trigger the crystalline phase, whereas low intensity milling resulted in amorphization. Moreover, in the process of the reaction–amorphization–crystallization sequence with low activation intensities, the amorphization–crystallization process was reversible by controlling the energy intensity of the mechanochemical action (i.e. the angular velocities of the device). According to kinetics studies, in a first approximation both the forward and the reverse reactions of amorphous–crystalline equilibrium were of first-order characteristics. For high activation intensities, further studies are necessary to correct the kinetic model.

5.3.2 Other Aurivillius ferroelectrics

5.3.2.1 Bismuth vanadate (BiV)

BiV ($Bi_2VO_{5.5}$) nanosized powders were synthesized from oxide mixture through mechanochemical activation using a planetary ball mill (Fretsh, Pulverisette 6) [53–55]. An agate vial and agate balls were used as the milling media. An intermediate phase $BiVO_4$ was present after activation for 4 h, which was complete after further milling until 16 h. After that, the designed phase BiV was formed, while the single-phase BiV was obtained after activation for 54 h. In this case, the intermediate phase $BiVO_4$ was always present, regardless of the starting materials used. This was because the barrier for the nucleation of $BiVO_4$ was lower than that of BiV, leading to ceramics with a relative density of 97% and an average grain size of about 2 μm. As a result, the BiV ceramics displayed high electrical properties.

Phase evolution of the mixture of Bi_2O_3–V_2O_5 as a result of high-energy mechanochemical activation was studied using a vibrating mill (Fritsch Pulverisette 0) [56]. Both agate and stainless steel media were used in this work. During the milling process, no intermediate phase was observed. After activation for 72 h, an amorphous product was obtained, instead of crystalline BiV. The absence of the crystalline BiV phase was readily ascribed to the higher energy density because of the vibrating mill and the heavier stainless steel media. The amorphous phase was crystallized at a very low temperature of 385 °C. The crystalline phase was γ-$Bi_2VO_{5.5}$, with particles having sizes in the range 100–200 nm.

Commercial powders of Bi_2O_3 and V_2O_5 were first mixed manually and then mechanochemically activated for times of up to three weeks [57]. The activation was conducted with a Fritsch vibrating mill (Pulverisette 0). Stainless steel vials and one 5 cm diameter steel ball were used as the milling media. After milling for one day, there was no obvious variation in phase composition, with just slight reduction in the peak density in the XRD patterns. Activation for two days led to essential refinement of the constituent oxides. As the milling time was increased to three days, the mixture was entirely amorphized.

The DTA curve of the amorphous product after thorough mechanochemical activation had a sharp exothermic peak at 385 °C on heating, while there were two weak exothermic peaks at 542 °C and 353 °C on cooling. The XRD results indicated that the as-activated powder was confirmed to be amorphous, while those calcined at temperatures lower than the exothermic peak on the heating curve were also amorphous. If the amorphous powder was calcined at temperatures of above 400 °C, it was crystallized to the γ-$Bi_2VO_{5.5}$ phase, which was stable up to 650 °C. During the cooling process, the γ phase was transferred to β-$Bi_2VO_{5.5}$ at 500 °C, which was further converted to α-$Bi_2VO_{5.5}$ at about 300 °C that was stable down to room temperature. This observation was different from the previous example, due to the stronger activation effect.

If the γ-$Bi_2VO_{5.5}$ phase, which was obtained by calcining the amorphous powder at 385 °C, was heated at a temperature below 500 °C no phase transition was observed. In other words, the γ-$Bi_2VO_{5.5}$ phase was stable up to 500 °C. However, if the same γ-$Bi_2VO_{5.5}$ phase was heated to 700 °C, phase transition from γ to β to α would occur. As claimed by the authors, that was the first report that the γ-$Bi_2VO_{5.5}$ phase was stable up to 500 °C without the use of any dopants. The phase transition process of the system was also evidenced by SEM observation results. A similar variation trend was observed for the oxide mixture of Bi_2O_3 and VO_2, in which Bi_2VO_5 was formed.

5.3.2.2 Alkaline earth bismuth titanate

In the Aurivillius family, alkaline earth bismuth titanate, $ABi_4Ti_4O_{15}$ (with A = Ca, Sr, Ba and Pb), has an orthorhombic structure, with $n = 4$ [58–60]. The mechanochemical process has been employed to prepare CBiT and SBiT based nanosized powders, thus leading to ceramics with high sinterability and promising electrical properties. CBiT nanosized powder was obtained directly from the commercial oxide mixture of CaO, Bi_2O_3 and TiO_2, by mechanochemical activation in N_2 for 30 h at room temperature [61]. The as-synthesized CBiT powder had an average particle size of about 50 nm. CBiT ceramics were fabricated with the nanosized powder after sintering at 1175 °C for 2 h, which had a relative density of 93.4% and a peak dielectric constant of 1049 near the Curie temperature of 774 °C at 1 MHz.

In a similar way, SBiT powder was prepared from a mixture of CaO, Bi_2O_3 and TiO_2 after mechanochemical activation for 20 h [62]. SBiT ceramics were derived from the powder after sintering at 1175 °C, with a relative density of 98% and a peak dielectric constant of 2770 at 100 kHz, near the Curie point of 539 °C. The SBiT ceramics exhibited a conduction activation energy of 0.22 eV at relatively low temperatures. The sample had a room temperature piezoelectric constant (d_{33}) of 24.0 p/CN. In these two cases, no intermediate phases, such as $Bi_{12}TiO_{20}$ and $Bi_4Ti_3O_{12}$, were present in the systems during the mechanochemical activation process.

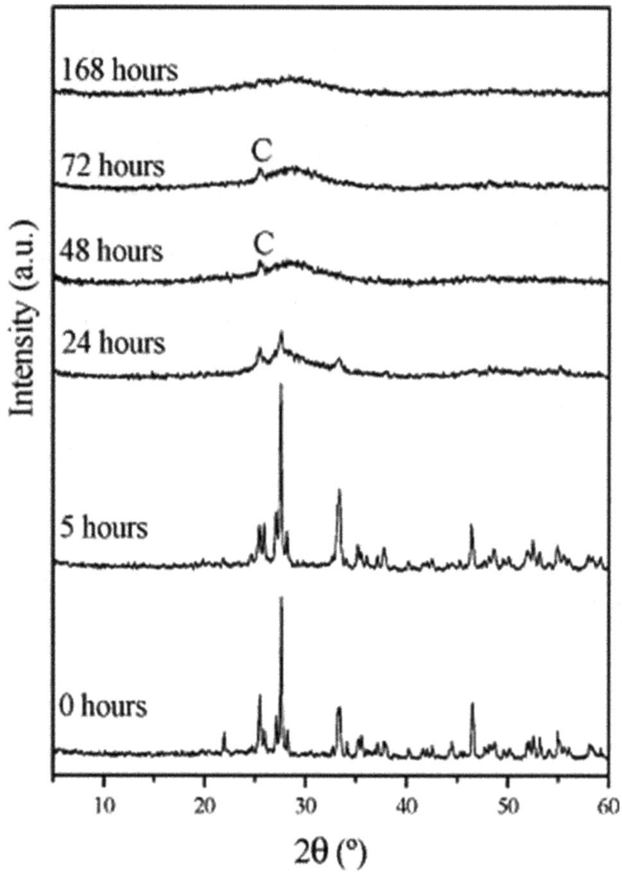

Figure 5.10. XRD patterns of the oxide mixture of Bi_2O_3, TiO_2 and $SrCO_3$ mechanochemically activated for different times with a vibrating mill (C:$SrCO_3$). Reproduced with permission from [63]. Copyright 2004 American Chemical Society.

In a separate study, amorphous SBiT powders, instead of crystalline phases, were obtained using both vibrating and planetary mills [63]. In addition, the conventional solid-state reaction process was also conducted to prepare the materials for comparison. The only difference was that $SrCO_3$, rather than SrO, was used as the starting material, together with Bi_2O_3 and TiO_2. For vibration milling, particle refinement of the starting components and amorphization of the systems were observed after activation for 24 h, but $SrCO_3$ was still present even after milling for 72 h. The mixtures were completely amorphized as the activation time was increased to 168 h. In comparison, activation for 12 h could reach amorphization when using the planetary mill. No matter what conditions were used, no no crystal phases were formed in the systems according to the XRD measurement results.

The XRD patterns of the oxide mixtures mechanochemically activated for different times, with vibrating and planetary mills, are shown in figures 5.10 and 5.11, respectively. In the case of a vibrating mill, obvious refinement in particle size

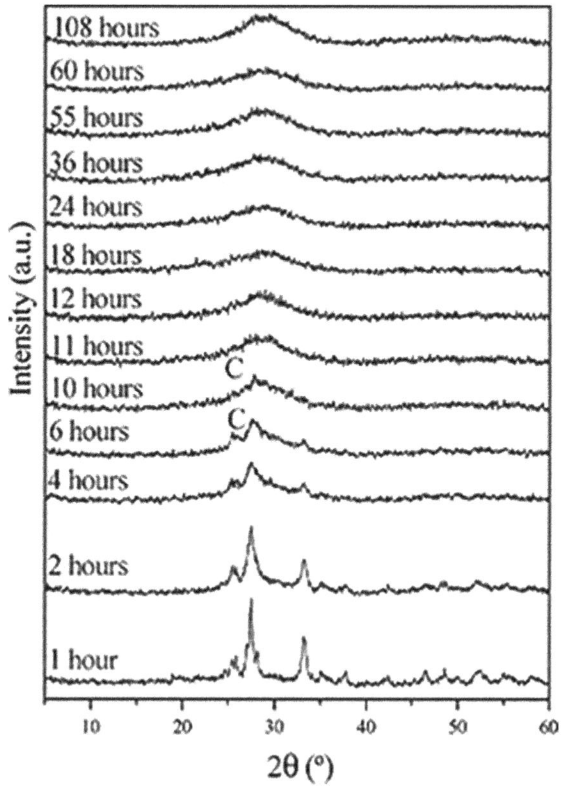

Figure 5.11. XRD patterns of the oxide mixture of Bi_2O_3, TiO_2 and $SrCO_3$ mechanochemically activated for different times with a planetary mill (C:$SrCO_3$). Reproduced with permission from [63]. Copyright 2004 American Chemical Society.

of the precursors was observed after milling for 24 h, while the peak of $SrCO_3$ was still present even when the activation proceeded for 48 h, although significant amorphization occurred. Complete amorphization required a long milling time of 168 h. In contrast, the planetary mill was much more efficient. The diffraction peaks of the constituents were minimized after milling for 4 h, while a complete amorphization was essentially achieved after activation for 11 h.

The thermogravimetry (TG) curve of the powder milled with a vibrating mill (VM) had three stages of weight loss, including water loss up to 266 °C, a second loss up to 485 °C and a third loss up to 671 °C. The corresponding DTA curve displayed two exothermic peaks at 376 °C and 451 °C. Figure 5.12 shows the high-temperature XRD patterns of the amorphous powder activated with a vibrating mill. A weak peak appeared at 200 °C corresponding to $SrCO_3$, which was present up to 600 °C. Therefore, the second and the third weight losses in the TG curve were due to the release of CO_2. A phase transition was observed at 400 °C, corresponding to the exothermic peak in the DTA curve at 376 °C, i.e. the conversion of the amorphous phase to a fluorite phase. More new peaks were detected at 500 °C, which corresponded to the second exothermic peak at 451 °C in the DTA curve, implying

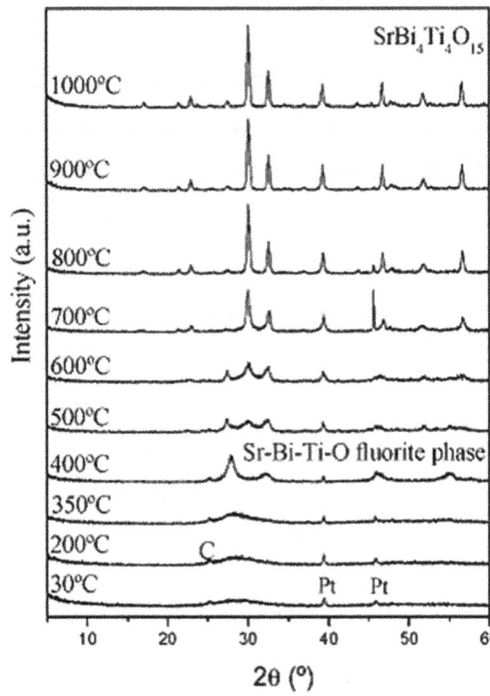

Figure 5.12. High-temperature XRD patterns of the amorphous precursor activated with a vibrating mill (VM) (C: SrCO3; Pt: platinum). Reproduced with permission from [63]. Copyright 2004 American Chemical Society.

the formation of the Aurivillius phase. As CO_2 was completely released, fluorite-type and SBiT phases coexisted until 600 °C. The fluorite-type phase finally converted to SBiT at 700 °C. A further increase in temperature to 1000 °C only enhanced the crystallization of SBiT.

A similar variation trend was observed in the TG/DTA responses of the powder milled with a planetary mill for 12 h. There were two exothermic peaks at 366 °C and 471 °C, along with an endothermic peak at 668 °C. The weight loss in the TG curve over 561 °C–693 °C corresponded to the endothermic peak in the DTA curve, due to the release of CO_2. However, the powder activated for 108 h with a planetary mill exhibited different thermal behavior. The TG curve had only one weight loss in the temperature range of 684 °C–823 °C. The DTA curve exhibited a special exothermic peak at 482 °C. Figure 5.13 shows the high-temperature XRD patterns of a sample milled for 12 h with a planetary mill. Similarly, the intermediate fluorite-type phase was detected at 350 °C, together with SBiT over 450 °C–600 °C, while the conversion to SBiT was finished at 700 °C.

However, slightly different from that in the vibrating sample and the planetary 12 h sample, the peak of $SrCO_3$ at low temperatures was not observed in the planetary 108 h sample. As a result, its TG curve had only one weight loss at a relatively higher temperature. Also, the fluorite-type phase was not present in this

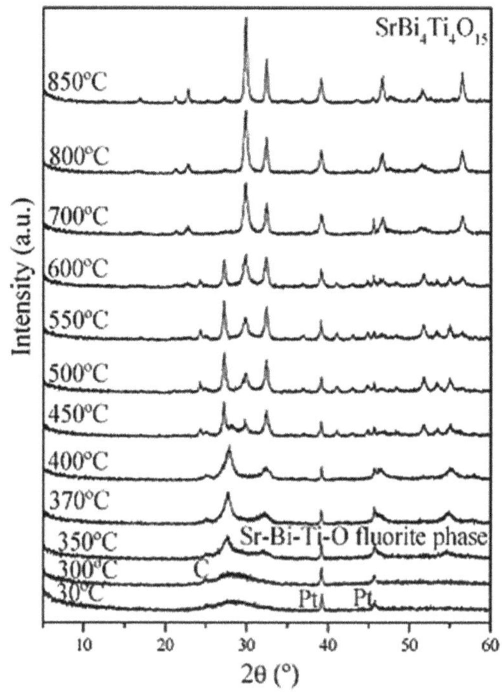

Figure 5.13. High-temperature XRD patterns of the amorphous precursor activated with a planetary mill for 12 h (PM12): (C) $SrCO_3$ and (Pt) platinum. Reproduced with permission from [63]. Copyright 2004 American Chemical Society.

case, suggesting that the SBiT phase was formed without the presence of the intermediate phase. Briefly, the phase formation of SBiT from the amorphous powders activated using the high-energy milling process occurred at much lower temperatures, compared to that required in the conventional solid-state reaction process, where the lowest temperature was 1000 °C.

Morphology variation and particle refinement were present during the phase evolution caused by the thermal annealing. The starting precursor powders all had particle sizes at the micron level. The powder milled with the vibrating mill displayed a uniform particle size distribution, with an average particle size of about 0.3 μm. After calcining at 400 °C, the particles were crystallized to a phase with tabular morphology and spherical aggregates. The fluorite crystal had a cubic symmetry, while the particle size was almost unchanged. After calcining at 550 °C for 1 h, the spherical grains were retained, while the average size was slightly increased to about 0.33 μm. Calcination at 600 °C for 5 h led to spherical particles with an average size of 0.37 μm, together with the presence of agglomerates. Both the morphology and size of the particles experienced an obvious change as the calcination temperature was raised at 800 °C, suggesting an increase in phase crystallinity, whereas the average particle size was increased to about 0.5 μm. The sample heated at 900 °C exhibited the layered morphology of the Aurivillius phases, the smaller particles were spherical. The disk-like particles had a thickness of about 0.22 μm, with

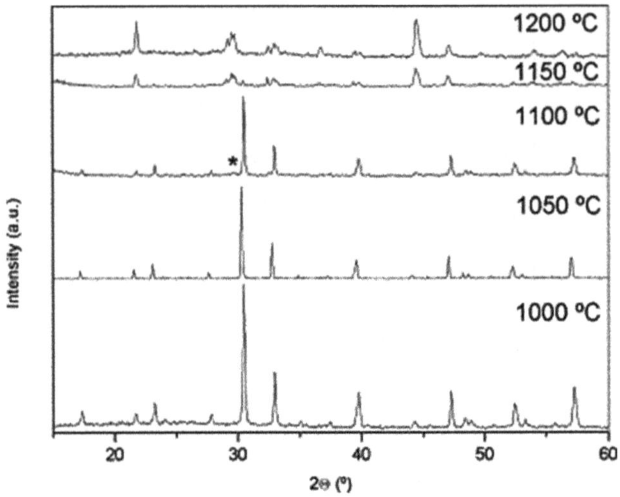

Figure 5.14. XRD patterns of the SBiT ceramics derived from $2Bi_2O_3 + 4TiO_2 + SrCO_3$ mechanochemically activated with a vibrating mill after sintering at different temperatures: (*) secondary phase. Reproduced with permission from [64]. Copyright 2008 Elsevier.

dimensions in the range 0.26–1.68 μm, which were increased to 0.65 μm and 0.94–2.99 μm, respectively, as the calcination temperature was further increased to 1000 °C. The characteristic of the Aurivillius phases became more pronounced.

The effect of the mechanochemical activation parameters on the electrical properties of SBiT ceramics has been studied by the same authors [64]. SBiT powders processed under different conditions were made into ceramics by optimizing the sintering temperature in order to achieve desired electrical properties. Commercial powders of Bi_2O_3, TiO_2 and $SrCO_3$/SrO were weighted in Ar to avoid the reaction with water vapor and CO_2. The powders were mixed according to the composition of SBiT with a quantity of 3 g for each part of the experiment.

Stainless steel vials and balls were used as the milling media, with both vibrating and planetary mills (Fritsch, Pulverizette models 0 and 6, respectively). In the case of the planetary mill, the rotation speed was 200 rpm. One ball with a diameter of 5 cm and five balls with diameters of 1 cm were used for the vibrating and planetary mills, respectively. When using a vibrating mill, the activation experiment was conducted for times of up to 168 h. If the planetary mill was used, different mechanical treatments were used to obtain ceramic precursors; the initial mixture was treated with milling times of 12–108 h and 10–90 h, respectively, for $SrCO_3$ and SrO used as the starting materials. All the powders were sintered at temperatures of 1000 °C–1200 °C for 5 h, at a heating rate of 3 °C min^{-1}.

Figure 5.14 shows the XRD patterns of the SBiT ceramics from the mechanochemically obtained powder with a vibrating mill and $SrCO_3$ as the starting material. The proper sintering temperatures were in the range 1000 °C–1050 °C, due to the formation of the Aurivillius phase and the absence of crystallographic orientation. Without the presence of texture, the ceramics would have high mechanical strengths. In addition, the sample sintered at 1100 °C contained a secondary phase at the surface, which was undesirable in terms of electrical properties. To remove the secondary phase, the sample should be polished.

The SBiT ceramics consisted of platelet shaped grains, which is a typical microstructure of Aurivillius materials. Obviously, a dense microstructure was achieved in the samples sintered at temperatures of 1050 °C and 1100 °C, with porosities of 3%–4% and grain sizes of about 1.5 μm. Figure 5.15 shows SEM images of the SBiT ceramics derived from the powders processed undert different conditions. Among the processing parameters, the type of mill had the most pronounced influence on the microstructure of the final ceramics. The grain size of the ceramics

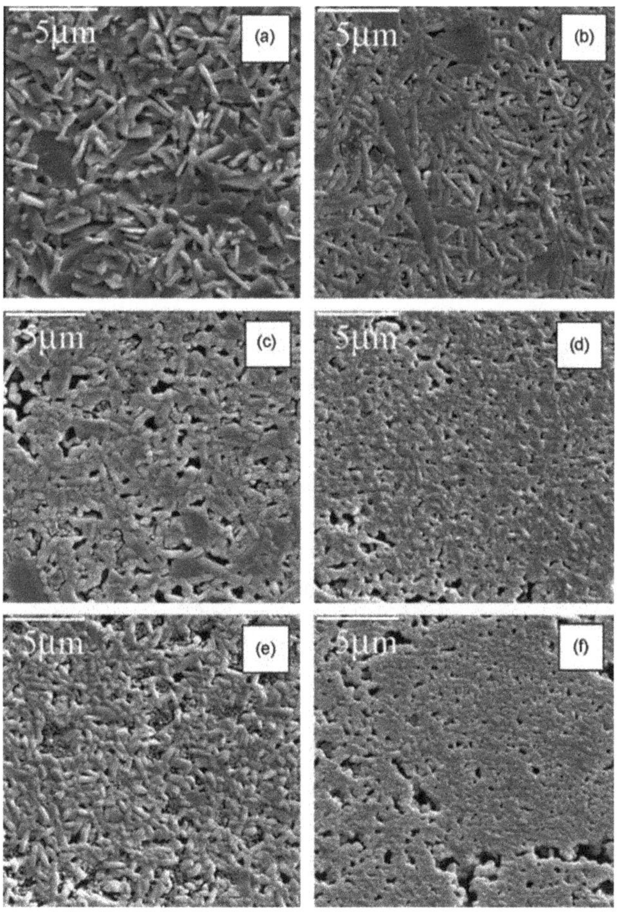

Figure 5.15. SEM images of the SBiT ceramics sintered at 1050 °C for 5 h derived from different mechanochemically activated precursors: (a) $2Bi_2O_3 + 4TiO_2 + SrCO_3$ milled for 168 h with a vibrating mill, (b) $2Bi_2O_3 + 4TiO_2 + SrO$ milled for 168 h with a vibrating mill, (c) $2Bi_2O_3 + 4TiO_2 + SrCO_3$ milled for 12 h with a planetary mill, (d) $2Bi_2O_3 + 4TiO_2 + SrCO_3$ milled for 108 h with a planetary mill, (e) $2Bi_2O_3 + 4TiO_2 + SrO$ milled for 12 h with a planetary mill and (f) $2Bi_2O_3 + 4TiO_2 + SrO$ milled for 90 h with planetary mill. Reproduced with permission from [64]. Copyright 2008 Elsevier.

from powders activated with the vibrating mill was relatively larger, compared to that when using the planetary mill.

Specifically, the SBiT ceramic sample from the powder of $SrCO_3$ with the vibrating mill after sintering at 1050 °C had an anomaly in the dielectric constant curve at about 503 °C, corresponding to the ferroelectric–paraelectric phase transition. A similar trend in dielectric properties was observed in other ceramic samples. Figure 5.16 shows representative dielectric constant curves of the ceramics derived from the powders activated under different conditions. For example, the dielectric properties of the samples, from the powders milled with a planetary mill, with $SrCO_3$ as the precursor, after sintering for 12 h and 108 h, had no significant difference, as seen in figure 5.16(a). Prolonged milling only led to a slight decrease in the peak dielectric constant. The two samples from the powders made with $SrCO_3$ and SrO as the precursor were essentially the same, as illustrated in figure 5.16(b). However, the effect of the type of mill was relatively significant, as observed in figure 5.16(c). The effect of milling conditions/parameters on electrical properties was actually a reflection of the effect on the microstructure of the final ceramics. In other words, when the mechanochemical activation method is used to synthesize materials, the processing parameters should be optimized.

$BaBi_4Ti_4O_{15}$ (BBiT) ceramics were fabricated from the powders that were synthesized using mechanochemical synthesis during intensive milling [65]. Commercial oxide powders of BaO, TiO_2 and Bi_2O_3 were mixed and mechanochemically treated for times of up to 6 h. Zirconium oxide vials and zirconium oxide balls with a diameter of d = 10 mm were used as the milling media. The ball-to-powder weight ratio was 20:1. A Fritsch Pulverisette 5 planetary ball mill was employed to conduct the mechanochemical activation. The basic disc rotation speed was 320 rpm and the rotation speed of disks with vials was 400 rpm. All experiments were carried out in air at room temperature. The as-activated powders were calcinated at 750 °C for 4 h and pressed into pallets with a cold isostatic press. BBiT ceramics were fabricated by sintering the green body pellets at 1130 °C for 1 h. The ceramics had a relative density of about 92% and exhibited relaxor ferroelectric characteristics.

5.3.2.3 Bismuth titanate niobate

Due to the weak mass diffusion, it is difficult to fabricate fully dense ceramics of Aurivillius compounds, including bismuth titanate niobite, Bi_3TiNbO_9 or BiTN [39, 66–68]. As a result, it is necessary to use special sintering facilities to develop these ceramics. Mechanochemical activation has been demonstrated to be an effective technique to synthesize Bi_3TiNbO_9 based powders, thus leading to ceramics with promising microstructural and electrical properties [5–7]. The as-synthesized powders were most likely of the amorphous phase.

The oxide mixture of $3Bi_2O_3 + Nb_2O_5 + 2TiO_2$ was mechanochemically activated for times of up to 336 h, leading to the formation of the amorphous phase after milling for 168 h [5]. The sample activated for 336 h was totally amorphized. The DTA curve revealed that there were two exothermic peaks at 370 °C and 560 °C, due to the formation of a metastable fluorite phase and the Aurivillius phase,

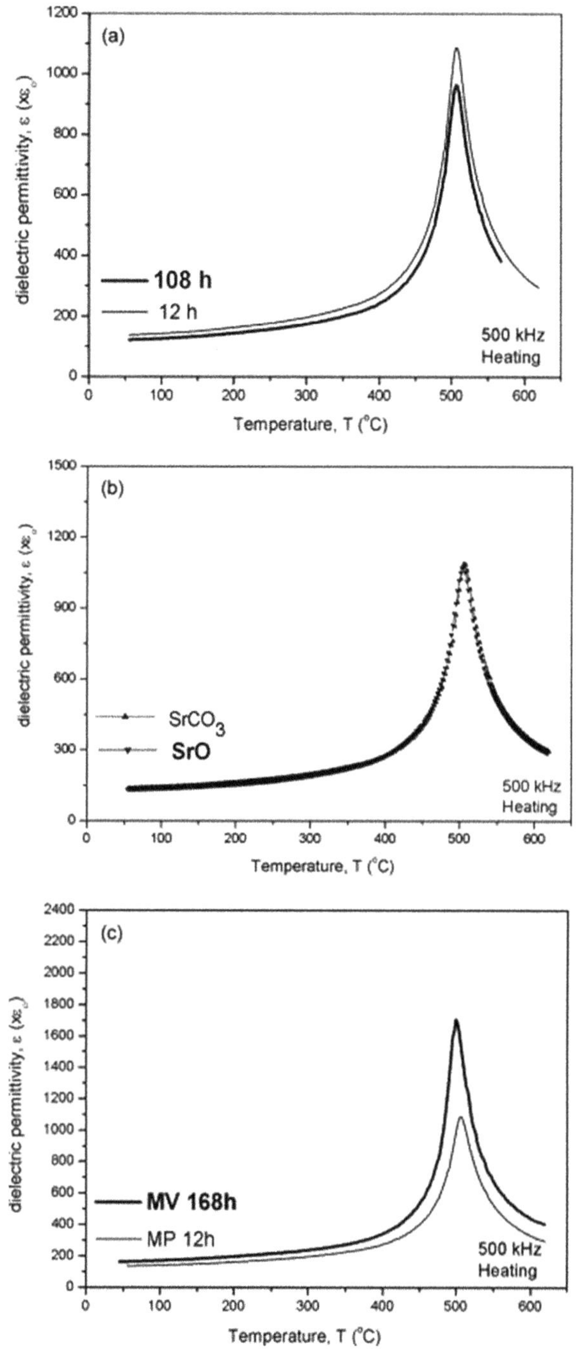

Figure 5.16. Dielectric constant curves (500 kHz) of SBiT ceramics derived from the powders with different mechanochemical activation parameters: (a) $2Bi_2O_3 + 4TiO_2 + SrCO_3$ milled for 12 and 108 h with a planetary mill, (b) $2Bi_2O_3 + 4TiO_2 + SrCO_3/SrO$ milled for 12 h with a planetary mill and (c) $2Bi_2O_3 + 4TiO_2 + SrCO_3$ milled for 12 h with a planetary mill and 168 h with a vibrating mill. Reproduced with permission from [64]. Copyright 2008 Elsevier.

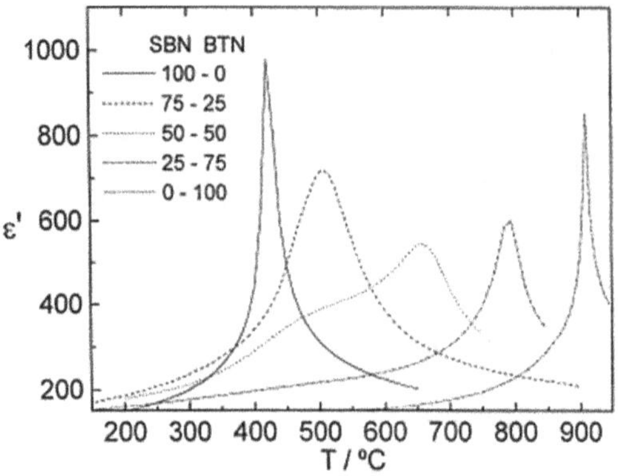

Figure 5.17. Dielectric constant of the SBN–BTN ceramics with different compositions as a function of temperature at 1.3 MHz. Reproduced with permission from [8]. Copyright 2001 Elsevier.

respectively. The sample could be fully crystallized after calcining at 600 °C. In comparison, the Aurivillius phase was crystallized from the unmilled powder at a high temperature of 1050 °C. BiTN ceramics could be derived from the amorphous powder by sintering at 1100 °C, which had lower porosity and thus higher electrical performance compared to those made using the conventional ceramic process.

Ceramics of compositions within the solid solution system $(SrBi_2Nb_2O_9)_{1-x}(Bi_3TiNbO_9)_x$ (SBN–BTN) have been obtained for $0 \leqslant x \leqslant 1$ [8–11]. Bi_2O_3, Nb_2O_5, $SrCO_3$ and TiO_2 were mixed according to the compositions of $(Bi_2SrNb_2O_9)_{1-x}(Bi_3TiNbO_9)_x$ ($x = 0$, 0.25. 0.50, 0.75 and 1.00) and then mechanochemically activated for 336 h with a Fritsch Pulverisette 0 type vibrating mill [8]. The mixtures were all completely amorphized after the long duration high-energy activation.

The solid solution ceramics had a relative density of about 98%, with promising piezoelectric properties. For example, the piezoelectric coefficient (d_{33}) versus temperature of the SBN–BTN ($x = 0.25$) ceramics exhibited a stable profile, making them suitable for high-temperature piezoelectric device applications. The dielectric profiles of all the ceramics with temperature were nearly the same, with the Curie temperature increasing almost linearly with increasing value of x, as shown in figure 5.17. Some of the samples displayed interesting ferroelectric and piezoelectric behaviors.

5.4 Other lead-free ferroelectric materials

Lithium niobate ($LiNbO_3$ or LN) has shown a ferroelectric effect, with various potential applications. High-energy mechanochemical milling has been used to prepare LN powder [69]. The starting materials were commercial powders of Nb_2O_5 and Li_2CO_3, which were milled for times in the range of 2–42 h. The milling was conducted with a Fritsch Pulverisette 5 type planetary mill, with stainless steel vials and balls as the milling media, with a ball-to-powder weight ratio of 10:1. According to the XRD results, the LN phase was present after activation for just 2 h, while the sample milled for 42 h contained LN as the dominant phase, along with a small fraction of amorphous phase and residual Nb_2O_5. The formation of crystalline LN was also supported by other measurement results.

Sodium niobate ($NaNbO_3$ or NN) was also obtained by using a mechanochemical activation process from the mixture of Na_2CO_3 and Nb_2O_5 [70]. Different to LN, the NN phase was not formed after milling for times of up to one month, when using a Fritsch Pulverisette 0 type vibrating mill. Instead, the sample contained crystalline Nb_2O_5 and amorphous Na_2CO_3. The crystalline NN phase was achieved by calcining the milled powder at a temperature of 600 °C, which was lower than that required in the conventional solid-state reaction process by 150 °C. The powder could be used to fabricate NN ceramics with high piezoelectric properties after sintering at 1000 °C for 2 h, when combined with a hot-pressing process.

$K_{0.5}Na_{0.5}NbO_3$ (KNN) nanosized ceramic powders were prepared using the mechanochemical activation method, combined with post-thermal calcination at a relatively low temperature [71]. The activation was started with a powder mixture of K_2CO_3, Na_2CO_3 and Nb_2O_5 for times of up to 32 h. The alkaline carbonate powders were predried at 200 °C for 1 h to remove water. The equipment was an SPEX 8000D shaker mill. The milling was conducted at a speed of 875 rpm. The sample milled for 32 h was calcined at 550 °C for 2 h to form phase pure KNN powder, with an orthorhombic crystal structure. Such a phase formation temperature was lower than that in the typical ceramic process by about 300 °C. KNN ceramics were obtained from the powder after sintering at 1100 °C for 2 h, with electrical properties comparable to the literature data.

KNN ceramics with high density, fine grains and uniform size distribution were derived from powders activated with mechanochemical milling [72]. The total time of the process was as short as 100 min, much shorter than the time of one day for the conventional solid-state reaction process. Moreover, the dielectric and ferroelectric properties of the KNN ceramics could be optimized by controlling the particle size of the as-activated powders. Specifically, the ceramics from the powder milled for 100 min exhibited high energy storage capacity, with an electrical energy storage density of 1.612 J cm^{-3} and a recoverable energy storage density of 0.431 J cm^{-3}, which was attributed to the high dielectric breakdown strength of 110 kV cm^{-1}.

The starting powders were mixed in the presence of high purity water in air with a zirconia vial and balls as the milling media. The ball-to-powder weight ratio was 8:1. The activation was then performed with a pulverisette P7™ Fritsch Vario planetary mill, with the sun wheel and the grinding jar rotating in opposite directions, at a speed ratio of 1:2. The rotational directions of the sun wheel and vial were reversed every 2 min with a rest interval of 8 min to prevent overheating. The rotation speeds were 300/600 rpm for times of up to 360 min. After milling, the slurry was freeze-dry calcined at 700 °C–850 °C for 9 h, resulting in precursors of KNN powders. KNN ceramics were derived from the powders after sintering at 1120 °C–1140 °C for 2 h.

Bismuth sodium barium titanate ceramics, with a composition of $(Bi_{0.5}Na_{0.5})_{0.94}Ba_{0.06}TiO_3$ (BNBT), as a lead-free piezoelectric material were developed from powders obtained using mechanochemical activation [73]. Starting powders of TiO_2, Bi_2O_3, BaO and Na_2CO_3 were mixed according to the designed compositions then activated in air at room temperature. The experiment was conducted with a Sepahan 84 D type planetary ball mill with a steel vial with a volume of 90 ml at a rotation speed of 180 rpm. The number of tempered steel balls with diameters of 20 mm and 8 mm was four and seven, respectively, corresponding to a ball-to-powder weight ratio of 20:1. The samples were milled for times of up to 360 h. The perovskite phase was formed after milling for 360 h. It seems that the time for perovskite phase formation was relatively long, probably because the milling configuration was not properly set, thus leading to a relatively low efficiency. For practical applications, it is necessary to increase the activation efficiency.

5.5 Multiferroic bismuth ferrite

Multiferroics are defined as multiple-effect ferroelectric, ferromagnetic and/or ferroelastic orderings present in one material [74–76]. They find potential applications in a wide range of areas, such as data storage, sensors, filters, attenuators and spintronics devices. Among various candidates for multiferroic materials, bismuth ferrite ($BiFeO_3$), with perovskite crystal structure, has drawn much attention due to its special ferroelectric and magnetic characteristics. For example, it was antiferromagnetic and ferroelectric up to temperatures of 370 °C and 825 °C, respectively [77, 78]. Although strong magnetic behavior and highly spontaneous polarizations have been observed in epitaxial $BiFeO_3$ thin films, both ceramics and single crystals of $BiFeO_3$ display a relatively low spontaneous polarization, insufficient for real applications [79, 80]. However, the reasons behind this observation have not been clarified. The preparation of high quality $BiFeO_3$ powder is a key factor to fabricate high performance $BiFeO_3$ ceramics.

$BiFeO_3$ nanosized powders were synthesized by employing the mechanochemical activation method [81]. Commercial oxides of Bi_2O_3 and Fe_2O_3 were mixed and activated mechanochemically with an SPEX 8000 shaker mill. The milling was conducted for times of up to 120 h. Figure 5.18 shows the XRD patterns of the starting oxides and the mixture after activation for different times. After milling for 10 h, all the diffraction peaks were absent in the XRD patterns, indicating that they were completely refined and the reaction to form $BiFeO_3$ started. After further milling for 70 h, the perovskite structured $BiFeO_3$ phase was detected. According to TEM results, the final $BiFeO_3$ power had particle sizes in the range 100 nm–150 nm. The authors did not make $BiFeO_3$ ceramics with the nanosized powder.

$BiFeO_3$ (BFO) and $Bi_{0.95}Eu_{0.05}FeO_3$ (BEFO) powders were synthesized using the mechanochemical activation process, in order to study their structural, microstructural and magnetic properties [82]. Commercial oxide powders of α-Fe_2O_3, α-Bi_2O_3 and Eu_2O_3 were thoroughly mixed and then mechanochemically activated with a Fritsch Pulverisette 6 type planetary ball mill, with a hardened steel vial and balls used as the milling media. The ball-to-powder weight ratio was about 30:1. The milling was carried out at a rotation speed of 32 rad s^{-1} for a time of 24 h. The as-synthesized powders were calcined at 973 K in air for different times of 1 h–24 h.

The effect of experimental parameters on the crystal structure, as well as the electronic and magnetic properties of $BiFeO_3$ prepared using high-energy mechanochemical activation, combined with calcination, was systematically studied [83].

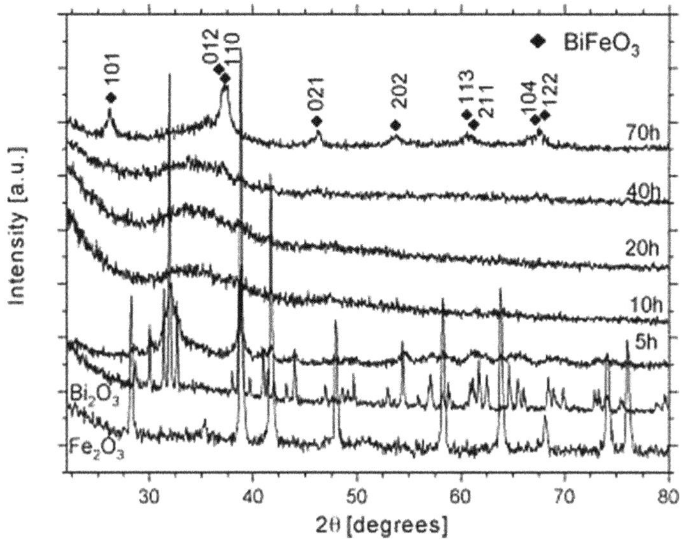

Figure 5.18. XRD patterns of the starting oxide mixture for $BiFeO_3$ before and after milling for different times. Reproduced with permission from [81]. Copyright 2007 Elsevier.

Oxide powders of Fe_2O_3 and Bi_2O_3 were mixed and activated for times of up to 13 h, followed by calcination at temperatures of 350 °C–750 °C for 2 h. High purity $BiFeO_3$ powder was obtained from the sample milled for 5 h and then calcined at 650 °C. The powder displayed a weak ferromagnetic effect, which was converted to antiferromagnetic order as the powder was pressed at 900 MPa and sintered at 800 °C for 2 h.

$Bi_{1-x}Yb_xFeO_3$ (x = 0.02, 0.05, 0.07) was for the first time prepared using mechanical activation followed by sintering. Bi_2O_3, Fe_2O_3 and Yb_2O_3 commercial powders were used as the starting materials [84]. The oxide mixture was mechanochemically milled for 6 h at room temperature and at 7 bar of O_2 pressure, with a Fritsch Pulverisette 7 planetary mill. A hardened steel vial with a volume of 80 ml and nine balls with diameters of 15 mm were used as the milling media. The ball-to-powder weight ratio was 20:1, while the rotation speed was 700 rpm. The as-obtained powders were pressed into pellets, which were sintered at 825 °C for 6 min in air at both heating and cooling rates of 10 °C min^{-1}. The ceramics possessed high resistivity and displayed antiferromagnetic properties at room temperature.

The same authors attempted to synthesize $Bi_{1-x}Sm_xFeO_3$ ($0.05 \leqslant x \leqslant 0.20$) using mechanochemical activation [85]. The processing parameters were nearly the same as those described above. Ceramic samples were prepared from the powders by sintering at 900 °C. Various analytical results confirmed that the $Bi_{1-x}Sm_xFeO_3$ samples were of single-phase perovskite structure. However, it was found that the electrical and magnetic properties of the $Bi_{1-x}Sm_xFeO_3$ ceramics were independent of the concentration of Sm substitution, such as resistivity and remnant magnetization.

Solid solutions of $BiFeO_3$ (BFO) and $Ba(Zr_{0.4}Ti_{0.6})O_3$ (BZT), with compositions of $(1 - x)$BFO–xBZT (x = 0, 0.15, 0.25, 0.40 and 0.50), were obtained using the

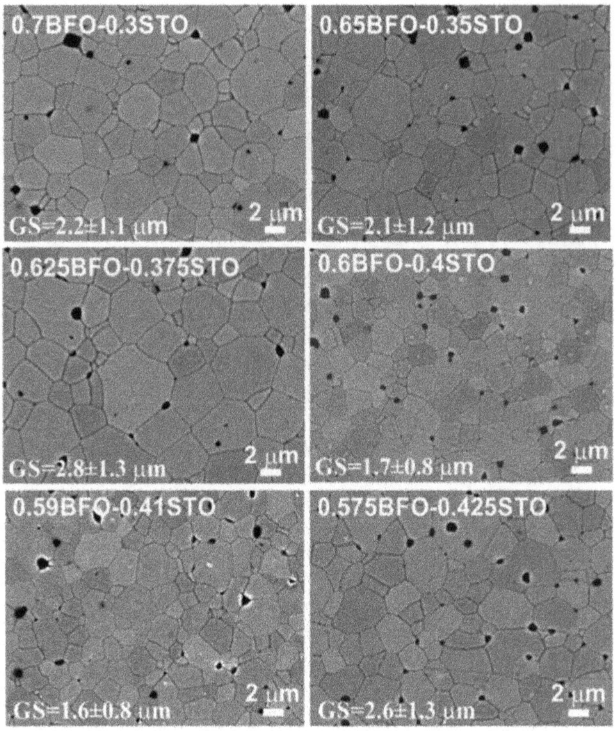

Figure 5.19. SEM images of the (x)BFO–(1 − x)STO ceramics after polishing and thermal etching, with their corresponding average grain sizes (GS) to be labeled. Reproduced with permission from [87]. Copyright 2019 Elsevier.

mechanochemical activation technique [86]. In addition, the solid-state reaction process was also used for comparison. However, the activation parameters, such as the type of mill, milling media, milling speed and ball-to-powder weight ratio, were not mentioned in the study. The degree of distortion of the two end components (i.e. BFO and BZT) in the $(Bi_{1-x}Ba_x)(Fe_{1-x}Zr_{0.6x}Ti_{0.4x})O_3$ (BBFZT) system was examined in terms of tolerance factors (t), which greatly deviated from the value (t = 1) of ideal perovskite. Both the dielectric constant and loss tangent of the BBFZT ceramics were decreased with increasing content of BZT in the solid solutions. As a consequence, the leakage current and electrical conductivity of the BBFZT were largely reduced due to the incorporation of BZT.

$BiFeO_3$–$SrTiO_3$ (BFO–STO) solid solution ceramics were developed from presynthesized powders that were mechanochemically treated, in order to enhance the microstructural and electrical properties of the materials [87]. The authors studied the compositions of (x)BFO–(1 − x)STO ceramics with $0.7 \geqslant x \geqslant 0.575$. Figure 5.19 shows SEM images of the (x)BFO–(1 − x)STO ceramics. They had relative densities of above 97% and grain sizes between 1.6 μm ($x = 0.59$) and 2.8 μm ($x = 0.625$). The BFO–STO ceramics possessed homogeneous microstructure and high densities due to the mechanochemical activation. High remanent polarizations of 30–50 μC cm^{-1} were observed for solid solution samples, while a maximum d_{33} value of 69 pC/N was achieved in the sample with $x = 0.625$.

The Bi_2O_3, Fe_2O_3, TiO_2 and $SrCO_3$ commercial powders were milled individually first and then thoroughly mixed with the designed compositions. The mixtures were calcined at 750 °C for 6 h twice, at both heating and cooling rates of 5 °C min^{-1}. After each round of calcination, the powders were milled with a Retsch PM400 planetary mill at a rotation speed of 200 rpm for 4 h, in the presence of ethanol. Polyethylene vials with yttria-stabilized-zirconia (YSZ) balls with a diameter of 3 mm were used as the milling media. All the powders were compacted and sintered at 1025 °C for 2 h at heating and cooling rates of 5 °C min^{-1}. According to the statement in the study, the experiment was not a high-energy activation, due to the low density of the milling media.

5.6 Conclusions

The effects of mechanochemical activation on the phase formation of barium titanate based and bismuth containing ferroelectric compounds are dependent on both the properties of the precursors and the processing parameters. In some cases, the target compounds are not directly formed during the mechanochemical process, although the phase formation temperatures are significantly reduced. The phase formation is only possible under proper conditions. For some bismuth containing compounds, amorphous phases are formed instead of crystalline phases. However, the amorphous powders can also be used to fabricate the corresponding ceramics. In addition, the mechanochemical process is a promising technique to prepare multiferoic bismuth ferrite ceramics, due to the low temperature requirements.

Acknowledgments

Shenzhen Technology University (SZTU) is acknowledged for the financial support of a start-up grant (2018) and also the Natural Science Foundation of Top Talent of SZTU (grant No. 2019 010 801 002).

References

[1] Xue J M, Wang J and Wan D M 2000 Nanosized barium titanate powder by mechanical activation *J. Am. Ceram. Soc.* **83** 232–4

[2] Abe O and Suzuki Y 1996 Mechanochemically assisted preparation of $BaTiO_3$ powder *Mater. Sci. Forum* **225** 563–8

[3] Kong L B, Ma J, Huang H, Zhang R F and Que W X 2002 Barium titanate derived from mechanochemically activated powders *J. Alloys Compd.* **337** 226–30

[4] Berbenni V, Marini A and Bruni G 2001 Effect of mechanical milling on solid state formation of $BaTiO_3$ from $BaCO_3$–TiO_2 (rutile) mixtures *Thermochim. Acta* **374** 151–8

[5] Castro A, Millan P, Pardo L and Jimenez B 1999 Synthesis and sintering improvement of Aurivillius type structure ferroelectric ceramics by mechanochemical activation *J. Mater. Chem.* **9** 1313–7

[6] Ricote J, Pardo L, Moure A, Castro A, Millan P and Chateigner D 2001 Microcharacterisation of grain-oriented ceramics based on Bi_3TiNbO_9 obtained from mechanochemically activated precursors *J. Eur. Ceram. Soc.* **21** 1403–7

[7] Moure A, Pardo L, Alemany C, Millan P and Castro A 2001 Piezoelectric ceramics based on Bi_3TiNbO_9 from mechano-chemically activated precursors *J. Eur. Ceram. Soc.* **21** 1399–402

[8] Jimenez B, Castro A, Pardo L, Millan P and Jimenez R 2001 Electric and ferro-piezoelectric properties of $(SBN)_{1-x}(BTN)_x$, ceramics obtained from amorphous precursors *J. Phys. Chem. Solids* **62** 951–8

[9] Pardo L, Castro A, Millan P, Alemany C, Jimenez R and Jimenez B 2000 $(Bi_3TiNbO_9)_x(SrBi_2Nb_2O_9)_{1-x}$ Aurivillius type structure piezoelectric ceramics obtained from mechanochemically activated oxides *Acta Mater.* **48** 2421–8

[10] Moure A, Castro A and Pardo L 2004 Improvement by recrystallisation of Aurivillius-type structure piezoceramics from mechanically activated precursors *Acta Mater.* **52** 945–57

[11] Moure A, Alemany C and Pardo L 2004 Electromechanical properties of SBN/BTN Aurivillius-type ceramics up to the transition temperature *J. Eur. Ceram. Soc.* **24** 1687–91

[12] Maurice A K and BR C 1987 Preparation and stoichiometry effects on microstructure and properties of high-purity $BaTiO_3$ *Ferroelectrics* **74** 61–75

[13] Bergström L, Shinozaki K, Tomiyama H and Mizutani N 1997 Colloidal processing of a very fine $BaTiO_3$ powder-effect of particle interactions on the suspension properties, consolidation, and sintering behavior *J. Am. Ceram. Soc.* **80** 291–300

[14] Clark I J, Takeuchi T, Ohtori N and Sinclair D C 1999 Hydrothermal synthesis and characterization of $BaTiO_3$ fine powder: precursors, polymorphism and properties *J. Mater. Chem.* **9** 83–91

[15] Brzozowski E and Castro M S 2000 Synthesis of barium titanate improved by modifications in the kinetics of the solid state reaction *J. Eur. Ceram. Soc.* **20** 2347–51

[16] Brzozowski E and Castro M S 2003 Lowering the synthesis temperature of high-purity $BaTiO_3$ powders by modification in the processing conditions *Thermochim. Acta* **389** 123–9

[17] Kodama S, Kido O, Suzuki H, Saito Y and Kaito C 2005 Characterization of nanoscale $BaTiO_3$ ultrafine particles prepared by gas evaporation method *J. Cryst. Growth* **282** 60–5

[18] Huang Y A *et al* 2017 Control of tetragonality via dehydroxylation of $BaTiO_3$ ultrafine powders *Ceram. Int.* **43** 16462–6

[19] Welham N J 1998 Mechanically induced reaction between alkaline earth metal oxides and TiO_2 *J. Mater. Res.* **13** 1607–13

[20] Stojanovic B D, Jovalekic C, Vukotic V, Simoes A Z and Varela J A 2005 Ferroelectric properties of mechanically synthesized nanosized barium titanate *Ferroelectrics* **319** 65–73

[21] Pavlovic V P *et al* 2007 Microstructural evolution and electric properties of mechanically activated $BaTiO_3$ ceramics *J. Eur. Ceram. Soc.* **27** 575–9

[22] Thakur O P, Feteira A, Kundys B and Sinclair D C 2007 Influence of attrition milling on the electrical properties of undoped-$BaTiO_3$ *J. Eur. Ceram. Soc.* **27** 2577–89

[23] Nath A K, Jiten C, Singh K C, Laishram R, Thakur O P and Bhattacharya D K 2010 Effect of ball milling time on the electrical and piezoelectric properties of barium titanate ceramics *Integr. Ferroelectr.* **116** 51–8

[24] Singh K C and Nath A K 2011 Barium titanate nanoparticles produced by planetary ball milling and piezoelectric properties of corresponding ceramics *Mater. Lett.* **65** 970–3

[25] Pavlovic V P *et al* 2011 Structural investigation of mechanically activated nanocrystalline $BaTiO_3$ powders *Ceram. Int.* **37** 2513–8

[26] Pavlovic V P, Nikolic M V, Pavlovic V B, Labus N, Zivkovic L and Stojanovic B D 2005 Correlation between densification rate and microstructure evolution of mechanically activated $BaTiO_3$ *Ferroelectrics* **319** 75–85
[27] Liu S B *et al* 2002 Preparation and characterization of $Ba_{1-x}Sr_xTiO_3$ thin films for uncooled infrared focal plane arrays *Mater. Sci. Eng.:* C **22** 73–7
[28] Ianculescu A *et al* 2007 Investigation of $Ba_{1-x}Sr_xTiO_3$ ceramics prepared from powders synthesized by the modified Pechini route *J. Eur. Ceram. Soc.* **27** 3655–8
[29] Curecheriu L P, Mitoseriu L and Ianculescu A 2009 Nonlinear dielectric properties of $Ba_{1-x}Sr_xTiO_3$ ceramics *J. Alloys Compd.* **482** 1–4
[30] Chou X J, Zhao Z Y, Du M X, Liu J and Zhai J W 2012 Microstructures and dielectric properties of $Ba_{1-x}Sr_xTiO_3$ ceramics doped with B_2O_3–Li_2O glasses for LTCC technology applications *J. Mater. Sci. Technol.* **28** 280–4
[31] Fuentes S, Chávez E, Padilla-Campos L and Diaz-Droguett D E 2013 Influence of reactant type on the Sr incorporation grade and structural characteristics of $Ba_{1-x}Sr_xTiO_3$ ($x = 0$–1) grown by sol–gel-hydrothermal synthesis *Ceram. Int.* **39** 8823–31
[32] Hungria T, Alguero M, Hungria A B and Castro A 2005 Dense, fine-grained $Ba_{1-x}Sr_xTiO_3$ ceramics prepared by the combination of mechanosynthesized nanopowders and spark plasma sintering *Chem. Mater.* **17** 6205–12
[33] Tusseau-Nenez S, Ganne J P, Maglione M, Morell A, Niepce J C and Pate M 2004 BST ceramics: effect of attrition milling on dielectric properties *J. Eur. Ceram. Soc.* **24** 3003–11
[34] Jain A, Panwar A K and Jha A K 2016 Structural, dielectric and ferroelectric studies of $Ba_{1-x}Sr_xTiO_3$ ceramics prepared by mechanochemical activation technique *J. Mater. Sci. Mater. Electron.* **27** 9911–9
[35] Jain A, Panwara A K and Jha A K 2016 Influence of milling duration on microstructural, electrical, ferroelectric and piezoelectric properties of $Ba_{0.9}Sr_{0.1}Zr_{0.04}Ti_{0.96}O_3$ ceramic *Ceram. Int.* **42** 18771–8
[36] Shrivastava V, Jha A K and Mendiratta R G 2005 Structural distortion and phase transition studies of Aurivillius type $Sr_{1-x}Pb_xBi_2Nb_2O_9$ ferroelectric ceramics *Solid State Commun.* **133** 125–9
[37] Henriques E I, Kim H J, Haluska M S, Edwards D D and Misture S T 2007 Solid solubility and electrical conduction mechanisms in 3-layer Aurivillius ceramics *Solid State Ionics* **178** 1175–9
[38] Diao C L, Zheng H W, Gu Y Z, Zhang W F and Fang L 2014 Structural and electrical properties of four-layer Aurivillius phase $BaBi_{3.5}Nd_{0.5}Ti_4O_{15}$ ceramics *Ceram. Int.* **40** 5765–9
[39] Wang Q, Wang C M, Wang J F and Zhang S J 2016 High performance Aurivillius-type bismuth titanate niobate (Bi_3TiNbO_9) piezoelectric ceramics for high temperature applications *Ceram. Int.* **42** 6993–7000
[40] Wong Y J, Hassan J, Chen S K and Ismail I 2017 Combined effects of thermal treatment and Er-substitution on phase formation, microstructure, and dielectric responses of $Bi_4Ti_3O_{12}$ Aurivillius ceramics *J. Alloys Compd.* **723** 567–79
[41] Wu B, Ma J, Wu W J and Chen M 2018 Evolution of microstructure and electrical properties of Aurivillius phase $(CaBi_4Ti_4O_{15})_{1-x}(Bi_4Ti_3O_{12})_x$ ceramics *Ceram. Int.* **44** 9168–73
[42] Dorrian J F, Newnham R E and Smith K K 1971 Crystal structure of $Bi_4Ti_3O_{12}$ *Ferroelectrics* **3** 17–27
[43] Cummings S E and Cross L E 1968 Electrical and optical properties of ferroelectric $Bi_4Ti_3O_{12}$ single crystals *J. Appl. Phys.* **39** 2268–74

[44] Shulman H S, Testorf M, Damjanovic D and Setter N 1996 Microstructure, electrical conductivity and piezoelectric properties of bismuth titanate *J. Am. Ceram. Soc.* **79** 3214–8

[45] Du X and Chen I W 1998 Ferroelectric thin films of bismuth-containing layered perovskites: part I, $Bi_4Ti_3O_{12}$ *J. Am. Ceram. Soc.* **81** 3253–9

[46] Jiang A Q, Li H G and Zhang L D 1998 Dielectric study in nanocrystalline $Bi_4Ti_3O_{12}$ prepared by chemical doprecipitation *J. Appl. Phys.* **83** 4878–83

[47] Lavado C and Stachiotti M G 2018 Fe^{3+}/Nb^{5+} co-doping effects on the properties of Aurivillius $Bi_4Ti_3O_{12}$ ceramics *J. Alloys Compd.* **731** 914–9

[48] Kong L B, Ma J, Zhu W and Tan O K 2001 Preparation of $Bi_4Ti_3O_{12}$ ceramics via a high-energy ball milling process *Mater. Lett.* **51** 108–14

[49] Lisoni J G, Millán P, Vila E, Martín de Vidales J L, Hoffmann T and Castro A 2001 Synthesis of ferroelectric $Bi_4Ti_3O_{12}$ by alternative routes: wet no-coprecipitation chemistry and mechanochemical activation *Chem. Mater.* **13** 2084–91

[50] Ng S H, Xue J M and Wang J 2002 Bismuth titanate from mechanical activation of a chemically coprecipitated precursor *J. Am. Ceram. Soc.* **85** 2660–5

[51] Stojanovic B D *et al* 2006 Mechanically activating formation of layered structured bismuth titanate *Mater. Chem. Phys.* **96** 471–6

[52] Zdujic M, Poleti D, Jovalekic C and Karanovic L 2006 The evolution of structure induced by intensive milling in the system $2Bi_2O_3{\cdot}3TiO_2$ *J. Non-Cryst. Solids* **352** 3058–68

[53] Shantha K, Subbanna G N and Varma K B R 1999 Mechanically activated synthesis of nanocrystalline powders of ferroelectric bismuth vanadate *J. Solid State Chem.* **142** 41–7

[54] Shantha K and Varma K B R 1999 Preparation and characterization of nanocrystalline powders of bismuth vanadate *Mater. Sci. Eng.* B **60** 66–75

[55] Shantha K and Varma K B R 2000 Characterization of fine-grained bismuth vanadate ceramics obtained using nanosized powders *J. Am. Ceram. Soc.* **83** 1122–8

[56] Ricote J, Pardo L, Castro A and Millan P 2001 Study of the process of mechanochemical activation to obtain Aurivillius oxides with $n = 1$ *J. Solid State Chem.* **160** 54–61

[57] Castro A, Millan P, Ricote J and Pardo L 2000 Room temperature stabilisation of γ-$Bi_2VO_{5.5}$ and synthesis of the new fluorite phase f-Bi_2VO_5 by a mechanochemical activation method *J. Mater. Chem.* **10** 767–71

[58] Zhao M L, Wang C L, Zhong W L, Zhang P L, Wang J F and Chen H C 2003 Dielectric and pyroelectric properties of $SrBi_4Ti_4O_{15}$-based ceramics for high-temperature applications *Mater. Sci. Eng.* B **99** 143–6

[59] Zhu J, Mao X Y and Chen X B 2004 Properties of vadadium-doped $SrBi_4Ti_4O_{15}$ ferroelectric ceramics *Solid State Commun.* **129** 707–10

[60] Pribošič I, Makovec D and Drofenik M 2001 Electrical properties of donor- and acceptor-doped $BaBi_4Ti_4O_{15}$ *J. Eur. Ceram. Soc.* **21** 1327–31

[61] Sim M H, Xue J M and Wang J 2004 Layer structured calcium bismuth titanate by mechanical activation *Mater. Lett.* **58** 2032–6

[62] Ng S H, Xue J M and Wang J 2002 High-temperature piezoelectric strontium bismuth titanate from mechanical activation of mixed oxides *Mater. Chem. Phys.* **75** 131–5

[63] Ferrer P, Iglesias J E and Castro A 2004 Synthesis of the Aurivillius phase $SrBi_4Ti_4O_{15}$ by a mechanochemical activation route *Chem. Mater.* **16** 1323–9

[64] Ferrer P, Algueró M and Castro A 2008 Influence of the mechanochemical conditions on the processing of $Bi_4SrTi_4O_{15}$ ceramics from submicronic powdered precursors *J. Alloys Compd.* **464** 252–8

[65] Bobić J D, Vijatović M M, Greičius S, Banys J and Stojanović B D 2010 Dielectric and relaxor behavior of $BaBi_4Ti_4O_{15}$ ceramics *J. Alloys Compd.* **499** 221–6
[66] Su D, Zhu J S, Xu Q Y, Liu J S and Wang Y N 2003 Transmission electron microscopy study on domain structures in Bi_3TiNbO_9 ceramics *Microelectron. Eng.* **66** 825–9
[67] Yuan J, Nie R, Chen Q, Xiao D Q and Zhu J G 2019 Structural distortion, piezoelectric properties, and electric resistivity of A-site substituted Bi_3TiNbO_9-based high-temperature piezoceramics *Mater. Res. Bull.* **115** 70–9
[68] Yi Z G, Li Y X, Yang Q B and Yin Q R 2008 La doping effects on intergrowth Bi_2WO_6–Bi_3TiNbO_9 ferroelectrics *Ceram. Int.* **34** 735–9
[69] de Figueiredo R S, Messai A, Hernandes A C and Sombra A S B 1998 Piezoelectric lithium niobate obtained by mechanical alloying *J. Mater. Sci. Lett.* **17** 449–51
[70] Castro A, Jiménez B, Hungría T, Moure A and Pardo L 2004 Sodium niobate ceramics prepared by mechanical activation assisted methods *J. Eur. Ceram. Soc.* **24** 941–5
[71] Singh R, Patro P K, Kulkarni A R and Harendranath C S 2014 Synthesis of nano-crystalline potassium sodium niobate ceramic using mechanochemical activation *Ceram. Int.* **40** 10641–7
[72] Chen B *et al* 2019 High-efficiency synthesis of high-performance $K_{0.5}Na_{0.5}NbO_3$ ceramics *Powder Technol.* **346** 248–55
[73] Amini R, Ghazanfari M R, Alizadeh M, Ardakani H A and Ghaffari M 2013 Structural, microstructural and thermal properties of lead-free bismuth–sodium–barium–titanate piezoceramics synthesized by mechanical alloying *Mater. Res. Bull.* **48** 482–6
[74] Wang K F, Liu J M and Ren Z F 2009 Multiferroicity: the coupling between magnetic and polarization orders *Adv. Phys.* **58** 321–448
[75] Cheong S W and Mostovoy M 2007 Multiferroics: a magnetic twist for ferroelectricity *Nat. Mater.* **6** 13–20
[76] Vaz C A F, Hoffman J, Ahn C H and Ramesh R 2010 Magnetoelectric coupling effects in multiferroic complex oxide composite structures *Adv. Mater.* **22** 2900–18
[77] Catalan G and Scott J F 2009 Physics and applications of bismuth ferrite *Adv. Mater.* **21** 2463–85
[78] Kumar M M, Palkar V R, Srinivas K and Suryanarayana S V 2000 Ferroelectricity in a pure $BiFeO_3$ ceramic *Appl. Phys. Lett.* **76** 2764–6
[79] Wang J *et al* 2003 Epitaxial $BiFeO_3$ multiferroic thin film heterostructures *Sci. Adv. Mater.* **299** 1719–22
[80] Ryu S, Kim J Y, Shin Y H, Park B G, Son J Y and Jang H M 2009 Enhanced magnetization and modulated orbital hybridization in epitaxially constrained $BiFeO_3$ thin films with rhombohedral symmetry *Chem. Mater.* **21** 5050–7
[81] Szafraniak I, Połomska M, Hilczer B, Pietraszko A and Kępiński L 2007 Characterization of $BiFeO_3$ nanopowder obtained by mechanochemical synthesis *J. Eur. Ceram. Soc.* **27** 4399–402
[82] Freitas V F, Grande H L C, de Medeiros S N, Santos I A, Cótic L F and Coelho A A 2008 Structural, microstructural and magnetic investigations in high-energy ball milled $BiFeO_3$ and $Bi_{0.95}Eu_{0.05}FeO_3$ powders *J. Alloys Compd.* **461** 48–52
[83] Pedro-García F, Sánchez-De Jesús F, Cortés-Escobedo C A, Barba-Pingarrón A and Bolarín-Miró A M 2017 Mechanically assisted synthesis of multiferroic $BiFeO_3$: Effect of synthesis parameters *J. Alloys Compd.* **711** 77–84

[84] Gil-González E, Perejón A, Sánchez-Jiménez P E, Hayward M A and Pérez-Maqueda L A 2017 Preparation of ytterbium substituted $BiFeO_3$ multiferroics by mechanical activation *J. Eur. Ceram. Soc.* **37** 945–54

[85] Gil-González E *et al* 2017 Characterization of mechanosynthesized $Bi_{1-x}Sm_xFeO_3$ samples unencumbered by secondary phases or compositional inhomogeneity *J. Alloys Compd.* **711** 541–51

[86] Choudhary R N P, Perez K, Bhattacharya P and Katiyar R S 2007 Structural and dielectric properties of mechanochemically synthesized $BiFeO_3$–$Ba(Zr_{0.6}Ti_{0.4})O_3$ solid solutions *Mater. Chem. Phys.* **105** 286–92

[87] Makarovic M, Bencan A, Walker J, Malic B and Rojac T 2019 Processing, piezoelectric and ferroelectric properties of $(x)BiFeO_3$–$(1-x)SrTiO_3$ ceramics *J. Eur. Ceram. Soc.* **39** 3693–702

Chapter 6

Ferrite ceramics (I)

Ling Bing Kong, Zhuohao Xiao, Xiuying Li, Shijin Yu, Wenxiu Que, Yin Liu, Tianshu Zhang, Kun Zhou and Hongfang Zhang

6.1 Introduction

Due to their high refractive index ($n = \sqrt{\mu'\varepsilon'}$, where $\mu'$and ε' are the real parts of relative permeability and permittivity, respectively) and impedance matching to free space ($Z = \eta_0\sqrt{\mu'/\varepsilon'} = \eta_0$, with $\varepsilon' = \mu'$, where η_0 is the impedance of free space), magneto-dielectric materials have potential electromagnetic (EM) fields [1–4]. If dielectric and magnetic loss tangents are sufficiently low, they could be used to reduce the dimensions of antennas. If the materials have high dielectric and magnetic loss tangents, they could be potential candidates as EM wave absorbers. EM wave absorbers have been extensively studied, while the development of magneto-dielectric materials, with low losses and matching permeability and permittivity, has been less explored.

For antenna miniaturization, the desired frequency bands include high frequency (HF, 3–30 MHz) and very high frequency (VHF, 30–90 and 100–300 MHz), because antennas working at these frequency bands usually have large physical sizes. Although materials for such purposes had been proposed a long time ago, the real achievement was only made more recently. Because ferrites are both magnetic and dielectric materials, they are the most promising candidates as magneto-dielectric materials. There are three groups of ferrites: (i) spinel, (ii) garnet and (iii) hexaferrite. In this chapter, the fabrication and characterization of magnesium ferrite ceramics, with close values of real permeability and permittivity, along with sufficiently low dielectric and magnetic loss tangents, will be presented and discussed. The rest of the examples will be covered in next chapter.

6.2 Mg–Cu–Co ferrite ceramics

Three groups of samples, $MgFe_{1.98}O_4$, $Mg_{1-x}Cu_xFe_{1.98}O_4$ (x = 0.01–0.30) and $Mg_{0.90-x}Co_xCu_{0.10}Fe_{1.98}O_4$ (x = 0.05, 0.10, 0.15 and 0.20), were studied [5, 6]. The samples were designed with Fe deficiency to prevent the formation of Fe^{2+} ions at

doi:10.1088/978-0-7503-2191-4ch6

high temperatures during sintering. Eventually, the purpose was to have samples with high DC resistivity and a low dielectric loss tangent. Copper oxide (CuO) was used as a sintering aid to reduce the sintering temperature of $MgFe_{1.98}O_4$, while cobalt oxide (CoO) was employed to adjust the magnetic properties. The ferrite ceramics were all prepared using conventional ceramic processing. First, oxides of the desired amounts according to the designated compositions were thoroughly mixed using a planetary ball mill, with tungsten carbide vials of 250 ml plus 100 balls 10 mm in diameter as the milling media. The milling was conducted for 2 h at a rotation speed of 200 rpm. The activated powders were calcined at 1000 °C for 2 h in air. The calcined powders were then compacted into pellets and rings. The green bodies were sintered at different temperatures for 2 h in air.

No reaction among the starting oxides was triggered by the mechanochemical activation due to the short milling time, while calcination was used to obtain the designed ferrite powders. Figure 6.1 shows the XRD patterns of the $Mg_{1-x}Cu_xFe_{1.98}O_4$ powders after they were calcined at 1000 °C for 2 h. All samples exhibited phase pure spinel structure, suggesting that spinel ferrites were formed at this temperature. The XRD patterns of samples of $Mg_{0.90-x}Co_xCu_{0.10}Fe_{1.98}O_4$ are depicted in figure 6.2, which were similar to those shown in figure 6.1.

Densification behaviors of selected $Mg_{1-x}Cu_xFe_{1.98}O_4$ samples ($x = 0.15$ and 0.20) are shown in figure 6.3. It was found that the introduction of CuO significantly promoted the densification process of the ferrite powders. Other samples also displayed similar shrinkage curves, with the shrinking onset temperatures shifting to lower temperatures as the content of CuO was increased. The maximum linear shrinkages and the temperatures of the peak shrinkage rate for the samples of $Mg_{1-x}Cu_xFe_{1.98}O_4$ are listed in table 6.1. The temperature for peak shrinkage rate was decreased monotonically from 1214 °C to 1055 °C as the content of CuO was increased from $x = 0.10$ to $x = 0.30$. In comparison, the peak temperature of $MgFe_{1.98}O_4$ was >1250 °C. Even though the temperature of the peak shrinkage rate decreased continuously with increasing concentration of CuO, the final linear shrinkage did

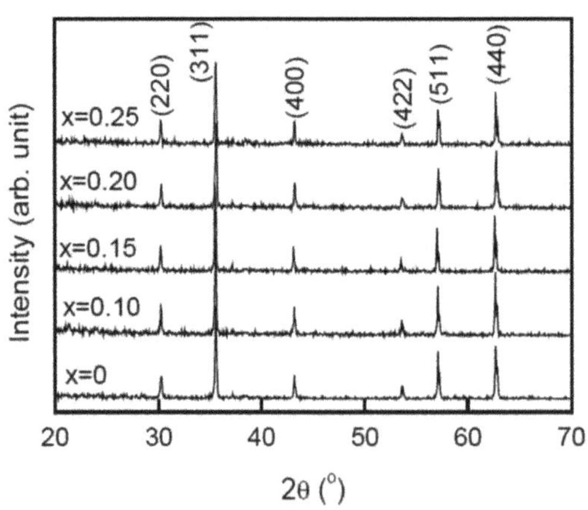

Figure 6.1. XRD patterns of the $Mg_{1-x}Cu_xFe_{1.98}O_4$ powders synthesized at 1000 °C for 2 h. Reproduced with permission from [5]. Copyright 2007 Wiley.

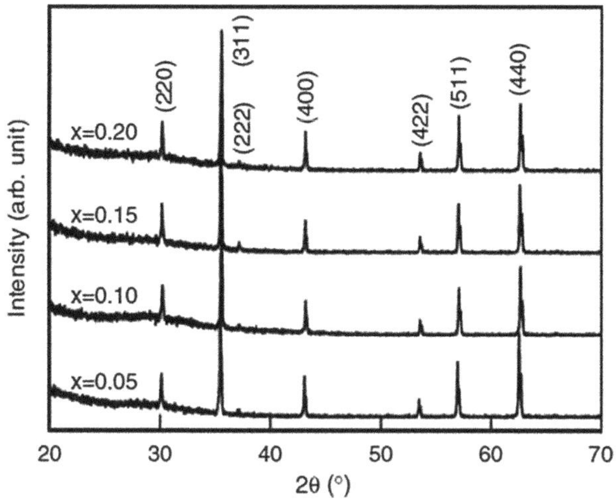

Figure 6.2. XRD patterns of the $Mg_{0.90-x}Co_xCu_{0.10}Fe_{1.98}O_4$ powders synthesized at 1000 °C for 2 h. Reproduced with permission from [5]. Copyright 2007 Wiley.

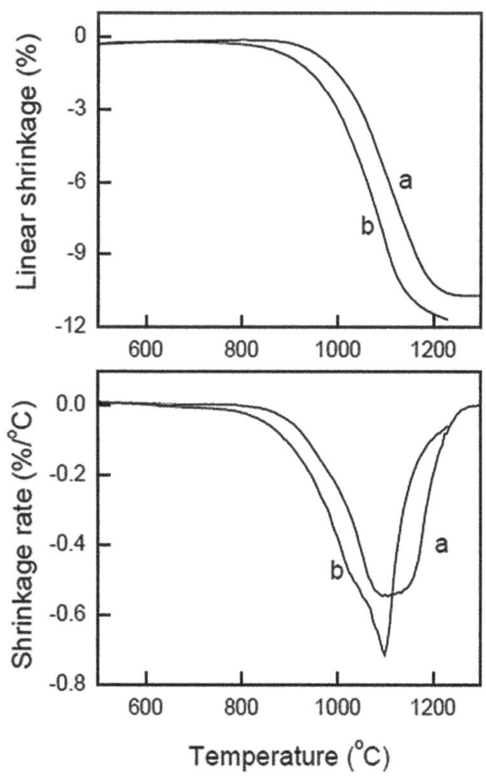

Figure 6.3. Densification curves of selected samples in the group of $Mg_{1-x}Cu_xFe_{1.98}O_4$: (a) x = 0.15 and (b) x = 0.20. Reproduced with permission from [5]. Copyright 2007 Wiley.

Table 6.1. Densification parameters of the $Mg_{1-x}Cu_xFe_{1.98}O_4$ samples. Reproduced with permission from [5]. Copyright 2007 Wiley.

x	Peak shrinkage rate (°C)	Max. linear shrinkage (%)
0	>1250 °C	−8.2
0.1	1214 °C	−10.1
0.15	1121 °C	−10.7
0.2	1098 °C	−11.7
0.25	1068 °C	−10.7
0.3	1055 °C	−10.3

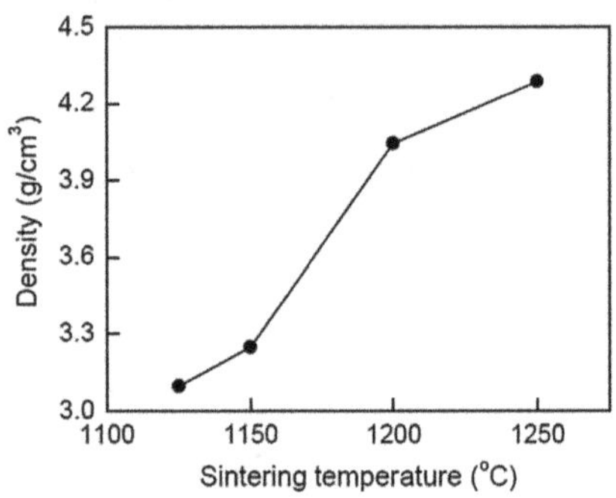

Figure 6.4. Density of the $MgFe_{1.98}O_4$ samples as a function of sintering temperature. Reproduced with permission from [5]. Copyright 2007 Wiley.

not follow this trend. It was increased from −8.2% to −11.7%, reached a maximum of −11.7% for the sample with $x = 0.20$ and then decreased to −10.3%.

Figure 6.4 shows the density of the $MgFe_{1.98}O_4$ samples as a function of sintering temperature. Obviously, $MgFe_{1.98}O_4$ could not be fully densified at temperatures <1200 °C. The sample sintered at 1200 °C had a density of 4.27 g cm^{-3}, which was even lower than 90% of the theoretical density of 4.523 g cm^{-3}. After sintering at 1250 °C for 2 h, the sample had a relative density of 94%. This temperature was too high in order to maintain a high resistivity, because Fe^{2+} could be readily formed at temperatures >1200 °C.

Densities of the groups of $Mg_{1-x}Cu_xFe_{1.98}O_4$ and $Mg_{0.90-x}Co_xCu_{0.10}Fe_{1.98}O_4$ ceramics are depicted in figures 6.5 and 6.6, respectively. Theoretical densities of $Mg_{1-x}Cu_xFe_{1.98}O_4$ could be simply calculated with those of $MgFe_{1.98}O_4$ and $CuFe_{1.98}O_4$ according to the linear relationship, because the solid solution characteristics of the materials. The calculated densities for the compositions of $x = 0.10$, 0.15, 0.20, 0.25 and 0.30 were 4.62, 4.66, 4.71, 4.76 and 4.8 g cm^{-3}, respectively. As observed in figure 6.4, $Mg_{1-x}Cu_xFe_{1.98}O_4$ ceramics could easily achieve relative densities >95%. For example, the sample with $x = 0.10$ had a relative density of 94% after sintering at 1100 °C. The sample with $x = 0.25$ was almost fully densified after sintering at 1000 °C for 2 h. This sintering temperature was lower than that of

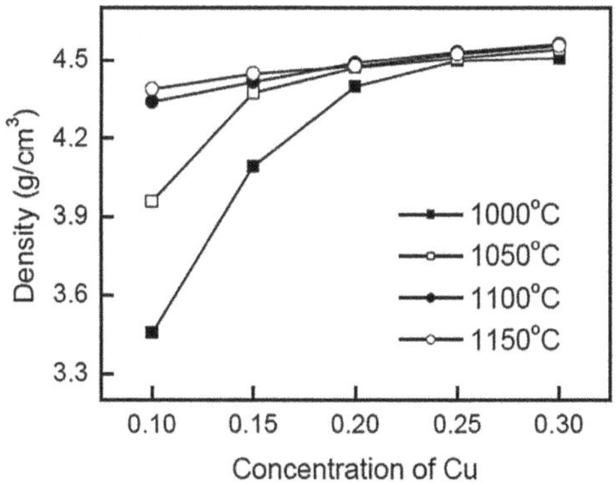

Figure 6.5. Densities of the $Mg_{1-x}Cu_xFe_{1.98}O_4$ ceramics as a function of Cu concentration after sintering at different temperatures. Reproduced with permission from [5]. Copyright 2007 Wiley.

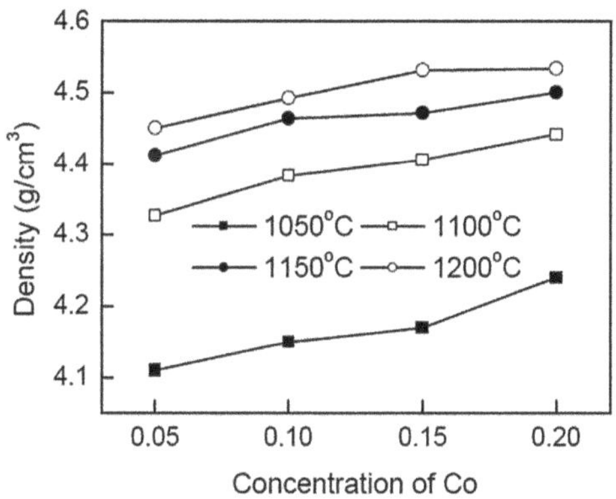

Figure 6.6. Densities of the $Mg_{0.90-x}Co_xCu_{0.10}Fe_{1.98}O_4$ ceramics as a function of Co concentration after sintering at different temperatures. Reproduced with permission from [5]. Copyright 2007 Wiley.

$MgFe_{1.98}O_4$ by more than 200 °C. The density measurement results were in a good agreement with the densification curves.

For the group of $Mg_{1-x}Cu_xFe_{1.98}O_4$ with $x \geqslant 0.20$, the densities of the samples sintered at 1150 °C were slightly lower than those of the samples sintered at 1100 °C. In addition, the bulk shrinkage was decreased slightly with increasing content of CuO after sintering at temperatures ⩾1100 °C. This observation was consistent with the linear shrinkage results, i.e. there is a critical concentration of CuO at which densification was maximized. This result could be understood by considering the densification mechanism involving the liquid phase.

As seen in figure 6.6, in the group of $Mg_{0.90-x}Co_xCu_{0.10}Fe_{1.98}O_4$, the density was increased almost linearly with increasing content of CoO simply because the density of $CoFe_2O_4$ (5.334 g cm^{-3}, JCPDS No. 2–1045) was higher than that of $MgFe_2O_4$. All the samples could be sintered to relative densities of ⩾92%, after sintering at 1050 °C. The $Mg_{0.90-x}Co_xCu_{0.10}Fe_{1.98}O_4$ samples demonstrated a similar variation trend in density to that of the $Mg_{0.90}Cu_{0.10}Fe_{1.98}O_4$ samples, as a function of sintering

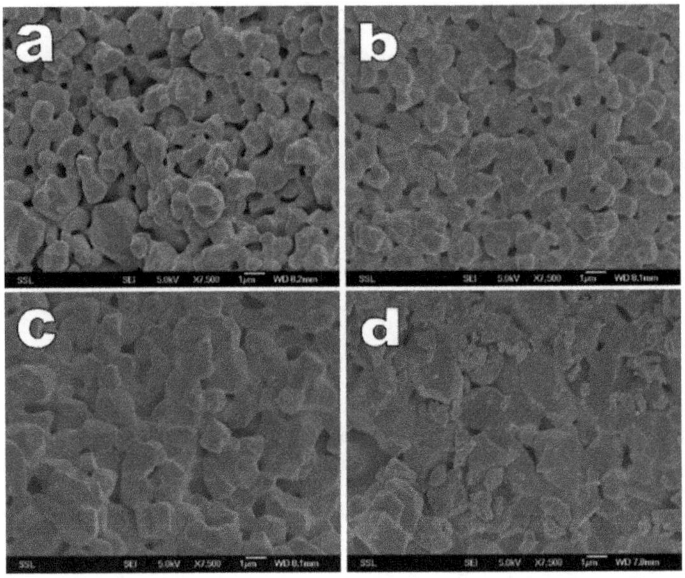

Figure 6.7. SEM images of the $MgFe_{1.98}O_4$ ceramics sintered for 2 h at different temperatures: (a) 1125 °C, (b) 1150 °C, (c) 1200 °C and (d) 1250 °C. Reproduced with permission from [5]. Copyright 2007 Wiley.

temperature. Therefore, the CuO played a dominant role in determining the densification of the $MgFe_{1.98}O_4$-based ceramics compared to CoO.

Figure 6.7 shows cross-sectional SEM images of the $MgFe_{1.98}O_4$ ceramics sintered at different temperatures for 2 h. Obviously, the $MgFe_{1.98}O_4$ sample sintered at 1125 °C was highly porous, with large pores uniformly distributed in the matrix. The porous microstructure was consistent with the density result, which was only 3.1 g cm^{-3}, corresponding to a relative density of 68%, as illustrated in figure 6.4. As the sintering temperature was increased to 1150 °C, a relatively dense microstructure was developed, but there were still a large number of pores to be observed. After sintering at 1200 °C for 2 h, the sample could reach a relative density of 90%, which meant that there was still about 10% porosity. In addition, the skeleton structures due to the presence of the point contacts indicated that the sintering process was carried out through a solid-state diffusion process. Therefore, it was concluded that $MgFe_{1.98}O_4$ has poor sinterability.

Along with a low densification rate, the $MgFe_{1.98}O_4$ sample also exhibited a relatively slow grain growth rate. Almost no grain growth was observed as the sintering temperature was increased from 1125 °C to 1150 °C. The average grain sizes of the samples sintered at 1150 °C, 1200 °C and 1250 °C were 1.5 μm, 1.9 μm and 2.8 μm, respectively. These values were smaller than those doped with CuO but sintered at lower temperatures by about an order of magnitude. The low grain growth rate of $MgFe_{1.98}O_4$ was a reflection of its poor densification behavior.

Figure 6.8 shows cross-sectional SEM images of the $Mg_{1-x}Cu_xFe_{1.98}O_4$ ceramics after sintering at 1000 °C for 2 h. Figure 6.9 shows cross-sectional SEM images of the $Mg_{0.90}Cu_{0.10}Fe_{1.98}O_4$ ceramics sintered at 1050 °C and 1150 °C. The average grain

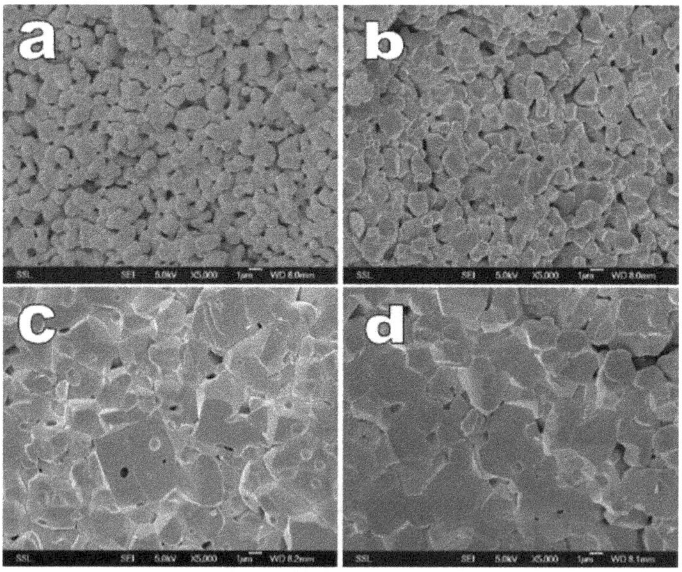

Figure 6.8. Cross-sectional SEM images of the $Mg_{1-x}Cu_xFe_{1.98}O_4$ ceramics sintered at 1000 °C for 2 h with different contents of CuO: (a) 0.10, (b) 0.15, (c) 0.20 and (d) 0.30. Reproduced with permission from [5]. Copyright 2007 Wiley.

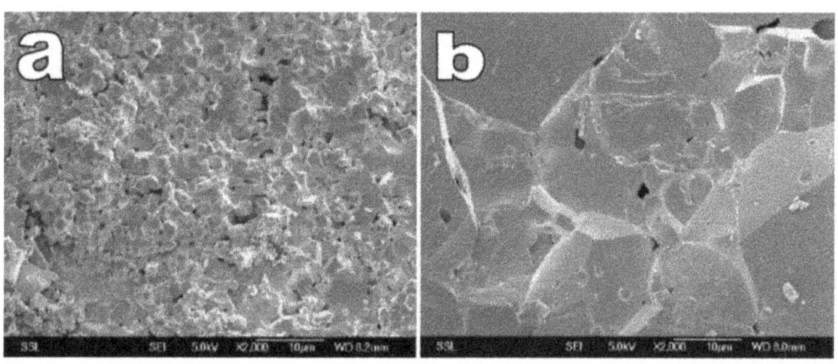

Figure 6.9. Cross-sectional SEM images of the $Mg_{0.90}Cu_{0.10}Fe_{1.98}O_4$ ceramics sintered for 2 h at different temperatures: (a) 1050 °C and (b) 1150 °C. Reproduced with permission from [5]. Copyright 2007 Wiley.

sizes of the samples sintered at different temperatures, as a function of the content of CuO, are depicted in figure 6.10. Compared to $MgFe_{1.98}O_4$, the $Mg_{1-x}Cu_xFe_{1.98}O_4$ samples exhibited a much higher densification rate and stronger grain growth tendency.

When comparing figures 6.7 and 6.8, one could easily find that the microstructures of the ferrite ceramics were significantly different with and without the introduction of CuO. First, as shown in figure 6.8(a), the $Mg_{0.90}Cu_{0.10}Fe_{1.98}O_4$ sample sintered at 1000 °C consisted of spherical grains. The point contact profile was absent in the samples doped with CuO, suggesting that the densification mechanism was varied. In this case, the liquid phase was formed during the sintering process of $Mg_{1-x}Cu_xFe_{1.98}O_4$. The average grain size of the $Mg_{0.90}Cu_{0.10}Fe_{1.98}O_4$

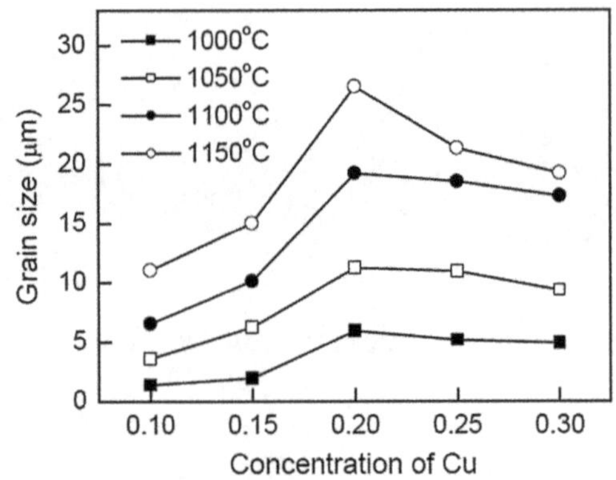

Figure 6.10. Average grain sizes of the $Mg_{1-x}Cu_xFe_{1.98}O_4$ ceramics as a function of the content of CuO after sintering at different temperatures. Reproduced with permission from [5]. Copyright 2007 Wiley.

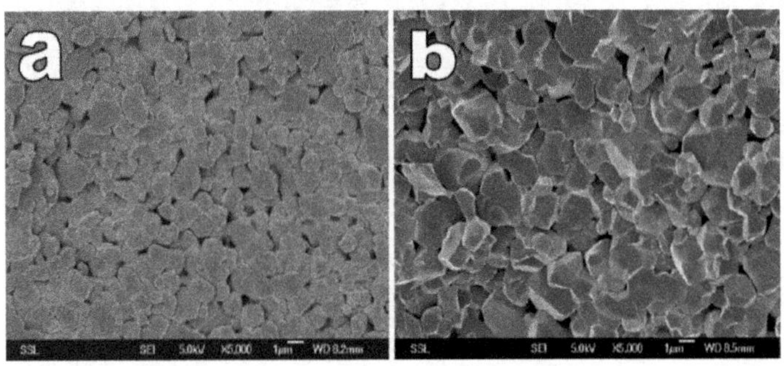

Figure 6.11. SEM images of the $Mg_{85}Co_{0.05}Cu_{0.10}Fe_{1.98}O_4$ ceramics sintered for 2 h at different temperatures: (a) 1050 °C and (b) 1100 °C. Reproduced with permission from [5]. Copyright 2007 Wiley.

sample sintered at 1000 °C was close to that of the $MgFe_{1.98}O_4$ sample sintered at 1125 °C. However, the samples with $x \geqslant 0.20$ possessed grains that were much larger than those of the $MgFe_{1.98}O_4$ sample sintered at 1250 °C. Figure 6.10 revealed that the average grain size of the $Mg_{0.90}Cu_{0.10}Fe_{1.98}O_4$ samples was monotonically increased with the content of Cu. The average grain size was maximized at the composition of $x = 0.20$.

Both the densification behavior and grain growth profile of the $Mg_{0.90-x}Co_x Cu_{0.10}Fe_{1.98}O_4$ samples were similar to those of the $Mg_{1-x}Cu_xFe_{1.98}O_4$ samples. Comparatively, the effect of CoO could be ignored. Figure 6.11 shows cross-sectional SEM images of the $Mg_{0.85}Co_{0.05}Cu_{0.10}Fe_{1.98}O_4$ ceramics sintered at different temperatures, while figure 6.12 depicts SEM images of the $Mg_{0.90-x}Co_x Cu_{0.10}Fe_{1.98}O_4$ sintered at 1200 °C with different contents of CoO. Figure 6.13 shows the average grain size of the samples as a function of sintering temperature. These results further confirmed that CuO played a dominant role in determining the densification and grain growth behaviors of the ferrite ceramics, compared to CoO.

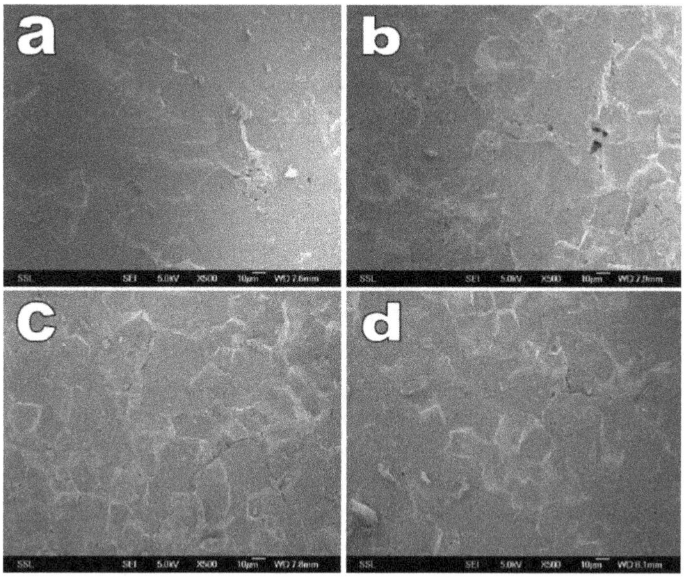

Figure 6.12. SEM images of the $Mg_{0.90-x}Co_xCu_{0.10}Fe_{1.98}O_4$ ceramics with different concentrations of CoO after sintering at 1200 °C for 2 h: (a) $x = 0.05$, (b) $x = 0.10$, (c) $x = 0.15$ and (d) $x = 0.20$. Reproduced with permission from [5]. Copyright 2007 Wiley.

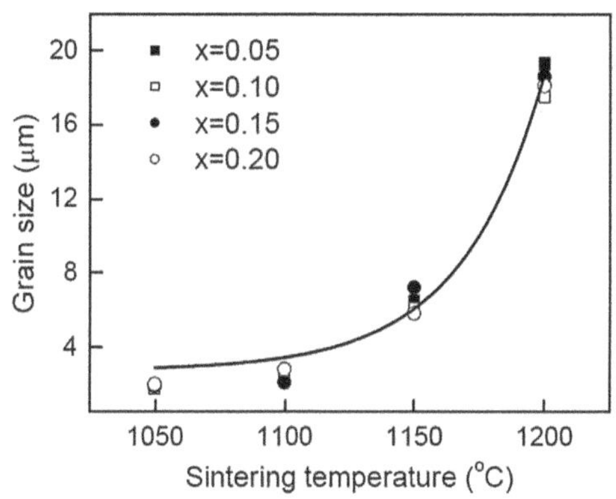

Figure 6.13. Grain size of the $Mg_{0.90-x}Co_xCu_{0.10}Fe_{1.98}O_4$ ceramics as a function of sintering temperature. Reproduced with permission from [5]. Copyright 2007 Wiley.

Due to the incorporation of CuO, $Mg_{1-x}Cu_xFe_{1.98}O_4$ experienced a different densification mechanism from $MgFe_{1.98}O_4$. It has been found that CuO is decomposed at about 1026 °C, forming Cu_2O with a low melting point of 1235 °C [5, 6]. Because Cu_2O was present, an eutectic Cu-rich liquid phase would be generated at temperatures that were lower than the melting point of Cu_2O. The Cu-rich liquid phase facilitates the liquid-phase sintering process. Generally, liquid-phase sintering has three stages: (i) rearrangement of grains/particles upon the formation of the liquid phase, (ii) the solution–precipitation process of solid phases and (iii) coalescence of the solid phase with the formation of skeletons [7].

Most likely, the three stages could not be clearly isolated, in particular between the rearrangement of grains/particles (densification) and the solution–precipitation process (grain growth). The temperature of rearrangement of grains/particles is much lower than that of the solid-state reaction sintering. The difference between liquid-phase sintering and solid-state reaction sintering was reflected by the difference in microstructure. Such a difference is well demonstrated by the SEM images of the $MgFe_{1.98}O_4$ and $Mg_{1-x}Cu_xFe_{1.98}O_4$ samples, as illustrated above in figures 6.6 and 6.7. The $MgFe_{1.98}O_4$ sintered at low temperatures contained pores, together with the rigid skeleton structure. In such a case, they were difficult to remove. In other words, densification was prohibited at low temperatures. The removal of the pores required the collapse of the skeleton structure, for which the temperature should be sufficiently high. In contrast, with the presence of the liquid phase, the grains/particles could be rearranged, and no skeleton structure was formed. As a result, the coalescence of pores became much easier, so that a rapid densification occurred at a relatively low temperature. Due to the difference in microstructure, these two groups of samples also have different electrical and dielectric properties, as discussed later [6].

Liquid-phase sintering also promoted grain growth due to enhanced mass transport [7–12]. As a result, the grain sizes of the $Mg_{1-x}Cu_xFe_{1.98}O_4$ samples were much larger than those of the $MgFe_{1.98}O_4$ samples. During liquid-phase sintering, the grains grow through the dissolution-precipitation process. Usually, smaller grains have higher ratios of surface area to volume, thus making them be at higher energy levels. As a result, the smaller the grains the more unstable they are. The smaller grains were preferentially dissolved into the liquid phase. As the dissolving concentration arrives at a critical value, these grains will be precipitated onto the larger grains. In other words, the larger grains will grow further at the expense of the smaller grains. In this case, the small grains need to cross the liquid-phase layer towards a certain large grain.

The results of figure 6.10 could be explained by considering the mechanism of liquid-phase sintering [13]. For the samples with $x < 0.20$, the content of the liquid phase was relatively low. As the content of CuO was increased, the coverage of the grains by the liquid-phase layer was enlarged. As a consequence, the rate of grain growth was increased gradually with increasing content of CuO. Once the critical value was arrived at, the surface of all grains was coated with the liquid-phase layer. In this case, further increase in the content of CuO resulted in thickening of the liquid-phase layers. In other words, the diffusion path of the precipitates was increased, so that it was more difficult for the smaller grains to approach the larger grains.

Therefore, after the critical content, the grain growth rate declined with increasing content of CuO. The fact that too high a content of the liquid phase retarded the densification of the samples to a certain degree is reflected by the densification behaviors and the density results, as discussed above. In addition, the grain size of the $Mg_{0.90-x}Co_xCu_{0.10}Fe_{1.98}O_4$ ceramics was slightly smaller than that of the

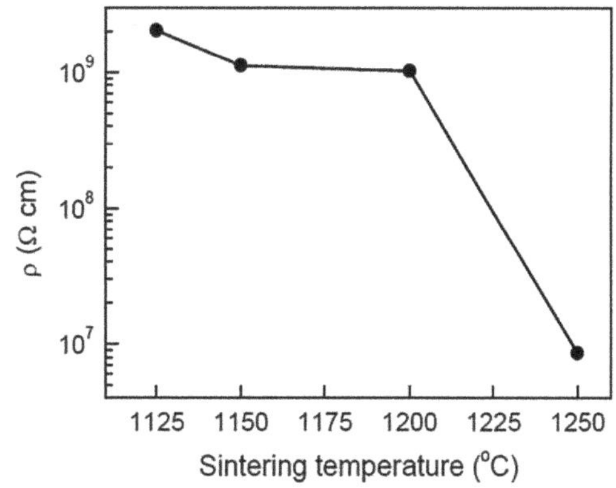

Figure 6.14. DC resistivities of the $MgFe_{1.98}O_4$ ceramics as a function of sintering temperature. Reproduced with permission from [6]. Copyright 2007 Wiley.

$Mg_{0.90}Cu_{0.01}Fe_{1.98}O_4$ samples after sintering at the same temperature, which suggested that the addition of CoO exerted a negative effect on the grain growth rate of the ferrite ceramics. However, the effect was very weak, due to the small concentrations of CoO used in the compositions.

Figure 6.14 shows the DC resistivities of the $MgFe_{1.98}O_4$ ceramics after sintering at different temperatures. The DC resistivity of the $MgFe_{1.98}O_4$ ceramics was reduced from 2×10^9 Ω·cm to 1×10^9 Ω·cm as the sintering temperature was raised from 1125 °C to 1200 °C. However, it was abruptly reduced to 8.6×10^6 Ω·cm as the sample was sintered at 1250 °C. The reduction was more than two orders of magnitude. The DC resistivities of the $MgFe_{1.98}O_4$ ceramics were very close to the values of other ferrite ceramics which could be found in the available literature [14–16].

The DC resistivities of the $Mg_{1-x}Cu_xFe_{1.98}O_4$ and $Mg_{0.90-x}Co_xCu_{0.10}Fe_{1.98}O_4$ ceramics are depicted in figures 6.15 and 6.16, respectively. The trend in DC resistivity of the $Mg_{1-x}Cu_xFe_{1.98}O_4$ versus the content of CuO was somehow complicated. For the samples sintered at 1000 °C, the DC resistivity was increased

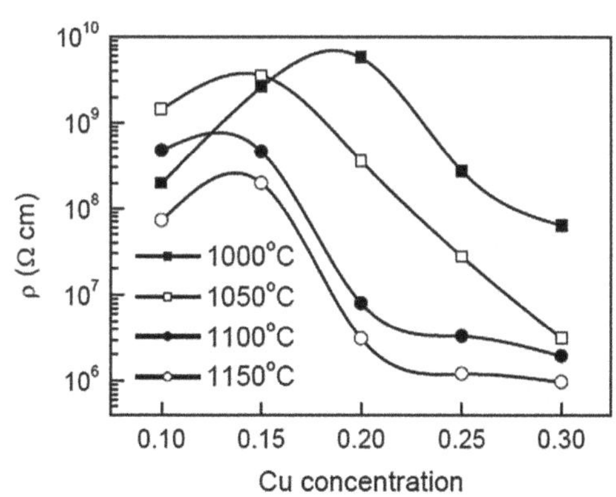

Figure 6.15. DC resistivities of the $Mg_{1-x}Cu_xFe_{1.98}O_4$ ceramics sintered at various temperatures as a function of the concentration of CuO. Reproduced with permission from [6]. Copyright 2007 Wiley.

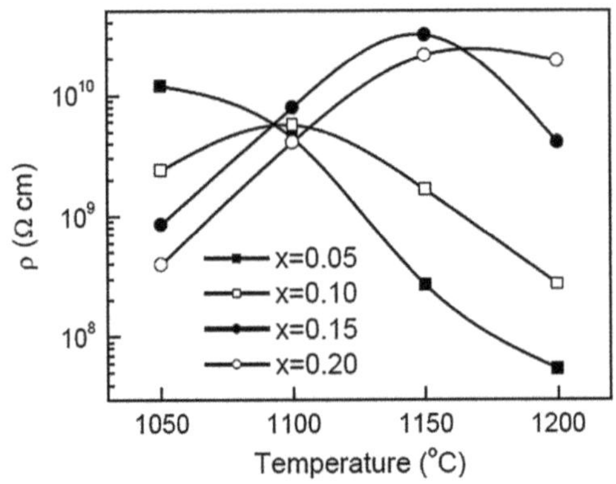

Figure 6.16. DC resistivities of the $Mg_{0.90-x}Co_xCu_{0.10}Fe_{1.98}O_4$ ceramics with different contents of CoO as a function of sintering temperature. Reproduced with permission from [6]. Copyright 2007 Wiley.

monotonically as the concentration was increased from $x = 0.10$ to $x = 0.20$. The resistivity value was maximized for the sample with $x = 0.20$. Thereafter, the resistivity declined with a further increase in the content of CuO. As the sintering temperature was increased, the critical content of CuO corresponding to the maximum resistivity was downshifted to the composition of $x = 0.15$. A similar variation trend was observed for other ferrite ceramics, such as Mg–Zn, Ni–Cu and Ni–Zn–Cu systems [14–16].

For the $Mg_{0.90-x}Co_xCu_{0.10}Fe_{1.98}O_4$ samples, the variations in DC resistivity with sintering temperature were different from one another. For example, the samples with $x = 0.05$ exhibited DC resistivity that was decreased with increasing temperature. In comparison, DC resistivity of the samples with $x = 0.20$ was increased as the sintering temperature was raised. However, for the samples with $x = 0.10$ and $x = 0.15$, the resistivities were maximized at temperatures of 1100 °C and 1150 °C, respectively. As for the effect of CoO, the DC resistivity of the samples sintered at 1050 °C was decreased with increasing content of CoO. Comparatively, the samples sintered at 1200 °C displayed an increasing trend with the content of CoO.

Figure 6.17 shows the complex permittivity curves of the $MgFe_{1.98}O_4$ ceramics after sintering at different temperatures. It was found that, if the samples were sintered at temperatures $\leqslant$1200 °C, both the real and imaginary permittivities were nearly constant over the frequency range. However, when the samples were sintered at 1250 °C for 2 h, both real and imaginary permittivities were exponentially increased versus frequency. At the same time, the values of permittivity were increased, compared to those of the samples sintered at low temperatures.

Figure 6.18 shows complex relative permittivity curves of the $Mg_{1-x}Cu_xFe_{1.98}O_4$ ceramics sintered at different temperatures. The sintering temperature had a similar effect on the variation trend of complex permittivity as for the $MgFe_{1.98}O_4$ samples. The complex permittivity curves of the $Mg_{1-x}Cu_xFe_{1.98}O_4$ ceramics sintered at 1150 °C displayed a similar variation trend to that of the $MgFe_{1.98}O_4$ sample sintered at 1250 °C, with sensitive responses to frequency. Those sintered at lower temperatures remained constant in the real and imaginary parts. Figure 6.19 depicts

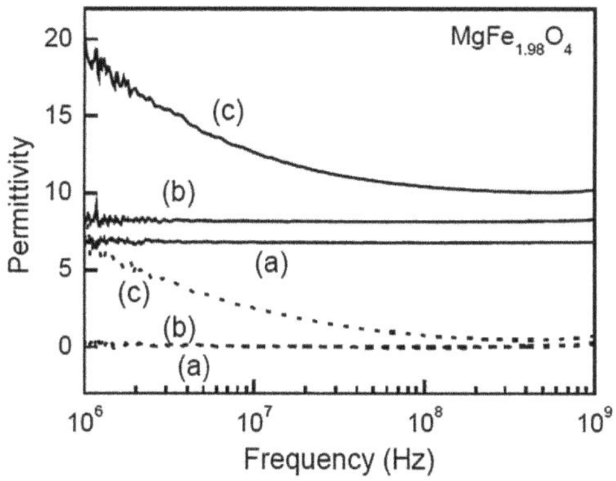

Figure 6.17. Complex permittivity curves of the $MgFe_{1.98}O_4$ ceramics after sintering for 2 h at different temperatures: (a) 1125 °C, (b) 1200 °C and (c) 1250 °C. Reproduced with permission from [6]. Copyright 2007 Wiley.

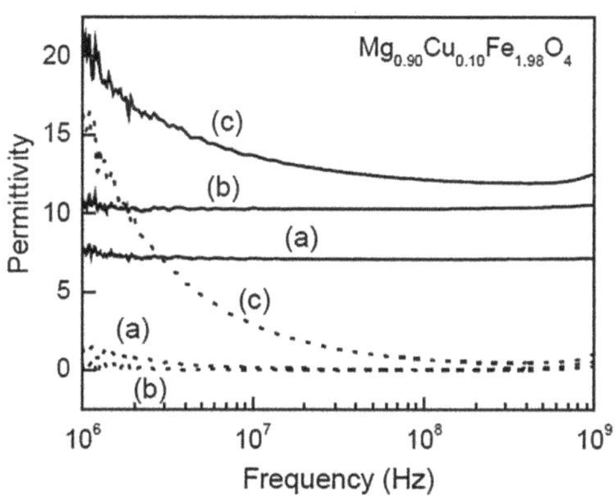

Figure 6.18. Complex permittivity curves of the $Mg_{0.90}Cu_{0.10}Fe_{1.98}O_4$ ceramics sintered at different temperatures: (a) 1000 °C, (b) 1100 °C and (c) 1150 °C. Reproduced with permission from [6]. Copyright 2007 Wiley.

the real permittivities of the $Mg_{1-x}Cu_xFe_{1.98}O_4$ ceramics, which were recorded at 10 MHz, as a function of sintering temperature. The real permittivity was monotonically increased with sintering temperature for this group of samples.

Figure 6.20 shows complex permittivity curves of the $Mg_{0.90-x}Co_xCu_{0.10}Fe_{1.98}O_4$ ceramics sintered at different temperatures. Real permittivities measured at 10 MHz, as a function of sintering temperature, are plotted in figure 6.21. At a given concentration of CoO, the real permittivity was increased with increasing sintering temperature. At a given sintering temperature, the value was decreased as the concentration of CoO was increased. Similarly, both real and imaginary permittivities were increased at low frequencies, which could be attributed to the dielectric loss related to the conduction of the samples.

The complex permeability curves of the $MgFe_{1.98}O_4$, $Mg_{0.90}Cu_{0.10}Fe_{1.98}O_4$ and $Mg_{0.85}Co_{0.05}Cu_{0.10}Fe_{1.98}O_4$ ceramics after sintering at different temperatures are

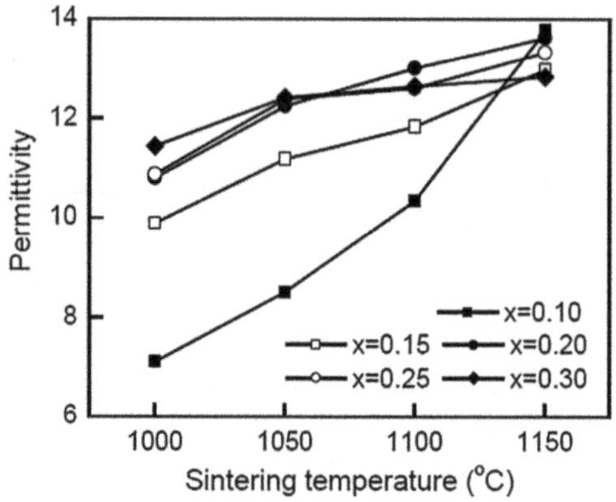

Figure 6.19. Real permittivity values measured at 10 MHz of the $Mg_{1-x}Cu_xFe_{1.98}O_4$ ceramics with different compositions as a function of sintering temperature. Reproduced with permission from [6]. Copyright 2007 Wiley.

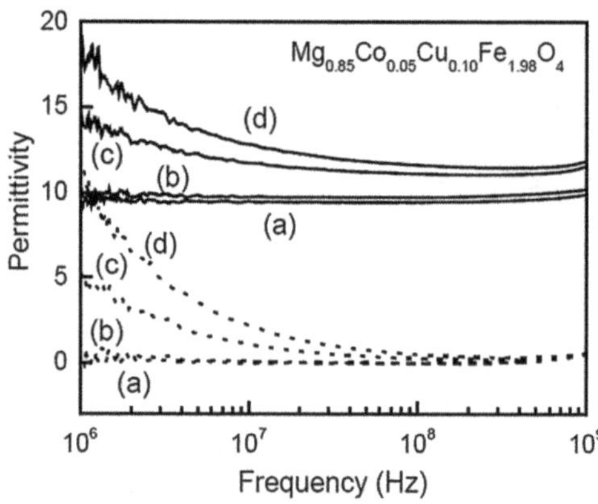

Figure 6.20. Complex permittivity curves of the $Mg_{085}Co_{0.05}Cu_{0.10}Fe_{1.98}O_4$ ceramics sintered at different temperatures: (a) 1050 °C, (b) 1100 °C, (d) 1150 °C and (d) 1200 °C. Reproduced with permission from [6]. Copyright 2007 Wiley.

illustrated in figures 6.22–6.24. In this case, the real permeability at 1 MHz was used to represent the static permeability (μ_0), because it was far below the resonance frequency [17, 18]. The resonance frequency (f_r) was that at which the magnetic loss peaked in the μ''–f curves. As demonstrated later, the resonance frequency of a given sample was dependent on both the intrinsic and extrinsic characteristics. Intrinsic properties include saturation magnetization, the gyromagnetic ratio and so on, while extrinsic properties include density and grain size, etc.

As shown in figure 6.22, the complex permeability curves of the $MgFe_{1.98}O_4$ samples possessed a similar profile. The real permeability (μ') remained constant until a certain frequency. After that, the real permeability started to increase. In addition, the imaginary permeability (μ'') was maintained at a low level concurrently with the real permeability. Similarly, a peak would be suddenly present. The

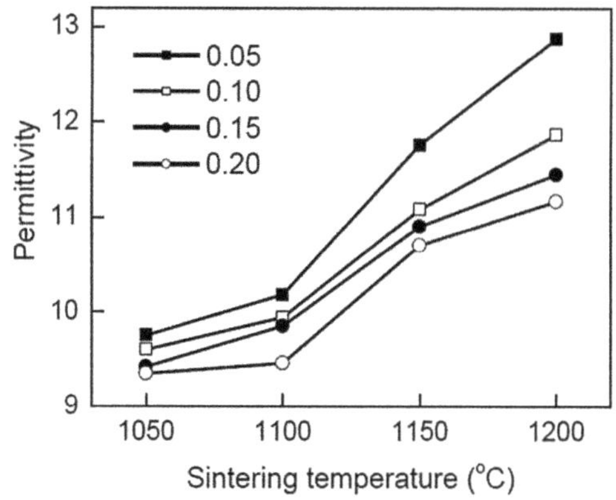

Figure 6.21. The real part of the relative permittivity values (at 10 MHz) of the $Mg_{0.90-x}Co_xCu_{0.10}Fe_{1.98}O_4$ ceramics with different contents of CoO as a function of sintering temperature. Reproduced with permission from [6]. Copyright 2007 Wiley.

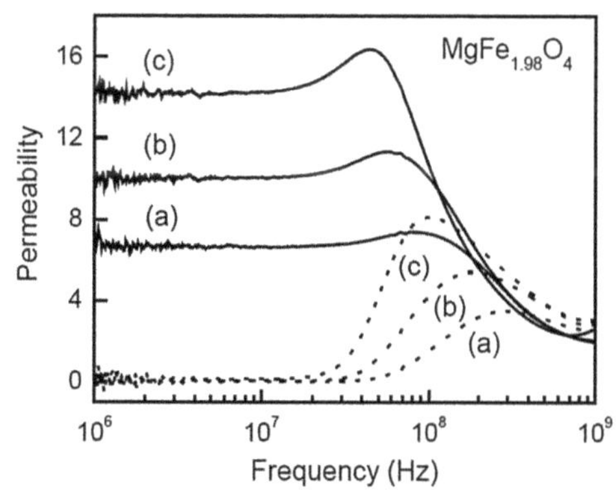

Figure 6.22. Complex relative permeability curves of $MgFe_{1.98}O_4$ ceramics sintering at different temperatures: (a) 1125 °C, (b) 1200 °C and (c) 1250 °C. Reproduced with permission from [6]. Copyright 2007 Wiley.

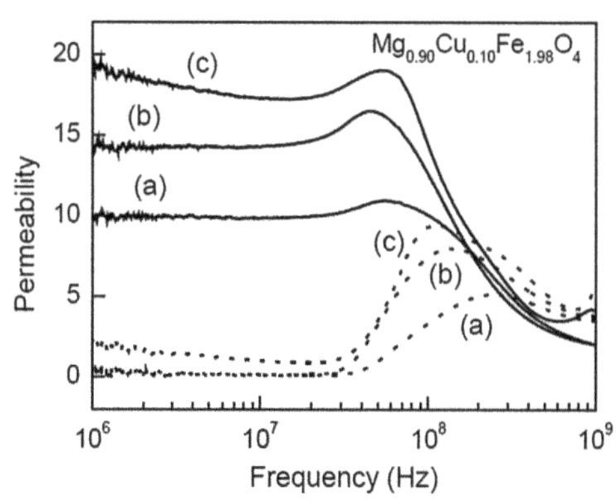

Figure 6.23. Complex relative permeability curves of the $Mg_{0.90}Cu_{0.10}Fe_{1.98}O_4$ ceramics sintered at different temperatures: (a) 1000 °C, (b) 1050 °C and (c) 1150 °C. Reproduced with permission from [6]. Copyright 2007 Wiley.

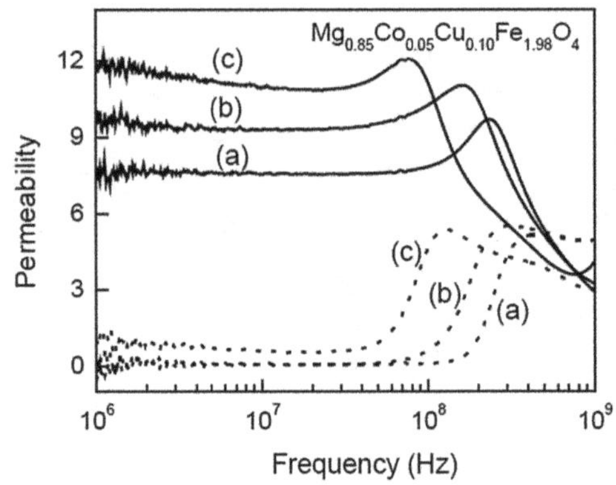

Figure 6.24. Complex permeability curves of the $Mg_{0.85}Co_{0.05}Cu_{0.10}Fe_{1.98}O_4$ ceramics sintered for 2 h at: (a) 1050 °C, (b) 1100 °C and (c) 1150 °C. Reproduced with permission from [6]. Copyright 2007 Wiley.

resonance frequencies were 260, 176 and 101 MHz, while the values of μ_0 were 6.6, 10 and 14.3, after the samples were sintered at 1125 °C, 1200 °C and 1250 °C, respectively. The imaginary permeability values at a low frequency of 10 MHz were 0.04, 0.06 and 0.11, corresponding to magnetic loss tangents given by $\tan\delta_\mu = \mu''/\mu'$, to be 5.9×10^{-3}, 6.2×10^{-3} and 8.1×10^{-3}, respectively. Therefore, the magnetic loss tangents of all the samples were lower than 10^{-2} at low frequencies, ensuring that they could meet the requirements for practical applications.

For the $Mg_{0.90}Cu_{0.10}Fe_{1.98}O_4$ ceramics, the complex permeability curves were similar, while the real permeabilities of the $Mg_{0.90}Cu_{0.10}Fe_{1.98}O_4$ ceramics were higher than those of the $MgFe_{1.98}O_4$ ceramics. The magnetic loss tangents were 8.9×10^{-3}, 9.6×10^{-3} and 5.8×10^{-2}. In this case, the sample sintered at 1150 °C had a magnetic loss tangent higher than 10^{-2}. Therefore, it could not be used for antenna miniaturization.

The addition of CoO resulted in a decrease in the real permeability and an increase in the resonance frequency of the ferrite ceramics, as shown in figure 6.24. In fact, the sample sintered at 1200 °C had real and imaginary permeability curves which were very close to those of the samples sintered at 1150 °C. In other words, the permeability of $Mg_{0.85}Co_{0.05}Cu_{0.10}Fe_{1.98}O_4$ was somehow saturated in the sample sintered at 1150 °C for 2 h. The real permeabilities at low frequencies were 7.62, 9.35 and 10.92, together with imaginary permeabilities of 0.51, 0.74 and 0.59. These corresponded to magnetic loss tangents of 6.6×10^{-3}, 7.9×10^{-3} and 5.4×10^{-2}, respectively. Therefore, the incorporation of CoO could also reduce the magnetic loss tangent at low frequencies. However, the effects of sintering temperature on the magnetic loss tangent in two groups of samples showed no significant differences. In other words, the function of CuO was stronger than that of CoO.

Figure 6.25 shows the impedance data of the $MgFe_{1.98}O_4$ ceramics in the frequency range 3–30 MHz (HF). The impedance of the sample sintered at 1125 °C was closest to 1, which is the impedance of free space. After sintering at 1150 °C, the impedance was increased to 1.1 and slightly dropped to 1.06 for the sample sintered at 1200 °C. The sample sintered at 1250 °C showed the largest change in impedance,

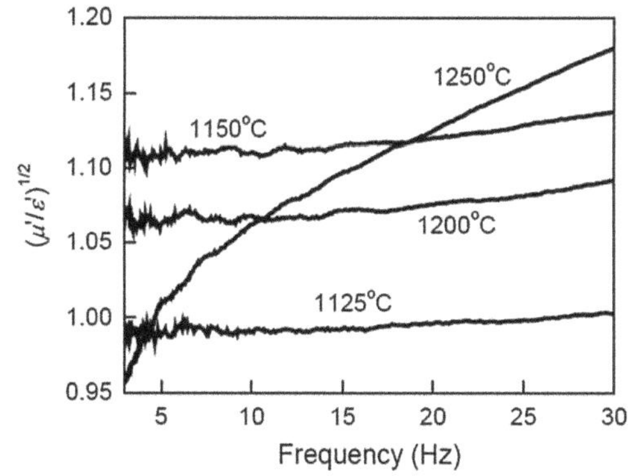

Figure 6.25. Impedance curves of the $MgFe_{1.98}O_4$ ceramics sintered at different temperatures. Reproduced with permission from [6]. Copyright 2007 Wiley.

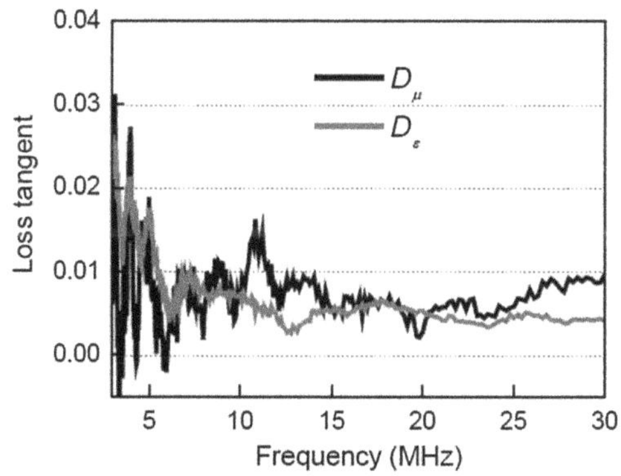

Figure 6.26. Magnetic and dielectric loss tangents of the $MgFe_{1.98}O_4$ ceramics sintered at 1125 °C for 2 h in the frequency range 3–30 MHz. Reproduced with permission from [6]. Copyright 2007 Wiley.

reflecting the large variation of the real permeability and permittivity. Therefore, the sintering temperature was an important processing parameter to obtain ferrite ceramics with the desired magnetic and dielectric properties in terms of real applications.

Figure 6.26 shows the magnetic and dielectric loss tangents of the $MgFe_{1.98}O_4$ sample sintered at 1125 °C in the frequency range 3–30 MHz. Both the magnetic and dielectric loss tangents were less than 10^{-2}, which ensured potential application in the HF band. Although some $Mg_{1-x}Cu_xFe_{1.98}O_4$ samples exhibited close values of real permeability and permittivity in the HF band, their dielectric loss tangents were too high for real applications. The impedance data of the $Mg_{0.85}Co_{0.05}Cu_{0.10}Fe_{1.98}O_4$ sample are shown in figure 6.27, while its dielectric and magnetic loss tangents are depicted in figure 6.28. The sample sintered at 1100 °C exhibited interesting magnetic and dielectric properties in the frequency range of 3–30 MHz. The $Mg_{0.85}Co_{0.05}Cu_{0.10}Fe_{1.98}O_4$ sample had higher real permeability and permittivity than the $MgFe_{1.98}O_4$ sample, thus having higher effectiveness in antenna miniaturization.

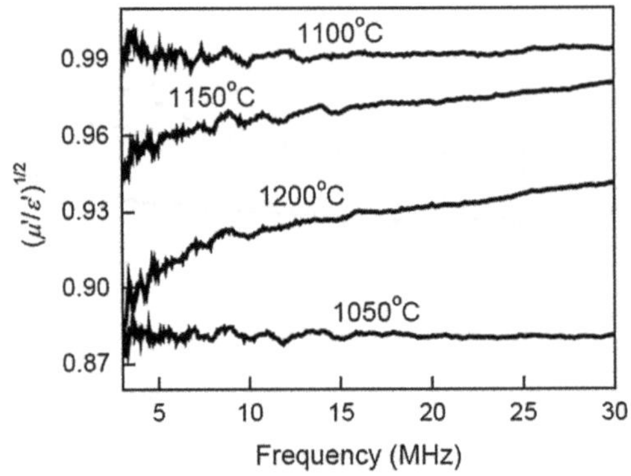

Figure 6.27. Impedance data of the $Mg_{0.85}Co_{0.05}Cu_{0.10}Fe_{1.98}O_4$ ceramics sintered at various temperatures for 2 h in the frequency range 3–30 MHz. Reproduced with permission from [6]. Copyright 2007 Wiley.

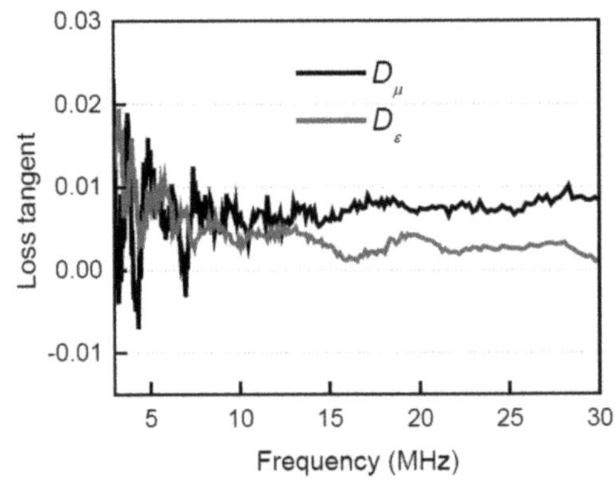

Figure 6.28. Magnetic and dielectric loss tangents of $Mg_{0.85}Co_{0.05}Cu_{0.10}Fe_{1.98}O_4$ ceramics sintered at 1100 °C for 2 h in the range 3–30 MHz. Reproduced with permission from [6]. Copyright 2007 Wiley.

DC resistivity is one of the most important parameters to reflect the quality of ferrite ceramics. For many applications, the DC resistivity should be as high as possible. Usually, the DC resistivity of ferrite ceramics is dependent on various factors, including composition, density, grain size and size distribution, microstructure uniformity and impurities. Ferrite ceramics consist of conductive grains that are isolated by highly resistive grain boundaries. Generally, it is difficult to achieve 100% densification, i.e. there are always pores with different sizes and shapes. Normally, the ferrites would have high DC resistivity if they have closed pores. However, open pores would adsorb impurities and water vapor, thus leading to low resistivity. The DC resistivities of ferrite ceramics are also dependent on the valence state of iron. Depending on the environment and conditions, Fe^{3+} could be reduced to Fe^{2+}. Once Fe^{2+} is formed, the resistivity will be reduced. If the content of Fe^{2+} reached 0.3%, the DC resistivity of the ferrites could be reduced by nearly two orders of magnitude [19].

The reduction in DC resistivity of the $MgFe_{1.98}O_4$ ceramics, as the sintering temperature was raised from 1125 °C to 1200 °C, was attributed to the increase in densification and grain size. After sintering at 1250 °C, the DC resistivity was abruptly decreased, because Fe^{2+} was formed during the high temperature sintering. The resistivity was reflected by the permittivity results.

The DC resistivities of ferrite ceramics have been widely studied [20, 21]. However, it was found that the DC resistivity of $Ni_{1-x}Cu_xFe_2O_4$ was monotonically decreased, with increasing content of CuO in the range $x = 0$–0.5 [20]. $Ni_{0.34}Zn_{0.66-x}Cu_xFe_2O_4$ had similar results [21]. However, the observations were explained differently. In the former case, the decrease in DC resistivity of $Ni_{1-x}Cu_xFe_2O_4$ was ascribed to the decreased porosity caused by the increased content of CuO [20]. In the other case, the porosity was increased with increasing content of CuO [21]. However, the samples had relatively large pores and the pores were mostly at the grain boundaries. As a result, the resistivity should not be influenced by the porosity.

In general, the DC resistivity of ferrite ceramics is increased with decreasing grain size and density, which could be used to explain the DC resistivities of the $MgFe_{1.98}O_4$ samples sintered at temperatures of 1125 °C–1200 °C. However, the variation in resistivity of the $Mg_{1-x}Cu_xFe_{1.98}O_4$ samples with compositions lower than the critical concentration could be explained. The critical concentration regarding the resistivity should be correlated to the microstructures. Although $MgFe_{1.98}O_4$ and $Mg_{1-x}Cu_xFe_{1.98}O_4$ were both porous, they possessed different microstructures. The $MgFe_{1.98}O_4$ samples exhibited a rigid skeleton structure due to the solid-state diffusion sintering mechanism. In comparison, the $Mg_{1-x}Cu_xFe_{1.98}O_4$ samples contained spherical grains owing to the liquid-phase sintering behavior.

As a result, the pores of the $MgFe_{1.98}O_4$ samples were closed, but those of the $Mg_{1-x}Cu_xFe_{1.98}O_4$ samples were open. In this case, the $MgFe_{1.98}O_4$ samples had relatively higher resistivities than the $Mg_{1-x}Cu_xFe_{1.98}O_4$ samples with $x = 0.10$–0.20. For the samples with concentrations over the critical point, full densification was achieved, so that the effect of pores disappeared. In addition, the higher the sintering temperatures, the lower the critical concentration of CuO. This was simply because high temperatures resulted in earlier densification.

The incorporation of CoO could increase the DC resistivity of the ferrite ceramics. For example, the samples sintered at 1050 °C and 1100 °C exhibited DC resistivities that were higher than those without the presence of CoO by almost one order of magnitude, because the conductive mechanism of the ferrite was changed. In the ferrites containing CoO, *p*-type conduction could be present through the hole migration between cobalt ions of Co^{2+} and Co^{3+} [22, 23]. Therefore, the normal *n*-type conduction of the ferrites was compensated, thus leading to an increase in DC resistivity. In the samples with relatively low contents of CoO, e.g. $x = 0.05$, the effect of CuO was much more pronounced. As a result, the DC resistivities of the $Mg_{0.85}Co_{0.05}Cu_{0.10}Fe_{1.98}O_4$ and $Mg_{0.90}Cu_{0.10}Fe_{1.98}O_4$ samples with sintering temperature were nearly the same. In addition, the temperature for the decline of the DC resistivity was increased as the content of CoO was increased. This suggested that the compensation effect due to the *p*-type conduction related to cobalt ions was gradually enhanced.

The main contributors to high frequency permittivity of ferrite single crystals are atomic and electronic polarizations. However, due to their polycrystalline characteristics, the permittivity of ferrite ceramics is also dependent on their microstructure, grain size, density and impurities. The dependence of permittivity of ferrite ceramics on grain size can be well described using the Maxwell–Wagner effect [19, 24]. In that model, ferrite ceramics are considered to consist of highly conductive grains and high insulating grain boundaries. Therefore, as the grain size of a ferrite ceramic is increased, the volume fraction of grain boundaries will be reduced, so that the permittivity will be increased. Similar to DC conductivity, the effect of pores on permittivity is dependent on the properties of the pores. Closed pores will lead to low permittivity, because the permittivity of vacuum/air is close to 1. In contrast, open pores might adsorb impurities, e.g. water vapor or dust, thus leading high permittivity, since the permittivity of water is quite high. Some defects in the crystal structure could also cause an increase in permittivity, as the presence of defects could increase polarizations.

If Fe^{2+} is formed in ferrite ceramics, their permittivity will be increased, because the polarization of Fe^{2+} is larger than that of Fe^{3+}. The Fe^{3+} ion has a stable d-shell configuration according to Hund's rule, so that the charge cloud possesses a spherical symmetry. In comparison, the Fe^{2+} ion has one more electron than Fe^{3+}. Therefore, the symmetry of the electron cloud is broken [19]. In other words, the formation of Fe^{2+} will increase the polarization of the ferrites, thus leading to high permittivity, as the concentration of Fe^{2+} ions is sufficiently high. In addition, the formation of Fe^{2+} corresponds to high conductivity or low resistivity, i.e. high permittivity is usually accompanied by low resistivity in ferrite ceramics.

The dielectric loss tangent of ferrite ceramics is a measure of the lag in polarization against the alternative electric fields, which is dependent on various parameters, such as the content of impurities, the defects in the crystal structure, high conductivity and so on [19]. Among them, the effect of conduction loss was strongest on the dielectric loss tangent of ferrite ceramics, in particular that caused by the electron hopping between Fe^{2+} and Fe^{3+} ions. Such a loss is mainly observed at low frequencies. In this regard, the high permittivity contributed by the formation of Fe^{2+} ions is not preferred, simply because such a high permittivity usually means that the materials will have very high dielectric loss tangents. High dielectric loss should be avoided in most device designs.

The variation trend in the complex permittivity of the $MgFe_{1.98}O_4$ ceramics as a function of sintering temperature can be understood by using an explanation of their DC resistivities. The slight increment in real permittivity as the sintering temperature was increased from 1125 °C to 1200 °C was attributed to the enhanced densification and grain growth of the samples. The abrupt increase in both real and imaginary permittivities at low frequencies was ascribed to the formation of Fe^{2+} ions during the sintering at the high temperatures. In this respect, ferrite ceramics should be sintered at as low a temperature as possible in order to avoid high dielectric loss tangents.

The variation trends in permittivity of the $Mg_{0.90}Cu_{0.10}Fe_{1.98}O_4$ and $MgFe_{1.98}O_4$ samples were different in two aspects. On one hand, the permittivity values of the

$Mg_{0.90}Cu_{0.10}Fe_{1.98}O_4$ samples were higher than those of the $MgFe_{1.98}O_4$ samples. This difference could be explained in terms of microstructures and grain growth rates. Obviously, the $Mg_{0.90}Cu_{0.10}Fe_{1.98}O_4$ samples exhibited higher densification and grain growth rates that the $MgFe_{1.98}O_4$ samples. As stated before, large grain size corresponded to a low fraction of grain boundaries, thus leading to high permittivity of ferrite ceramics [25]. $Mg_{0.90}Cu_{0.10}Fe_{1.98}O_4$ had a larger grain size and the grains were more conductive, compared to $MgFe_{1.98}O_4$. As a result, the $Mg_{0.90}Cu_{0.10}Fe_{1.98}O_4$ sample had relatively higher permittivity.

On the other hand, both the permittivity values and dielectric loss tangent were quickly increased at the low frequency end, as the sintering temperature was raised from 1200 °C to 1250 °C and 1100 °C to 1150 °C, for the $MgFe_{1.98}O_4$ and $Mg_{1-x}Cu_x Fe_{1.98}O_4$ samples, respectively. It was related to the presence of Fe^{2+} ions. In other words, Fe^{2+} was formed at a lower temperature in the $Mg_{1-x}Cu_xFe_{1.98}O_4$ samples, owing mainly to the addition of CuO. The formation mechanism of Fe^{2+} ions should be employed to interpret this observation. During the sintering at high temperatures, Fe^{2+} ions were formed due to the reduction of Fe^{3+} ions, if local oxygen was deficient because of the dissociation from the crystal lattice. Although the Fe^{2+} ions could be recovered to Fe^{3+} ions during the cooling process, the oxidation could be completely achieved, thus leaving Fe^{2+} ions in the ferrite ceramics at room temperature. Due to its low densification rate, the $MgFe_{1.98}O_4$ ceramics were still of certain porosity after sintering at 1200 °C so that oxygen gas in the air could approach the Fe^{2+} ions for oxidation. In comparison, the densification of the samples doped with CuO was already finished after sintering at 1150 °C. As a result, the Fe^{2+} ions could not be completely oxidized, because it was more difficult for oxygen gas to diffuse in.

At the low frequency of 10 MHz, the real permittivity of the $Mg_{0.90}Cu_{0.10}Fe_{1.98}O_4$ ceramics was increased as the content of CuO and sintering temperature were increased. The dielectric loss tangent was minimized for the samples with $x = 0.10$ and $x = 0.15$ after they were sintered at 1100 °C and 1050 °C, respectively. The increase in real permittivity could be attributed to the increase in grain size and the enhancement in densification, due to the doping of CuO. The trend of the dielectric loss tangent was actually close to the DC resistivities. In other words, the dielectric loss tangent of ferrite ceramics was most likely caused by the conduction loss. There was no critical content of CuO for real permittivity, which might be ascribed to the fact that the effect of porosity on real permittivity was much weaker than that on DC resistivity.

The permittivity of the $Mg_{0.90-x}Co_xCu_{0.10}Fe_{1.98}O_4$ ceramics was concurrently dependent on the contents of CuO and CoO. For example, the increase in real and imaginary permittivities at low frequencies was present in the $Mg_{0.85}Co_{0.05}Cu_{0.10} Fe_{1.98}O_4$ sample sintered at 1150 °C and the $Mg_{0.80}Co_{0.10}Cu_{0.10}Fe_{1.98}O_4$ sample sintered at 1200 °C. However, such an observation was not present in the samples with compositions of $x \geqslant 0.15$. As the sintering temperature was fixed, the permittivity of the $Mg_{0.90-x}Co_xCu_{0.10}Fe_{1.98}O_4$ sample was nearly linearly increasing with increasing content of CoO. In comparison, the effect of sintering temperature on the complex permeability of the three groups of samples was stronger than those of the contents of CuO and CoO.

As shown in figures 6.22–6.24, the magnetic properties of the samples exhibited several interesting characteristics as a function of sintering temperature. First, the static permeability (μ_0) of the samples was significantly increased. Second, the resonance frequency (f_r) was downshifted with increasing sintering temperature. Furthermore, a low content of CoO could lead to a great increase in f_r. In addition, the real permeability decreased accordingly.

The effect of pores on the permeability of ferrite materials was complicated. With the presence of pores, magnetic poles could be produced on the surface of ferrite grains when an external magnetic field is applied. As a result, demagnetizing fields were generated, thus resulting in a reduction in the real permeability. The magnitude of the demagnetizing field was dependent on the grain size and grain boundary properties of the ferrite ceramics. According to the magnetic circuit model [25], the static permeability (μ_0) could be derived from

$$\mu_0 = \frac{\mu_i\left(1 + \dfrac{\delta}{D}\right)}{1 + \mu_i\dfrac{\delta}{D}}, \tag{6.1}$$

where μ_i is the intrinsic static permeability of the materials that are defect-free, while D and δ are the average grain size and the thickness of the grain boundaries which include pores and all non-magnetic phases. With this equation, the variation trend in μ_0 as a function of sintering temperature could be explained. As the sintering temperature was increased, both the grain size (D) and density (ρ) were increased, thus leading to a reduction in the ratio of δ/D. As a consequence, static permeability would be increased. The ratio of δ/D could be obtained with the measured densities of the samples, according to the following equation [26]:

$$\frac{\delta}{D} = \left(\frac{\rho_i}{\rho}\right)^{1/3} - 1, \tag{6.2}$$

where ρ_i is the theoretical density of the materials. In this case, the values of 4.52 g cm^{-3}, 4.62 g cm^{-3} and 4.65 g cm^{-3} were used for the $MgFe_{1.98}O_4$, $Mg_{0.90}Cu_{0.10}Fe_{1.98}O_4$ and $Mg_{0.85}Co_{0.05}Cu_{0.10}Fe_{1.98}O_4$ samples, respectively. According to equation (6.1), the static permeability (μ_0) of the samples sintered at different temperature could be derived. Figure 6.29 shows the fitting results as the dashed lines. Clearly, the fitting results were in good agreement with the measurement data of the three groups of ferrite ceramics.

Generally, the relationship between resonance frequency (f_r) and static permeability ($\mu_0 - 1$) for spinel ferrite ceramics could be described according to the Snoek-like law, given by

$$(\mu_0 - 1)f_r = C. \tag{6.3}$$

At natural resonance, there is $C = \frac{2}{3}\gamma M_s$, with γ and M_s being the gyromagnetic ratio and saturation magnetization, respectively [27]. At wall resonance, there is $C = C_1 M_s^2$, with C_1 being a constant dependent on the size of the domain and the thickness of the domain wall. As a result, as the sintering temperature was increased,

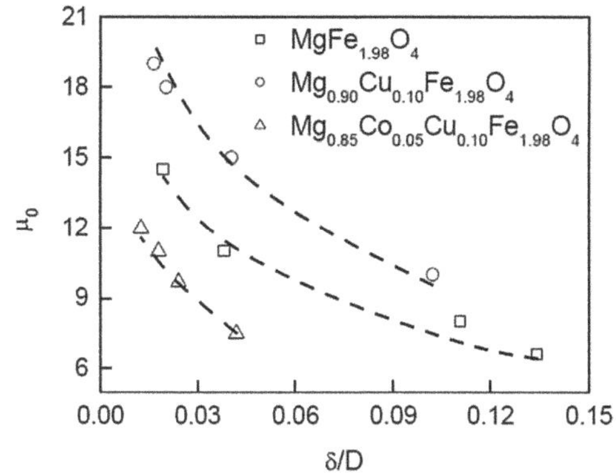

Figure 6.29. Static permeability of the three groups of ferrite ceramics, as a function of δ/D. Reproduced with permission from [6]. Copyright 2007 Wiley.

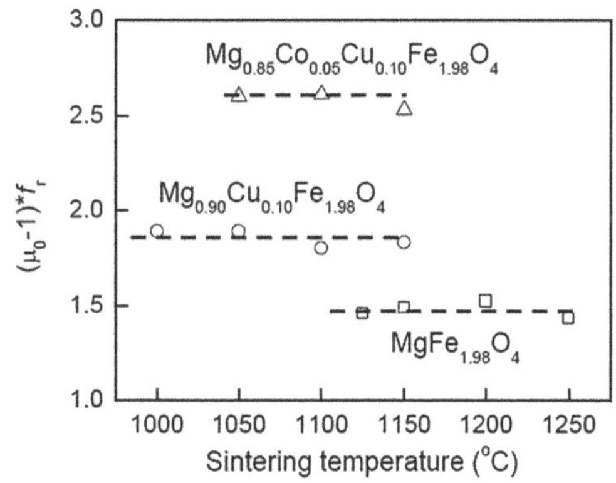

Figure 6.30. The product $[(\mu_0 - 1)*f_r]$ of the three groups of ferrite ceramics as a function of sintering temperature. Reproduced with permission from [6]. Copyright 2007 Wiley.

the grain size increased, thus leading to high static permeability. Accordingly, the resonance frequency was reduced. Figure 6.30 shows products of $(\mu_0 - 1)$ and f_r of the three groups of ferrite ceramics as a function of sintering temperature. In each group, the product was a constant for the samples sintered at different temperatures.

It has been demonstrated that the doping of CoO with a low concentration could result in enhanced microwave properties of ferrites [28]. Co^{2+} ions have a very high positive magnetocrystalline anisotropy. In contrast, the magnetocrystalline anisotropy of most cations in spinel ferrites is negative. Therefore, incorporation of Co^{2+} could alter the magnetocrystalline anisotropy of a given ferrite. For example, the magnetocrystalline anisotropy constant (K_1) of $(Ni_{0.2}Cu_{0.2}Zn_{0.6})_{1.02}Fe_{1.98}O_4$ was -2.7×10^4 erg cm^{-3}, while that of $(Ni_{0.2}Cu_{0.2}Zn_{0.6})_{0.97}Co_{0.05}Fe_{1.98}O_4$ became $+16.3 \times 10^4$ erg cm^{-3}, where the concentration of Co^{2+} was only 5 mol% [28].

As the absolute value of magnetocrystalline anisotropy is increased, the permeability will decrease and the resonance frequency will increase accordingly. For example, the f_r of the $Mg_{0.90}Cu_{0.10}Fe_{1.98}O_4$ sample was about 150 MHz, while the sample with the introduction of Co^{2+} had a resonance frequency of 400 MHz, but they were sintered at the same temperature, as seen in figures 6.24 and 6.23. Once the resonance frequency is increased, the low magnetic loss tangent could be achieved at higher frequencies. As a result, the ferrite ceramics could be used for high frequency applications. Generally, increasing the resonance frequency of magnetic materials is always a challenge.

6.3 Bi_2O_3 doped $MgFe_{1.98}O_4$ ferrite ceramics

$MgFe_{1.98}O_4$ based ceramics doped with Bi_2O_3 as a sintering aid have been explored to develop magneto-dielectric materials [29]. The concentration of Bi_2O_3 was up to 10 wt%, while the sintering temperature was in the range 900 °C–1150 °C. The starting materials were commercial oxide powders without any treatment. The powders were mixed and milled with WC milling media. The activated mixtures were directly used to prepare the ferrite ceramics. The phase composition, microstructural evolution, grain growth behavior, and electrical, dielectric and magnetic properties will be presented and discussed, with the aim to identify candidates with promising magneto-dielectric properties.

Figure 6.31 shows the XRD patterns of the $MgFe_{1.98}O_4$ samples doped with different contents of Bi_2O_3, which were sintered at 1000 °C for 2 h. All samples were of a single phase of spinel structure, while Bi_2O_3 was not clearly detected. In other words, the phase formation of $MgFe_{1.98}O_4$ was not influenced by the presence of Bi_2O_3. The absence of Bi_2O_3 in the XRD patterns could be attributed to the fact the corresponding volume and molar concentrations were only about 5% and 4.3%, respectively, for the weight content of 10% Bi_2O_3, as listed in table 6.2. By using high-energy activation, the starting oxide powders were homogeneously mixed and significantly refined. As a result, the phase formation of the ferrite was enhanced and the reaction temperature was decreased.

Figure 6.32 shows the densities of the $MgFe_{1.98}O_4$ ceramics doped with different concentrations of Bi_2O_3 as a function of sintering temperature. Theoretical densities of the samples with different contents of Bi_2O_3 were obtained from the theoretical densities of $MgFe_2O_4$ (4.523 g cm^{-3}, JCPDS No. 17–464) and Bi_2O_3 (8.93 g cm^{-3}, JCPDS No. 2–498), which are listed in table 6.3.

Full densification of $MgFe_{1.98}O_4$ could not be achieved when the sintering temperature was ⩽1150 °C. The sample sintered at 1150 °C had a density of 3.93 g cm^{-3}, corresponding to a relative density of 87%. In comparison, the samples doped with 0.5 wt% and 1 wt% Bi_2O_3 exhibited a relative density of about 95% after sintering at 1150 °C. The relative density of the sample doped with 2 wt% Bi_2O_3 was 96.1% after sintering at 1000 °C. When the concentration of Bi_2O_3 was ⩾3%, the samples were fully densified even after sintering at a low temperature of 950 °C. For fully densified samples, the density increased with increasing content of Bi_2O_3, because the density of Bi_2O_3 was higher than that of $MgFe_{1.98}O_4$. In other words, Bi_2O_3 was an effective sintering aid to promote the densification of $MgFe_{1.98}O_4$.

Cross-sectional SEM images of the undoped $MgFe_{1.98}O_4$ samples and those doped with 0.5 wt%, 1 wt% and 2 wt% Bi_2O_3, sintered at different temperatures, are

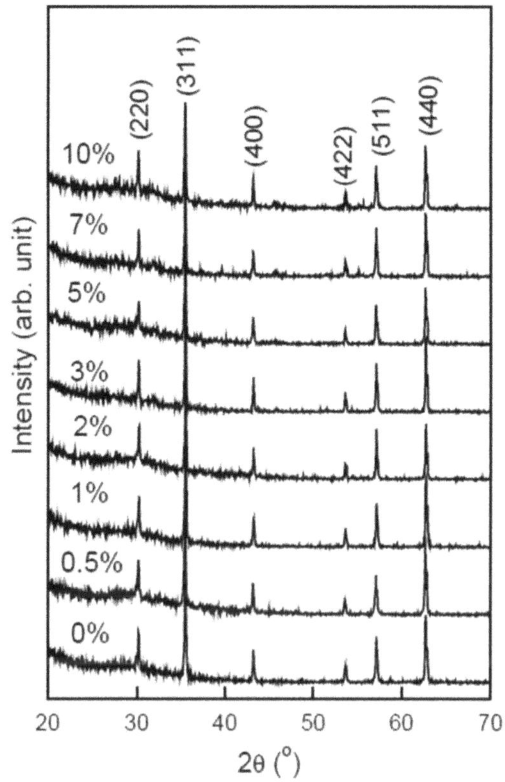

Figure 6.31. XRD patterns of the samples doped with different concentrations of Bi_2O_3 after sintering at 1000 °C for 2 h.

Table 6.2. Concentrations of Bi_2O_3 in weight, volume and molar percentages of the sample.

wt%	0.5	1	2	3	5	7	10
vol%	0.25	0.51	1.02	1.52	2.54	3.55	5.08
mol%	0.21	0.43	0.86	1.29	2.15	3.01	4.29

depicted in figures 6.33–6.36. Figure 6.37 shows cross-sectional SEM images of the $MgFe_{1.98}O_4$ ceramics with high concentrations of Bi_2O_3 in the range 3–10 wt% sintered at 1100 °C for 2 h. Figure 6.38 shows the average grain sizes of the samples, derived from the SEM images, as a function of sintering temperature.

As seen in figure 6.33, the $MgFe_{1.98}O_4$ ceramics without the doping of Bi_2O_3 possessed porous microstructures after sintering at a temperature $\leqslant$1100 °C. They consisted of loosely packed spherical grains, thus corresponding to relatively low densities. In addition, the grain growth rate was quite low with increasing sintering temperature. The sample sintered at 1100 °C exhibited an average grain size of about 1 μm. After sintering at 1150 °C for 2 h, the sample was densified to a certain degree.

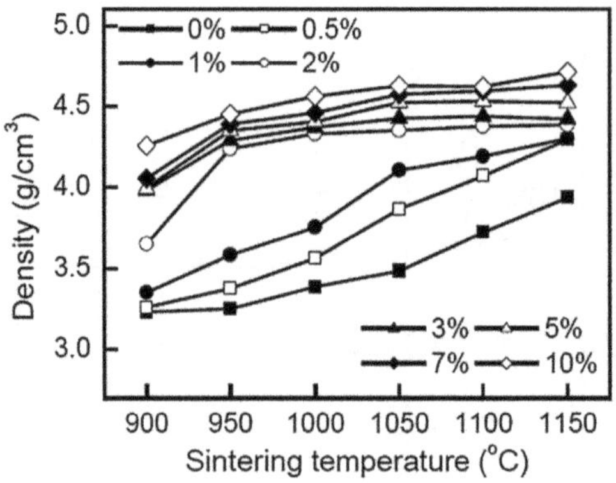

Figure 6.32. Densities of the $MgFe_{1.98}O_4$ ceramics with different contents of Bi_2O_3 as a function of sintering temperature.

Table 6.3. Theoretical densities of the $MgFe_{1.98}O_4$ ferrite ceramics with different concentrations of Bi_2O_3.

wt%	0.5	1	2	3	5	7	10
ρ_d (g cm^{-3})	4.53	4.54	4.56	4.59	4.63	4.67	4.73

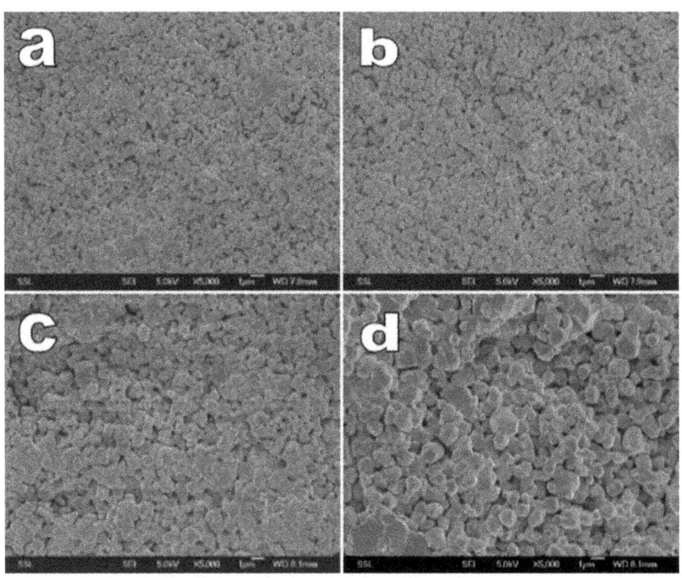

Figure 6.33. Cross-sectional SEM images of the undoped $MgFe_{1.98}O_4$ ceramics sintered for 2 h at different temperatures: (a) 1000 °C, (b) 1050 °C, (c) 1100 °C and (d) 1150 °C.

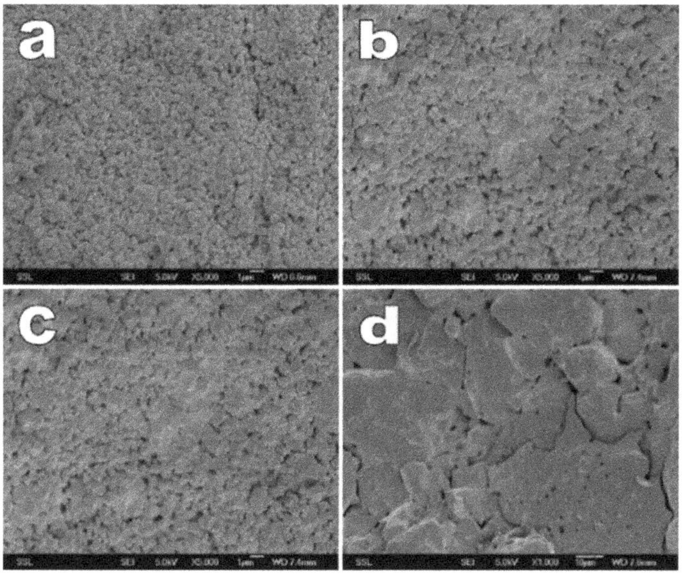

Figure 6.34. Cross-sectional SEM images of the $MgFe_{1.98}O_4$ ceramics doped with 0.5 wt% Bi_2O_3 after sintering for 2 h at different temperatures: (a) 1000 °C, (b) 1050 °C, (c) 1100 °C and (d) 1150 °C.

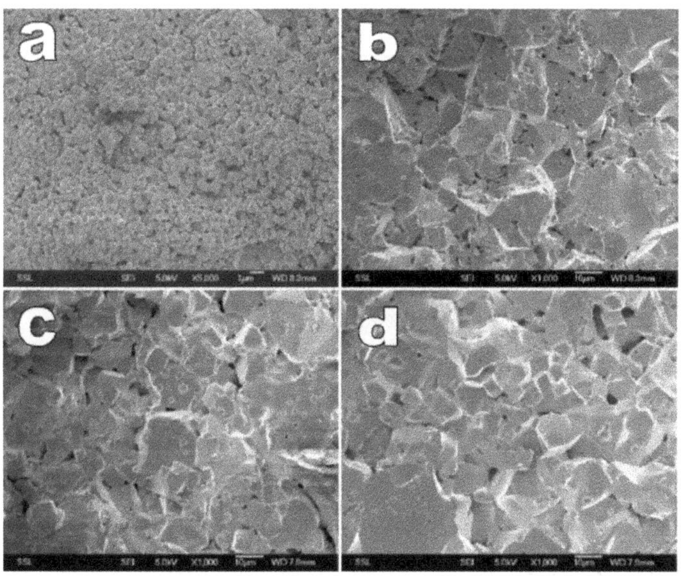

Figure 6.35. Cross-sectional SEM images of the $MgFe_{1.98}O_4$ ceramics doped with 1 wt% Bi_2O_3 after sintering for 2 h at different temperatures: (a) 1000 °C, (b) 1050 °C, (c) 1100 °C and (d) 1150 °C.

At the same time, the average grain size was increased by about two times compared to that of the sample sintered at 1100 °C. In addition, more faceted grains were present in the sample sintered at 1150 °C. Also, the fracture of all the samples was of an inter-grain profile, suggesting weak strength of the grain boundaries.

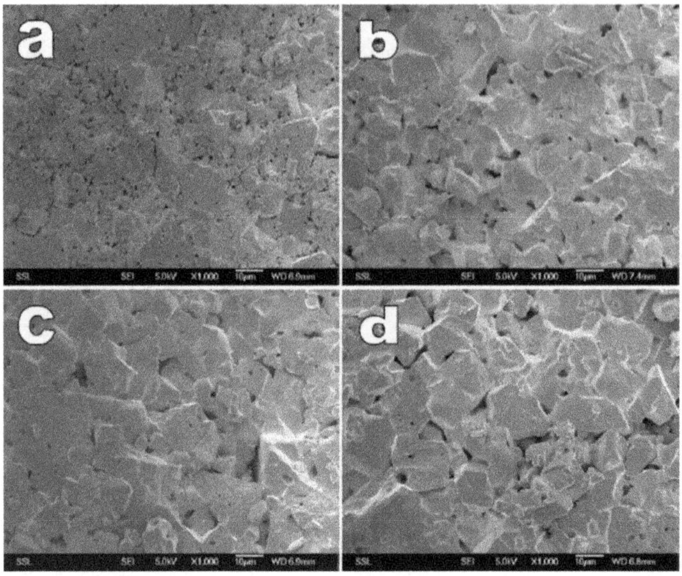

Figure 6.36. Cross-sectional SEM images of the $MgFe_{1.98}O_4$ ceramics doped with 2 wt% Bi_2O_3 after sintering for 2 h at different temperatures: (a) 950 °C, (b) 1000 °C, (c) 1050 °C and (d) 1100 °C.

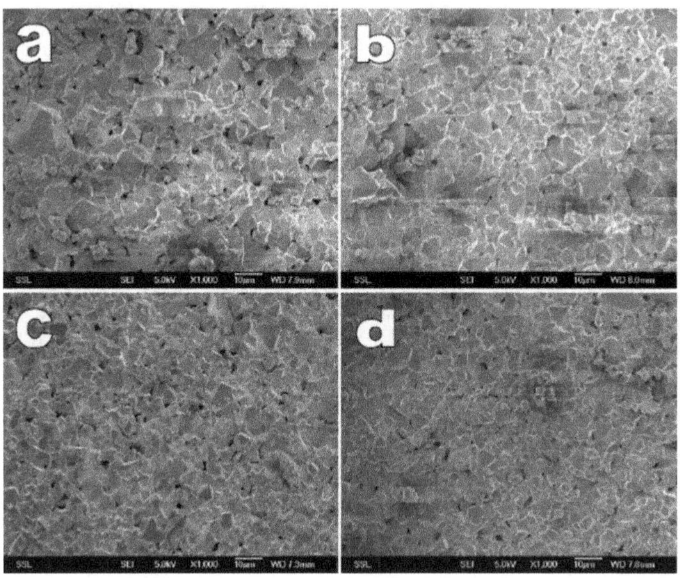

Figure 6.37. Cross-sectional SEM images of the $MgFe_{1.98}O_4$ ceramics sintered at 1100 °C doped with different concentrations of Bi_2O_3: (a) 3 wt%, (b) 5 wt%, (c) 7 wt% and (d) 10 wt%.

The doping of 0.5 wt% Bi_2O_3 had no significant effect on the densification behavior and microstructure features of the $MgFe_{1.98}O_4$ ceramics, after they were sintered at temperatures ⩽1100 °C, as illustrated in figures 6.34(a)–(c). The samples doped with 0.5 wt% Bi_2O_3 exhibited average grain sizes that were just slightly larger

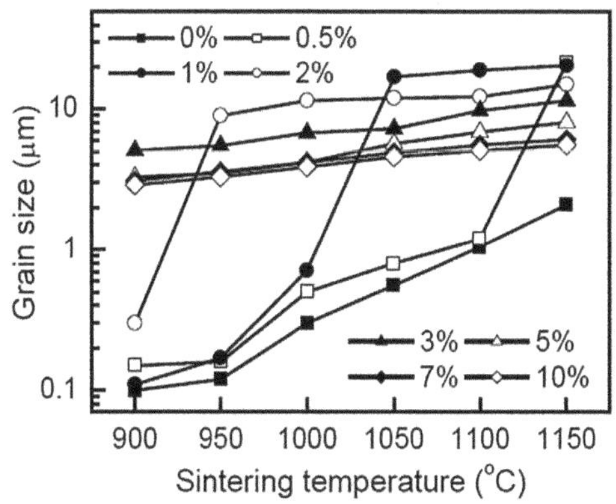

Figure 6.38. Average grain sizes of the $MgFe_{1.98}O_4$ ceramics doped with different concentrations of Bi_2O_3 as a function of sintering temperature.

than those without 0.5 wt% Bi_2O_3, because several large grains were present in the doped samples, thus enlarging the average grain sizes. In other words, the microstructure of the samples doped with 0.5 wt% Bi_2O_3 was not homogeneous, simply because the concentration of the dopant was too low. In this case, Bi_2O_3 was isolated in the sample, within which area large grains were present. However, the sample sintered at 1150 °C displayed a distinctive variation in microstructure and an abrupt increment in grain size, as revealed in figure 6.34(d). The average grain size was increased by a factor of more than ten times and the sample experienced intra-grain fracture.

The samples doped with 1 wt% and 2 wt% Bi_2O_3 followed a similar microstructure evolution and grain growth behavior. However, the abrupt change in microstructure was observed at lower temperatures, as demonstrated in figures 6.35 and 6.36. The critical sintering temperatures were 1050 °C and 950 °C for the samples doped with 1 wt% and 2 wt% Bi_2O_3, respectively. In other words, the abrupt microstructure variation temperature was decreased by about 100 °C as the concentration of Bi_2O_3 was doubled. For the samples with high contents of Bi_2O_3, the averaged grain size decreased with increasing concentration of Bi_2O_3.

In addition, more intra-grain fractures were observed, which indicated that the strength of grain boundaries was enhanced with increasing content of Bi_2O_3. As shown in figures 6.35(b) and 6.36(a), pores were present both at grain boundaries and inside the grains. A typical bimodal size distribution was observed in the sample doped with 2 wt% Bi_2O_3 after sintering at 950 °C, as revealed in figure 6.36(a). It suggested that the larger grains grew at the expense of the smaller grains. In this case, because the grains grew at a too high a rate, the pores could not escape, so they were trapped at grain boundaries and inside the grains.

The bimodal grain size distribution was not observed in the samples with Bi_2O_3 concentrations $\geqslant$3%, simply because such a microstructure could be present at temperatures lower than 900 °C. In addition, the reduction in the average grain size of the samples with increasing concentration of Bi_2O_3 was more pronounced in the

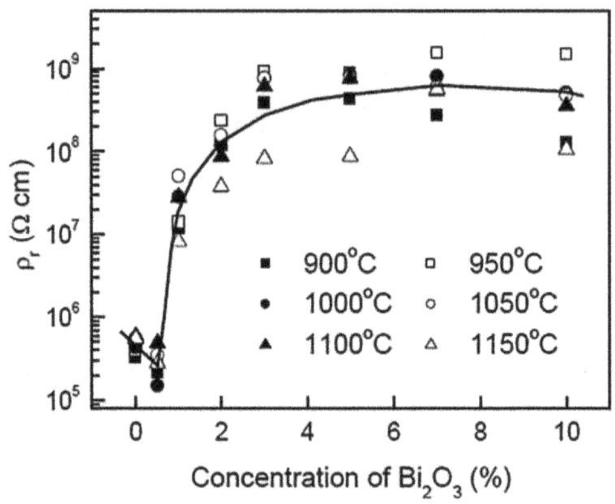

Figure 6.39. DC resistivity values of the $MgFe_{1.98}O_4$ ceramics sintered at different temperatures as a function of the concentration of Bi_2O_3. The solid line is a guide for the eyes. Reproduced with permission from [29]. Copyright 2007 Elsevier.

samples doped with high concentrations of Bi_2O_3, as demonstrated in figure 6.38, which will be discussed later.

Figure 6.39 shows the DC resistivities of the $MgFe_{1.98}O_4$ ceramics, sintered at different temperatures, as a function of the concentration of Bi_2O_3. It was observed that the DC resistivity of the samples doped with Bi_2O_3 was increased rapidly as the concentration of Bi_2O_3 was increased from 0.5 wt% to 1 wt%. The increasing rate of resistivity was slowed down with a concentration of 1–3 wt%. Above 3 wt%, the resistivity remained almost unchanged. It was quite interesting to speculate that the DC resistivities of the $MgFe_{1.98}O_4$ samples doped with 0.5 wt% Bi_2O_3 were slightly decreased, compared to those of the sample without doping of Bi_2O_3. For the ceramics without doping and doped with low concentrations of Bi_2O_3, the DC resistivity was increased with increasing sintering temperature up to 1100 °C. The resistivity declined in the ferrite ceramic samples sintered at 1150 °C.

Complex permittivity curves of the $MgFe_{1.98}O_4$ ceramics without the doping of Bi_2O_3 are shown in figure 6.40, while those of the three groups of samples doped with Bi_2O_3 at concentrations $\leqslant 2\%$ are depicted in figures 6.41–6.43. Two types of curves were observed for all the ferrite ceramic samples. Group one was represented by the curves shown in figures 6.43(a)–(c), in which both real and imaginary permittivities were kept nearly constant in the frequency range of the test. Samples with these kinds of permittivity curves had a relatively low dielectric loss tangent. Group two was characterized by exponential curves at low frequencies. Samples with high dielectric loss tangents should be avoided, if the ferrite ceramics were proposed for antenna applications. The real permittivity values and the loss tangents recorded at 1 MHz are shown in figures 6.44 and 6.45, respectively. Figure 6.46 depicts the real permittivity values (collected at 100 MHz) of all samples, as a function of sintering temperature.

The dielectric properties of the $MgFe_{1.98}O_4$ samples without doping of Bi_2O_3 after sintering at temperatures $\leqslant 1100$ °C were nearly the same, while their real permittivity was slightly increased with increasing sintering temperature. However,

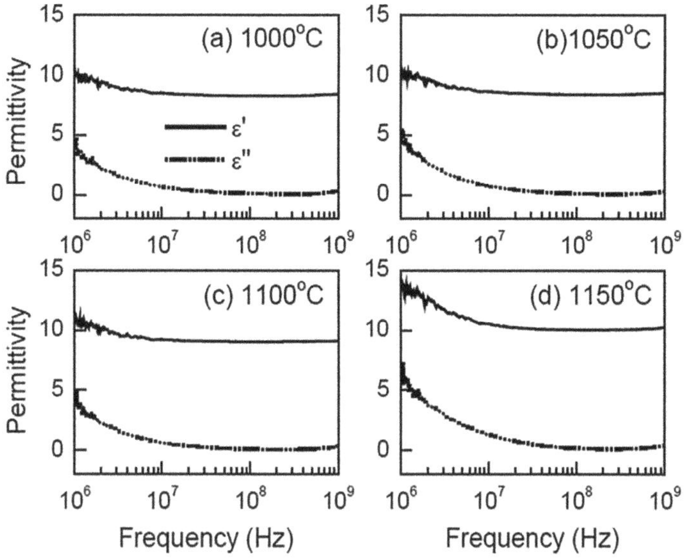

Figure 6.40. Complex relative permittivity curves of the $MgFe_{1.98}O_4$ ceramics without dopant after sintering at different temperatures. Reproduced with permission from [29]. Copyright 2007 Elsevier.

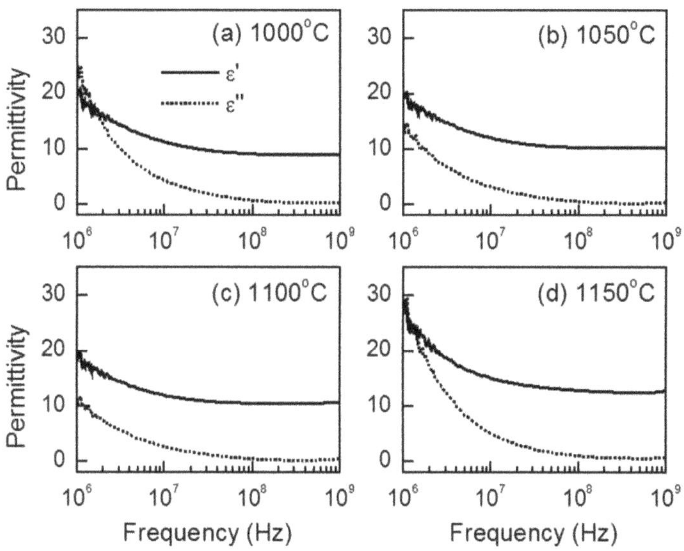

Figure 6.41. Complex relative permittivity curves of the $MgFe_{1.98}O_4$ ceramics doped with 0.5 wt% Bi_2O_3 after sintering at different temperatures. Reproduced with permission from [29]. Copyright 2007 Elsevier.

after sintering at 1150 °C, both the real and imaginary permittivities were largely increased. The introduction of 0.5 wt% Bi_2O_3 obviously had a negative effect on the dielectric properties of the $MgFe_{1.98}O_4$ ceramics. Although both the real and imaginary permittivities were increased, the dielectric loss tangent was also increased, in particular for the samples sintered at high temperatures. The samples

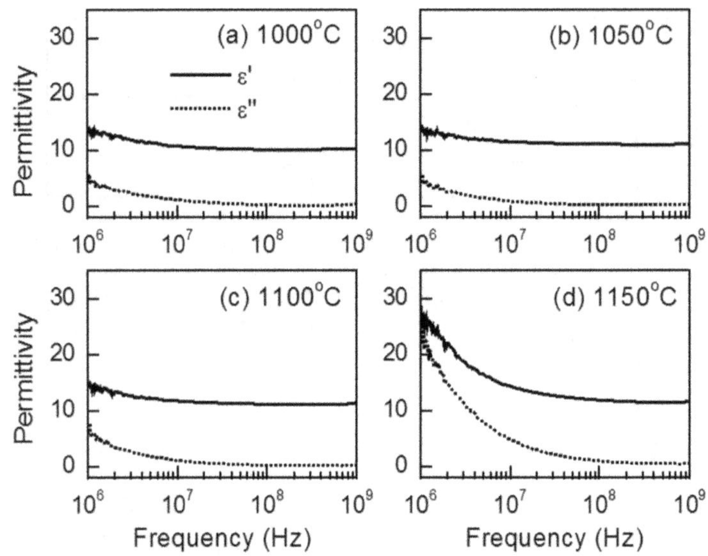

Figure 6.42. Complex relative permittivity curves of the $MgFe_{1.98}O_4$ ceramics doped with 1 wt% Bi_2O_3 after sintering at different temperatures. Reproduced with permission from [29]. Copyright 2007 Elsevier.

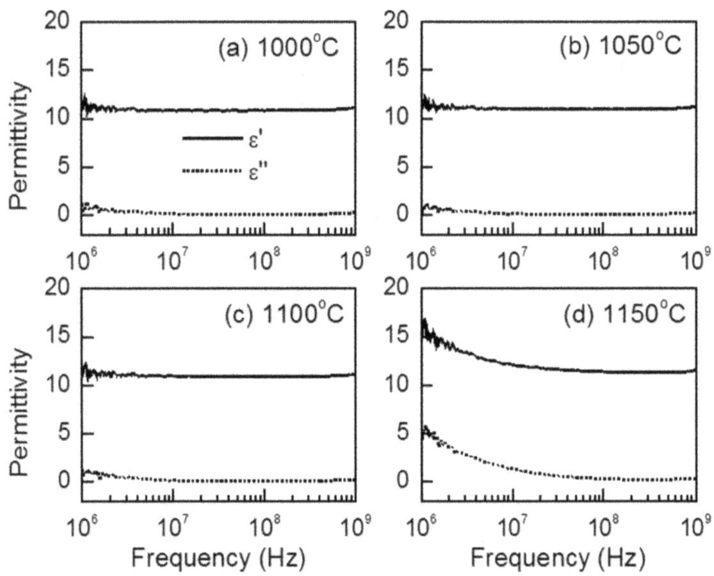

Figure 6.43. Complex relative permittivity curves of the $MgFe_{1.98}O_4$ ceramics doped with 2 wt% Bi_2O_3 after sintering at different temperatures. Reproduced with permission from [29]. Copyright 2007 Elsevier.

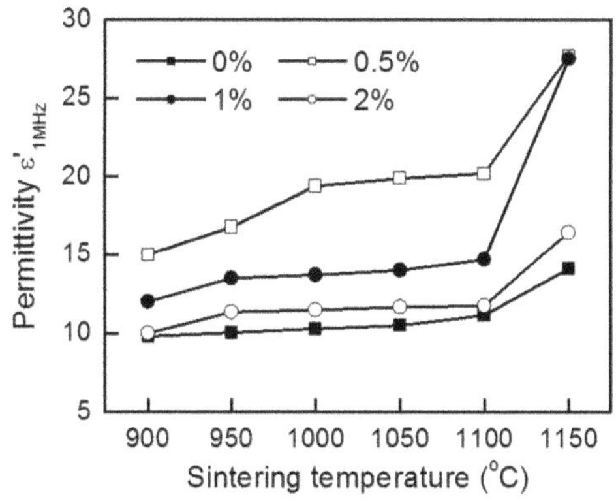

Figure 6.44. Values of the real part of permittivity (recorded at 1 MHz) of the $MgFe_{1.98}O_4$ ceramics, doped with Bi_2O_3 at low concentrations ⩽2%, as a function of sintering temperature. Reproduced with permission from [29]. Copyright 2007 Elsevier.

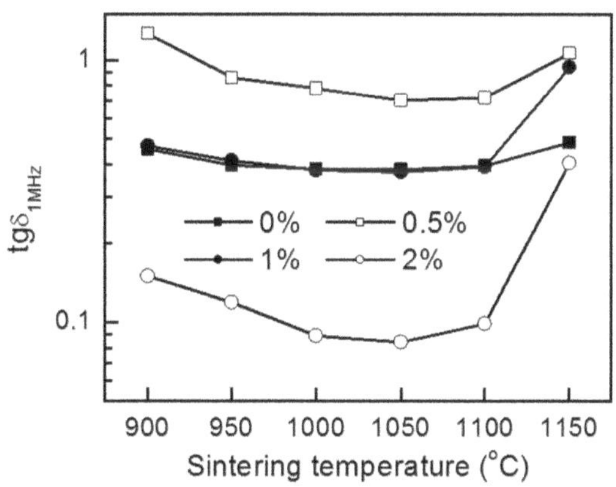

Figure 6.45. Dielectric loss tangents (collected at 1 MHz) of the $MgFe_{1.98}O_4$ ceramics, doped with Bi_2O_3 at low concentrations ⩽2%. Reproduced with permission from [29]. Copyright 2007 Elsevier.

doped with 1 wt% Bi_2O_3 exhibited similar dielectric properties to those without doping, when the sintering temperature was ⩽1100 °C. However, after sintering at 1150 °C, an extremely high dielectric loss tangent was observed. At a fixed concentration of Bi_2O_3, the dielectric loss tangent was decreased gradually with increasing sintering temperature, in the temperature range 950 °C–1050 °C.

The dielectric properties of the ferrite ceramics were significantly improved in the samples doped with 2 wt% Bi_2O_3. As observed in figure 6.43, both the real and imaginary permittivities were nearly constant in the whole range of frequency, if the samples were sintered at temperatures ⩽1100 °C. However, after sintering at 1150 °C, a high dielectric loss feature was present, i.e. the real and imaginary permittivities were increased at low frequencies. Further increase in the content of Bi_2O_3 resulted in almost no variation in the dielectric properties. For example, the samples doped with 3 wt% Bi_2O_3 possessed nearly the same dielectric curves as the

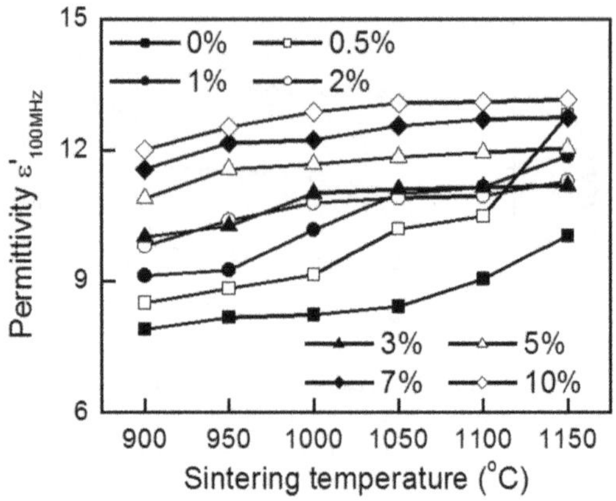

Figure 6.46. Values of the real part of permittivity (recorded at 100 MHz) of the $MgFe_{1.98}O_4$ ceramics, doped with different concentrations of Bi_2O_3, as a function of sintering temperature. Reproduced with permission from [29]. Copyright 2007 Elsevier.

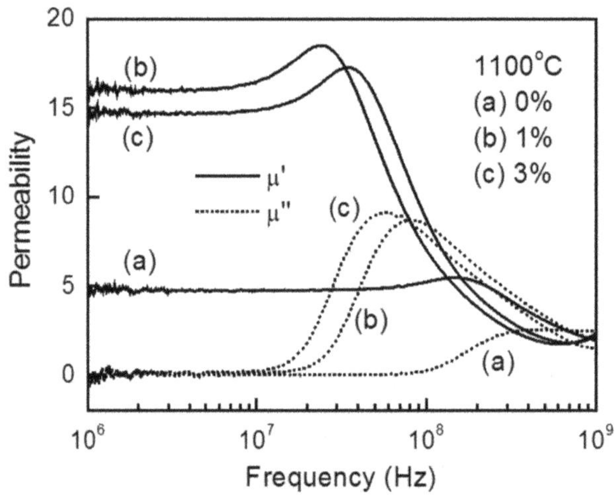

Figure 6.47. Complex relative permeability curves of selected $MgFe_{1.98}O_4$ ceramics doped with different contents of Bi_2O_3 after sintering at 1100 °C. Reproduced with permission from [29]. Copyright 2007 Elsevier.

samples with 2 wt% Bi_2O_3. However, a slightly higher content of Bi_2O_3 could suppress the dielectric loss tangent of the samples after sintering at high temperatures. Too high a content of Bi_2O_3 produced an increase in dielectric loss tangent. Therefore, the optimal concentration of Bi_2O_3 was 2–3 w%, in terms of achieving a minimum dielectric loss tangent for the $MgFe_{1.98}O_4$ ferrite ceramics.

Figure 6.47 shows complex relative permeability curves of selected $MgFe_{1.98}O_4$ ceramics doped with different contents of Bi_2O_3 after sintering at 1100 °C. Similarly, the real permeability at a low frequency (1 MHz) was used to stand in for the static permeability (μ_0). The magnetic loss tangent is defined as $tg\delta_\mu = \mu'/\mu'$. The magnetic loss tangent of the samples at frequencies far below the resonance frequency could be lower than 10^{-2}. Therefore, to develop ferrite ceramics for high frequency applications, it is necessary to increase the resonance frequency, which is always a challenge. In addition, it was found that the magnetic loss tangent at low frequencies was not influenced by the content of Bi_2O_3.

Figure 6.48 shows static permeabilities of the ferrite ceramics doped with different concentrations of Bi_2O_3, as a function of sintering temperature. For the samples without doping and doped with high levels of Bi_2O_3, the static permeability was increased almost linearly as the sintering temperature was increased. However, for samples doped with relatively low concentrations of Bi_2O_3, there was a sharp increase in static permeability, while the temperature at which the static permeability started to rise gradually decreased with increasing content of Bi_2O_3. For example, the temperatures for the static permeability increase were 1100 °C, 1000 °C and 900 °C for the samples doped with 0.5 wt%, 1 wt% and 2 wt% Bi_2O_3, respectively. In fact, this observation was closely related to densification and grain growth behavior in the samples.

Nevertheless, the presence of Bi_2O_3 could significantly enhance the magnetic properties of the $MgFe_{1.98}O_4$ ceramics. For example, the static permeability of the samples doped with 3 wt% Bi_2O_3 was larger than that of the samples without the doping of Bi_2O_3 by about three times, as they were sintered at the same temperature. However, too high a concentration of Bi_2O_3 led to a reduction in static permeability, which was simply because Bi_2O_3 is a non-magnetic material.

Bi_2O_3 has four polymorphic crystal structures, i.e. monoclinic (α), tetragonal (β), face-centered cubic (δ) and body-centered cubic (γ). Among them, γ-Bi_2O_3 is stable up to the melting point, together with the cations of Fe^{3+}, Co^{2+} and Ni^{2+} [14, 30, 31]. Because Bi_2O_3 has a low melting point of 820 °C, the liquid phase was formed during the sintering at high temperatures [30]. The liquid phase is accumulated at the grain boundaries, which suggests that the ferrite grains were wetted. As a result, the liquid-phase sintering process was triggered, which is

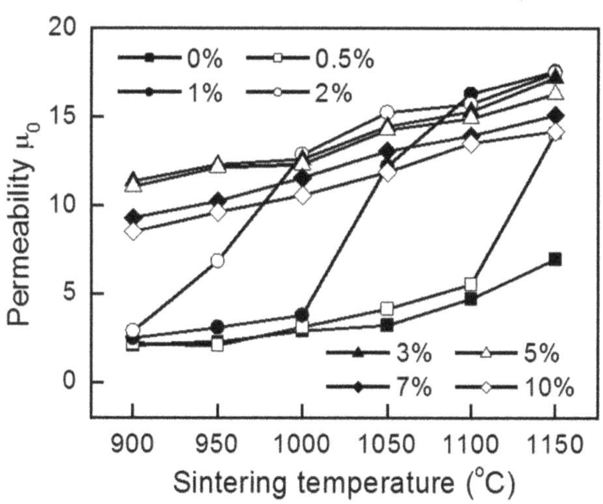

Figure 6.48. Static permeability values of the $MgFe_{1.98}O_4$ ceramics doped with different concentrations of Bi_2O_3 as a function of sintering temperature. Reproduced with permission from [29]. Copyright 2007 Elsevier.

significantly different from the solid-state reaction sintering process, as described above.

During the liquid-phase sintering, there are three mechanisms at the different quantities of the liquid phase [30, 32]. When the content of Bi_2O_3 is too low, the quantity of the liquid phase is not sufficient to cover all the ferrite grains, and the sintering is actually governed by the solid-state diffusion. Above a critical content, a continuous layer of liquid phase is formed, then the solution–precipitation process takes place. In other words, the items are dissolved in the liquid layer, followed by precipitation to realize mass transport. In this case, all the ferrite grains are fully covered by the liquid phase. However, when the content of Bi_2O_3 is too high, the thickness of the liquid phase is largely increased. As a consequence, the dissolved items should 'swim' through the thick liquid layers.

The liquid phase promotes the densification process and increases the grain growth of the ferrite ceramics. At a relatively low temperature, due to the formation of the liquid-phase layer, the particles are rearranged, thus facilitating early densification [32]. The liquid phase offers capillary pressure, thus providing an extra driving force for the sintering process. Furthermore, the effect of the liquid phase on sintering behavior and grain growth is highly dependent on the sintering temperature. This is simply because the viscosity of the liquid phase is decreased with increasing temperature, so the wetting function will be enhanced.

In addition, with increasing temperature the grain size will be increased, so that the volumetric fraction of the grain boundary is decreased. As a consequence, the liquid layer will be thickened. The thickness of the liquid-phase layer could be estimated according to the brick wall model (BWM) [30, 33], in which the ferrite grains are treated as cubic. Before the occurrence of the early stage sintering, the sample had an average grain size of 1 μm. Therefore, it was used to obtain the thickness of the liquid-phase layer as Bi_2O_3 was just melting. The thicknesses of the samples doped with different contents of Bi_2O_3 are listed in table 6.4. It has been reported that the dissolution–reprecipitation mechanism would take place if the liquid-phase layer reached the critical thickness of ~1.7 nm [30].

Obviously, the liquid-phase layer of the sample doped with 1 wt% Bi_2O_3 reached the critical thickness. Therefore, for the $MgFe_{1.98}O_4$ without doping and the sample doped with 0.5% Bi_2O_3, when they were sintered at temperatures of <1150 °C, the densification and grain growth occurred through the solid-state sintering, i.e. the diffusion of items was via the crystal lattice. In comparison, for the samples doped with 1–2 wt% Bi_2O_3, the sintering mechanism was the solution–precipitation process. Finally, the densification and grain growth of the samples with even higher

Table 6.4. Thicknesses of the liquid-phase layer for the samples with different contents of Bi_2O_3 during the early stage of sintering. Reproduced with permission from [29]. Copyright 2007 Elsevier.

wt%	0.5	1	2	3	5	7	10
t (nm)	0.85	1.69	3.38	5.06	8.41	11.7	16.7

contents of Bi_2O_3 were controlled by the diffusion rate of the items through the liquid layer which had a much greater thickness.

The grain growth of ceramics follows the kinetics, which can be described as

$$D_t^n - D_0^n = k_0 t \exp(-Q/RT), \tag{6.4}$$

where D_t is the average grain size at time t, D_0 is the initial grain size, n is the kinetic grain growth exponent, Q is the apparent activation energy, k_0 is the coefficient, and R and T are the ideal gas constant and absolute temperature [30, 34, 35]. At a given time and temperature, the final grain size is dependent on the activation energy (Q), which is related to the specific grain growth. In other words, the grain growth rate is a function of the activation energy, i.e. the lower the activation energy, the higher the grain growth rate will be.

As a liquid layer is present, the grain growth rate can be expressed by

$$dG/dt = 2DSM\sigma/kT\rho_d\delta(G/G_0 - 1), \tag{6.5}$$

where D is the diffusion coefficient of the solid items in the liquid phase, S is the dissolution limit of a flat surface, σ is the solid–liquid surface energy, ρ_d is the density of the solid, δ is the thickness of the liquid layers and G_0 is the critical grain size to determine the dissolving or growing of the grains [34]. In this case, the grain growth rate is closely related to the thickness of the liquid-phase layer, since the mass transport to maintain the grain growth should be across the liquid-phase layer.

As a result, the sample without the doping of Bi_2O_3 encountered the highest activation energy, thus presenting the slowest grain growth rate, as evidenced by the data shown in figures 6.33 and 6.39. For the sample doped with 0.5 wt% Bi_2O_3, the content of the liquid phase was too low to display the effect on grain growth of $MgFe_{1.98}O_4$, as they were sintered at temperatures $\leqslant$1100 °C. In fact, a continuous liquid layer was not formed in this case. However, by comparing figures 6.33 and 6.44, it was found that the grains of samples without doping were much more uniform than those of the samples doped with 0.5 wt% Bi_2O_3, as mentioned earlier. There were occasionally very large grains in the background of the smaller grains. This suggested that the small quantity of the liquid phase was isolated, thus having only a localized effect as the sintering temperature was sufficiently high (e.g. 1150 °C).

It was also possible that, as the grain size reached a critical level, the liquid phase could cover the all the grains and then liquid-phase sintering occurred. As shown in figure 6.34(d), when sintering at 1150 °C, the critical grain size might have been achieved to ensure the liquid-phase sintering, thus leading to rapid grain growth. As the content of Bi_2O_3 was further increased, the temperature required to form the continuous liquid-phase layer was decreased. For the samples doped with 1 wt% and 2 wt% Bi_2O_3, the corresponding temperatures were 1050 °C and 950 °C, respectively. In comparison, for the samples doped with Bi_2O_3 at concentrations $\geqslant$3%, densification and grain growth were observed at 950 °C. However, in this case, the grain growth was inversely proportional to the thickness of the liquid-phase layer. As a consequence, the average grain size was slightly decreased with increasing concentration of Bi_2O_3 for samples sintered at the same temperature.

The DC resistivity of ferrite ceramics is closely related to their microstructure. The DC resistivities of the $MgFe_{1.98}O_4$ ceramics presented above will be discussed by considering their grain growth and densification behaviors. As stated above, ferrite ceramics consist of highly conductive grains and highly insulating grain boundaries, in the form of a three-dimensional series–parallel network. The grains and grain boundaries of the ferrite ceramics without dopant are nearly the same or similar in composition, although the grain boundaries could be amorphous and contain structural and compositional defects. There could also be impurities or contaminations at the grain boundaries. In this case, the DC resistivities of ferrite ceramics are mainly dependent on the quantity and quality of the grain boundaries, where the quantity is reflected by the average grain size. Therefore, the decrease in the DC resistivity for the samples with high concentrations of Bi_2O_3 ($\geqslant$3%) was mainly attributed to the increase in the average grain size, as the sintering temperature was increased in the range 950 °C–1100 °C.

Among various other factors, porosity has an important effect on the DC resistivity of ferrite ceramics [36, 37]. It is expected that closed/isolated pores would increase DC resistivity, because air/vacuum has high insulation characteristics. Open pores could a provide conduction path for electricity by trapping impurities, thus leading to reduction in DC resistivity. The undoped $MgFe_{1.98}O_4$ samples and those doped with low concentrations of Bi_2O_3 (e.g. $\leqslant$2 wt%) had low densities, after they were sintered at relatively low temperatures. In this case, they exhibited low mechanical strengths and contained a large number of open pores. Therefore, there were two competitive factors to determine the DC resistivity, and enhanced densification decreased grain size. Because the positive effect of enhanced densification was stronger than the negative effect of increased grain size, the DC resistivities were increased with increasing sintering temperature, as the samples were sintered at temperatures $\leqslant$1100 °C.

As discussed earlier, the formation of Fe^{2+} ions usually reduces the DC resistivity of ferrite ceramics due to the increased conduction [38]. It has been found that the formation of Fe^{2+} ions could be enhanced in a ferrite doped with Bi_2O_3. Because Bi^{3+} could be oxidized to Bi^{5+}, the Bi^{5+} ion tended to substitute Fe^{3+} ions at the B-site of the spinel structure. As a result, Fe^{2+} ions would be produced to balance the electrical charges [30]. In other words, the addition of Bi_2O_3 could reduce the DC resistivity of ferrite ceramics.

However, the abrupt decline in the DC resistivity of the $MgFe_{1.98}O_4$ samples after sintering at 1150 °C could not be explained using the production of Fe^{2+} ions, because the temperature was still not sufficiently high to form Fe^{2+}. It was also not confirmed that Bi^{5+} ions were formed in the samples. Moreover, Bi^{5+} ions are only stable at high temperatures. For example, the temperature at which Bi^{5+} ions could be formed was 1340 °C [30]. Instead, the highest temperature was 1150 °C in this case. Moreover, if the statement was true, the effect should be proportional to the concentration of Bi_2O_3, which was not observed in all the samples. Therefore, the variation in the DC resistivity and the dielectric properties of the $MgFe_{1.98}O_4$ samples sintered at 1150 °C was completely irrelevant to Fe^{2+} ions. In fact, the liquid phase played a significant role in determining the electrical and dielectric properties of the $MgFe_{1.98}O_4$ ferrite ceramics.

The addition of Bi_2O_3 actually altered the properties of the grain boundaries of the $MgFe_{1.98}O_4$ ferrite ceramics. The Bi-rich grain boundaries had been observed in ZnO–Bi_2O_3-based ceramic varistors [33, 39]. The Bi-rich phase with a continuous network structure was left after the ZnO grains were etched out with acid [39]. It was found the three-dimensional network structure promoted oxygen transport in the ZnO–Bi_2O_3-based ceramics. This enhanced oxygen transport could be an important factor to improve the electrical properties of the ferrite samples. With the oxygen transport, the formation of Fe^{2+} ions could be suppressed to a certain degree, thus maintaining a high DC resistivity.

According to the Koops model [40], ferrite ceramics are composed of conductive grains, with a thickness of d_2 and resistivity of $\rho_{r,\ 2}$, which were isolated by an insulating layer, with a thickness of d_1 and resistivity of $\rho_{r,\ 1}$ [41]. If it is assumed that $x = d_1/d_2 \ll 1$ and $\rho_{r,1} \gg \rho_{r,2}$, the resistivity could be represented by [42]

$$\rho_r = \frac{d_1}{d_2}\rho_{r,\ 1} + \rho_{r,\ 2} = x\rho_{r,\ 1} + \rho_{r,\ 2}. \tag{6.6}$$

The grain resistivity ($\rho_{r,\ 2}$) can be considered to be a constant, because no new phase was formed between Bi_2O_3 and $MgFe_{1.98}O_4$. Therefore, the DC resistivities would be just dependent on the values of x and $\rho_{r,\ 1}$. As mentioned previously, the grain size of the samples was just slightly decreased, as the concentration of Bi_2O_3 was varied in the range 0.5–3 wt%. With increasing content of Bi_2O_3, because both x and $\rho_{r,\ 1}$ were increased, the DC resistivity was increased accordingly. However, as the content of Bi_2O_3 was further increased, the DC resistivity was saturated.

However, the DC resistivity of the samples doped with 0.5 wt% could not be explained in this way. A liquid phase was first formed during heating of the green bodies above the melting point of Bi_2O_3 [14, 30, 31]. The liquid layer covered and wetted the grains. After that, densification and grain growth immediately started upon the formation of the liquid phase. After the sintering process was finished, the Bi_2O_3-rich liquid phase was retracted from the two-grain boundaries to triple- and multiple-grain junctions during the cooling process. The amount of the liquid phase left in the two-grain boundaries was determined by the concentration of Bi_2O_3.

It has been observed that Ca- and Si-based impurities were likely dissolved in the Bi_2O_3-rich liquid phase at the two-grain boundaries [30, 40]. As a result, there should be a critical quantity for the liquid phase, below which all the liquid phase would flow into the triple- and multiple-grain junctions. In other words, it seemed that the two-grain boundaries were cleaned by the liquid phase and it was entirely retracted. The complete retraction of the liquid phase would take away all the impurities at the same time. Similar grain boundaries had been reported in Mn–Zn ferrite ceramics using TEM [30]. It was due to the presence of such clean grain boundaries that the samples doped with 0.5 wt% Bi_2O_3 had lower resistivities, compared to the samples without the dopant. In this case, both x and ρ_1 in equation (6.6) were minimized, thus leading to the lowest DC resistivity. At the same time, the abrupt decrease in

DC resistivity of the samples sintered at 1150 °C followed a similar mechanism, in particular for the sampled doped with relatively low contents of Bi_2O_3.

The dielectric loss tangents of ferrite ceramics include electron polarization losses, ion vibration losses, deformation losses and ionic migration losses. The electron polarization losses determine the absorption and color of the materials in the visible range. The losses of ion vibration and deformation are properties at the infrared band. As a result, the dielectric losses of the ferrite ceramics in the frequency range here would be attributed to ion migration. In addition, conduction losses could contribute to a large degree if they are present. If Fe^{2+} ions are not formed, the conduction loss is mainly related to the porous microstructure of the ferrite ceramics.

Polarization induced by conduction loss has a dispersion at the frequencies 10–100 kHz, while it could also be extended to high frequencies <100 MHz [43]. In this regard, the dielectric properties of the samples with high loss tangents, as shown in figures 6.40–6.43, were mainly related to the conduction loss because they had relatively low densities. In comparison, the contribution of the atomic and electronic polarizations to the real and imaginary permittivities of the ferrites were kept constant over the frequency range from 1 MHz to 1 GHz. In other words, if there is no conduction loss, both the real and imaginary permittivities will be unchanged in the frequency range of interest, while the dielectric loss tangent will be sufficiently low. Therefore, the key strategy to develop ferrite ceramics with a low dielectric loss tangent is to realize their densification at relatively low temperatures, so that the materials are fully densified and the formation of Fe^{2+} is effectively avoided.

Similarly, static dielectric permittivity (ε_s), which could represent the relative permittivity (ε_r) of ferrite ceramics (e.g. yttrium iron garnet or YIG), can be given by [44]

$$\varepsilon_s = \left(\frac{d_1 + d_2}{d_1}\right)\varepsilon_i, \tag{6.7}$$

where ε_i is the relative permittivity at infinite frequency. In this case, the permittivities at 1 MHz and 100 MHz could be used to stand for ε_s and ε_i, respectively. However, this formula is only applicable to the samples with high dielectric loss tangents (i.e. those doped with low concentrations of Bi_2O_3), while those with low loss tangents (i.e. those doped with high concentrations of Bi_2O_3) could be interpreted with this equation. Accordingly, the extremely high level of ε'_{1MHz} observed in the samples doped with 0.5 wt% Bi_2O_3 was quickly understood. As stated previously, the cleaned two-grain boundaries corresponded to minimized d_1 in the samples doped with 0.5 wt% Bi_2O_3. This explained well the extremely high permittivity, as seen in figure 6.41. In addition, the dielectric loss tangent of this group of samples was the highest, as illustrated in figure 6.42.

The effect of sintering temperature on the permittivity (ε'_{100MHz}), as illustrated in figure 6.43, could be described using the Maxwell–Wagner effect [38], i.e. the permittivity was increased with increasing grain size, while the grain size was increased with increasing sintering temperature. At a fixed sintering temperature, the

$\varepsilon'_{100\text{MHz}}$ was increased as the concentration of Bi_2O_3 was increased, which was most likely attributed to the higher permittivity of Bi_2O_3 than that of $MgFe_{1.98}O_4$.

The difference between ε_s and ε_i, i.e. $\Delta\varepsilon = \varepsilon_s - \varepsilon_i = \varepsilon'_{1\text{MHz}} - \varepsilon'_{100\text{MHz}}$, can be used to distinguish the contribution of conduction polarization to the dielectric properties of the ferrite ceramics. Because the conduction was only caused by the porous microstructure, the use of high sintering temperatures could effectively reduce the conduction loss by improving the densification of the materials. Therefore, with increasing sintering temperature, the value of $\varepsilon'_{1\text{MHz}}$ (ε_s) was decreased, since the level of $\Delta\varepsilon$ was lower already. However, the result shown in figure 6.41 was different. In this case, the permittivity was increased, implying that the reduction in $\varepsilon'_{1\text{MHz}}$ (ε_s) was compensated by the increase in $\varepsilon'_{100\text{MHz}}$ (ε_i).

The high permittivity of the sample doped with 0.5 wt% Bi_2O_3 after sintering at 1150 °C was due to the special characteristics of the liquid phase. In this case, the liquid phase was entirely accumulated at the grain junctions, thus resulting in cleaned two-grain boundaries. This conclusion was supported by the fact that the phenomenon was observed for the sample without doping and the samples doped with high concentrations of Bi_2O_3. Similarly, the lowest DC resistivities of the samples doped with 0.5 wt% Bi_2O_3 were also related to the cleaned two-grain boundaries. As the content of Bi_2O_3 was increased, the quantity of the liquid phase increased. As a consequence, the two-grain boundaries cannot be cleaned, because the junctions cannot take all the liquid phase.

Similarly, an attempt was made to use equations (6.1) and (6.2) to describe the magnetic properties of the $MgFe_{1.98}O_4$ ceramics. Because the samples doped with Bi_2O_3 at concentrations $\geqslant$3% were nearly completely densified after sintering at temperatures of 900 °C – 1150 °C, equation (6.2) was applicable to the samples without doping and those doped with relatively low concentrations of Bi_2O_3 (i.e. 0.5 wt%, 1 wt% and 2 wt%). Bi_2O_3 is non-magnetic, but it was involved in the calculation of the theoretical densities of the ceramic samples. However, it is found that the data of the four groups of samples could not be fitted with equation (6.1), as illustrated in figure 6.49.

It was found that the static permeabilities of the samples that had low densities were all much lower than the values derived from the equation. However, several data were close to the dotted line, such as that of the 0.5 wt% sample after sintering at 1150 °C, those doped with 1 wt% Bi_2O_3 after sintering at temperatures of at 1050 °C – 1150 °C and the samples doped with 2 wt% dopant after sintering at 1000 °C – 1150 °C. Careful inspection revealed that these samples had pretty high densification and quite large grains, as seen in figures 6.33–6.36. According to magnetic theory, the magnetization of ferrite ceramics is related to both spin rotation and domain-wall motions. The contribution of domain-wall motion is dependent on grain size. For magnetic materials, there is a critical size for the presence of domain walls.

As a result, the low static permeabilities of the samples, which were observed in figure 6.49, compared to the calculated values, were attributed to their immature domain walls. In fact, the sample doped with 2 wt% Bi_2O_3 after sintering at 950 °C

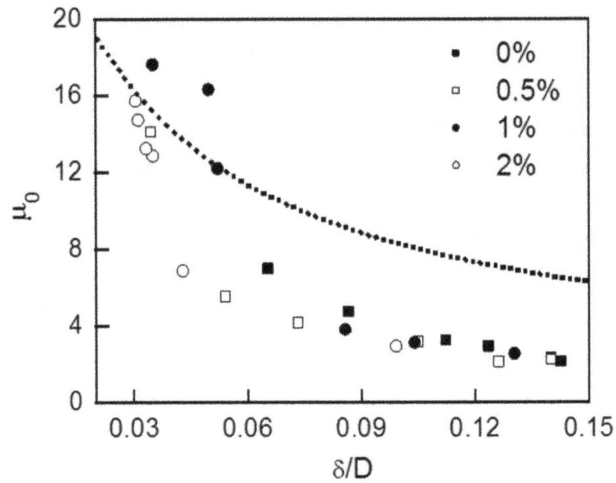

Figure 6.49. Static permeability versus δ/D for the $MgFe_{1.98}O_4$ ceramics with Bi_2O_3 at low concentrations ⩽2%. The dot line is calculated from equation (6.1) with static permeability $\mu_i = 30$. Reproduced with permission from [29]. Copyright 2007 Elsevier.

possessed essentially small sized grains as a matrix, in which large grains were occasionally observed. Therefore, it was not surprising that this sample exhibited a static permeability that was lower than the calculated value. In other words, in the magnetic circuit model, the calculation of the δ/D ratio just used the measured density, while the critical grain size was not considered.

For the samples doped with Bi_2O_3 contents ⩾3 wt%, the increase in the static permeability as a function of sintering temperature could be explained by considering the variation in grain size. It was simply because all the samples had been nearly completely densified. The increase in grain size resulted in more domain walls, thus leading to high static permeability.

The effects of Bi_2O_3 on the magnetic properties of ferrites have been widely reported in the literature. For example, it was reported that the grain size, microstructure and permeability of $Mn_{0.52}Zn_{0.44}Fe_{2.04}O_4$ could be modified with Bi_2O_3 [30]. In this case, the permeability was maximized when the doping concentration of Bi_2O_3 was 0.03%. Above that optimal concentration, the permeability started to decline. The achievement of the maximized permeability was attributed to the removal of impurities from grain boundaries. However, as the grain boundaries were cleaned, the coercivity was reduced. As a result, the permeability began to decrease. It is important to note that a relatively high sintering temperature of 1340 °C was used in that study. Because of the high sintering temperature, Bi_2O_3 would be volatilized, in particular, the concentration of Bi_2O_3 was too high. Generally, the volatilization of Bi_2O_3 occurs at about 1200 °C [14]. In addition, the non-magnetic characteristics of Bi_2O_3 and the reduction in density could result in a reduction in magnetization, thus leading to a decline of the static permeability.

Bi_2O_3 has also been employed to enhance the magnetic properties of $Ni_{0.8}Zn_{0.2}Fe_2O_4$ ceramics, which were synthesized using a wet-chemical method [31].

In this study, a maximum permeability of 75 was obtained at the doping concentration of 1%, while the values of the sample without doping and the sample doped with 6% were 37 and 48, respectively. The reduction in permeability at high contents of Bi_2O_3 was ascribed to the addition of the non-magnetic Bi_2O_3 accumulated at the grain boundaries, because no difference in grain size or density was observed among the samples. However, the decrease in permeability could not be quantized with the non-magnetic phase.

In a separate study, Ni–Cu–Zn ferrite ceramics were doped with Bi_2O_3 with concentrations of up to 1 wt% [31]. In the doping concentration range of 0.375 – 0.5 wt%, maximum permeability was observed. The presence of the maximum permeability was attributed to the formation of larger grains in the ferrite ceramics. These inconsistent results reported in the available literature suggest that not only is the concentration of Bi_2O_3 an important factor in affecting the magnetic properties of ferrite ceramics, but also the properties of the materials and the synthetic conditions have significant effects. Therefore, when citing data from the literature, it is necessary to make all these conditions clear.

Figure 6.50 shows the static permeability of the samples sintered at 1150 °C, as a function of the concentration of Bi_2O_3, revealing the effect of Bi_2O_3 as a non-magnetic phase. It was found that the relationship between the static permeability and the content of Bi_2O_3 was linear. However, as the extrapolated permeability at zero concentration was used to evaluate the effect of the non-magnetic phase, the data were higher than the designed concentrations of Bi_2O_3. Therefore, the effect of grain size should also be taken into account when interpreting the magnetic properties of the ferrite ceramics as a function of the concentration of Bi_2O_3.

It was observed that real permeability was higher than the real permittivity in the systems of $MgFe_{1.98}O_4$ doped with Bi_2O_3. In this regard, to achieve magneto-dielectric properties with matching permeability and permittivity in the low

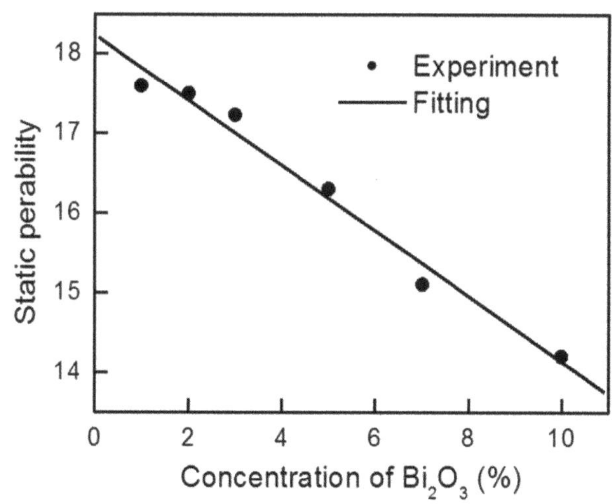

Figure 6.50. Static permeability versus concentration of Bi_2O_3 for the $MgFe_{1.98}O_4$ ceramics sintered at 1150 °C. Reproduced with permission from [29]. Copyright 2007 Elsevier.

frequency range (3–30 MHz), the permeability should be reduced by the addition of CoO [23].

According to the results discussed above, 3 wt% Bi_2O_3 was the optimal doping concentration in order to maintain a sufficiently low dielectric loss tangent of the $MgFe_{1.98}O_4$ ferrite ceramics. To obtain magneto-dielectric materials based on $MgFe_{1.98}O_4$, CoO was used to modify the magnetic properties of $MgFe_{1.98}O_4$ [45]. The sample fabrication was similar to that used to prepare other samples, as presented earlier. Briefly, the starting oxide powders were mixed according to the composition of $Mg_{1-x}Co_xFe_{1.98}O_4$, with x values of up to 0.05. The concentration of Bi_2O_3 was 3 wt%. The mixtures were activated, compacted and then sintered at different temperatures.

XRD results indicated that all the samples had a phase pure spinel structure. Also, the addition of 3 wt% Bi_2O_3 had no effect on the phase composition of the $Mg_{1-x}Co_xFe_{1.98}O_4$ ceramics. In addition, the densification, microstructural evolution and grain growth behavior of the ferrite ceramics were only influenced by the sintering temperature, while the effect of CoO could be neglected. The average grain size of the samples sintered at 900 °C was <1 μm, but it was abruptly increased to about 5 μm after sintering at 950 °C. As the sintering temperature was raised from 950 °C to 1150 °C, the average grain size was increased monotonically to 12 μm. After sintering at 1000 °C all the samples had been fully densified. The densification and grain growth characteristics of the $Mg_{1-x}Co_xFe_{1.98}O_4$ ceramics could be explained in terms of liquid-phase sintering due to the low melting point of Bi_2O_3, as discussed above [29].

Figure 6.51 shows complex permittivity curves of selected $Mg_{1-x}Co_xFe_{1.98}O_4$ ($x = 0.04$) ceramics sintered at different temperatures. Both the real and imaginary permittivities of the sample sintered at 950 °C were strongly dependent on the frequency, in particular at the frequencies of <100 MHz. This dependence was largely weakened in the sample after sintering at 1000 °C. Figure 6.52 shows real permittivity values of the samples with $x = 0.04$, collected at 1 MHz and 100 MHz, as a function of sintering temperature. At 1 MHz, the real permittivity was decreased significantly from 12.5 to 10.7 as the sintering temperature was increased from 900 °C to 950 °C. Above 950 °C, the real permittivity was monotonically increased up to the sintering temperature of 1150 °C. The real permittivity at 100 MHz had a similar variation trend with sintering temperature.

The highly frequency-dependent real and imaginary permittivities of the sample sintered at 900 °C, as shown in figure 6.51, implied that the conduction loss was very high, thus leading to a high dielectric loss tangent. The high conduction loss could be attributed to the low densification of the sample. Therefore, the high permittivities were caused by external contributions. Once the external contributions are present, the ferrites can have a high dielectric loss tangent. In comparison, the permittivity related to atomic and electronic polarizations is intrinsic, which is independent of frequency in the frequency range from 1 MHz to 1 GHz. In this regard, to achieve a low dielectric loss tangent it is necessary to eliminate the external contributions. In other words, low temperature densification would be an effective way to develop ferrite ceramics with low dielectric loss.

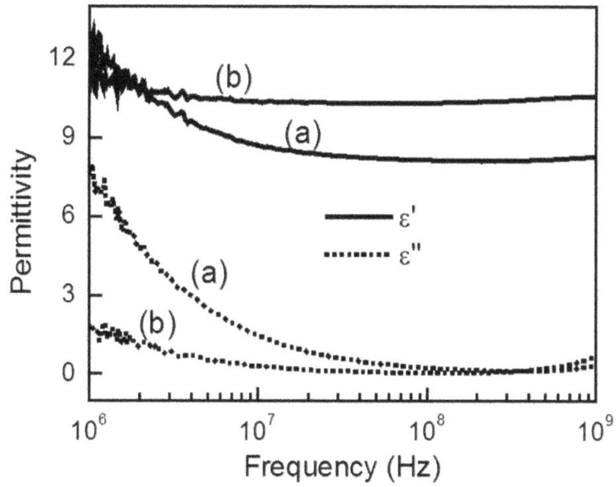

Figure 6.51. Representative relative complex permittivity curves of the $Mg_{0.96}Co_{0.04}Fe_{1.98}O_4$ ceramics sintered at different temperatures: (a) 900 °C and (b) 1000 °C. Reproduced with permission from [45]. Copyright 2008 IEEE.

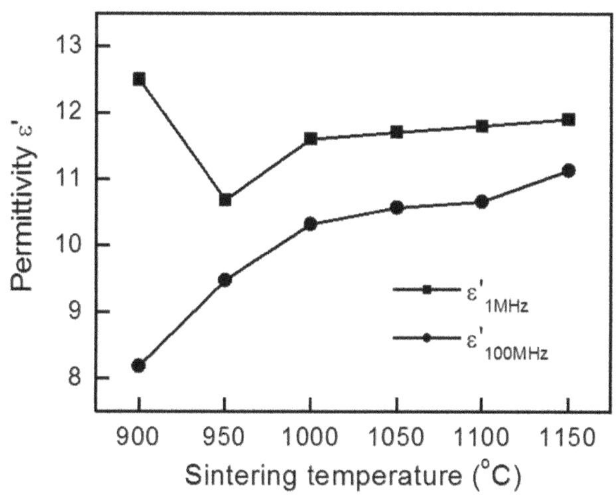

Figure 6.52. Real permittivity values (recorded at 1 MHz and 100 MHz) of the $Mg_{0.96}Co_{0.04}Fe_{1.98}O_4$ ceramics as a function of sintering temperature. Reproduced with permission from [45]. Copyright 2008 IEEE.

This is the reason why we proposed using a sintering aid to fabricate ferrite ceramics with a sufficiently low dielectric loss tangent. According to previous studies, 1 wt% Bi_2O_3 was sufficient to achieve high sinterability of the $MgFe_{1.98}O_4$ ceramics in terms of densification, while 3 wt% must be used to ensure a low dielectric loss tangent. In fact, the decrease in real permittivity (1 MHz) as the sintering temperature was reduced from 900 °C to 950 °C was attributed to the absence of conduction loss. After that, the increases in real permittivity with increasing sintering temperature were related to the grain size [19].

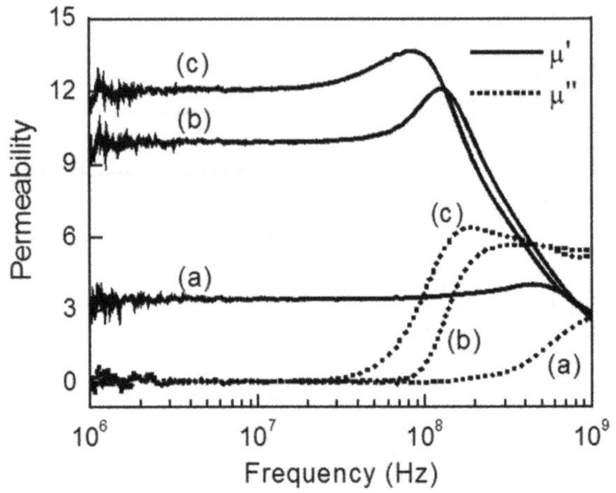

Figure 6.53. Representative complex relative permeability curves of the $Mg_{0.96}Co_{0.04}Fe_{1.98}O_4$ ceramics sintered at different temperatures: (a) 900 °C, (b) 1000 °C and (c) 1100 °C. Reproduced with permission from [45]. Copyright 2008 IEEE.

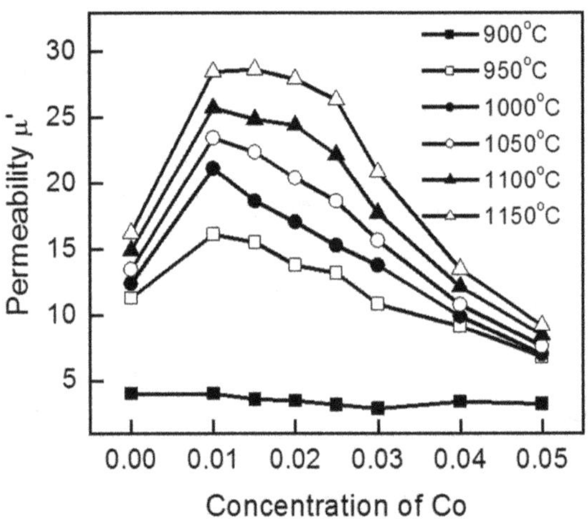

Figure 6.54. Real permeability values (collected at 1 MHz) of the $Mg_{1-x}Co_xFe_{1.98}O_4$ ceramics as a function of the concentration of CoO. Reproduced with permission from [45]. Copyright 2008 IEEE.

Figure 6.53 shows complex permeability curves of selected $Mg_{0.96}Co_{0.04}Fe_{1.98}O_4$ samples after sintering at different temperatures. The complex permeability curves of the $Mg_{0.96}Co_{0.04}Fe_{1.98}O_4$ ceramics were similar to those presented above. Figure 6.54 depicts the static permeability values of the ceramics with different concentrations of CoO as a function of sintering temperature. At a fixed concentration of CoO, the static permeability was increased continuously with increasing sintering temperature. At a fixed sintering temperature, except for 900 °C, a maximum static permeability was observed in terms of the concentration of CoO. The samples sintered at 900 °C had not been fully densified, so their magnetic ordering was not well formed.

Because the samples sintered at 900 °C were not fully densified and their grains were not matured, their data were not included. Figure 6.55 shows a μ_0–δ/D curve of the samples with a composition of $x = 0.04$. The data could be fitted with equation (6.1), leading to a value of μ_i of 31. This value was just slightly higher than the μ_0 of the sample sintered at 1150 °C. It was suggested that the sample after sintering at 1150 °C was nearly matured after sintering at this temperature. Higher sintering temperatures were not tested in order to avoid the formation of Fe^{2+} ions.

Figure 6.54 shows static permeabilities of the $Mg_{1-x}Co_xFe_{1.98}O_4$ ceramics, which were sintered at different temperatures, as a function of the concentration of CoO. The variation trend could be explained similarly, as discussed previously. The permeability of polycrystalline ferrite ceramics is related to the domain-wall motions and pin rotations [46]. The permeability due to spin rotation is given by

$$\mu_0 - 1 = kM_s^2/|K_1|, \tag{6.8}$$

where M_s is the saturation magnetization and K_1 is the magnetocrystalline anisotropy constant. According to the single-ion model, the magnetic ions in ferrite crystals occupy the interstitials formed by close packing of anions, because anions are much larger than the cations. As a result, the total magnetic anisotropy is the summation of all individual magnetic ions [47].

In spinel $Mg_{1-x}Co_xFe_{1.98}O_4$, Mg^{2+} and Fe^{3+} ions exhibited negative magnetocrystalline anisotropies, while the Co^{2+} ion displayed a positive anisotropy with a large value. Therefore, the incorporation of a low quantity of CoO could change the total magnetocrystalline anisotropy from a negative value to a positive one. In this case, the absolute anisotropy would go through zero, so that the absolute value of magnetocrystalline anisotropy could be minimized. $Mg_{1-x}Co_xFe_2O_4$ is a solid solution of $MgFe_2O_4$ and $CoFe_2O_4$, so that the magnetocrystalline anisotropy

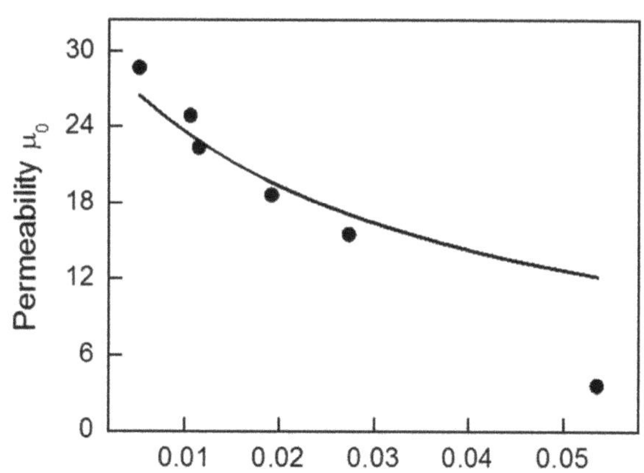

Figure 6.55. Static permeability of the $Mg_{1-x}Co_xFe_{1.98}O_4$ samples with $x = 0.04$ as a function of δ/D. Reproduced with permission from [45]. Copyright 2008 IEEE.

constant can be estimated accordingly. The magnetocrystalline anisotropy constants of $MgFe_2O_4$ and $CoFe_2O_4$ are -3.75×10^4 erg cm^{-3} and 180×10^4 erg cm^{-3}, respectively [48, 49]. Obviously, the value would be changed from negative to positive as the concentration of CoO was increased. Therefore, the sample with $x = 0.02$ had a magnetocrystalline anisotropy of near zero. In other words, samples around this composition would display maximized permeability. In fact, the experimental result was $x = 0.01$. Nevertheless, additional considerations should be taken into account when using this model to explain the experimental results of the materials.

It has been further reported that Co^{2+} ions could also influence the stability of the domain walls. The ordering of Co^{2+} ions at the *B*-site of spinel crystal could induce local uniaxial anisotropy (K_u), which was more pronounced if there were cation vacancies. The substitution of Co in ferrite is usually accompanied by the presence of *p*-type conduction, because of the oxidation of Co^{2+} to Co^{3+}. For every Co^{3+}, there would be a cation vacancy. As a result, domain walls would be immobilized. Experimental results of thermoelectric power indicate that the concentration of cation vacancies in $(Ni_{0.2}Cu_{0.2}Zn_{0.6})_{1.02-x}Co_xFe_{1.98}O_4$ ceramics was increased with increasing content of CoO [28].

In other words, the localized anisotropy related to the cation vacancies was increased with increasing concentration of CoO. Because this extra effect was sufficiently strong, maximized permeability was not observed in the $(Ni_{0.2}Cu_{0.2}Zn_{0.6})_{1.02-x}Co_xFe_{1.98}O_4$ ceramics, with $x = 0.005$–0.05. However, according to the above mentioned equation, a permeability peak should be present in the samples with $x = 0.02$–0.03. Furthermore, the domain walls could be immobilized by the Co^{3+} ions that acted as spin centers. As a consequence, the permeability was decreased at even lower concentrations of CoO. In this regard, the concentration of CoO for the presence of the maximum permeability was shifted to $x = 0.01$ in experiment from $x = 0.02$ in theory.

Figure 6.56 shows the representative magneto-dielectric properties of the $Mg_{1-x}Co_xFe_{1.98}O_4$ ceramics. In this case, the sample with $x = 0.04$, which was sintered at 1000 °C for 2 h, exhibited promising magneto-dielectric properties, with close values of real permeability and permittivity of about 10 over the HF frequency band of 3–30 MHz. The impedance of the sample was >97%, matching the free space. The magnetic loss tangent was lower than 10^{-2} over the whole HF frequency band, whereas the dielectric loss tangent suitable for real application was above 15 MHz. The high dielectric loss tangent below 15 MHz could be caused by either intrinsic or extrinsic factors, which deserves further study.

6.4 Conclusions and perspectives

Various ferrite ceramics have been developed with promising magneto-dielectric properties. These materials have very close values of real permittivity and real permeability, when their dielectric and magnetic loss tangents are sufficiently low. As a result, they have potential applications in reducing the physical dimensions of antennas in the HF and VHF bands.

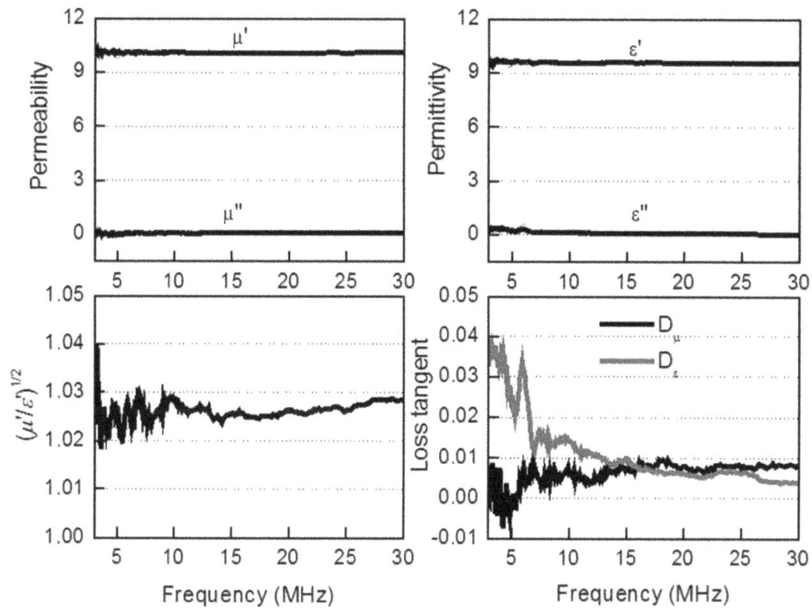

Figure 6.56. Magneto-dielectric properties of the $Mg_{0.96}Co_{0.04}Fe_{1.98}O_4$ ceramics sintered at 1000 °C for 2 h. Reproduced with permission from [45]. Copyright 2008 IEEE.

Ferrites are selected for the development of magneto-dielectric materials, because they have both magnetic and dielectric characteristics. However, ferrites without the presence of sintering aids could only be fully densified at relatively high temperatures (e.g. >1200 °C). Usually, after sintering at high temperatures, Fe^{2+} ions could be formed, due to the localized deficiency of oxygen. Once Fe^{2+} ions are present, the ferrite ceramics will have low resistivity or high conductivity. As a result, the dielectric loss tangent will be very high. However, low temperature cannot ensure the densification of the ferrite ceramics. In other words, they will have high porosity. The porous ferrite ceramics could be contaminated or adsorb water vapor, thus leading to a high dielectric loss tangent. Therefore, the first step is to solve the issue of this high dielectric loss tangent. In order to achieve this, it is necessary to use sintering aids to promote the densification of ferrites. Ferrite ceramics densified at low temperatures avoid the formation of Fe^{2+} ions and prevent contamination and water adsorption. As a result, the dielectric loss tangent could be effectively reduced.

It has been shown that Bi_2O_3 is a promising candidate for a sintering aid for ferrite ceramics, mainly because it has a relatively low melting point and does not react with most ferrites. The low melting point of Bi_2O_3 ensured the formation of a liquid phase, thus triggering liquid-phase sintering. Therefore, low temperature densification was achieved. Because Bi_2O_3 has no reaction with ferrite, it has no effect on the phase composition and thus the magnetic and dielectric properties of the ferrite ceramics.

The second step is to modify the magnetic properties of the ferrite ceramics, so that the permeability and real permittivity are equal to each other or very close, so as to ensure their impedance matching with free space. This is simply because Co^{2+} has a positive contribution to the magnetocrystalline anisotropy constant, while nearly all other cations have a negative contribution. Therefore, the magnetic properties can be adjusted by substituting Co for the cations in a given ferrite.

Until now, the magneto-dielectric materials reported in the available literature have mainly been spinel ferrites. One issue of spinel ferrites is their relatively low working frequencies. Currently, spinel ferrites could have matching real permeability and permittivity up to 100 MHz. To further increase the work frequencies, it is necessary to increase the resonance frequencies of the ferrites. In this case, barium ferrites with a hexagonal crystal structure should be used, because hexagonal ferrites have higher resonance frequencies than spinel ferrites. However, hexagonal ferrites have even higher sintering temperatures. Therefore, it is still a challenge to develop magneto-dielectric materials with high working frequencies, which could be a research topic for future studies.

Acknowledgments

The Shenzhen Technology University (SZTU) is acknowledged for the financial support of a start-up grant (2018) and also the Natural Science Foundation of Top Talent of SZTU (grant no. 2019010801002).

References

[1] Mosallaei H and Sarabandi K 2004 Magneto-dielectrics in electromagnetics: concept and applications *IEEE Trans. Anten. Propag.* **52** 1558–67

[2] Buell K, Mosallaei H and Sarabandi K 2006 A substrate for small patch antennas providing tunable miniaturization factors *IEEE Trans. Microw. Theor. Technol* **54** 135–46

[3] Colburn J S and Rahmat-Samii Y 1999 Patch antennas on externally perforated high dielectric constant substrates *IEEE Trans. Anten. Propag* **47** 1785–94

[4] Hansen R C and Burke M 2000 Antennas with magneto-dielectrics *Microwave Opt. Technol. Lett.* **26** 75–8

[5] Kong L B, Li Z W, Lin G Q and Gan Y B 2007 Magneto-dielectric properties of Mg–Cu–Co ferrite ceramics: I. Densification behavior and microstructure development *J. Am. Ceram. Soc.* **90** 3106–12

[6] Kong L B, Li Z W, Lin G Q and Gan Y B 2007 Magneto-dielectric properties of Mg–Cu–Co ferrite ceramics: II. Electrical, dielectric and magnetic properties *J. Am. Ceram. Soc.* **90** 2104–12

[7] Stevenson A J, Kupp E R and Messing G L 2011 Low temperature, transient liquid phase sintering of B_2O_3–SiO_2-doped Nd:YAG transparent ceramics *J. Mater. Res.* **26** 1151–58

[8] Svoboda J, Riedel H and Gaebel R 1996 A model for liquid phase sintering *Acta Mater.* **44** 3215–26

[9] Dong W M, Jain H and Harmer M P 2005 Liquid phase sintering of alumina, I. Microstructure evolution and densification *J. Am. Ceram. Soc.* **88** 1702–07

[10] Dong W M, Jain H S and Harmer M P 2005 Liquid phase sintering of alumina, II. Penetration of liquid phase into model microstructures *J. Am. Ceram. Soc.* **88** 1708–13

[11] Dong W M, Jain H S and Harmer M P 2005 Liquid phase sintering of alumina, III. Effect of trapped gases in pores on densification *J. Am. Ceram. Soc.* **88** 1714–19

[12] German R M, Suri P and Park S J 2009 Review: liquid phase sintering *J. Mater. Sci.* **44** 1–39
[13] Rezlescu N, Rezlescu E, Popa P D, Craus M L and Rezlescu L 1998 Copper ions influence on the physical properties of a magnesium–zinc ferrite *J. Magn. Magn. Mater.* **182** 199–206
[14] Murbe J and Topfer J 2006 Ni–Cu–Zn ferrites for low temperature firing: II. Effects of powder morphology and Bi_2O_3 addition on microstructure and permeability *J. Electroceram.* **16** 199–205
[15] Wang S F, Wang Y R, Yang T C K, Chen C F, Lu C A and Huang C Y 2000 Densification and magnetic properties of low-fire NiCuZn ferrites *J. Magn. Magn. Mater.* **220** 129–38
[16] Ram S, Krishnan H, Rai K N and Narayan K A 1989 Magnetic and electrical properties of Bi_2O_3 modified $BaFe_{12}O_{19}$ hexagonal ferrite *Japan. J. Appl. Phys.* **28** 604–8
[17] Nakamura T 2000 Snoek's limit in high-frequency permeability of polycrystalline Ni–Zn, Mg–Zn and Ni–Zn–Cu spinel ferrites *J. Appl. Phys.* **88** 348–53
[18] Tsutaoka T 2003 Frequency dispersion of complex permeability in Mn–Zn and Ni–Zn spinel ferrites and their composite materials *J. Appl. Phys.* **93** 2789–96
[19] Verma A and Dube D C 2005 Processing of nickel–zinc ferrites via the citrate precursor route for high-frequency applications *J. Am. Ceram. Soc.* **88** 519–23
[20] Hoque S M, Choudhury M A and Islam M F 2002 Characterization of Ni–Cu mixed spinel ferrite *J. Magn. Magn. Mater.* **251** 292–303
[21] Low K O and Sale F R 2002 Electromagnetic properties of gel-derived NiCuZn ferrites *J. Magn. Magn. Mater.* **246** 30–5
[22] Jonker G H 1959 Analysis of the semiconducting properties of cobalt ferrite *J. Phys. Chem. Solids* **9** 165–75
[23] Nakamura T and Hatakeyama K 2000 Complex permeability of polycrystalline hexagonal ferrites *IEEE Trans. Magn.* **36** 3415–17
[24] Chang H F 1984 Modeling of electrical response for semiconducting ferrite *J. Appl. Phys.* **56** 1831–37
[25] Johnson M T and Visser E G 1990 A coherent model for the complex permeability in polycrystalline ferrites *IEEE Trans. Magn.* **26** 1987–89
[26] Nakamura T, Tsutaoka T and Natakeyama K 1994 Frequency dispersion of permeability in ferrite composite materials *J. Magn. Magn. Mater.* **138** 319–28
[27] Snoek J L 1948 Dispersion and absorption in magnetic ferrites at frequencies above one Mc/s *Physica* **14** 207–17
[28] Byun T Y, Byeon S C, Hong K S and Kim C K 1999 Factors affecting initial permeability of Co-substituted Ni–Zn–Cu ferrites *IEEE Trans. Magn.* **35** 3445–47
[29] Kong L B, Li Z W, Lin G Q and Gan Y B 2007 Electrical and magnetic properties of magnesium ferrite ceramics doped with Bi_2O_3 *Acta Mater.* **55** 6561–72
[30] Drofenik M, Znidarsic A and Makovec D 1998 Influence of the addition of Bi_2O_3 on the grain growth and magnetic permeability of MnZn ferrites *J. Am. Ceram. Soc.* **81** 2841–48
[31] Kumar P S A, Sainkar S R, Shrotri J J, Kulkami S D, Deshpande C E and Date S K J 1998 Particle size dependence of rotational responses in Ni–Zn ferrite *J. Appl. Phys.* **83** 6864–66
[32] Jorgensen P J and Bartlett R W J 1973 Liquid-phase sintering of $SmCo_5$ *J. Appl. Phys.* **44** 2876–80
[33] Wong J 1980 Sintering and varistor characteristics of ZnO–Bi_2O_3 ceramics *J. Appl. Phys.* **51** 4453–59
[34] Lay K W 1968 Grain growth in UO_2–Al_2O_3 in the presence of a liquid phase *J. Am. Ceram. Soc.* **51** 373–76

[35] Senda T and Bradt R C 1991 Grain growth of zinc oxide during the sintering of zinc oxide–antimony oxide ceramics *J. Am. Ceram. Soc.* **74** 1296–302
[36] Verma A, Goel T C, Mendiratta R G and Gupta R G 1999 High-resistivity nickel–zinc ferrites by the citrate precursor method *J. Magn. Magn. Mater.* **192** 271–76
[37] Shinde T J, Gadkari A B and Vasambekar P N 2008 DC resistivity of Ni–Zn ferrites prepared by oxalate precipitation method *Mater. Chem. Phys.* **111** 87–91
[38] Verma A and Dube D C 2005 Processing of nickel–zinc ferrites via the citrate precursor route for high-frequency applications *J. Am. Ceram. Soc.* **88** 519–23
[39] Clarke D R 1999 Varistor ceramics *J. Am. Ceram. Soc.* **82** 485–502
[40] Brooks K G, Rerta Y and Amarakoon V R W 1992 Effect of Bi_2O_3 on impurity ion distribution and electrical resistivity of Li–Zn ferrites *J. Am. Ceram. Soc.* **75** 3065–69
[41] Koops C G 1951 On the dispersion of resistivity and dielectric constant of some semi-conductors at audiofrequencies *Phys. Rev.* **83** 121–24
[42] Miles P A, Westphal W B and Von Hippel A 1957 Dielectric spectroscopy of ferromagnetic semiconductors *Revi. Mod. Phy* **29** 279–307
[43] Abdeen A M 1998 Electric conduction in Ni–Zn ferrites *J. Magn. Magn. Mater.* **185** 199–206
[44] Larsen P K and Metselaar R 1973 Electric and dielectric properties of polycrystalline yttrium iron garnet: space-charge-limited currents in an inhomogeneous solid *Phys. Rev.* B **8** 2016–25
[45] Kong L B, Li Z W, Lin G Q and Gan Y B 2008 $Mg_{1-x}Co_xFe_{1.98}O_4$ ceramics with promising magneto-dielectric properties as potential candidate for antenna miniaturizations *IEEE Trans. Magn.* **44** 559–65
[46] Van Uitert L G, Schafer J P and Hogan C L 1954 Low-loss ferrites for applications at 4000 millicycles per second *J. Appl. Phys.* **25** 925
[47] Slonczewski J C 1958 Origin of magnetic anisotropy in $Co_xFe_{3-x}O_4$ *J. Appl. Phys.* **39** 448–9
[48] Darby M I and Issac E D 1974 Magnetocrystalline anisotropy of ferro- and ferrimagnetics *IEEE Trans. Magn.* **10** 259–304
[49] Bozorth R M, Tilden E F and Williams A J 1955 Anisotropy and magnetostriction of some ferrites *Phys. Rev.* **99** 1788–98

IOP Publishing

Functional Ceramics Through Mechanochemical Activation

Ling Bing Kong

Chapter 7

Ferrite ceramics (II)

Ling Bing Kong, Zhuohao Xiao, Xiuying Li, Shijin Yu, Wenxiu Que, Yin Liu, Tianshu Zhang, Kun Zhou and Hongfang Zhang

7.1 Li-ferrite ceramics

$Li_{0.50}Fe_{2.50}O_4$ ceramics, doped with Bi_2O_3 at concentrations of 0–5 wt% and modified by cobalt (Co) with a composition of $Li_{0.50-0.50x}Co_xFe_{2.50-0.50x}O_4$ (x = 0.01–0.07) and sintered at different temperatures for 2 h, have been studied in order to obtain magneto-dielectric materials with matching permeability and permittivity, for antenna miniaturization [1–3]. The starting materials were commercial Li_2CO_3 and oxide powders. The powders were mixed and activated using high-energy ball milling with WC as the milling medium. The activated mixtures were calcined at 800 °C. The calcined powders were compacted and sintered at different temperatures. The resultant ferrite ceramics were characterized using XRD, SEM and a high-frequency materials' property analyzer. The magnetic and dielectric properties of the ferrite ceramics were measured over the frequency range 1 MHz—1 GHz.

Figure 7.1 shows XRD patterns of the $Li_{0.50}Fe_{2.50}O_4$ ceramics doped with different concentrations of Bi_2O_3, sintered at 850 °C for 2 h. No reaction occurred between Bi_2O_3 and the spinel $Li_{0.50}Fe_{2.50}O_4$, according to the XRD results. After sintering at higher temperatures, the XRD patterns were almost the same. Therefore, Bi_2O_3 had no effect on the phase formation of the spinel $Li_{0.50}Fe_{2.50}O_4$. The XRD patterns for the samples doped with different contents of CoO are shown in figure 7.2 and the representative samples sintered at different temperatures are depicted in figure 7.3. In both cases, the content of Bi_2O_3 was 3 wt%. Also, the spinel phase of $Li_{0.50-0.50x}Co_xFe_{2.50-0.50x}O_4$ was readily formed for all the compositions and after sintering at different temperatures.

Figure 7.4 shows densification curves of the $Li_{0.50}Fe_{2.50}O_4$ samples doped with different concentrations of Bi_2O_3, with the densification parameters listed in table 7.1. The addition of Bi_2O_3 greatly enhanced the densification behaviors of the ferrites. The sample without doping shrank at about 900 °C, while the shrinkage rate peaked at

doi:10.1088/978-0-7503-2191-4ch7

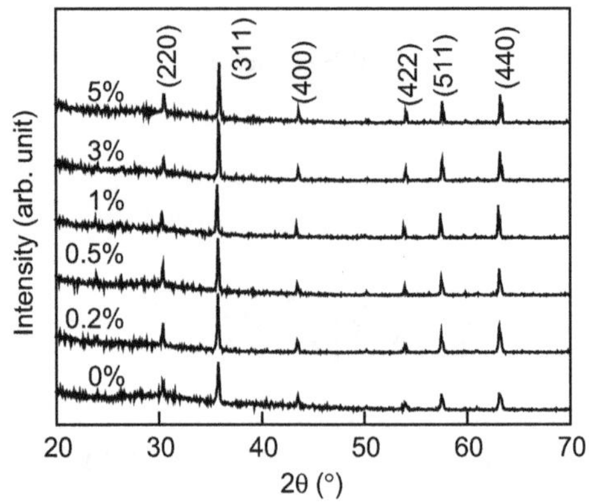

Figure 7.1. XRD patterns of the $Li_{0.50}Fe_{2.50}O_4$ samples doped with different concentrations of Bi_2O_3, after sintering at 850 °C for 2 h. Reproduced with permission from [1]. Copyright 2008 Elsevier.

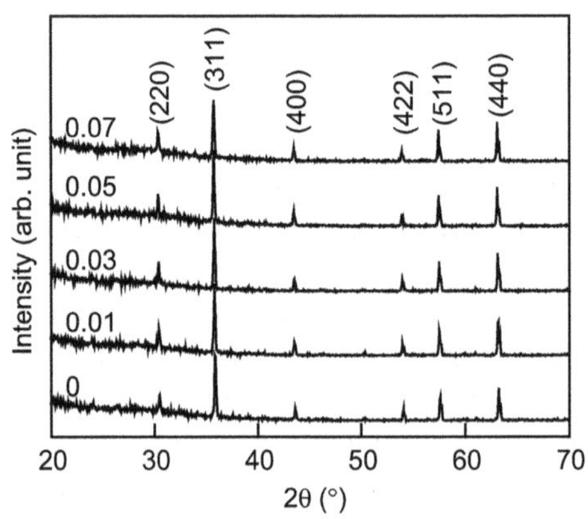

Figure 7.2. XRD patterns of the $Li_{0.50-0.50x}Co_xFe_{2.50-0.50x}O_4$ samples with different concentrations of CoO and 3 wt% Bi_2O_3, after sintering at 850 °C for 2 h. Reproduced with permission from [1]. Copyright 2008 Elsevier.

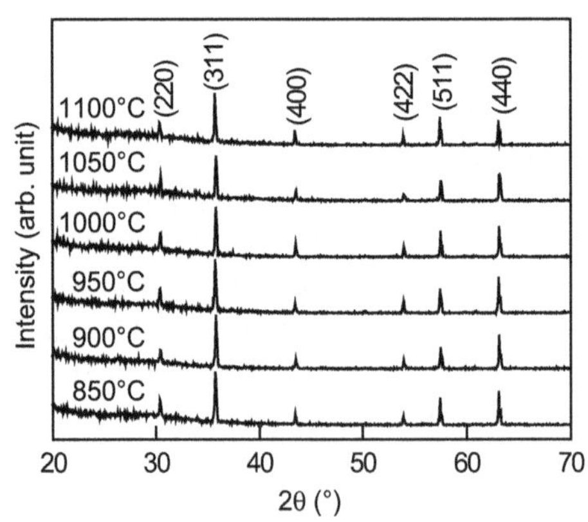

Figure 7.3. XRD patterns of the $Li_{0.50-0.50x}Co_xFe_{2.50-0.50x}O_4$ samples ($x = 0.07$) doped with 3 wt% Bi_2O_3 sintered at different temperatures for 2 h. Reproduced with permission from [1]. Copyright 2008 Elsevier.

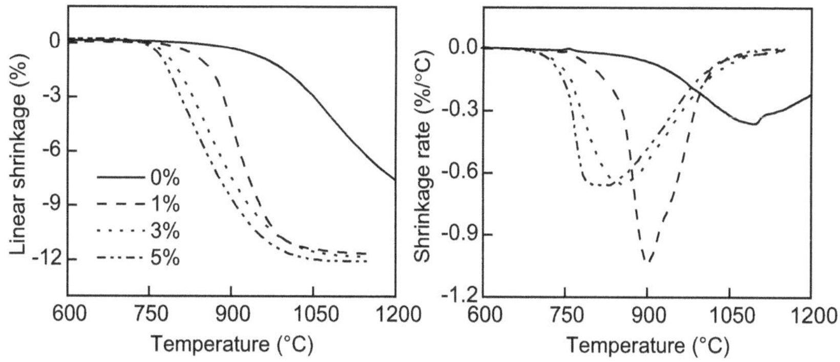

Figure 7.4. Sintering behaviors of the $Li_{0.50}Fe_{2.50}O_4$ powders with different concentrations of Bi_2O_3. Reproduced with permission from [1]. Copyright 2008 Elsevier.

Table 7.1. Densification parameters of the $Li_{0.50}Fe_{2.50}O_4$ powders with different contents of Bi_2O_3. Reproduced with permission from [1]. Copyright 2008 Elsevier.

Bi_2O_3 concentration (%)	0	1	3	5
Start shrinkage temperature (°C)	900	760	690	690
Max. shrinkage rate temperature (°C)	1100	900	855	813
Max. shrinkage rate (%/°C)	0.36	1.03	0.64	0.66
Final linear shrinkage (%)	7.4	11.5	11.7	11.9

about 1100 °C. The corresponding maximum shrinkage rate was 0.36%/°C, while the final linear shrinkage was 7.4%. After doping with 1 wt% Bi_2O_3, the temperature for the shrinkage to start was decreased to about 760 °C. In addition, the peak of shrinkage rate was reduced to about 900 °C. Therefore, the shrinkage temperature was decreased by about 200 °C when the $Li_{0.50}Fe_{2.50}O_4$ was doped with 1 wt% Bi_2O_3. The maximum shrinkage rate was increased to ~1.03%/°C, about three times that of the sample without doping. At the same time, the final linear shrinkage was 11.5%. However, as the concentration of Bi_2O_3 was further increased, the maximum shrinkage rate slightly decreased, compared to that of the sample doped with 1 wt% Bi_2O_3, although their final linear shrinkages were close.

The enhancement in densification behaviors of $Li_{0.50}Fe_{2.50}O_4$ due to the addition of Bi_2O_3 was also reflected by the densities of the samples. Figure 7.5 shows the densities of the $Li_{0.50}Fe_{2.50}O_4$ ceramics doped with different contents of Bi_2O_3 as a function of sintering temperature. The theoretical densities of the samples with different concentrations of Bi_2O_3 were derived from the theoretical densities of $Li_{0.50}Fe_{2.50}O_4$ (JCPDF No. 74–1191) and Bi_2O_3, which were 4.658 g cm^{-3} and 8.9 g cm^{-3}, respectively. The theoretical densities are summarized in table 7.2.

The density of the $Li_{0.50}Fe_{2.50}O_4$ ceramics without the doping of Bi_2O_3 was increased monotonically with increasing sintering temperature. The relative density was increased from 69.3% to 92.1%, as the sintering temperature was raised from

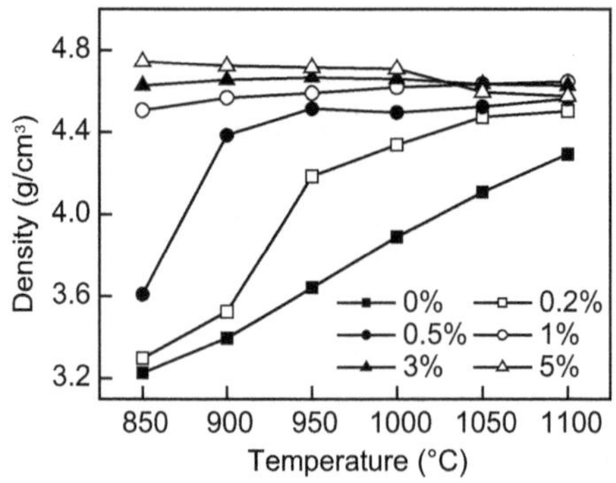

Figure 7.5. Densities of the $Li_{0.50}Fe_{2.50}O_4$ ferrite ceramics doped with different concentrations of Bi_2O_3 as a function of sintering temperature. Reproduced with permission from [1]. Copyright 2008 Elsevier.

Table 7.2. Theoretical densities of systems of $Li_{0.50}Fe_{2.50}O_4$ and Bi_2O_3. Reproduced with permission from [1]. Copyright 2008 Elsevier.

Bi_2O_3 (wt%)	0	0.2	0.5	1	3	5
D_{th} (g cm^{-3})	4.658	4.662	4.669	4.680	4.723	4.766

850 °C to 1100 °C. This suggested that $Li_{0.50}Fe_{2.50}O_4$ could not be fully densified at temperatures <1100 °C. In comparison, an abrupt increase in density was observed in the samples doped with 0.2 wt% and 0.5 wt% Bi_2O_3. The relative densities were increased from 75.6% to 89.9% and 77.3% to 93.9% as the sintering temperature was increased from 900 °C to 950 °C and 850 °C to 900 °C for the samples with 0.2 wt% and 0.5 wt% Bi_2O_3, respectively. The sharp increase in density was an indicator of the effect of the sintering aid.

The sample doped with 1 wt% Bi_2O_3 already had a relative density of 95.4% after sintering at a temperature as low as 850 °C, while the sample doped with 5 wt% Bi_2O_3 reached a value of 99.6%. After sintering at 1100 °C, the relative density of the former increased to 98.4%, whereas that of the latter slightly decreased to 96.1%. The relative density of the sample doped with 3 wt% Bi_2O_3 increased from 98.8% to 99.7% as the sintering temperature was raised from 850 °C to 950 °C. Finally, its relative density reduced to 98.8% after sintering at 1100 °C. For the reduction in density of the samples doped with high concentrations of Bi_2O_3, possible reasons included the swelling of the samples because of the presence of too much liquid phase and the volatilization of Li at certain temperatures.

Figure 7.6 shows the densities of the $Li_{0.50-0.50-x}Co_xFe_{2.50-0.50x}O_4$ samples doped with 3 wt% Bi_2O_3 as a function of sintering temperature. Clearly, the variation trend in density was similar to that of the $Li_{0.50}Fe_{2.50}O_4$ counterparts. This observation suggested that CoO had almost no effect on the densification of $Li_{0.50}Fe_{2.50}O_4$. It was not surprising, because CoO and Fe_2O_3 have similar physical properties. The incorporation of CoO only altered the magnetic properties of the ferrite ceramics.

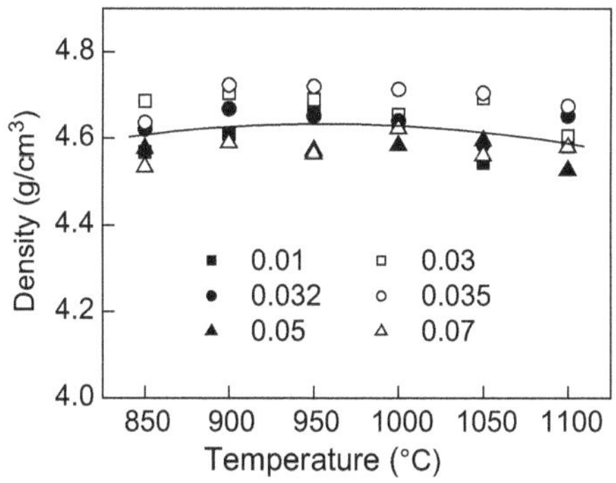

Figure 7.6. Densities of the $Li_{0.50-0.50x}Co_xFe_{2.50-0.50x}O_4$ ceramics doped with 3 wt% Bi_2O_3 as a function of sintering temperature. Reproduced with permission from [1]. Copyright 2008 Elsevier.

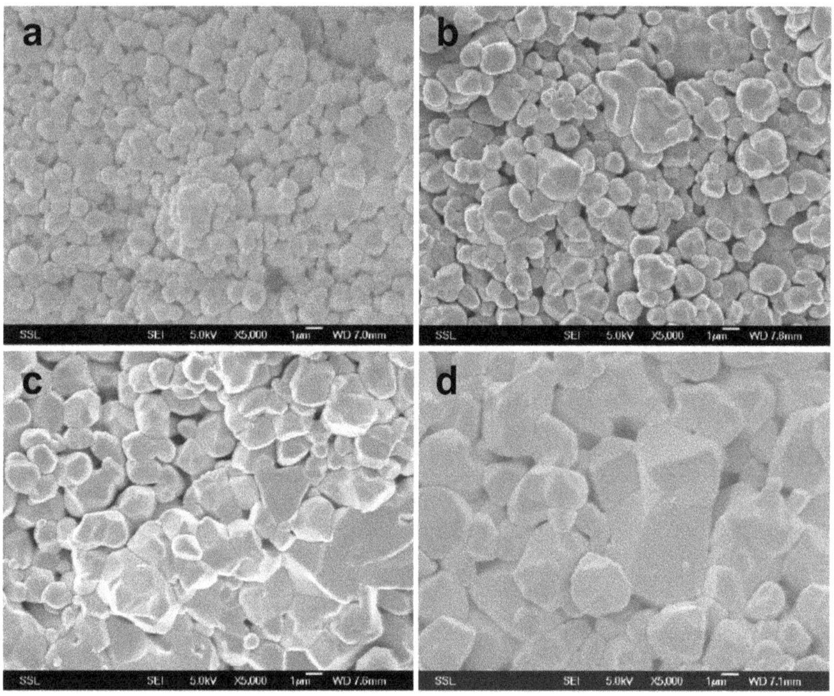

Figure 7.7. SEM images of the $Li_{0.50}Fe_{2.50}O_4$ ceramics sintered at different temperatures: (a) 950 °C, (b) 1000 °C, (c) 1050 °C and (d) 1100 °C. Reproduced with permission from [1]. Copyright 2008 Elsevier.

Figure 7.7 shows cross-sectional SEM images of the $Li_{0.50}Fe_{2.50}O_4$ ceramics after sintering at different temperatures. All the samples exhibited a porous microstructure, while an inter-grain fracture was clearly observed. In addition, the grains of the ceramics became more and more regular in morphology and the average grain size was increased gradually with increasing sintering temperature. This is a typical profile of a densification process following the mechanism of solid-state diffusion through the crystal lattices.

Cross-sectional SEM images of the selected samples doped with 0.2 wt% and 0.5 wt% Bi_2O_3 are depicted in figures 7.8 and 7.9. The sample doped with 0.2 wt% after sintering at 900 °C exhibited a similar microstructure to that of the $Li_{0.50}Fe_{2.50}O_4$ after sintering at 950 °C, both of which displayed spherical grains. However, a much denser microstructure was achieved after the sample was sintered at 950 °C for 2 h. This observation was consistent with the density result,

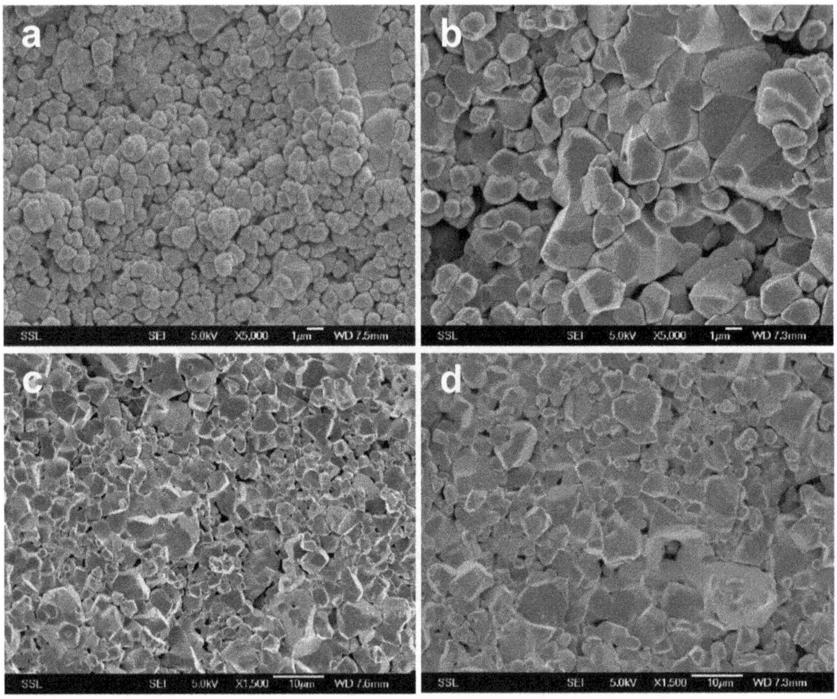

Figure 7.8. SEM images of the $Li_{0.50}Fe_{2.50}O_4$ ceramics with 0.2 wt% Bi_2O_3 sintered at different temperatures: (a) 900 °C, (b) 950 °C, (c) 1000 °C and (d) 1050 °C. Reproduced with permission from [1]. Copyright 2008 Elsevier.

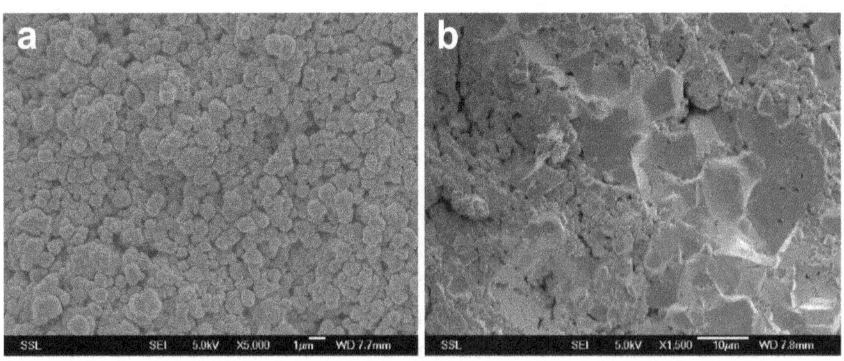

Figure 7.9. SEM images of the $Li_{0.50}Fe_{2.50}O_4$ ceramics with 0.5 wt% Bi_2O_3 sintered at different temperatures: (a) 850 °C and (b) 900 °C. Reproduced with permission from [1]. Copyright 2008 Elsevier.

corresponding to an abrupt increase in relative density from 75.6% to 89.9%. In addition, the samples sintered at temperatures ⩽950 °C encountered an inter-grain fracture, while intra-grain fracture was present in the SEM image of the sample after sintering at 1000 °C. The intra-grain fracture became more and more significant in the samples sintered at higher temperatures, as demonstrated in figure 7.8(d). In other words, the grain boundaries of the samples were dependent on the sintering temperature, i.e. the strength of the grain boundaries was gradually increased with increasing sintering temperature.

For samples doped with 0.5% Bi_2O_3, spherical grains were only observed after sintering at 850 °C. At the same time, the microstructure was tremendously changed after the sample was sintered at 900 °C, exhibiting a bimodal grain size distribution, i.e. the large grains were embedded in a matrix of smaller grains, as discussed for other ferrite materials in previous sections. This microstructure suggested that larger grains grew by consuming the smaller ones. The microstructures of the samples doped with 0.5 wt% after sintering at temperatures ⩾950 °C were similar to those of the samples doped with higher concentrations of Bi_2O_3. Correspondingly, the temperature at which the rapid variation in microstructure occurred was in good agreement with the density data. However, the such a temperature was reduced compared to that observed in the sample doped with 0.2 wt% Bi_2O_3, clearly confirming the positive effect of the sintering aid on densification of the ferrite ceramics.

Cross-sectional SEM images of selected $Li_{0.50}Fe_{2.50}O_4$ samples doped with 1 wt% and 3 wt% Bi_2O_3, after sintering at 850 °C and 1100 °C for 2 h, are illustrated in figures 7.10 and 7.11, respectively. It was noted that the samples doped with 5 wt% Bi_2O_3 had no significant variation in microstructure. They shared a common feature, i.e. fully dense microstructure and strong intra-grain fracture. For these samples, densification had been achieved after they were sintered at 850 °C, while a further increase in sintering temperature resulted only in grain growth. The presence of the liquid phase at the grain boundaries made them stronger, thus leading to intra-grain fracture.

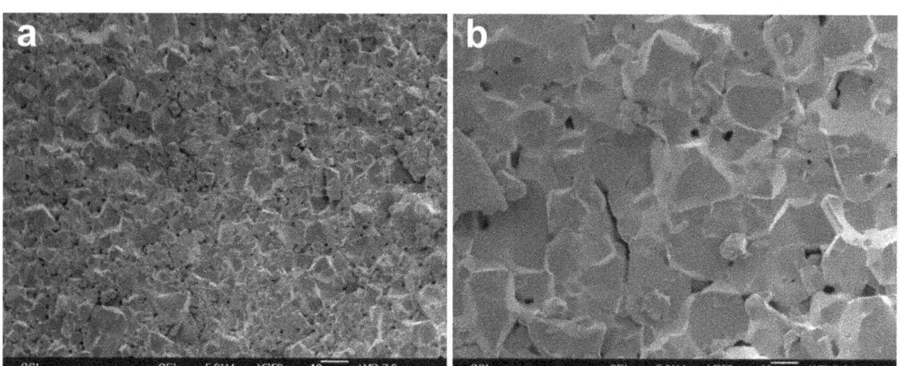

Figure 7.10. SEM images of the $Li_{0.50}Fe_{2.50}O_4$ ceramics doped with 1 wt% Bi_2O_3 sintered at different temperatures: (a) 850 °C and (b) 1100 °C. Reproduced with permission from [1]. Copyright 2008 Elsevier.

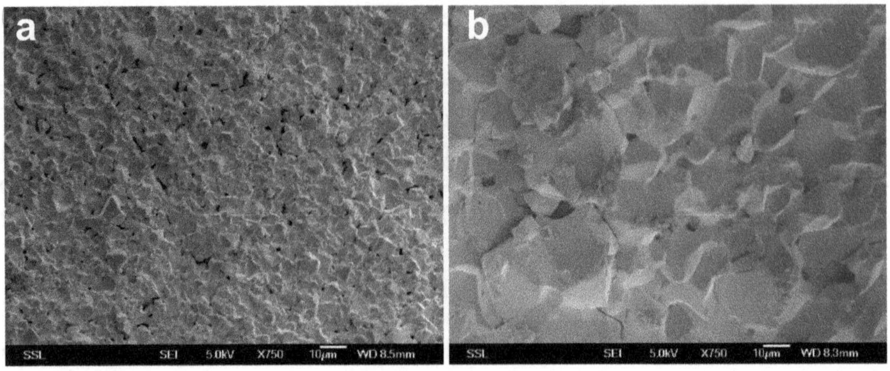

Figure 7.11. SEM images of the $Li_{0.50}Fe_{2.50}O_4$ ceramics doped with 3 wt% Bi_2O_3 sintered at different temperatures: (a) 850 °C and (b) 1100 °C. Reproduced with permission from [1]. Copyright 2008 Elsevier.

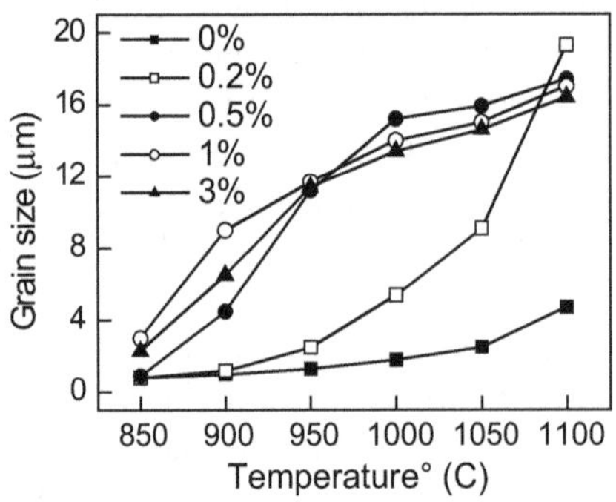

Figure 7.12. Grain sizes of the $Li_{0.50}Fe_{2.50}O_4$ samples doped with different contents of Bi_2O_3 as a function of sintering temperature. Reproduced with permission from [1]. Copyright 2008 Elsevier.

Figure 7.12 shows the average grain sizes of the samples doped with different contents of Bi_2O_3, which were derived from the SEM images, as a function of sintering temperature. Obviously, Bi_2O_3 enhanced the densification and also promoted the grain growth of the $Li_{0.50}Fe_{2.50}O_4$ ferrite. The grain size of the $Li_{0.50}Fe_{2.50}O_4$ samples without the doping of Bi_2O_3 was only increased from about 1 μm to about 5 μm as the sintering temperature was increased from 850 °C to 1100 °C. However, with the addition of 0.2 wt% Bi_2O_3 the grain growth rate started to accelerate at 900 °C. The difference in grain size between the two groups of samples was gradually enlarged with increasing sintering temperature. After sintering at 1100 °C, the sample doped with 0.2 wt% Bi_2O_3 had an average grain size about four times that of the sample without doping. Although they showed varying trends in grain growth, they should have different growth mechanisms, in particular at the higher sintering temperatures.

In addition, the sample doped with 0.2 wt% Bi_2O_3 after at 1100 °C exhibited the largest average grain size among all the samples, as seen in figure 7.12. A rapid grain growth was also observed in the samples doped with 0.5 wt% Bi_2O_3. However, the grain growth was slowed down when the sintering temperature was over 950 °C, although those samples possessed even larger grain sizes. This could be attributed to the fact that the grain growth mechanism did not change as the concentration of Bi_2O_3 was increased. For the samples with Bi_2O_3 at concentrations of 1–5 wt%, the average grain size demonstrated a slight decrease with increasing content of Bi_2O_3, at a given sintering temperature.

The $Li_{0.50-0.50x}Co_xFe_{2.50-0.50x}O_4$ samples have a similar microstructure and grain growth behavior to the $Li_{0.50}Fe_{2.50}O_4$ samples doped with 3 wt% Bi_2O_3. Cross-sectional SEM images of selected $Li_{0.50-0.50x}Co_xFe_{2.50-0.50x}O_4$ ceramics are shown in figures 7.13 and 7.14. For the samples sintered at low temperatures (e.g. 850 °C, figure 7.13), the average grain size was slightly decreased with increasing concentration of CoO, while both the microstructure and average grain size were nearly the same for the samples after sintering at high temperatures (e.g. 1100 °C, figure 7.14).

Table 7.3 lists the thicknesses of the liquid phase layer at the grain boundaries in the samples doped with different concentrations of Bi_2O_3 at the early stage of sintering. The thicknesses are according to the brick wall model (BWM) [15], with an assumption that the grains are of an ideal cubic morphology. Also, the average grain size of the $Li_{0.50}Fe_{2.50}O_4$ sample without doping after sintering at 850 °C was used for the calculation, because Bi_2O_3 could have been melting at this temperature. As mentioned previously, there is a critical thickness of ~1.7 nm for the liquid phase layer to ensure the occurrence of the dissolution–reprecipitation densification

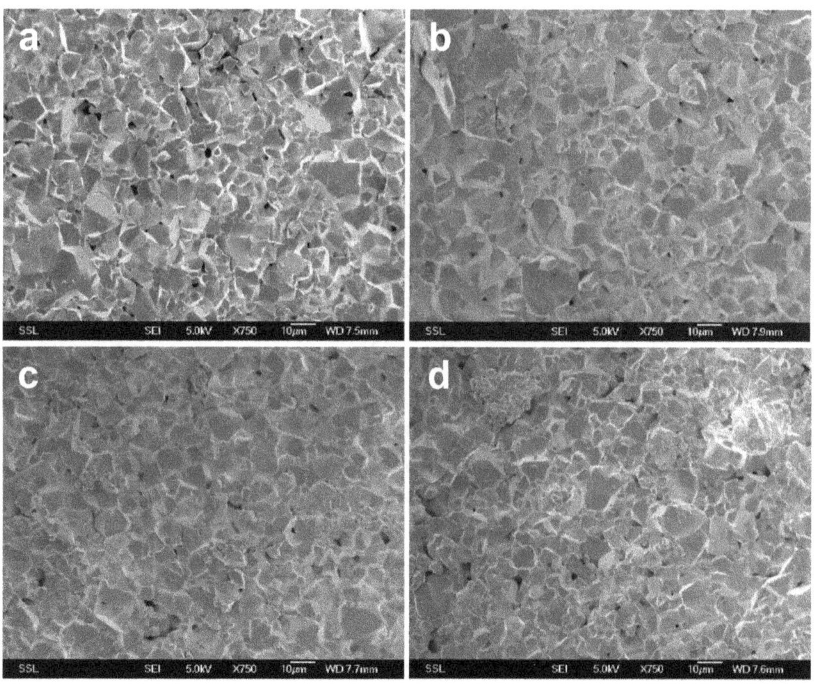

Figure 7.13. SEM images of the $Li_{0.50-0.50x}Co_xFe_{2.50-0.50x}O_4$ ceramics sintered at 950 °C with different concentrations of CoO: (a) $x = 0.01$, (b) $x = 0.03$, (c) $x = 0.05$ and (d) $x = 0.07$. Reproduced with permission from [1]. Copyright 2008 Elsevier.

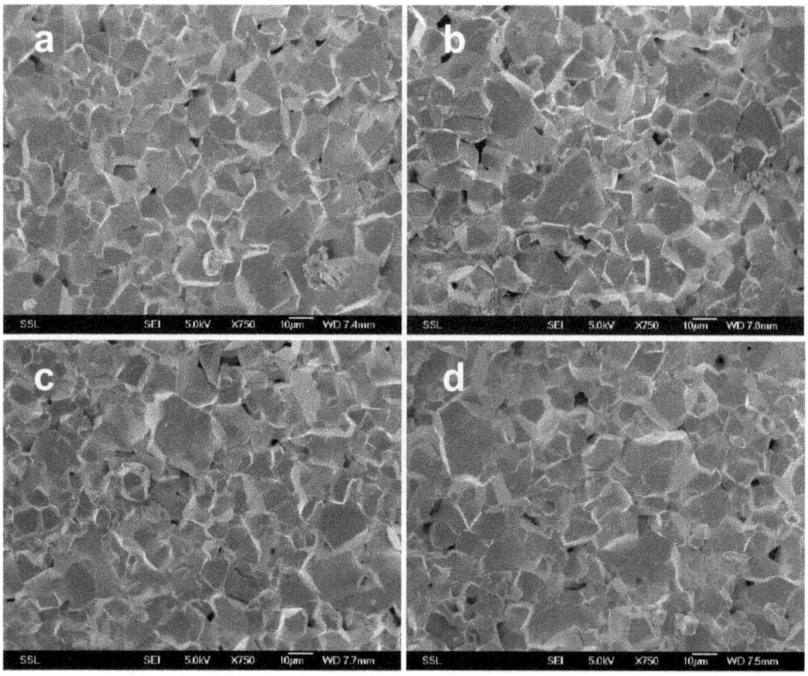

Figure 7.14. SEM images of the $Li_{0.50\text{-}0.50x}Co_xFe_{2.50\text{-}0.50x}O_4$ ceramics sintered at 1050 °C with different concentrations of CoO: (a) $x = 0.01$, (b) $x = 0.03$, (c) $x = 0.05$ and (d) $x = 0.07$. Reproduced with permission from [1]. Copyright 2008 Elsevier.

Table 7.3. Estimated thickness of the Bi_2O_3-rich liquid phase layer in the ferrite ceramics at the early stage of sintering. Reproduced with permission from [1]. Copyright 2008 Elsevier.

Bi_2O_3 (wt%)	0.2	0.5	1	3	5
t (nm)	0.35	0.87	1.74	5.21	8.65

mechanism. In this respect, the critical thickness of the liquid phase layer was reached when the content of Bi_2O_3 was 1 wt%. This result was roughly consistent with the experimental results, because 1 wt% Bi_2O_3 was sufficiently high to maintain full densification of the $Li_{0.50}Fe_{2.50}O_4$ ceramics in the sintering temperature range.

Since grain size was increased as the sintering temperature was raised, the thickness of the liquid phase layer would be increased due to the reduction in the volumetric fraction of the grain boundary. Therefore, the quantity of Bi_2O_3 to maintain the critical thickness of the liquid phase layer was reduced accordingly. Moreover, the decrease in viscosity of the liquid phase with increasing sintering temperature might require a thinner critical thickness, as stated in the previous section.

Interestingly, the average grain size of the sample doped with 0.2 wt% Bi_2O_3 after sintering at 1000 °C was about 5.4 μm. With this average grain size, the quantity of

Bi_2O_3 needed to sustain the critical thickness of the liquid phase layer was about 0.18 wt%, which was very close to the content of 0.2 wt% in the real experiment. However, the thickness of the liquid phase layer in the sample doped with 0.5 wt% Bi_2O_3 after sintering at 900 °C was much thicker than the critical thickness of 1.7 nm, if its average grain size was taken to be 4.5 μm. Accordingly, the mechanisms of the $Li_{0.50}Fe_{2.50}O_4$ + Bi_2O_3 system can be schematically described, as illustrated in figure 7.15, in order to qualitatively explain the densification and grain growth behaviors of the samples as a function of the content of Bi_2O_3 and the sintering temperature.

The $Li_{0.50}Fe_{2.50}O_4$ samples without the doping of Bi_2O_3 were densified through a solid-state reaction sintering mechanism, in which both the densification and grain growth took place through the diffusion of the atoms in the ferrite crystal lattice. Therefore, a high shrinking start temperature, a low shrinkage rate and a low final linear shrinkage were encountered because this sintering mechanism requires a relatively high activation energy. With the presence of Bi_2O_3 the liquid phase was generated at a certain temperature during the sintering process, because of the low melting point of Bi_2O_3. The formation of the liquid phase layer triggered the ferrite grains/particles to rearrange at relatively low temperatures, thus promoting the sintering procedure and grain growth of the materials.

It was noted that the sample doped with 1 wt% Bi_2O_3 had a liquid phase layer with a thickness that was almost the critical thickness, so that the liquid phase sintering took place in a swift way. In comparison, the liquid phase layers of the samples doped with 3 wt% and 5 wt% Bi_2O_3 were much thicker than the critical level. As a result, these two samples became less smart in densification, as reflected by the shrinkage behaviors as demonstrated in figure 7.4. In this respect, 1 wt% Bi_2O_3 was the optimal concentration in terms of densification. Interestingly, this difference was observed in the microstructure and density of the sintered samples, mainly because the density measurement and microstructure were static results, with a lower heating rate and longer dwelling time. Comparatively, the densification

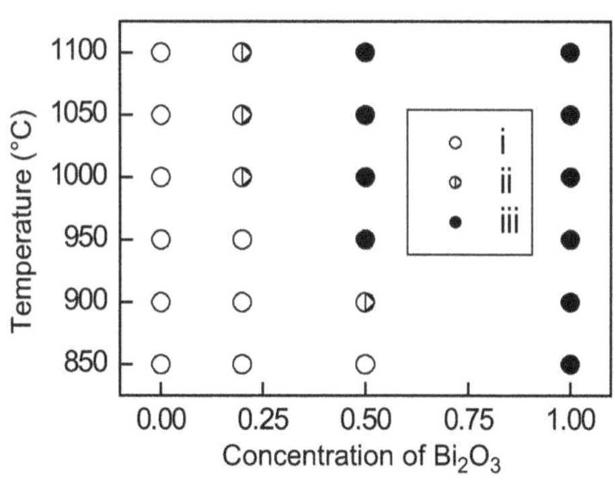

Figure 7.15. Schematic diagram of sintering mechanisms for the samples doped with different concentrations of Bi_2O_3: (i) solid-state diffusion through crystal lattices, (ii) solution–reprecipitation and (iii) diffusion through the liquid phase layer. Reproduced with permission from [1]. Copyright 2008 Elsevier.

behavior reflected a dynamic process, with a higher heating rate and without dwelling, which was more like an *in situ* monitoring.

By observing the grain sizes of the samples sintered at 1100 °C, it was noted that the grain size was maximized in the sample doped with 0.2 wt% of Bi_2O_3. Further increase in the content of Bi_2O_3 led to a gradual reduction in the average grain size, similar to the variation trend of the $MgFe_{1.98}O_4$ samples, as presented before. However, such a phenomenon was more significant in $Li_{0.50}Fe_{2.50}O_4$ than in $MgFe_{1.98}O_4$. The main reason could be that $Li_{0.50}Fe_{2.50}O_4$ has a lower sintering temperature than $MgFe_{1.98}O_4$. It was highly suspected that Li_2CO_3 could have been engaged in the formation of the liquid phase, because it is also a popular sintering aid for various ceramic materials [4–6].

As stated earlier, the liquid phase layer of the sample doped with 0.2 wt% Bi_2O_3 reached the critical thickness at 1000 °C. Therefore, the grain growth and densification mechanism were changed from solid-state diffusion to the solution–reprecipitation process. Once the sintering mechanism was switched to solution–reprecipitation, both the grain growth and densification were accelerated. For the sample doped with 0.5 wt%, the switching temperature was reduced to 900 °C. As the content of Bi_2O_3 was increased to 1 wt%, the critical thickness requirement was invalid, because the quantity of the liquid phase was too much, so the solution–reprecipitation mechanism started to work in the very beginning.

With the progress of densification and grain growth, the thickness of the liquid phase layer was gradually increased, so that the mechanism was transient from the solution–reprecipitation process to diffusion through the liquid phase layer. As a result, the grain growth rate became limited by the thickness of the liquid phase layer. This was simply because the diffusion path through the liquid phase layer to facilitate the grain growth was lengthened. The sample doped with 0.2 wt% Bi_2O_3 possessed the largest grain size, which could be attributed to the effect of liquid phase sintering, with the thickness of the liquid phase layer being the most suitable for grain growth of the ferrite, as long as the temperature was sufficiently high.

The $Li_{0.50-0.50x}Co_xFe_{2.50-0.50x}O_4$ samples are solid solutions of $Li_{0.50}Fe_{2.50}O_4$ and $CoFe_2O_4$. $CoFe_2O_4$ has a poorer sinterability than $Li_{0.50}Fe_{2.50}O_4$, because the starting materials contained Li_2CO_3, as mentioned above. At low sintering temperatures (e.g. 850 °C), the effect of Bi_2O_3 still had not started, and the role of $CoFe_2O_4$ in effecting the grain growth was somehow dominant. As a result, the average grain size was slightly decreased as the concentration of CoO was gradually increased. However, at high temperatures (e.g. 1100 °C) the effect of the liquid phase became dominant, such that the effect of CoO was surpassed.

Figure 7.16 shows the DC resistivities of the $Li_{0.50}Fe_{2.50}O_4$ ceramics sintered at different temperatures as a function of the concentration of Bi_2O_3. The resistivity was minimized in the samples doped with 0.2 wt% Bi_2O_3. As the content of Bi_2O_3 was raised from 0.2 wt% to 5 wt%, the resistivity increased by about three orders of magnitude. At a fixed doping concentration, the resistivity was maximized at an intermediate sintering temperature, with the exact temperature being dependent on the concentration of Bi_2O_3. Figure 7.17 depicts the DC resistivities of the

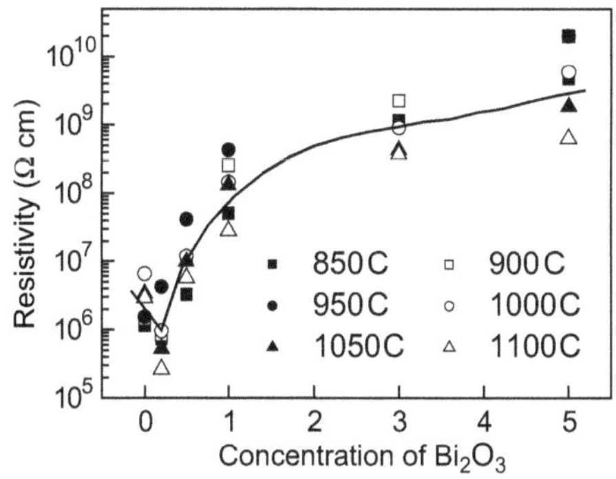

Figure 7.16. DC resistivities of the $Li_{0.50}Fe_{2.50}O_4$ ceramics sintered at different temperatures as a function of the concentration of Bi_2O_3. The solid line is a guide for the eyes. Reproduced with permission from [2]. Copyright 2008 Elsevier.

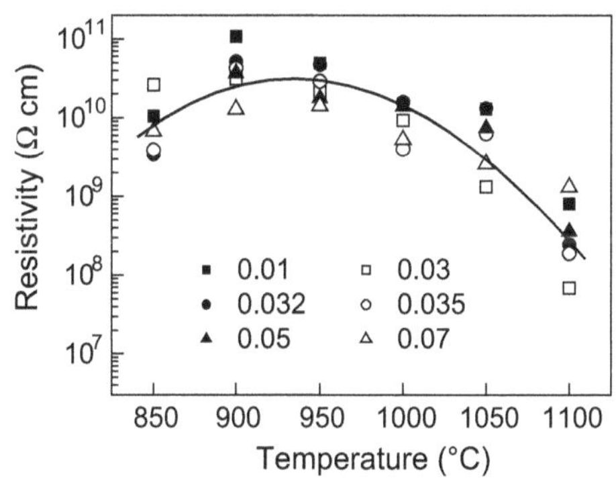

Figure 7.17. DC resistivities of the $Li_{0.50-0.50x}Co_xFe_{2.50-0.50x}O_4$ ceramics with different concentrations of CoO as a function of sintering temperature. The solid line is a guide for the eyes. Reproduced with permission from [2]. Copyright 2008 Elsevier.

$Li_{0.50-0.50x}Co_xFe_{2.50-0.50x}O_4$ ceramics as a function of sintering temperature, indicating that the maximum values were observed in the temperature range of 900 °C – 950 °C.

Complex relative permittivity curves of the $Li_{0.50}Fe_{2.50}O_4$ ceramics without doping of Bi_2O_3, after sintering at different temperatures, are illustrated in figure 7.18. Figure 7.19 shows the real permittivity values (collected at 1 MHz and 100 MHz) of the samples as a function of sintering temperature. The value at 1 MHz was decreased slightly with increasing sintering temperature in the range 850 °C – 950 °C. Above 950 °C, it was gradually and slowly increased. In comparison, the real permittivity values at 100 MHz were monotonically increased in the whole range of sintering temperature.

Figure 7.20 depicts loss tangents at 1 MHz of the $Li_{0.50}Fe_{2.50}O_4$ ceramics as a function of sintering temperature. It was found that both the real and imaginary permittivity values were decreased with a frequency below the frequency point of about 50 MHz, above which they remained nearly constant. The slight increase in

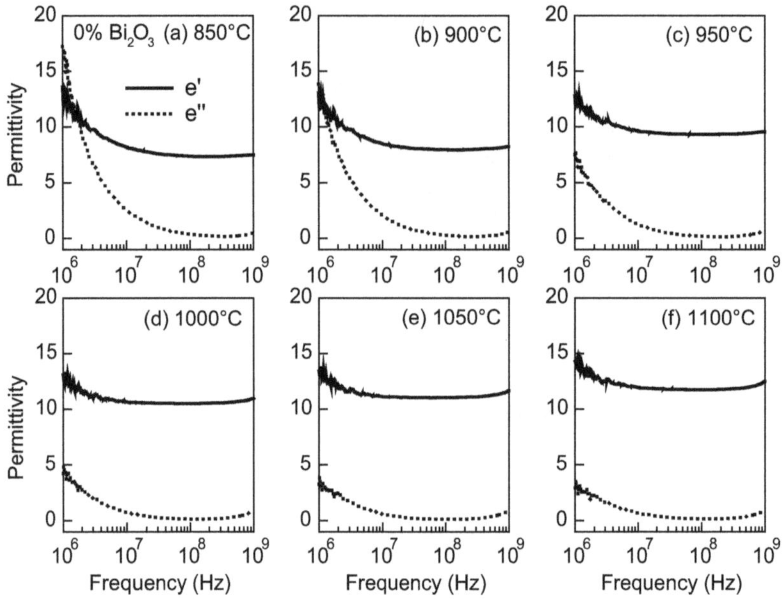

Figure 7.18. Complex relative permittivity curves of the $Li_{0.50}Fe_{2.50}O_4$ ceramics without doping of Bi_2O_3 after sintering at different temperatures. Reproduced with permission from [2]. Copyright 2008 Elsevier.

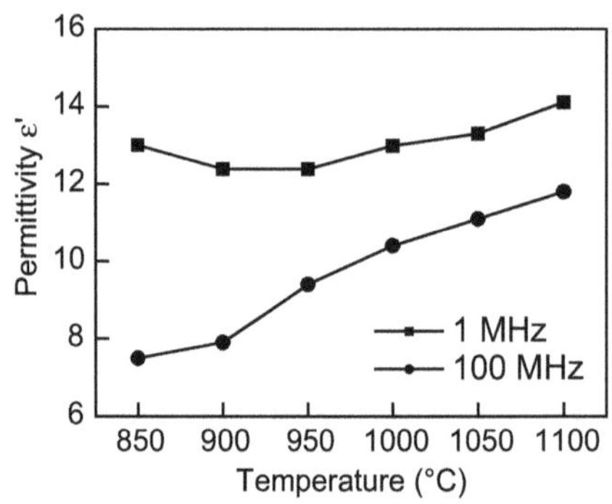

Figure 7.19. Real permittivity values (at 1 MHz and 100 MHz) of the $Li_{0.50}Fe_{2.50}O_4$ ceramics with the doping of Bi_2O_3 as a function of sintering temperature. Reproduced with permission from [2]. Copyright 2008 Elsevier.

permittivity at the high-frequency end (above 0.5 GHz) was caused by the instrument, and was not of the intrinsic properties of the materials. As the sintering temperature was increased, the increase of the real and imaginary permittivity values in the low frequency range was weakened, so that the dielectric loss tangent was reduced accordingly.

The dielectric properties of the $Li_{0.50}Fe_{2.50}O_4$ samples doped with low concentrations of Bi_2O_3 were very sensitive to the sintering temperature. Complex relative permittivity curves of the $Li_{0.50}Fe_{2.50}O_4$ samples doped with 0.2 wt% and 0.5 wt%

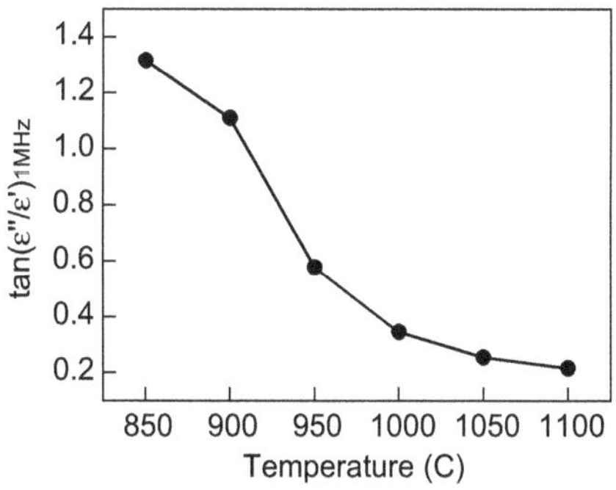

Figure 7.20. Dielectric loss tangents (at 1 MHz) of the $Li_{0.50}Fe_{2.50}O_4$ ceramics without the doping of Bi_2O_3 as a function of sintering temperature. Reproduced with permission from [2]. Copyright 2008 Elsevier.

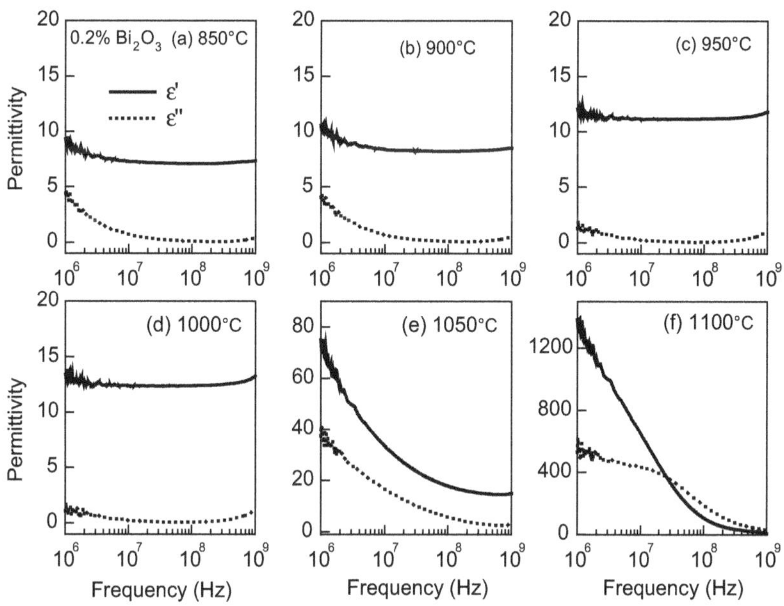

Figure 7.21. Complex relative permittivity curves of the $Li_{0.50}Fe_{2.50}O_4$ ceramics doped with 0.2 wt% Bi_2O_3 after sintering at different temperatures. Reproduced with permission from [2]. Copyright 2008 Elsevier.

Bi_2O_3 after sintering at different temperatures are depicted in figures 7.21 and 7.22, respectively. The real permittivity values collected at 1 MHz and 100 MHz are demonstrated in figure 7.23, while their dielectric loss tangents collected at 1 MHz are plotted in figure 7.24. The real permittivity values had an abrupt increase for both samples, occurring at 1050 °C and 950 °C for the samples doped with 0.2 wt% and 0.5 wt% Bi_2O_3, respectively. In comparison, the magnitude of the increase was much higher in the sample doped with 0.2 wt% Bi_2O_3, so this sample also had a much higher dielectric loss tangent.

The dielectric loss tangent of the $Li_{0.50}Fe_{2.50}O_4$ ceramics was largely reduced as the content of Bi_2O_3 was increased. Complex relative permittivity curves of the

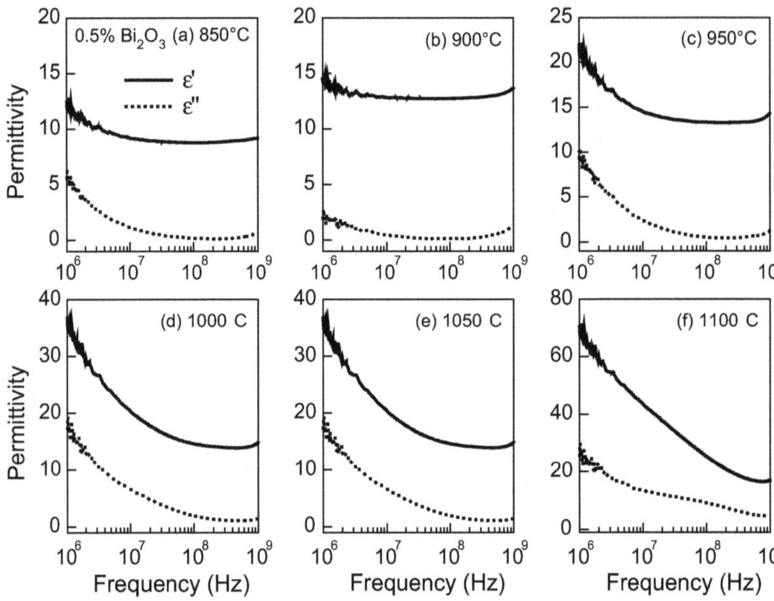

Figure 7.22. Complex relative permittivity curves of the $Li_{0.50}Fe_{2.50}O_4$ ceramics doped with 0.5 at% Bi_2O_3 after sintering at different temperatures. Reproduced with permission from [2]. Copyright 2008 Elsevier.

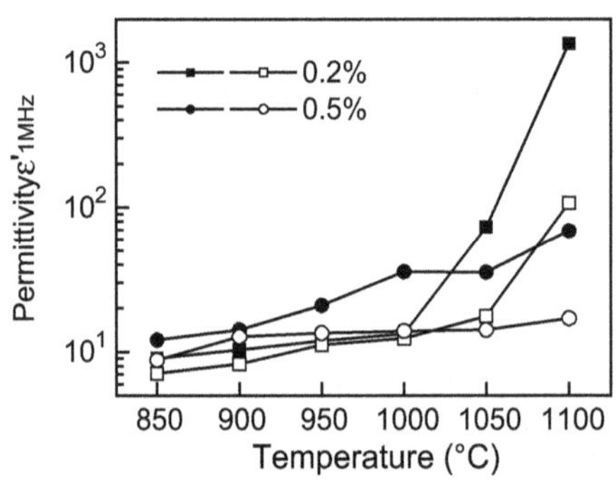

Figure 7.23. Real permittivity values (at 1 MHz) of the $Li_{0.50}Fe_{2.50}O_4$ ceramics (with 0.2 wt% and 0.5 wt% Bi_2O_3) as a function of sintering temperature. Reproduced with permission from [2]. Copyright 2008 Elsevier.

$Li_{0.50}Fe_{2.50}O_4$ samples doped with 1 wt% and 3 wt% Bi_2O_3 are depicted in figures 7.25 and 7.26, respectively. Dielectric properties of the samples doped with 5 wt% Bi_2O_3 were almost the same as those of the samples with 3 wt% Bi_2O_3, while the former had slightly lower dielectric loss tangents. To further demonstrate their differences in dielectric properties, the real permittivity and dielectric loss tangent values (at 1 MHz) of the $Li_{0.50}Fe_{2.50}O_4$ samples doped with high concentrations of Bi_2O_3 are demonstrated in figures 7.27 and 7.28, respectively.

For the $Li_{0.50}Fe_{2.50}O_4$ samples without doping and those doped with lower concentrations of Bi_2O_3 their low resistivity values after sintering at low temperatures were attributed to their poor densification or porous microstructure, due to

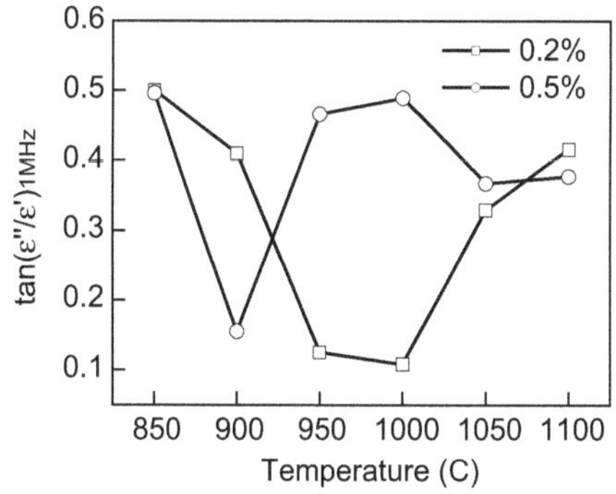

Figure 7.24. Dielectric loss tangents (at 1 MHz) of the $Li_{0.50}Fe_{2.50}O_4$ ceramics (with 0.2 wt% and 0.5 wt% Bi_2O_3) as a function of sintering temperature. Reproduced with permission from [2]. Copyright 2008 Elsevier.

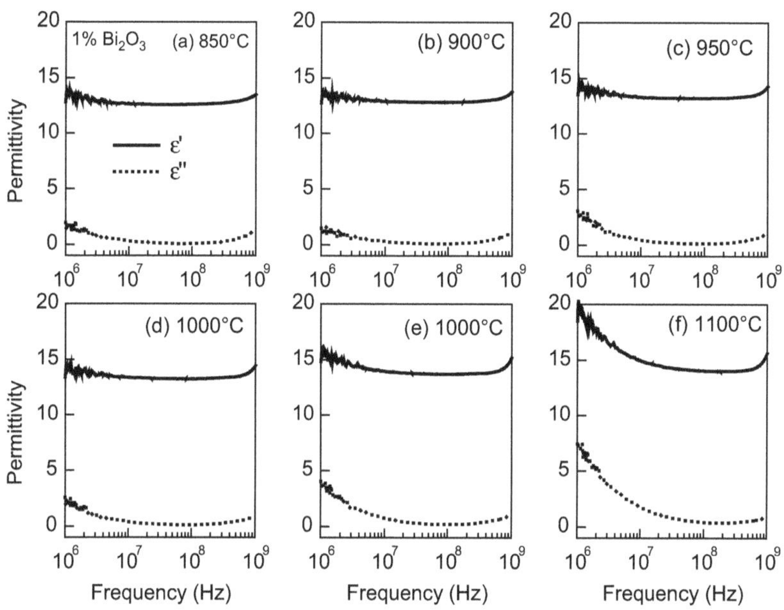

Figure 7.25. Complex relative permittivity of the $Li_{0.50}Fe_{2.50}O_4$ ceramics doped with 1 wt% Bi_2O_3 after sintering at different temperatures. Reproduced with permission from [2]. Copyright 2008 Elsevier.

the reasons discussed previously. In this case, the porous microstructure consisted of a large number of open pores, thus adsorbing impurities and water vapor. As a consequence, they had relatively low resistivities.

Except for the samples doped with 0.2 wt% Bi_2O_3, the variation in DC resistivity of the samples could generally be described using equation (6.6). The grain resistivity (ρ_2) was assumed to be constant with respect to the doping level of Bi_2O_3, because no reaction took place between Bi_2O_3 and the $Li_{0.50}Fe_{2.50}O_4$ ferrite. As a consequence, the DC resistivities of the samples were dependent on x and ρ_1. In addition, the average grain size of the samples was only very slightly decreased with increasing

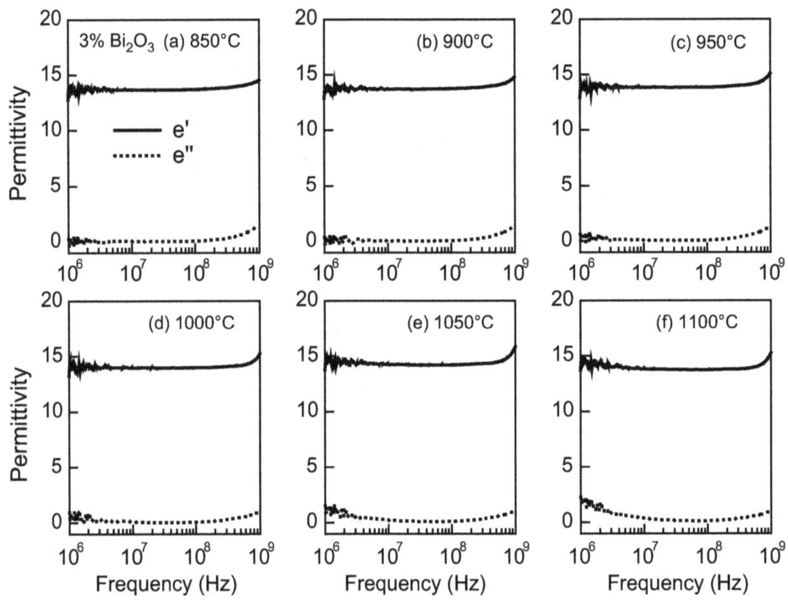

Figure 7.26. Complex relative permittivity curves of the $Li_{0.50}Fe_{2.50}O_4$ ceramics doped with 3 wt% Bi_2O_3 after sintering at different temperatures. Reproduced with permission from [2]. Copyright 2008 Elsevier.

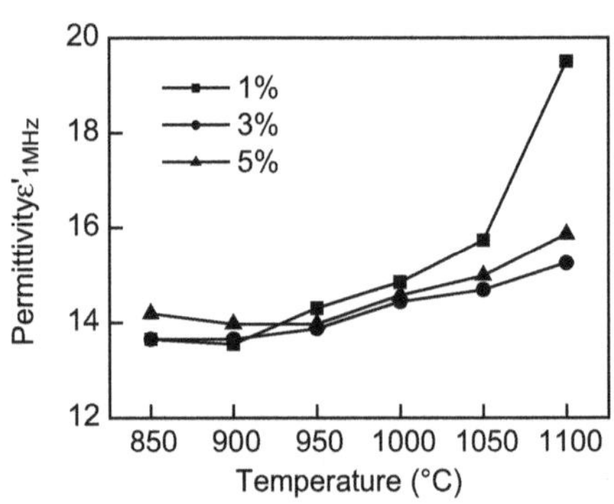

Figure 7.27. Real permittivity values (at 1 MHz) of the $Li_{0.50}Fe_{2.50}O_4$ samples doped with high concentrations of Bi_2O_3 as a function of sintering temperature. Reproduced with permission from [2]. Copyright 2008 Elsevier.

content of Bi_2O_3 from 0.5 wt% to 5 wt%, in particular for those sintered at temperatures ⩾950 °C, as presented earlier [1]. Therefore, as the content of Bi_2O_3 was raised, both x and ρ_1 were increased, thus leading to an increase in the DC resistivity.

Similarly, the lowest DC resistivity of the samples doped with 0.2 wt% was related to the clean two-grain boundaries. For $MgFe_{1.98}O_4$, the concentration of Bi_2O_3 for the presence of the lowest DC resistivity was 0.5 wt%. The DC resistivities were increased by about one order of magnitude in the samples with the substitution of Co for Li. This observation was even significant for the samples sintered at

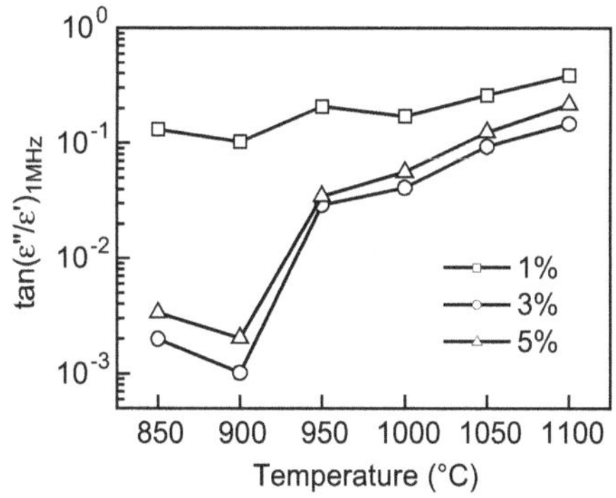

Figure 7.28. Dielectric loss tangents (at 1 MHz) of the $Li_{0.50}Fe_{2.50}O_4$ ceramics doped with high concentrations of Bi_2O_3 (1–3 wt%) as a function of sintering temperature. Reproduced with permission from [2]. Copyright 2008 Elsevier.

Table 7.4. Differences in the real permittivity of the samples between the data recorded at 1 MHz and 100 MHz. Reproduced with permission from [2]. Copyright 2008 Elsevier.

	850 °C	900 °C	950 °C	1000 °C	1050 °C	1100 °C
0%	5.51	4.49	3.01	2.62	2.21	2.32
0.2%	2.01	2.04	0.75	0.89	54.6	1250
0.5%	3.32	1.44	8.51	21.5	22.6	51.6
1%	1.05	0.71	1.12	1.61	2.04	5.53

temperatures in the range 900 °C–950 °C, which was ascribed to the presence of *p*-type conduction caused by the formation of Co^{3+} ions.

The dependent profile of permittivity on the grain size of the ferrite ceramics could be interpreted using the Maxwell–Wagner effect [7, 8]. The difference in permittivity between 1 MHz and 100 MHz was used to distinguish the contributions from the conduction and the atomic/electronic polarization. The differences in permittivity between 1 MHz and 100 MHz of some samples are summarized in table 7.4. For $Li_{0.50}Fe_{2.50}O_4$ samples without doping of Bi_2O_3 the difference decreased monotonically from 5.51 to 2.21 as the sintering temperature was increased from 850 °C to 1050 °C, while it increased slightly to 2.32 as the sintering temperature was further raised to 1100 °C.

This observation was consistent with the increase in density with sintering temperature above 1050 °C. The small increase in the value of the sample sintered at 1100 °C could be related to the production of Fe^{2+} ions. The continuous increase in the permittivity at 100 MHz could be understood according to the Maxwell–Wagner effect [9], i.e. the increase in grain size. If the 100 MHz value was taken as intrinsic permittivity, the difference in the permittivity would be extrinsic. According to dielectric theory, ferrites have a very low dielectric loss tangent over the frequency range from 1 MHz to 1 GHz. Therefore, the dielectric loss tangent of the $Li_{0.50}Fe_{2.50}O_4$

ceramics without doping was caused by the conduction loss. The loss tangent was decreased with increasing sintering temperature and the conduction loss was gradually reduced. Therefore, $Li_{0.50}Fe_{2.50}O_4$ ferrite ceramics without the doping of Bi_2O_3 could not be processed to have a dielectric loss tangent $<10^{-2}$.

In terms of the effect of Bi_2O_3 on dielectric properties, the $Li_{0.50}Fe_{2.50}O_4$ ceramics could be divided into two categories: the low concentration category (i.e. 0.2–0.5 wt %) and high concentration category (i.e. 1–5 wt%). The difference in real permittivity between 1 MHz and 100 MHz for the sample doped with 0.2 wt% Bi_2O_3 decreased with increasing sintering temperature from 850 °C to 950 °C. However, after sintering at 1000 °C the value was very small, which suggested that the samples sintered at 950 °C and 1000 °C would have a low dielectric loss tangent. Then, the difference was abruptly increased after the sample was sintered at temperatures ⩾1050 °C. An extremely high value of 1250 was observed in the sample after sintering at 1100 °C. In addition, the sample sintered at 1100 °C displayed a very high real permittivity at 1 MHz.

The minimum value of difference in permittivity between 1 MHz and 100 MHz for the sample doped with 0.5 wt% Bi_2O_3 was present after sintering at 900 °C, with a value about two times that of the sample doped with 0.2 wt% Bi_2O_3 sintered at 950 °C. However, the difference in permittivity between 1 MHz and 100 MHz of the sample doped with 0.5 wt% Bi_2O_3 was much lower than that of the sample with 0.2 wt% Bi_2O_3, as the sintering temperature was increased to 1050 °C and 1100 °C. In summary, the difference in permittivity between 1 MHz and 100 MHz was concurrent with the dielectric loss tangent, as revealed in figure 7.16. In other words, to achieve a low dielectric loss tangent of ferrite ceramics is to eliminate the extrinsic contribution, i.e. making fully densified samples at sufficiently low temperatures, so as to avoid porous microstructures and prevent the formation of Fe^{2+} ions.

Figure 7.29 shows the complex relative permittivity curves of the $Li_{0.50}Fe_{2.50}O_4$ ceramics without doping of Bi_2O_3 after sintering at different temperatures, while those of the samples doped with different contents of Bi_2O_3 after sintering at 900 °C are depicted in figure 7.30. Figure 7.31 illustrates the static permeability data of the $Li_{0.50}Fe_{2.50}O_4$ ferrite ceramics as a function of sintering temperature. For the

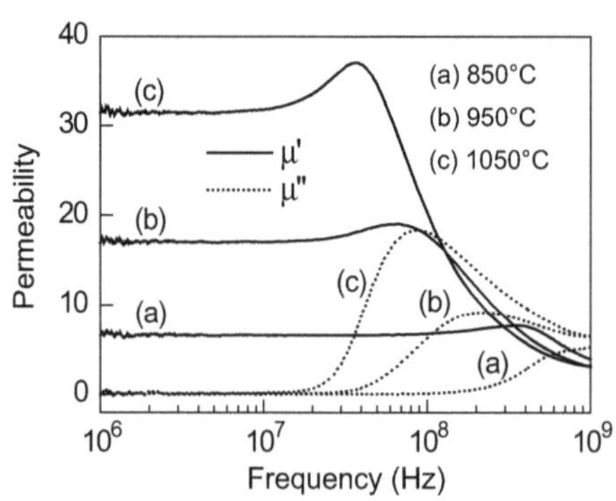

Figure 7.29. Complex permeability curves of the $Li_{0.50}Fe_{2.50}O_4$ ceramics sintered at different temperatures for 2 h. Reproduced with permission from [3]. Copyright 2008 Elsevier.

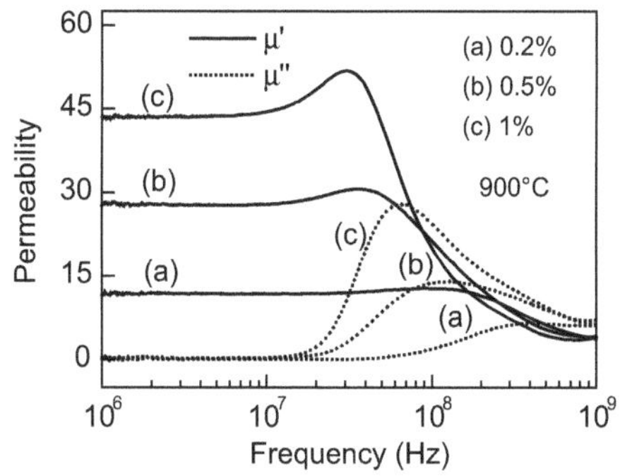

Figure 7.30. Complex permeability curves of the $Li_{0.50}Fe_{2.50}O_4$ ceramics doped with different concentrations of Bi_2O_3, after sintering at 900 °C for 2 h. Reproduced with permission from [3]. Copyright 2008 Elsevier.

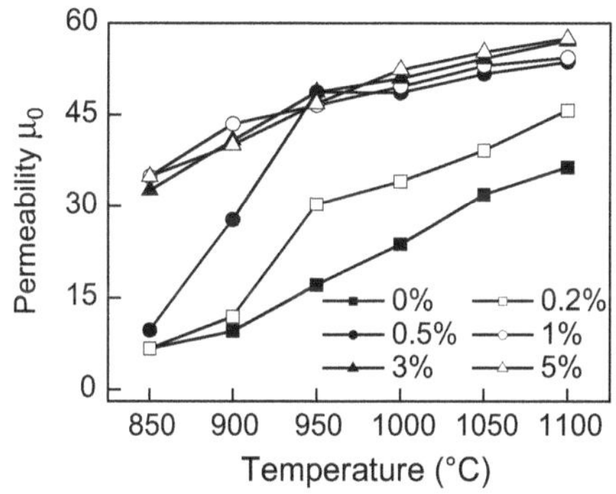

Figure 7.31. Real permeability values (at 1 MHz) of the $Li_{0.50}Fe_{2.50}O_4$ ceramics doped with different concentrations of Bi_2O_3 as a function of sintering temperature. Reproduced with permission from [3]. Copyright 2008 Elsevier.

$Li_{0.50}Fe_{2.50}O_4$ samples with doping with Bi_2O_3, static permeability monotonically increased as the sintering temperature was increased. The static permeability of the samples doped with 0.2 wt% Bi_2O_3 was very close to that of the $Li_{0.50}Fe_{2.50}O_4$ sample without the addition of Bi_2O_3 if the sintering temperature was ⩽900 °C. Once the sintering temperature was increased to ⩾950 °C, the static permeability of the samples doped with 0.2 wt% Bi_2O_3 was rapidly increased. The values were constantly larger than those of the samples without the sintering aid.

The static permeability of the samples doped with 0.5 wt% Bi_2O_3 linearly increased with the sintering temperature in the range 850 °C–950 °C. After that, the values were close to those of the samples with higher concentrations of Bi_2O_3. Above 1 wt %, the effect of Bi_2O_3 on the static permeability was very weak. The variation trend in the static permeability of the $Li_{0.50}Fe_{2.50}O_4$ ceramics was analogous to those of the density and the average grain size as a function of sintering temperature, as discussed previously.

Applying equation (6.1) to the samples without doping and with 0.2 wt% and 0.5 wt%, the data are shown in figure 7.32. In particular, the data of the two samples

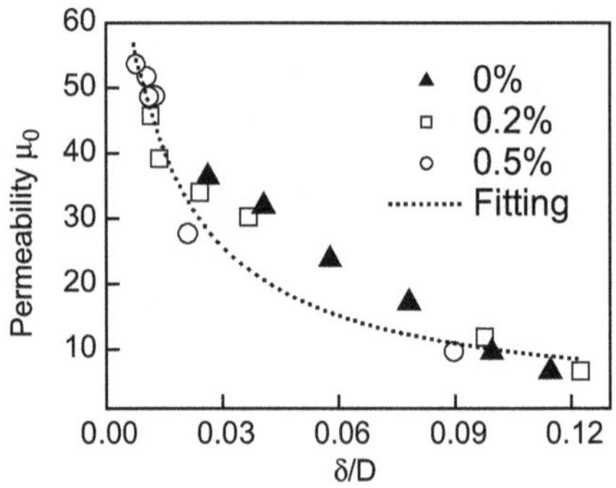

Figure 7.32. Static permeability versus δ/D of the $Li_{0.50}Fe_{2.50}O_4$ ceramic samples without doping and those doped with low concentrations of Bi_2O_3. Reproduced with permission from [3]. Copyright 2008 Elsevier.

doped with 0.2 wt% and 0.5 wt% Bi_2O_3 yielded intrinsic state permeabilities of about 100. However, the data of the samples without doping of Bi_2O_3 could not be fitted, which indicated that the variation in their permeability did not follow the magnetic circuit model. It was highly possible that the ratio (δ/D) calculated using the measured densities of the samples was not a real reflection of the contribution of the increase in grain size as a function of the sintering temperature.

The variation in static permeability of the samples with higher concentrations of Bi_2O_3 as a function of sintering temperature could be solely ascribed to the increase in their average grain size. Figure 7.33 shows the static permeability data of all the $Li_{0.50}Fe_{2.50}O_4$ samples as a function of grain size. As stated before, the permeability of polycrystalline ferrite ceramics is contributed to by both the spin rotation and domain wall motions. The contribution due to domain motion is almost linearly proportional to the grain size, although the grain size must be over a critical value in the range 0.1–1 μm, while the contribution of spin rotation is not related to grain size, according to Globus's model [10].

The critical size of a single domain of ferrites should be smaller than 1 μm. Therefore, for the samples with average grain sizes >2.5 μm, their permeability due to domain wall displacement would be dominant. A linear relation has been derived from the data, i.e. $\mu_0 = 28.7 + 1.7D$. Similar phenomena have also been found in other ferrites [11]. For the samples with an average grain size <2.5 μm, the permeability was quickly increased with increasing grain size. In this case, the grains could be classified into two groups, depending on whether the sizes were below or above the critical size for a single domain. According to the data shown in figure 7.33, the rotation permeability was about 6, which was much smaller than the permeability due to the wall displacement. The rapid increase in the static permeability could be attributed to the fact that the fraction of smaller grains in the samples was reduced, while that of the larger grains was increased.

The substitution of Li with Co modified the magnetic properties of the ferrite ceramics, if the concentration of CoO was $\geqslant x = 0.01$. Figure 7.34 shows representative typical complex permeability curves of the $Li_{0.50-0.50x}Co_xFe_{2.50-0.50x}O_4$ ceramics sintered at 900 °C with different concentrations of CoO. Representative static

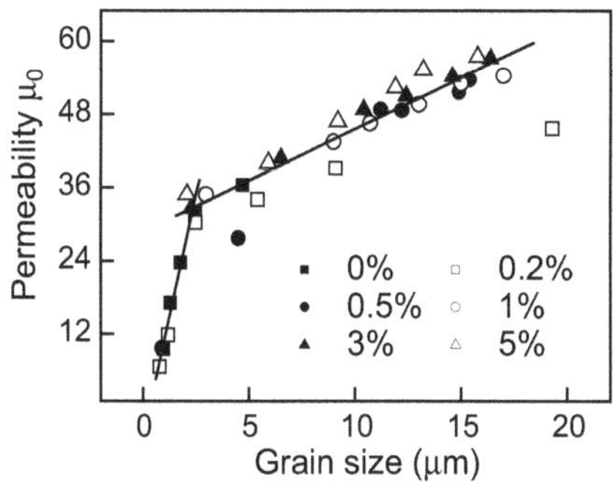

Figure 7.33. Static permeability versus grain size of the $Li_{0.50}Fe_{2.50}O_4$ ceramics doped with different contents of Bi_2O_3, after sintering at different temperatures. The solid line is a guide for the eyes. Reproduced with permission from [3]. Copyright 2008 Elsevier.

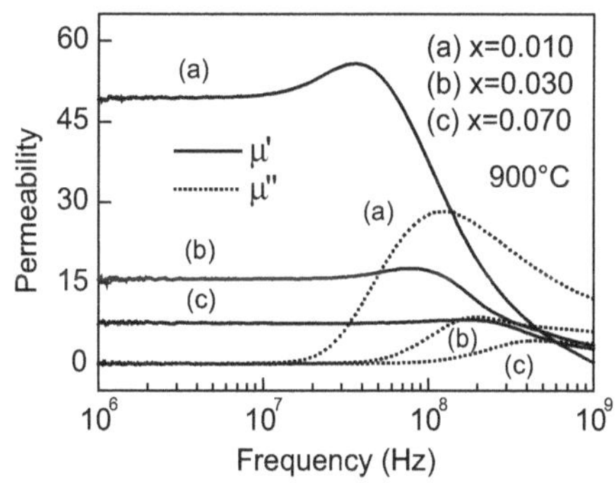

Figure 7.34. Complex permeability curves of the $Li_{0.50-0.50x}Co_xFe_{2.50-0.50x}O_4$ ceramics with different concentrations of CoO after sintering at 900 °C for 2 h. Reproduced with permission from [3]. Copyright 2008 Elsevier.

permeability values of the $Li_{0.50-0.50x}Co_xFe_{2.50-0.50x}O_4$ ceramics sintered at different temperatures as a function of the concentration of CoO are illustrated in figure 7.35. At a fixed composition, the permeability was increased as the sintering temperature was increased, which was similar to those of the $Li_{0.50}Fe_{2.50}O_4$ ceramics doped with 3 wt% Bi_2O_3. At a fixed sintering temperature, the static permeability was increased with increasing concentration of CoO up to $x = 0.01$, maximized in the sample with $x = 0.01$. After that, the permeability was quickly decreased.

Similarly, Li^+ and Fe^{3+} ions contributed to negative magnetocrystalline anisotropy in the ferrite. The magnetocrystalline anisotropy constants of the $Li_{0.50-0.50x}Co_xFe_{2.50-0.50x}O_4$ samples are listed in table 7.5, which were derived from -8.5×10^4 erg cm^{-3} and 180×10^4 erg cm^{-3} for $Li_{0.50}Fe_{2.50}O_4$ and $CoFe_2O_4$ [12]. Obviously, the value was varied from negative to positive as the substituting concentration of CoO was gradually increased from $x = 0.035$ to $x = 0.05$. A minimum absolute value was observed in the sample at $x = 0.05$, which was expected to possess the maximized permeability value. However, the experimental peak value was observed in the sample with $x = 0.01$. The difference between the theoretical and

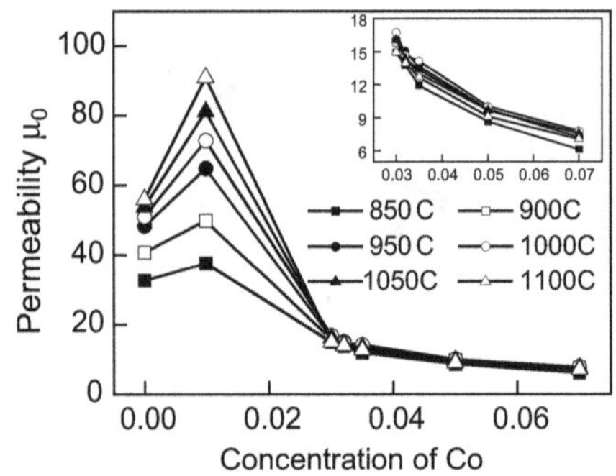

Figure 7.35. Real permeability values (at 1 MHz) of the $Li_{0.50-0.50x}Co_xFe_{2.50-0.50x}O_4$ ceramics sintered at different temperatures as a function of the concentration of CoO. The inset shows the detailed curves of real permeability versus the concentration of CoO in the range $x = 0.03–0.07$. Reproduced with permission from [3]. Copyright 2008 Elsevier.

Table 7.5. Magnetocrystalline anisotropy constants of the $Li_{0.50-0.50x}Co_xFe_{2.50-0.50x}O_4$ samples. Reproduced with permission from [3]. Copyright 2008 Elsevier.

x	0.01	0.03	0.032	0.035	0.05	0.07
K_1 (10^4 erg cm^{-3})	−6.62	−2.85	−2.47	−1.90	0.93	4.69

the experimental results could be explained similarly to that for the $MgFe_{1.98}O_4$ ferrites.

After the magnetic properties were modified, potential candidates with magneto-dielectric properties could be obtained using the $Li_{0.50-0.50x}Co_xFe_{2.50-0.50x}O_4$ ferrite ceramics. Figure 7.36 shows the magneto-dielectric properties of the representative $Li_{0.50-0.50x}Co_xFe_{2.50-0.50x}O_4$ ($x = 0.032$) ceramics which were sintered at 900 °C for 2 h. The samples substituted with CoO at concentrations of $x = 0.030 - 0.035$ are listed in table 7.6. The impedance of the sample was well matched to that of free space. At the same time, both the magnetic and dielectric tangents were below the desired value of 10^{-2}. Therefore, these ferrite ceramics could find potential applications for the antennas over the HF band (3–30 MHz). The real permeability and permittivity values of the $Li_{0.50-0.50x}Co_xFe_{2.50-0.50x}O_4$ ceramics were higher than those of the Mg–Cu–Co [13, 14] and Mg–Co ferrite ceramics, as presented previously [1–3].

7.2 Ni–Zn–Co ferrite ceramics

In this section the fabrication and characterization of Ni–Zn–Co ferrite ceramics with promising magneto-dielectric properties over 30–90 MHz derived from high-energy ball milled precursors containing iron (Fe) powder will be discussed [15]. Fe powder was used to compensate for the volume shrinkage due to its reaction with oxygen during the sintering process, so that it was possible to achieve near net-shape

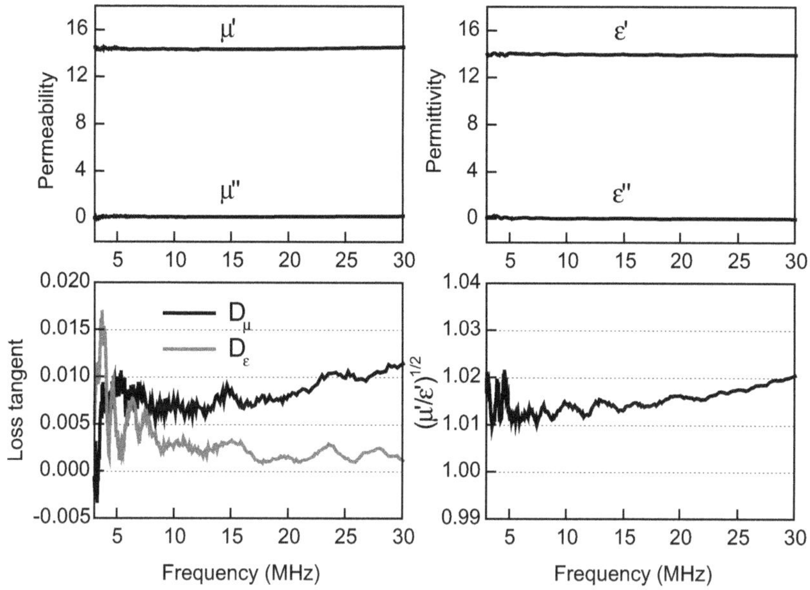

Figure 7.36. Magneto-dielectric properties (over 3–30 MHz) of the $Li_{0.50-0.50x}Co_xFe_{2.50-0.50x}O_4$ (x = 0.032) ceramics sintered at 900 °C for 2 h. Reproduced with permission from [3]. Copyright 2008 Elsevier.

Table 7.6. Magneto-dielectric parameters of the $Li_{0.50-0.50x}Co_xFe_{2.50-0.50x}O_4$ ceramics over 3–30 MHz. Reproduced with permission from [3]. Copyright 2008 Elsevier.

x	T (°C)	μ' (at 15 MHz)	ε' (at 15 MHz)	D_μ	D_ε	$(\mu'/\varepsilon')^{1/2}$
0.030	850	14.7	14.2	0.005–0.008	0.013–0.003	1.01–1.02
0.032	850	13.7	13.8	0.005–0.005	0.025–0.005	1.00–1.09
0.032	900	14.2	13.8	0.010–0.011	0.015–0.001	1.01–1.02
0.035	900	12.9	13.6	0.002–0.010	0.015–0.001	0.97–0.98
0.035	950	13.5	13.6	0.012–0.020	0.015–0.002	1.00–1.01

processing. In other words, the final ceramics would have nearly unshrunk dimensions compared to the green bodies.

Commercial Fe (99+% purity), Fe_2O_3 (99% purity), MnO_2 (98% purity) and Co_3O_4 (99+% purity) powders were selected as the starting materials. The ferrite composition was $Ni_{0.70}Zn_{0.25}Co_{0.05}Fe_{1.90}Mn_{0.02}O_4$. The effect of the relative content of Fe and Fe_2O_3 was studied, as expressed as the percentages of Fe_2O_3 [100%Fe_2O_3/(Fe_2+Fe_2O_3)] being 10%, 20%, 30% and 50%. The starting powders were blended and then activated using high-energy ball milling for 12 h. The high-energy milling was performed using a Retsch PM400 type planetary ball milling system. A 250 ml tungsten carbide vial and 100 tungsten carbide balls with diameters of 10 mm were adopted as a milling medium. The milling speed was fixed at 200 rpm. The activated powders were shaped and sintered at 800 °C for 8 h. A sample with the same composition was fabricated using the conventional ceramic process with all-oxide precursors. The oxide powders were mixed and calcined at 1000 °C for 2 h. The calcined powder was then sintered at 1250 °C for 2 h. Therefore, it was possible to compare the ferrite ceramics with micro- and nano-scale sizes.

Figure 7.37 shows the linear shrinkage curve of the sample with 10% Fe_2O_3. Shrinkage was maximized at about 500 °C, which was caused by the oxidation of the polymer binder in the green body. After that, expansion was started, which was slowed down above about 800 °C. Finally, the sample started to shrink at about 1050 °C. However, the linear size was still larger than the initial size at 1200 °C. The expansion was directly related to the oxidation of the Fe. With increasing content of Fe_2O_3 the maximum expansion was decreased. According to the densification behavior, 800 °C was selected as the sintering temperature to fabricate ferrite ceramics. The XRD patterns indicated that the samples were all of single spinel structure.

Figure 7.38 shows cross-sectional SEM images of the nano-sized samples and the microsized sample. The sample from the mixture containing 10% Fe_2O_3 consisted of large grains, while the grain size distribution was in the range 0.5–1.5 μm. The samples with higher contents of Fe_2O_3 had no significant difference in microstructure, with average grain sizes of 100–200 nm. The samples also contained nano-sized pores with a relatively uniform distribution. With increasing content of Fe_2O_3, the pores adopted a more and more uniform distribution.

The densities of the samples were measured using the weight/volume ratio, while the final linear expansion was calculated from the diameters before and after sintering. It was found that the density was decreased with increasing content of Fe_2O_3 powder in the precursor mixture. This observation was understandable, since the higher the content of Fe_2O_3 in the precursor mixture, the more difficult it is for the samples to be sintered. All the sintered samples retained nearly the same dimensions as their green bodies, suggesting that near net-shape fabrication of ferrite ceramics was realized. Because the final ceramics had no variation in physical dimensions, it was more convenient for the design in some applications.

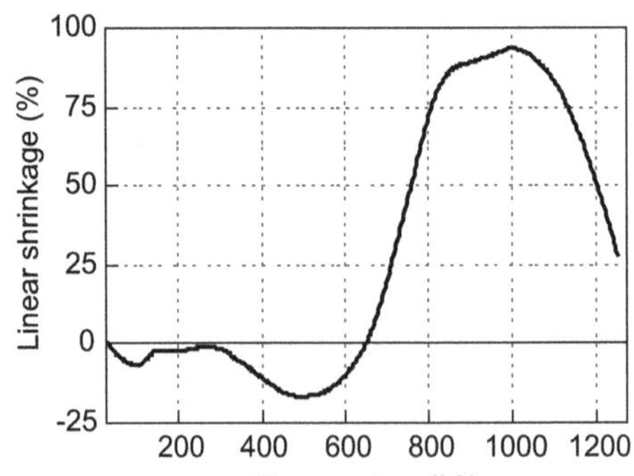

Figure 7.37. Linear shrinkage of the sample containing 10% Fe_2O_3. Reproduced with permission from [15]. Copyright 2018 Elsevier.

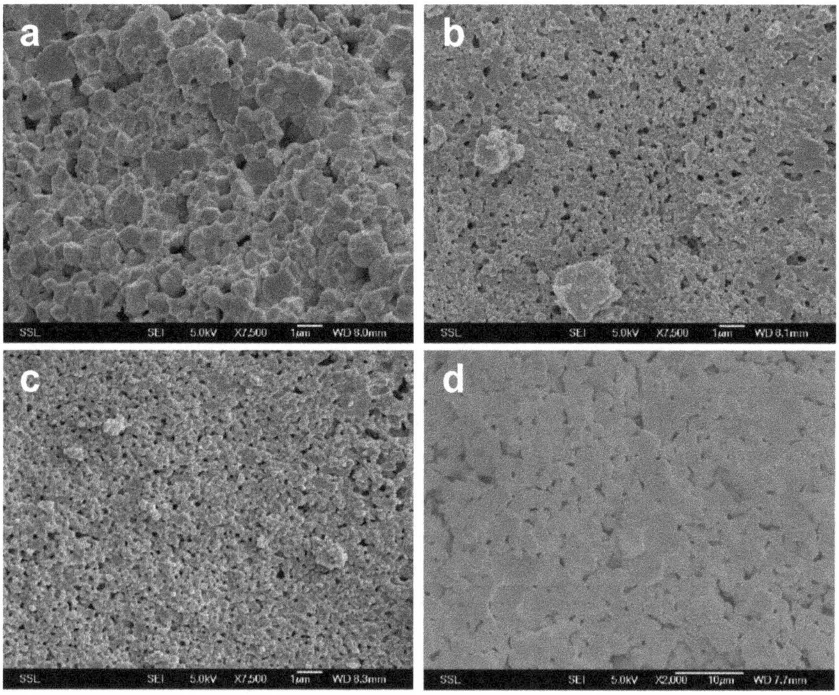

Figure 7.38. Cross-sectional SEM images of the samples with different compositions after sintering at 800 °C: (a) 10% Fe_2O_3, (b) 30% Fe_2O_3, (c) 50% Fe_2O_3 and (d) microsized sample. Reproduced with permission from [15]. Copyright 2018 Elsevier.

Complex permittivity and permeability curves of the ferrite ceramics are depicted in figures 7.39 and 7.40, respectively. The samples from the mixtures containing 10% and 20% Fe_2O_3 possessed very high permittivity values, together with a high dielectric loss tangent, in particular for the sample with 10% Fe_2O_3. It was highly possible that Fe was partially oxidized to Fe^{3+} instead of Fe^{3+}. Also, Fe would react with Fe_2O_3 to form Fe_3O_4, which has a similar spinel crystal structure to the nominal ferrite. Therefore, the XRD patterns revealed no secondary phases.

As mentioned previously, the presence of Fe^{2+} ions resulted in high permittivity, because the polarization of Fe^{2+} is stronger than that of Fe^{3+}. This observation was consistent with the magnetic properties of the samples. The sample with 10% Fe_2O_3 exhibited a complex permeability curve analogous to those of composites consisting of magnetic metal powders and a polymer matrix. The sample from the mixture containing 50% Fe_2O_3 displayed promising magneto-dielectric properties over the frequency band of 30–90 MHz, as shown in figure 7.41. It had close values of real permeability and permittivity of about 10, together with sufficiently low magnetic and dielectric loss tangents.

It was found that equation (6.1) was valid for data of μ_0 with different compositions. As the content of Fe_2O_3 was increased, both the grain size (D) and density (ρ) were increased, thus leading to an increment in the ratio of δ/D. As a consequence, the static permeability was reduced. The ratio of δ/D was derived from

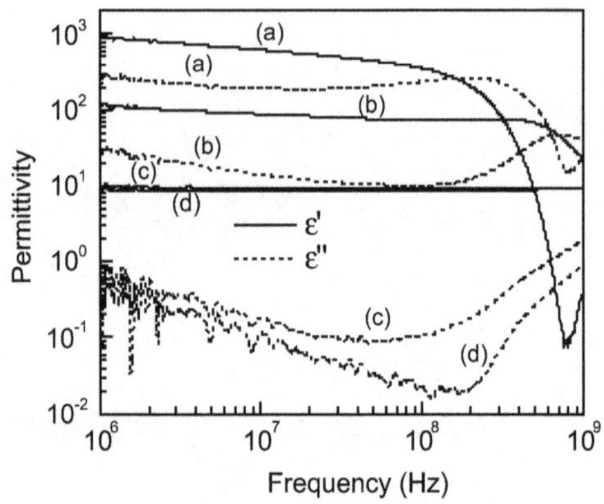

Figure 7.39. Complex relative permittivity curves of the ferrite ceramics with different contents of Fe_2O_3 in the precursor mixtures: (a) 10%, (b) 20%, (c) 30% and (d) 50%. Reproduced with permission from [15]. Copyright 2018 Elsevier.

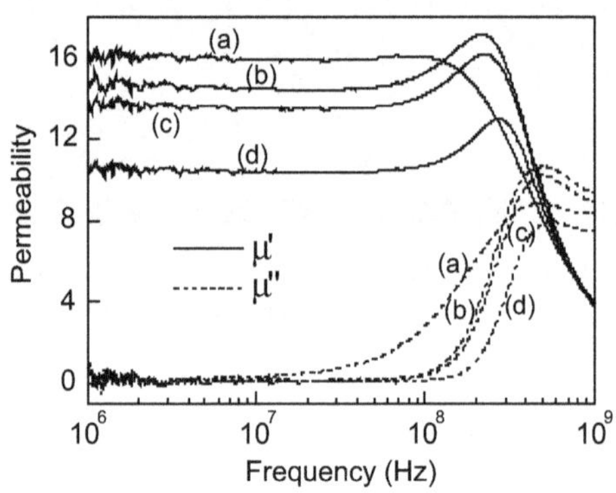

Figure 7.40. Complex permeability curves of the ferrite ceramics with different contents of Fe_2O_3 in the precursor mixtures: (a) 10%, (b) 20%, (c) 30% and (d) 50%. Reproduced with permission from [15]. Copyright 2018 Elsevier.

the densities of the samples with equation (6.2). It was observed that the simulated intrinsic static permeability was in good agreement with the value of the microsized sample, as demonstrated in figure 7.42. Figure 7.43 shows the curve-fitted data as the dashed-lines, consistent with the experimental values of the ferrite ceramics.

7.3 Effect of processing

$Li_{0.50}Fe_{2.50}O_4$ ceramics, with the addition of 3 wt% Bi_2O_3 as a sintering aid, were fabricated using a two-step ceramic processing. Commercial powders of Li_2CO_3 (99% purity), Fe_2O_3 (99% purity) and Bi_2O_3 (99% purity) were utilized as the starting materials. Two routes were used to examine the effect of Bi_2O_3 addition on the microstructure, densification, and dielectric and magnetic properties of the final $Li_{0.50}Fe_{2.50}O_4$ ceramics. In route one, Bi_2O_3 was mixed with the starting components, followed by high-energy activation and calcination. In route two, the

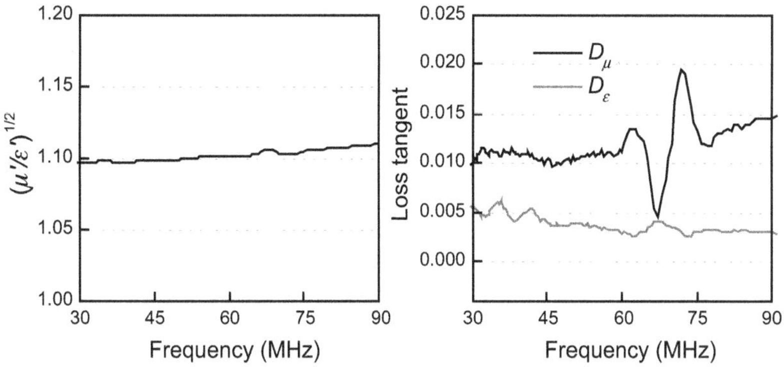

Figure 7.41. Magneto-dielectric properties of the sample from the mixture containing 50% Fe_2O_3 over 30–90 MHz. Reproduced with permission from [15]. Copyright 2018 Elsevier.

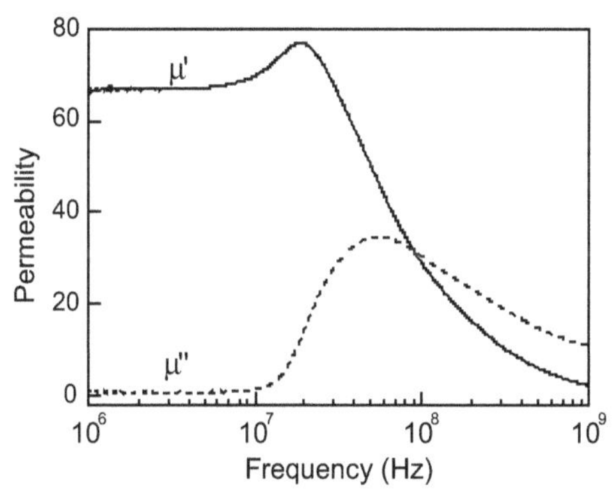

Figure 7.42. Complex relative permeability curves of the microsized sample. Reproduced with permission from [15]. Copyright 2018 Elsevier.

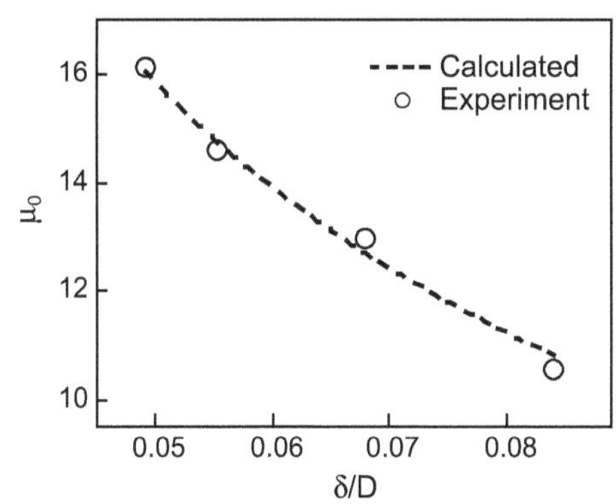

Figure 7.43. Static permeability versus δ/D of the samples. Reproduced with permission from [15]. Copyright 2018 Elsevier.

component precursors were mixed and activated first, followed by calcination. The calcined powder was mixed with Bi_2O_3 followed by compaction and sintering. The activation was conducted using high-energy ball milling for 2 h, with a Retsch PM400 type planetary ball mill and WC milling media. The activated powders were calcined at 900 °C for 2 h. The calcined powders were milled for a second time for 1 h. After the milling, the powders were compacted and then sintered at –1150 °C for 2 h.

Figure 7.44 shows the XRD patterns of the two samples with the Bi_2O_3 added before and after the calcination step. Obviously, there was no significant difference in the XRD patterns between the two groups of samples, suggesting that Bi_2O_3 had almost no effect on the phase formation of the $Li_{0.50}Fe_{2.50}O_4$ ferrite. This observation was similar to those discussed previously, simply because no reaction occurs between Bi_2O_3 and $Li_{0.50}Fe_{2.50}O_4$, as Bi_2O_3 only acted as a sintering aid to promote densification of the materials. Therefore, the addition of Bi_2O_3 had no effect on the phase composition of the ferrite, but it would exhibit a relatively pronounced influence on the microstructure, grain growth and thus the final properties of the $Li_{0.50}Fe_{2.50}O_4$ ceramics.

Figure 7.45 shows the densities of the two groups of samples as a function of sintering temperature, revealing a quite significant difference. If Bi_2O_3 was added before calcination, the density of the samples was monotonically increased from about 4.2 g cm^{-3} to about 4.5 g cm^{-3} as the sintering temperature was increased from 850 °C to 1100 °C. In contrast, if Bi_2O_3 was added after the calcination step, the sample was nearly fully densified after sintering at 850 °C. The density was only slightly increased with sintering temperature up to 1000 °C. Further increase in sintering temperature led to a slight reduction in the density of the samples. However, after sintering at 1150 °C, the two groups of samples had the same density.

The effect of the Bi_2O_3 addition sequence was also reflected in the difference in microstructure between the two groups of ceramics. SEM images of the samples sintered at different temperatures, from the mixtures with Bi_2O_3 added before and after the calcination step, are depicted in figures 7.46 and 7.47. A quick examination indicated that the difference in microstructure was consistent with that in density, as

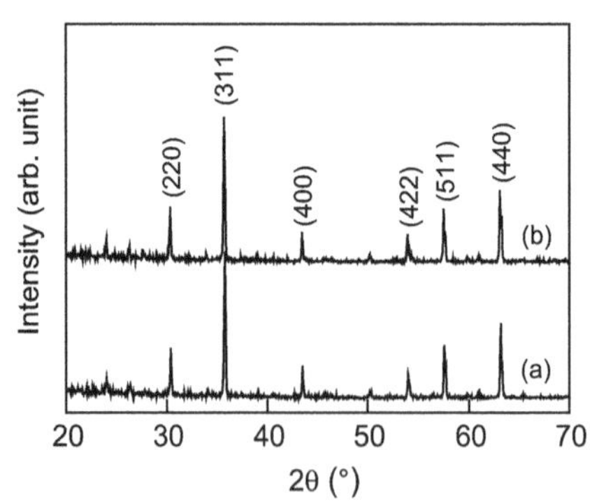

Figure 7.44. XRD patterns of the samples sintered at 850 °C for 2 h with Bi_2O_3 being added before (a) and after (b) calcination.

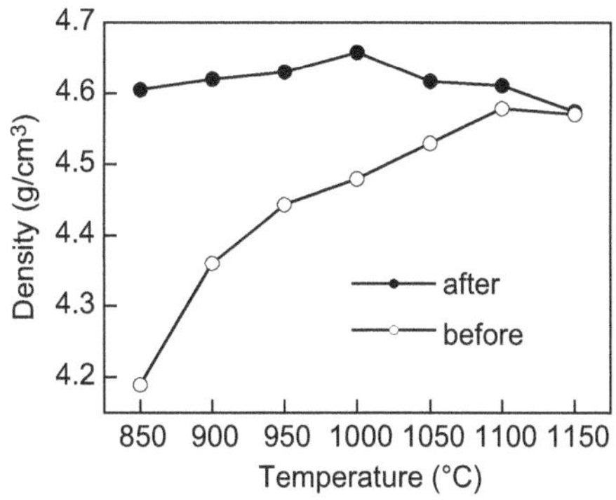

Figure 7.45. Measured density of the ferrite ceramics as a function of sintering temperature.

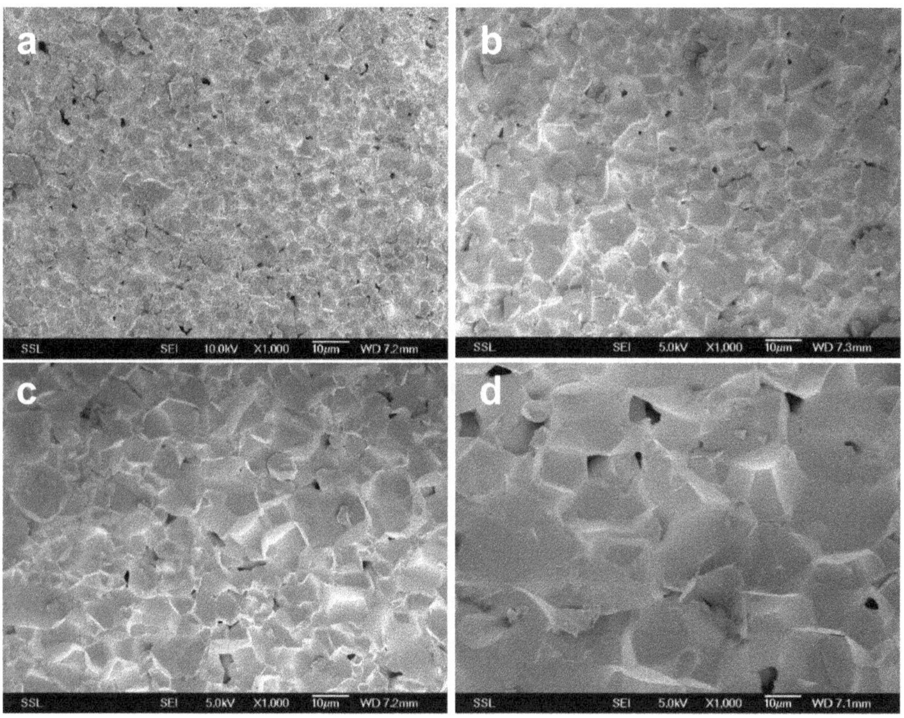

Figure 7.46. SEM images of the samples with Bi_2O_3 addition after calcination, sintered at different temperatures for 2 h: (a) 850 °C, (a) 900 °C, (a) 950 °C and (a) 1150 °C.

seen in figure 7.45. The samples with Bi_2O_3 added before calcination contained more pores, in particular those samples sintered at low temperatures, as illustrated in figures 7.46(a) and 7.47(b). With increasing sintering temperature, their difference in porosity was gradually weakened. After sintering at 1150 °C, the two groups of samples shared almost the same microstructure. In addition, the addition of Bi_2O_3 after the calcination step resulted in a relatively higher grain growth rate.

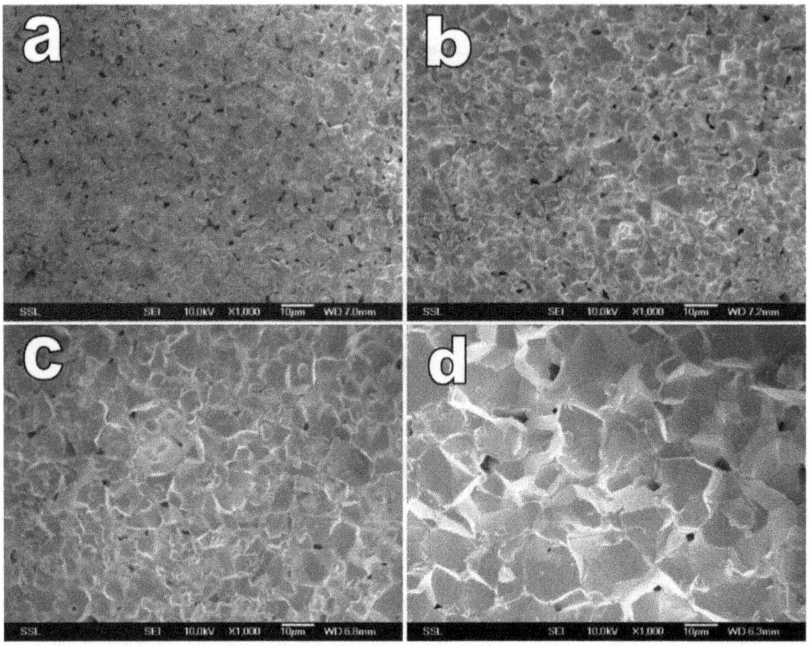

Figure 7.47. SEM images of the samples with Bi_2O_3 addition before calcination, sintered at different temperatures for 2 h: (a) 850 °C, (a) 900 °C, (a) 950 °C and (a) 1100 °C.

The differences in density and microstructure between the two groups of samples could be explained as follows. When the mixture of Li_2CO_3, Fe_2O_3 and Bi_2O_3 was mixed and activated with high-energy milling, the three components were homogenized at the nanometer scale, due to the high-energy impacts. In this case, during the sintering process, Li_2CO_3 and Fe_2O_3 reacted to form the $Li_{0.50}Fe_{2.50}O_4$ phase. As mentioned previously, Bi_2O_3 was not involved in the reaction. Instead, Bi_2O_3 was molten to form a liquid phase above its melting point, which promoted the densification of the sample. However, because the Bi_2O_3 was homogeneously distributed in the mixtures, it was difficult for it to accumulate to form the liquid phase. As a result, the effect of Bi_2O_3 on densification of the ferrite was largely weakened. In comparison, when Bi_2O_3 was added after the calcination step, the $Li_{0.50}Fe_{2.50}O_4$ phase was formed so that Bi_2O_3 was mixed with $Li_{0.50}Fe_{2.50}O_4$. Because Bi_2O_3 was not subject to high-energy activation, it was easier for it to form the liquid phase. As a consequence, the liquid phase participated in the densification of the sample.

Due to the differences in densification and microstructure development, the dielectric and magnetic properties of the two groups of ferrite ceramics were accordingly different. Figure 7.48 shows complex permittivity curves of the two groups of samples after sintering at 850 °C for 2 h. Obviously, the sample from the mixture with Bi_2O_3 added after the calcination step had comparatively higher real permittivity, while their imaginary values were very close. Therefore, the former

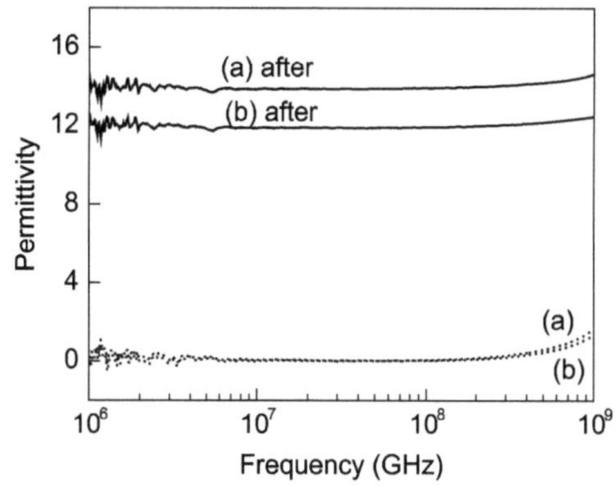

Figure 7.48. Complex relative permittivity curves of the ferrite ceramics sintered at 850 °C for 2 h: (a) after and (b) before.

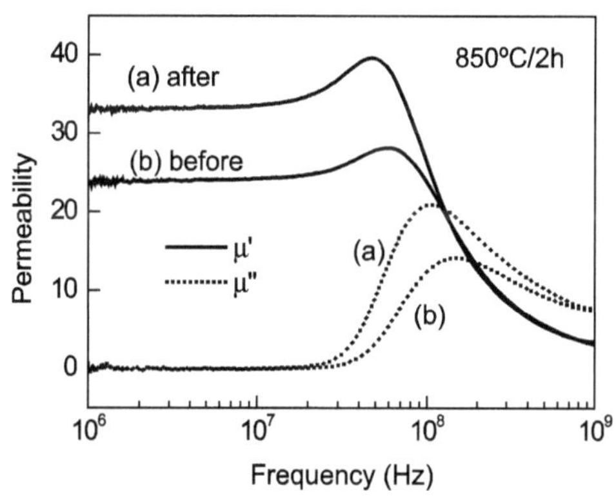

Figure 7.49. Complex relative permeability curves of the ferrite ceramics sintered at 850 °C for 2 h.

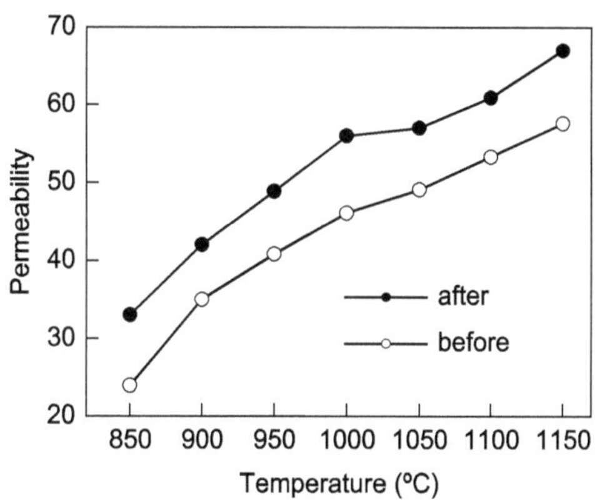

Figure 7.50. Static permeability of the ferrite ceramics as a function of sintering temperature.

would have a lower dielectric loss tangent. As discussed previously, high densification and large grain size correspond to high permittivity.

Figure 7.49 shows complex permeability curves of the two groups of samples after sintering at 850 °C for 2 h, while the static permeability values as a function of sintering temperature are plotted in figure 7.50. As expected, if Bi_2O_3 was added after the calcination step, the samples had higher real permeability. Specifically, the difference in static permeability between the two groups of samples was always about 10 in the range of the sintering temperature. This study indicated that the processing sequence could be used to control the densification, microstructure and thus the properties of ferrite ceramics.

Acknowledgments

Shenzhen Technology University (SZTU) is acknowledged for the financial support of a start-up grant (2018) and also the Natural Science Foundation of Top Talent of SZTU (grant no. 2019010801002).

References

[1] Teo M L S, Kong L B, Li Z W, Lin G Q and Gan Y B 2008 Development of magneto-dielectric materials based on Li-ferrite ceramics: I. Densification behavior and microstructure development *J. Alloys Compd.* **459** 557–66

[2] Teo M L S, Kong L B, Li Z W, Lin G Q and Gan Y B 2008 Development of magneto-dielectric materials based on Li-ferrite ceramics: II. DC resistivity and complex relative permittivity *J. Alloys Compd.* **459** 567–75

[3] Kong L B, Teo M L S, Li Z W, Lin G Q and Gan Y B 2008 Development of magneto-dielectric materials based on Li-ferrite ceramics: III. Complex relative permeability and magneto-dielectric properties *J. Alloys Compd.* **459** 576–82

[4] You H W and H K J 2006 Low temperature sintering of Li_2CO_3 added $(Ba,Sr)TiO_3$ ceramics *Integr. Ferroelectr.* **86** 59–65

[5] Vuong L D and Gio P D 2013 Effect of Li_2CO_3 addition on the sintering behavior and physical properties of PZT–PZN–PMnN ceramics *Inter. J. Mater. Sci. Appl.* **2** 89–93

[6] Guan S B, Yang H B, Zhao Y Z and Zhang R 2018 Effect of Li_2CO_3 addition in $BiFeO_3$–$BaTiO_3$ ceramics on the sintering temperature, electrical properties and phase transition *J. Alloys Compd.* **735** 386–93

[7] Verma A and Dube D C 2005 Processing of nickel–zinc ferrites via the citrate precursor route for high-frequency applications *J. Am. Ceram. Soc.* **88** 519–23

[8] Miles P A, Westphal W B and Von Hippel A 1957 Dielectric spectroscopy of ferromagnetic semiconductors *Rev. Mod. Phys.* **29** 279–307

[9] Jonker G H 1959 Analysis of the semiconducting properties of cobalt ferrite *J. Phys. Chem. Solids* **9** 165–75

[10] Globus A 1977 Some physical considerations about the domain wall size theory of magnetization mechanisms *J. Physique Coll.* **38** C1–15

[11] Bera J and Roy P K 2005 Effect of grain size on electromagnetic properties of $Ni_{0.7}Zn_{0.3}Fe_2O_4$ ferrite *Physica* B **363** 128–32

[12] Bozorth R M, Tilden E F and Williams A J 1955 Anisotropy and magnetostriction of some ferrites *Phys. Rev.* **99** 1788–98

[13] Kong L B, Li Z W, Lin G Q and Gan Y B 2007 Magneto-dielectric properties of Mg–Cu–Co ferrite ceramics: I. Densification behavior and microstructure development *J. Am. Ceram. Soc.* **90** 3106–12

[14] Kong L B, Li Z W, Lin G Q and Gan Y B 2007 Magneto-dielectric properties of Mg–Cu–Co ferrite ceramics: II. Electrical, dielectric and magnetic properties *J. Am. Ceram. Soc.* **90** 2104–12

[15] Xiao Z H *et al* 2018 Low temperature sintered magneto-dielectric ferrite ceramics with near net-shape derived from high-energy milled powders *J. Alloys Compd.* **751** 28–33

IOP Publishing

Functional Ceramics Through Mechanochemical Activation

Ling Bing Kong

Chapter 8

Mullite ceramics (I)

Ling Bing Kong, Zhuohao Xiao, Xiuying Li, Shijin Yu, Wenxiu Que, Yin Liu, Tianshu Zhang, Kun Zhou and Hongfang Zhang

8.1 Introduction

Mullite is an aluminum silicate with a composition of $3Al_2O_3 \cdot 2SiO_2$, which is the only stable phase under ambient conditions in the Al_2O_3–SiO_2 binary system [1]. Mullite and mullite-based composite ceramics are among the most important structural ceramics due to their room temperature and high-temperature mechanical properties [2–8]. They also have potential applications as substrates for electronic devices and solar cells [9–15] and window materials in mid-IR bands [16–18].

Various methods have been reported to synthesize mullite powder and fabricate mullite ceramics. The conventional ceramic processing required extremely high temperatures >1650 °C and >1500 °C for pressureless sintering and the hot-pressing method, respectively. To reduce the fabrication temperatures of mullite ceramics and composites, wet-chemical processes were developed to synthesize mullite powders with small particles and thus high reactivity [19–24].

With this background, the current authors attempted to use the mechanochemical activation process to reduce the temperatures of mullite phase formation and the fabrication of mullite ceramics. Initially, we wished to obtain fully dense mullite ceramics at relatively low sintering temperatures. However, the experimental results indicated that dense mullite ceramics could be achieved from a highly refined oxide mixture using high-energy mechanochemical activation. Instead, anisotropic grain growth occurred, so the mullite ceramics obtained in this way were highly porous. SEM observation revealed that the porous mullite ceramics consisted of well-developed mullite whiskers. It was then understood that the refined oxide powders resulted in a low temperature mullite phase formation. Due to its anisotropic crystal structure, mullite strongly tends to grow anisotropically as long as the grain growth takes place in unrestrictive environments. In comparison, during the conventional process of mullite ceramic fabrication, densification occurred before the mullite

doi:10.1088/978-0-7503-2191-4ch8

phase formation so anisotropic grain growth was prohibited. Therefore, the mullite ceramics were fully densified with spherical grains.

In other words, mechanochemical activation offers a new way to synthesize mullite whiskers directly from oxide mixtures through the conventional ceramic process, without the requirement for chemical processing methods and the use of toxic fluorides. Consequently, the mullite whiskers produced in this way will be much cheaper than those made with most traditional methods. Due to their important applications as a reinforcing element, it is highly desirable to synthesize mullite whiskers or fibers [25–32]. Therefore, this new method is of both technological and academic significance.

This chapter aims to provide an overview of the progress in processing and characterizing mullite ceramics produced from mixtures of oxides with mechanochemical activation and various oxide dopants. The obtained results will be discussed by considering the interrelationships among mullite phase formation, ceramic densification and anisotropic grain growth in order to offer a guide or reference for future studies on this interesting and important material. In addition, due to limitations of time and facilities, most of the results have been deeply or quantitatively characterized. Therefore, it is strongly suggested that one should pay more attention to studying and understanding the underlying mechanisms that govern the formation of mullite ceramics with a wide range of microstructures and grain morphologies. The mullitization, densification and anisotropic grain growth of mullite, as well as the effects of transitional oxides, are described in this chapter, while the effects of other oxides will be presented in the next chapter.

8.2 Mullitization, densification and anisotropic grain growth

An early attempt was made to prepare dense mullite ceramics from oxide mixture activated using a mechanochemical activation process, because it was expected that a refined precursor powder would have higher sinterability [33]. Commercial powders of SiO_2 and Al_2O_2 were used as the starting materials. The powders were mixed according to the stoichiometric composition of mullite ($3Al_2O_3 \cdot 2SiO_2$). The mechanochemical activation was conducted with a Retsch PM400 type planetary mill at room temperature in air for 10 h. A 250 ml tungsten carbide (WC) vial and 100 WC balls with diameters of 1 cm were utilized as the milling media, while the milling speed was 200 rpm. The milling was interrupted for 5 min every 25 min. The milled powders were calcined at different temperatures for 4 h.

Figure 8.1 shows the XRD patterns of the mixture of Al_2O_3 and SiO_2 before and after mechanochemical activation for 4 h. The diffraction peaks of the mixture before activation belonged to Al_2O_3 with a corundum structure. The broad peak over 20°–30° was related the amorphous silica. After activation for 10 h, all the diffraction peaks of Al_2O_3 were broadened and weakened, suggesting that the Al_2O_3 powder was significantly refined due to the high-energy mechanochemical activation. In addition, WC was detected in the mechanochemically activated mixture. In addition, no new phases consisting of Al_2O_3 and SiO_2 were observed, indicating that

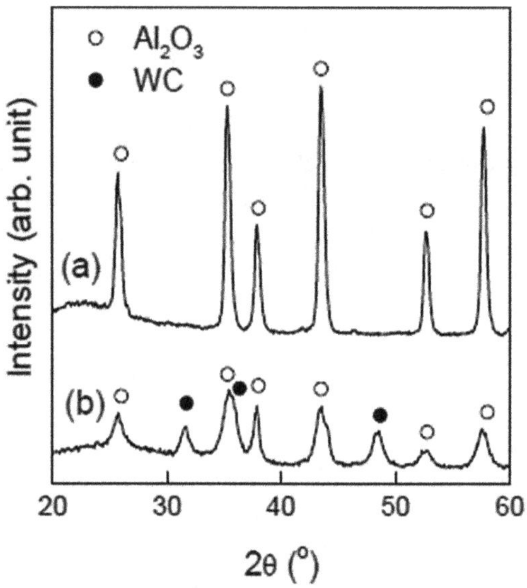

Figure 8.1. XRD patterns of the mixture before and after mechanochemical activation for 10 h. Reproduced with permission from [33]. Copyright 2002 Wiley.

no reaction between the two components was triggered by the mechanochemical activation.

The presence of WC resulted from abrasion of the milling media. Despite its strong diffraction peaks in the XRD pattern, the real content in terms of weigh percentage was not very high, because almost no weight change was observed for the WC balls after the milling experiment. For example, it was found that the WC contamination due to the abrasion of the WC vials and balls after milling for the long time of 110 h was less than 2 at% when the mixture of Fe_2O_3 and ZrO_2 was studied [34]. The diffraction peaks of the WC phase in the XRD pattern were attributed to its high degree of crystallization.

Figure 8.2 shows the XRD patterns of samples from the oxide mixture without mechanochemical activation after calcining at different temperatures. After calcination at 900 °C for 4 h only peaks of Al_2O_3 were present, implying that mullite was not formed at this temperature. As the temperature was increased to 1000 °C a new phase, cristobalite, was detected due to the crystallization of the amorphous SiO_2. The phase compositions reflected in the XRD patterns were almost unchanged below 1200 °C. After calcination at 1300 °C for 4 h, mullite was present as single-phase, indicating that the phase formation of mullite occurred at this temperature. A further increase in calcination temperature resulted in no change in XRD patterns.

In comparison, the formation temperature of mullite phase from the mixture with mechanochemical activation was tremendously reduced. Figure 8.3 shows XRD patterns of the samples from the mulled mixture after calcining at different temperatures. After calcining at 900 °C, Al_2O_3 was the only phase, while WC was absent. The absence of the WC phase suggested that it was oxidized into WO_3 with poor crystallization and thus weak diffraction peaks. This result supported the above

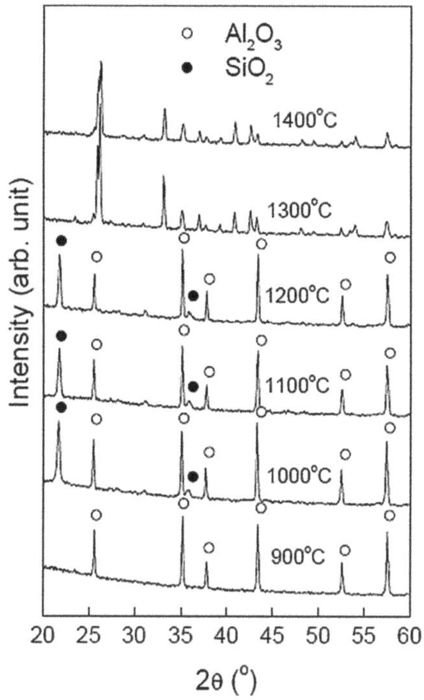

Figure 8.2. XRD patterns of the mixture without mechanochemical activation after calcination at different temperatures for 4 h. Reproduced with permission from [33]. Copyright 2002 Wiley.

statement that the content of WC in the milled mixture was relatively low. After calcining at 1000 °C for 4 h, strong diffraction peaks of the mullite phase were observed, while cristobalite and Al_2O_3 became minor phases. Mullite was nearly single-phase in the sample calcined at 1100 °C, while the mullite phase formation was complete at 1200 °C. Therefore, it could be concluded that the phase formation temperature of mullite was decreased by about 300 °C as a result of the mechanochemical activation. The enhanced mullite formation behavior of the activated mixture was ascribed to the refinement of the Al_2O_3 and SiO_2 after the high-energy mechanochemical activation.

Figure 8.4 shows the relative densities of the samples from the mixtures with and without mechanochemical activation as a function of sintering temperature. For the samples from the mixture without mechanochemical activation, the relative density was increased monotonically in the temperature range 900 °C – 1200 °C. A high densification with a relative density >95% was achieved after sintering at 1200 °C for 4 h. It was found that the mullite phase was still not visible at this temperature, as shown in figure 8.2. In other words, the mixture experienced densification before the phase formation of mullite, which was attributed to transient viscous sintering due to the presence of the liquid phase [35–38].

In contrast, the relative density of the samples from the activated powder was not over 70% when the phase formation of mullite was observed at 1000 °C.

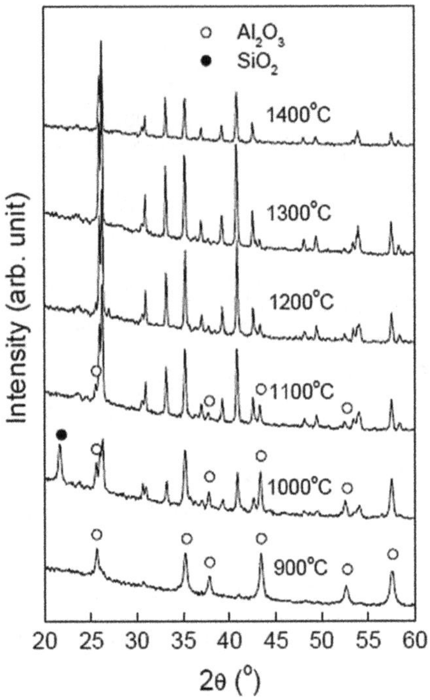

Figure 8.3. XRD patterns of the mixture with mechanochemical activation for 10 h after calcination at different temperatures for 4 h. Reproduced with permission from [33]. Copyright 2002 Wiley.

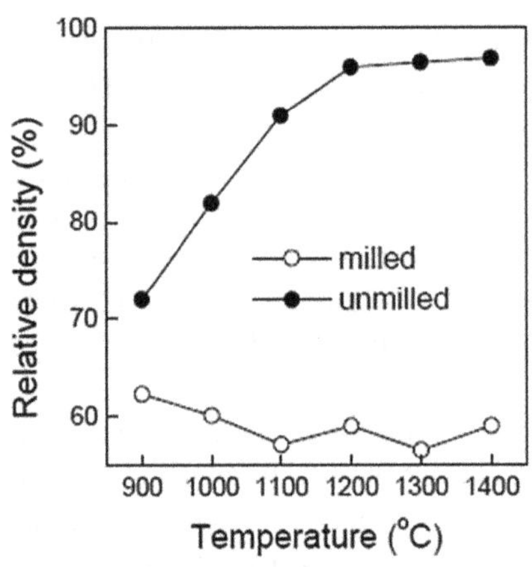

Figure 8.4. Densities of the samples from the mixture before and after mechanochemical activation for 10 h as a function of sintering temperature. Reproduced with permission from [33]. Copyright 2002 Wiley.

Furthermore, the relative density decreased slightly with sintering temperature up to 1100 °C. As shown in figure 8.5, highly faceted whiskers were obtained in the samples from the activated mixture. Due to the presence of the whiskers, their interlocking behavior prevented the samples from densifying. In comparison, no whiskers were observed in the samples from the mixture without mechanochemical

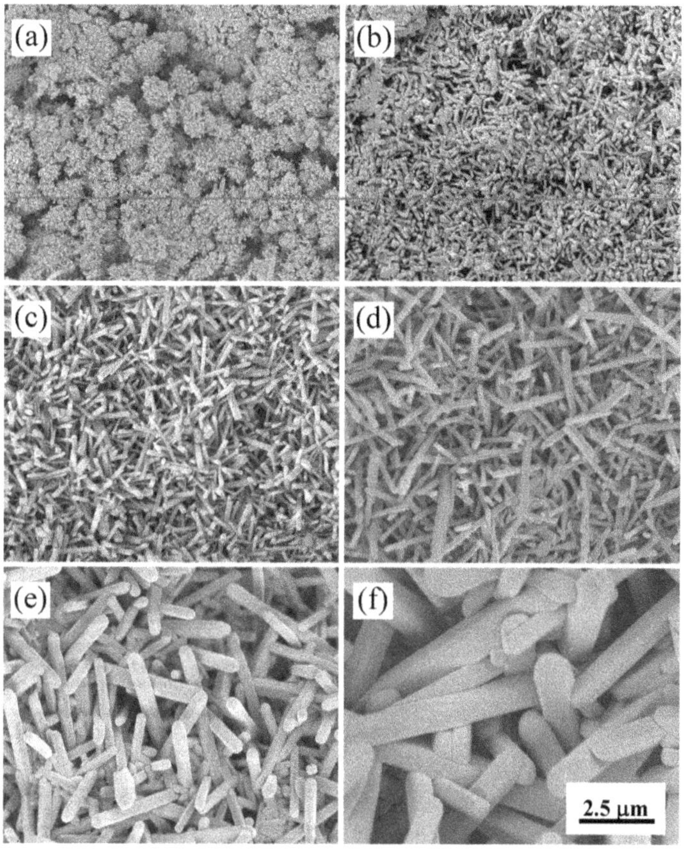

Figure 8.5. SEM images of the samples derived from the mixed powder milled for 10 h, after calcination at different temperatures for 4 h: (a) 900 °C, (b) 1000 °C, (c) 1100 °C, (d) 1200 °C, (e) 1300 °C and (f) 1400 °C. Reproduced with permission from [33]. Copyright 2002 Wiley.

activation. Figure 8.6 shows an SEM image of the sample from the unmilled powder after sintering at 1200 °C. The length, thickness and aspect ratio of the mullite whiskers were all increased with increasing sintering temperature.

Initially, the formation of the mullite whiskers from the activated oxide mixture was explained simply as the template effect of Al_2O_3. This was because nanosized whisker-like particles were observed in the sample after treatment at 900 °C, as shown in figure 8.5, while mullite still had not formed at this temperature, as illustrated in figure 8.3. In addition, the presence of the WO_3 derived from the WC contamination was also considered to have contributed to the formation of the mullite whiskers, which was also confirmed by XRD analysis. Figure 8.7 shows the XRD pattern of the sample after calcining at 700 °C for 2 h, in which WO_3 was detected.

This explanation was actually not correct. Mullite has an orthorhombic crystal structure, with a = 0.754 56 nm, b = 0.768 98 nm and c = 0.288 42 nm. This anisotropic crystal structure allows anisotropic grain growth, i.e. the formation of

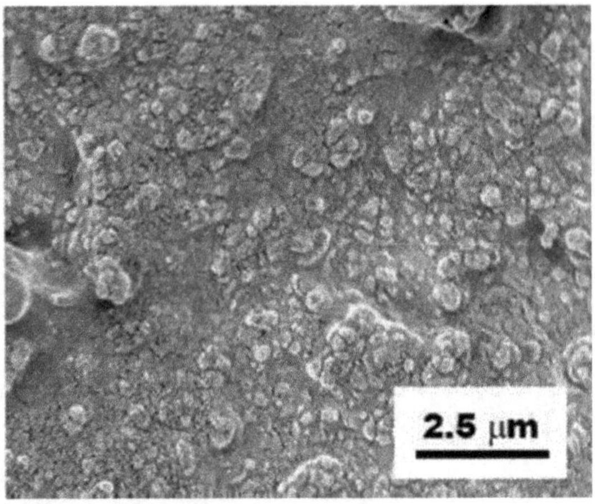

Figure 8.6. XRD patterns of the mixture before and after mechanochemical activation for 10 h. Reproduced with permission from [33]. Copyright 2002 Wiley.

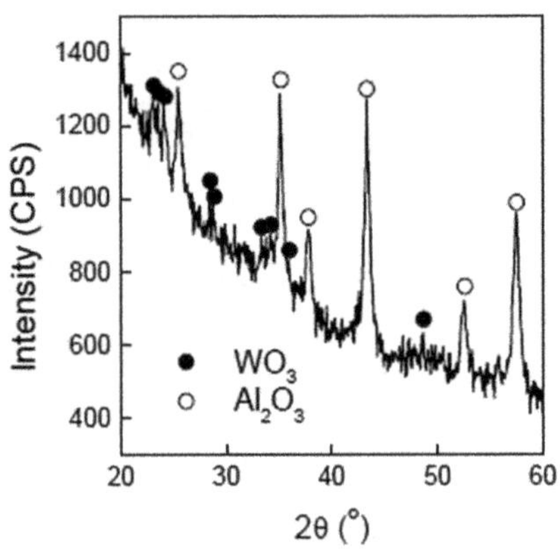

Figure 8.7. XRD patterns of the mixture before and after mechanochemical activation for 10 h. Reproduced with permission from [33]. Copyright 2002 Wiley.

mullite whiskers, if a free growth environment is present. For the mixture without mechanochemical activation, phase formation was too high due to the low reactivity of the reactants. However, due to the presence of SiO_2 eutectic liquid could be formed at relatively low temperatures. The formation of the liquid phase would promote the densification of the sample though a transient viscous flow sintering process. In this case, densification occurred before phase formation. After sintering at sufficiently high temperatures, the mullite phase was formed. Because the samples had been fully densified, the anisotropic grain growth of mullite was prohibited. In other words, once the sample is fully densified, mullite whiskers cannot be formed.

In contrast, after mechanochemical activation, the two reactants were highly activated, so that the phase formation temperature of mullite was largely reduced. As a result, the mullite phase was detected in the sample after sintering at 1000 °C, while densification had not started. As mentioned above, mullite has a tendency of anisotropic grain growth and mullite whiskers were formed accordingly. Once whiskers were present, it was difficult for the samples to be densified, due to the interclocking effect of the anisotropic mullite grains. As shown in figures 8.2 and 8.5(a), no mullite phase was detected in the XRD pattern, whereas short whisker-like particles were visible in the SEM image. Therefore, it was strongly believed that the mullite phase was actually formed at 900 °C, but it was only present at the surface of the sample and the content of mullite was too low for the XRD to detect. Further analysis should be conducted to confirm this explanation.

8.3 The effects of doping with transitional oxides and milling media

TiO_2 has been shown to exhibit a strong effect on anisotropic grain growth and densification of mullite ceramics, which were made from the mixture of Al_2O_3 and SiO_2 powders treated with mechanochemical activation [39]. Interestingly, anisotropic grain growth was only observed in the activated mixture of the two oxides. The anisotropic grain growth and densification of the samples were competitive with each other, which was closely related to the concentration of TiO_2. At low concentrations, TiO_2 promoted mullite phase formation and thus led to anisotropic grain growth. Too high a content of TiO_2 resulted in preferential densification.

Al_2O_3 (AKP-30, Japan), SiO_2 (Alfa Aesar-Johnson Matthey Company, USA) and TiO_2 (99.5+% purity, Aldrich Chemical Company Inc., USA) were mixed according to the composition of mullite ($3Al_2O_3{\cdot}2SiO_2$) together with TiO_2 with concentrations 1–10 wt% using the conventional ball milling process, followed by high-energy mechanochemical activation. The Al_2O_3 powder consisted of the θ and γ phases with particle sizes of tens of nanometers, while SiO_2 was quartz with high crystallinity and a large particle size. The milling parameters of mechanochemical activation were similar to those discussed above. The powders with and without mechanochemical activation were compacted and sintered at temperatures of 1100 °C – 1500 °C for 5 h, at heating and cooling rates of 10 °C/min. Some sintered samples were polished and thermally etched at temperatures that were lower than the corresponding sintering temperatures by 50 °C.

As expected, the phase formation of mullite from the mixture of Al_2O_3 and SiO_2 without high-energy activation was as high as 1400 °C, while the process was not completed after sintering at 1500 °C. This mullitization temperature was much higher than that discussed above, because fine SiO_2 powder was used there, whereas the SiO_2 powder was coarse grained quartz in this case. The addition of a small quantity of TiO_2 boosted the phase formation of mullite from the mixture of Al_2O_3 and SiO_2. In the sample with 5 wt% TiO_2, mullite phase formation was observed at 1300 °C, which was lower than that without the doping of TiO_2 by 100 °C. Also, the phase formation of mullite was finished at 1500 °C. Further increase in the content of TiO_2 exhibited a negative effect on mullite phase formation. For the mixture of Al_2O_3 and SiO_2 with mechanochemical activation, the optimal level of TiO_2 was 1 wt% in terms of mullite phase formation.

Figure 8.8 shows the XRD patterns of the samples from the activated mixture of Al_2O_3 and SiO_2 without the addition of TiO_2, after sintering at different

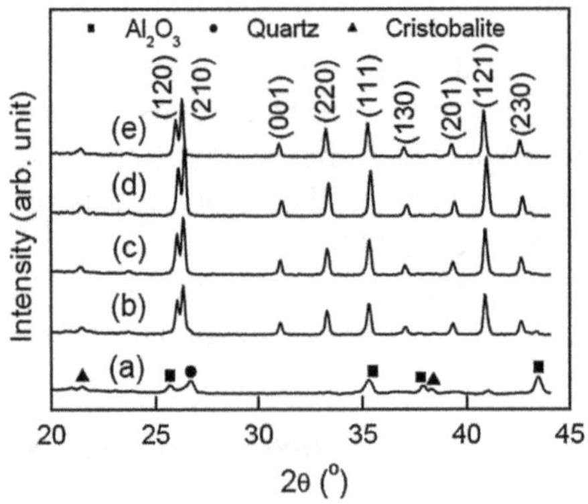

Figure 8.8. XRD patterns of the mechanochemically activated oxide mixture without TiO_2 after sintering for 5 h at different temperatures: (a) 1100 °C, (b) 1200 °C, (c) 1300 °C, (d) 1400 °C and (e) 1500 °C. Reproduced with permission from [39]. Copyright 2010 Elsevier.

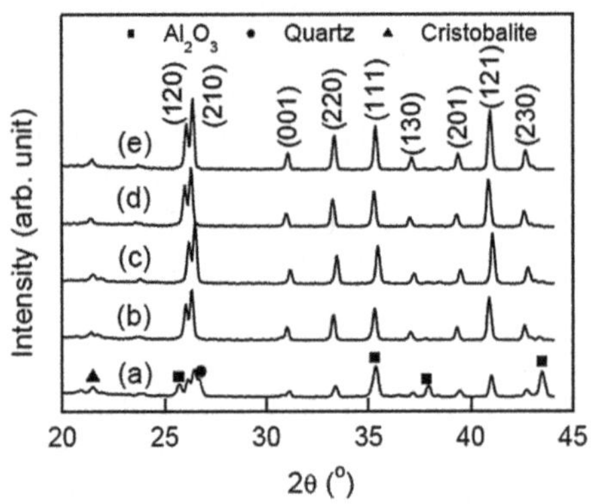

Figure 8.9. XRD patterns of the activated mixture with 1 wt% TiO_2 after sintering for 5 h at different temperatures: (a) 1100 °C, (b) 1200 °C, (c) 1300 °C, (d) 1400 °C and (e) 1500 °C. Reproduced with permission from [39]. Copyright 2010 Elsevier.

temperatures for 5 h. High-energy activation promoted the reaction between Al_2O_3 and SiO_2. The mullite phase was not present in the sample sintered at 1100 °C. Quartz and α-Al_2O_3 were the major phases, while weak peaks of cristobalite were observed, suggesting that a phase transition took place for quartz and alumina at this temperature. It was concluded that mechanochemical activation enhanced the reactivity of the two oxides. After sintering at 1200 °C the phase formation of mullite was nearly complete. In other words, the temperature of mullite phase formation was reduced by 200 °C – 300 °C as a result of the mechanochemical activation. The significantly reduced mullite phase formation temperature was ascribed to the refined oxide powders.

Figure 8.9 shows XRD patterns of the samples with 1 wt% TiO_2 after sintering at different temperatures. Obviously, the presence of 1 wt% TiO_2 enhanced the mullite phase formation by reducing the reaction temperature. As demonstrated in figure 8.9(a), the sample sintered at 1100 °C contained a pretty high level of mullite.

However, the mullite phase formation behavior was not further enhanced as the concentration of TiO_2 was further increased.

Figure 8.10 shows the XRD patterns of the samples with different contents of TiO_2 after sintering at 1200 °C for 5 h. As the concentration of TiO_2 was increased from 1 wt% to 3 wt%, diffraction peaks of the precursor oxides were present. This implied that the optimal content of TiO_2 was 1 wt% from the mullite phase formation point of view. This concentration was lower than that observed in the samples without mechanochemical activation. This was mainly because the mechanochemical activation increased the homogeneity of the powders, so that the effectiveness of TiO_2 was enhanced. In addition, TiO_2 was detected in the sample with 3 wt% TiO_2. Furthermore, the diffraction peaks of Al_2O_3 were increased as the content of TiO_2 was increased.

Figure 8.11 shows the relative densities of the samples from the activated mixture after sintering at 1300 °C – 1500 °C as a function of the content of TiO_2. The relative density of the samples sintered at 1300 °C was increased monotonically with increasing content of TiO_2. It was worth mentioning that the relative density was never higher than 80%. The variation in relative density of the samples sintered at 1400 °C was similar, while the undoped sample had a slightly higher density than that with 1 wt% TiO_2. For those sintered at 1500 °C, the relative density was increased in a linear way when the content of TiO_2 was in the range 1–5 wt%.

Moreover, the sample with 5 wt% TiO_2 was almost fully densified. In comparison, the sample without the addition of TiO_2 had a low relative density of 80%. This result indicated that the introduction of TiO_2 would promote the densification of the mullite ceramics, as long as its content was sufficiently high. In contrast, the oxide mixture without mechanochemical activation always displayed a lower densification rate compared to the mechanochemically activated sample, no matter whether the doping of TiO_2 was used or not. Therefore, high-energy mechanochemical activation played a significant role in accelerating the densification of the mullite ceramics from oxide mixtures.

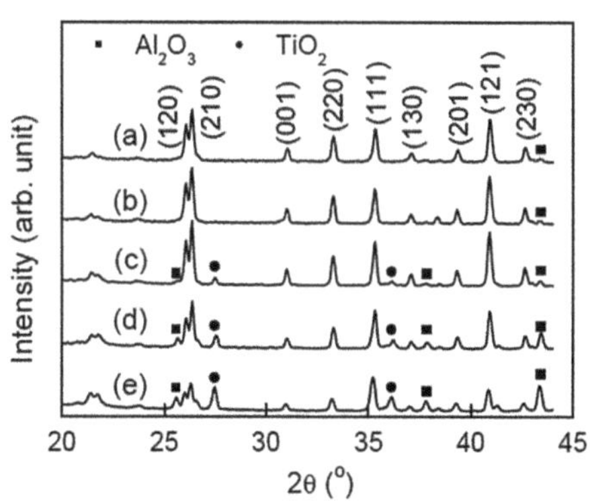

Figure 8.10. XRD patterns of the activated mixture after sintering at 1200 °C for 5 h with different contents of TiO_2: (a) 0 wt%, (b) 1 wt%, (c) 3 wt%, (d) 5 wt% and (e) 7 wt%. Reproduced with permission from [39]. Copyright 2010 Elsevier.

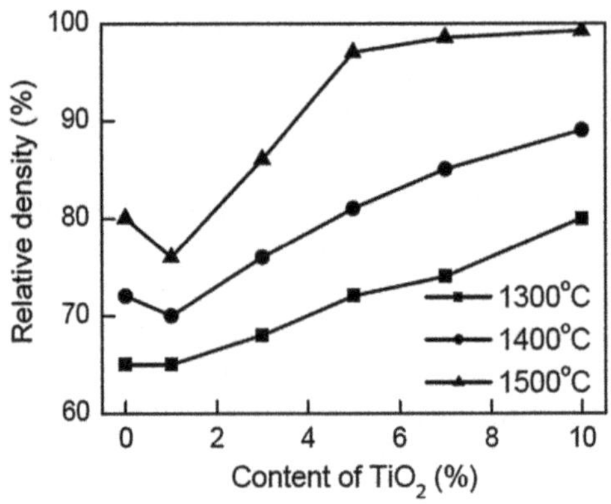

Figure 8.11. Relative densities of the samples derived from the activated mixtures as a function of concentration of TiO_2. Reproduced with permission from [39]. Copyright 2010 Elsevier.

Figure 8.12 depicts SEM images of the samples sintered at 1200 °C for 5 h, which were derived from the mixtures doped with different concentrations of TiO_2. The presence of TiO_2 had a strong effect on the anisotropic grain growth of mullite from the activated oxide mixtures. The sample from the powder without the presence of TiO_2 possessed grains with a short whisker morphology, where the average length of the whiskers was about 1 μm and the thickness was about 0.5 μm. This suggested that anisotropic grain growth was triggered in this sample after sintering at 1200 °C. Obviously, the anisotropic grain growth was enhanced if 1 wt% TiO_2 was introduced. In addition, the size of the mullite grains was significantly increased.

As the content of TiO_2 was increased to 3 wt%, the mullite grains were further enlarged. However, the grain size distribution was much wider, compared to the sample doped with 1% TiO_2. For instance, grains with a thickness of >1 μm were observed in this sample, as shown in figure 8.12(c). The uniformity of the mullite grains was reduced in the sample doped with 5 wt% TiO_2. The 7% TiO_2 sample exhibited a similar profile to the sample with 5 wt% TiO_2, while more irregular grains were present and some of them had extremely large sizes, as illustrated in figures 8.12(d) and (e). No whiskers were observed in the sample doped with 10% TiO_2.

This observation implied that the optimal content of TiO_2 was 1 wt% in terms of achieving anisotropic grain growth with uniform size distribution. It was more pronounced for the samples sintered 1000 °C, as shown in figure 8.13. Mullite whiskers were already formed in the sample doped with 1 wt% TiO_2, while the sample without TiO_2 and those with higher contents of TiO_2 were just loose compact powders with nearly spherical particles. Similar variation in anisotropic grain growth was demonstrated in the samples sintered at higher temperatures, as revealed in figure 8.14. The sample with 5 wt% TiO_2 was characterized by the presence of mullite grains with a nearly square morphology, indicating a reduction in the grain aspect ratio.

Figure 8.15 shows SEM images of selected samples after polishing and thermal etching. Clearly, the sample with 3 wt% TiO_2 had a larger aspect ratio that the

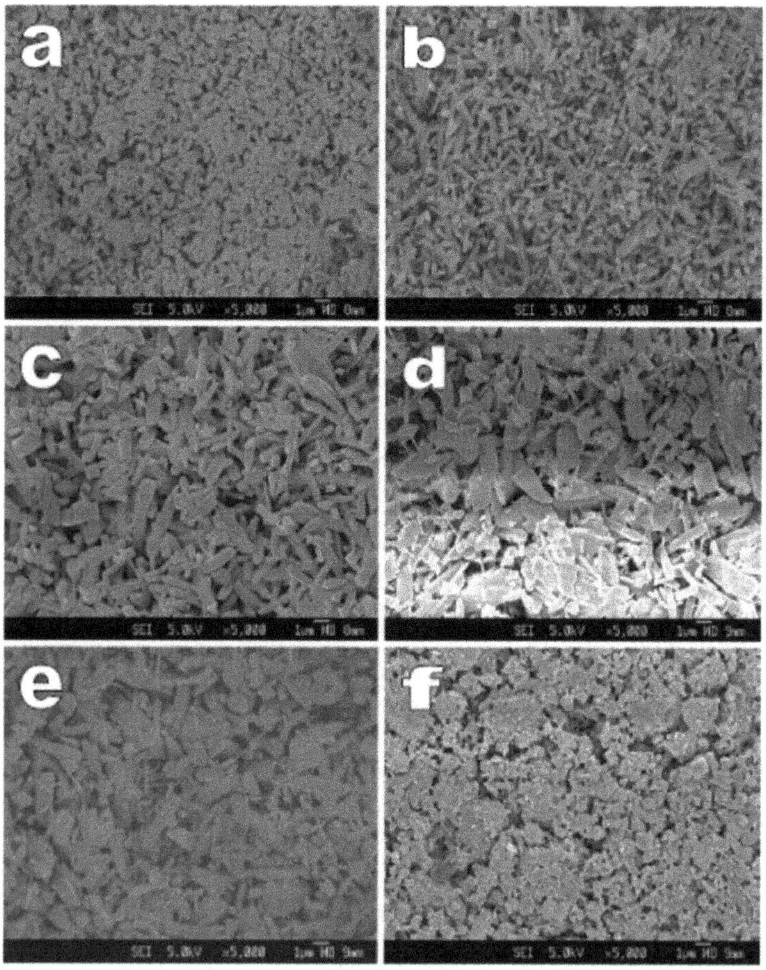

Figure 8.12. SEM images of the samples derived from the activated mixtures sintered at 1200 °C for 5 h with different contents of TiO_2: (a) 0 wt%, (b) 1 wt%, (c) 3 wt%, (d) 5 wt%, (e) 7 wt% and (f) 10 wt%. Reproduced with permission from [39]. Copyright 2010 Elsevier.

sample with 5 wt% TiO_2. In addition, the sample with a higher content of TiO_2 exhibited a denser microstructure. Therefore, this observation further confirmed that anisotropic grain growth and densification of mullite ceramics were competitive when the mullite ceramics were derived from mixtures of Al_2O_3 and SiO_2. The microstructural characterization results were consistent with the relative densities of the samples, as observed in figure 8.11. In this case, 5 wt% TiO_2 resulted in full densification of the sample after sintering at 1500 °C. This densification temperature was much lower than those required when using the conventional ceramic processing.

Figure 8.16 shows densification curves of the mixtures with mechanochemical activation and different contents of TiO_2. It was found that a sharp expansion

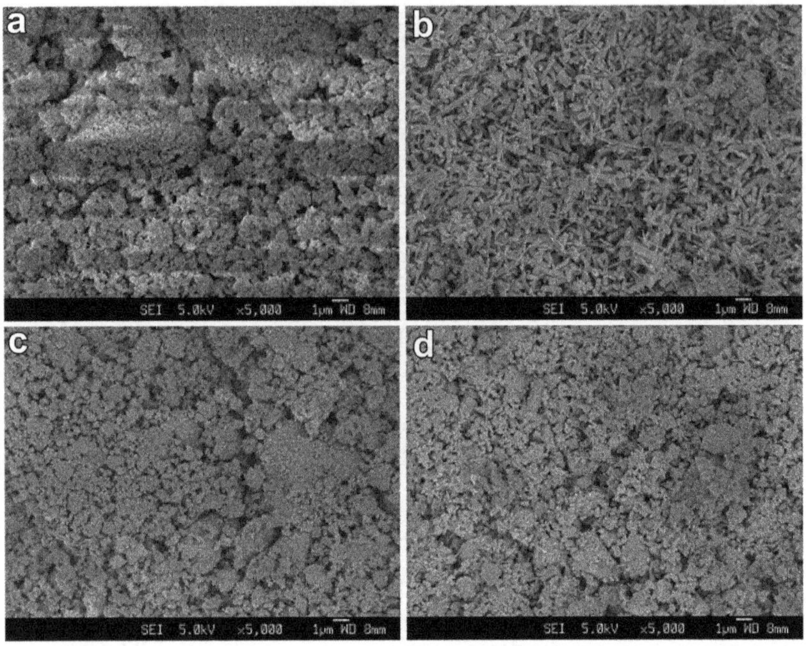

Figure 8.13. SEM images of the samples derived from the milled powders with different contents of TiO_2 after sintering at 1100 °C for 5 h: (a) 0 wt%, (b) 1 wt%, (c) 3 wt% and (d) 5 wt%.

occurred at about 600 °C and a sudden shrinking took place at about 1200 °C. The magnitude of the expansion decreased and the degree of the shrinkage increased as the content of TiO_2 was increased. However, the sudden shrinkage was not present in the samples without and with 1 wt% TiO_2. In contrast, weak expansion was observed at the temperature of 1200 °C. It has been reported that TiO_2 could accelerate the densification of mullite ceramics [40–42], which was attributed to the formation of the liquid phase at high temperature, thus leading to liquid phase sintering.

The precursor oxides and the dopant TiO_2 were highly refined due to the high-energy mechanochemical activation. Therefore, it was believed that the formation of a transient liquid phase would occur at much lower temperatures. As a consequence, rapid shrinkage took place at the relatively low temperature of about 1200 °C in the mechanochemically activated TiO_2-doped mixtures of Al_2O_3 and SiO_2. As the content of TiO_2 was too low, the liquid phase could not be formed due to the insufficient amount of the source. This was the reason the sample with 1 wt% TiO_2 exhibited similar behavior to the sample without TiO_2. Because the sintering curve of the sample with 1 wt% TiO_2 was overlapped with that of the undoped sample, it is excluded in figure 8.16. At the same time, a slight shrinkage at about 1500 °C was observed in all the samples. The expansion at 600 °C could be related to the phase transformation of the starting oxides, which should be clarified through adoption as a topic of future studies. The degree of densification was increased monotonically in

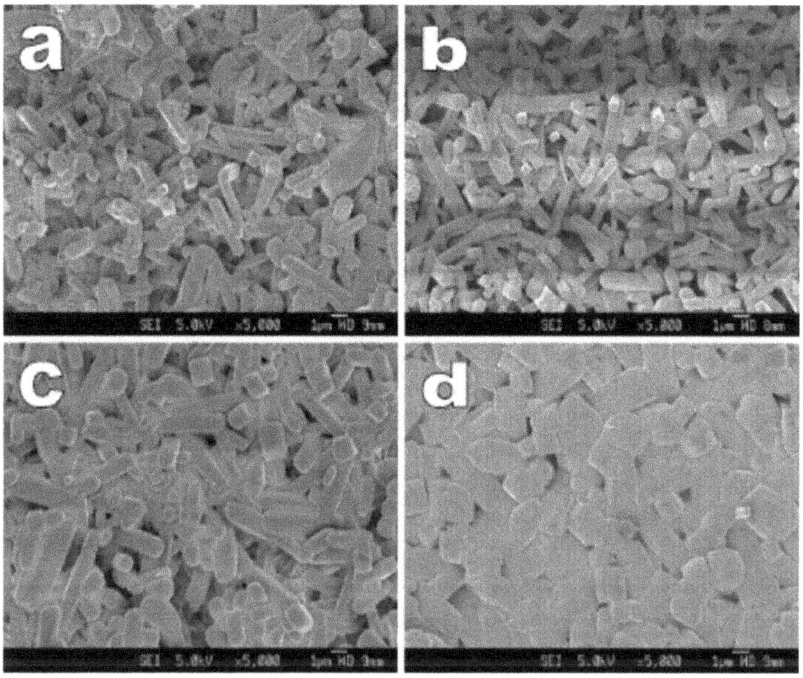

Figure 8.14. SEM images of the samples derived from the activated powders sintered at 1500 °C for 5 h with different contents of TiO_2: (a) 0 wt%, (b) 1 wt%, (c) 3 wt% and (d) 5 wt%. Reproduced with permission from [39]. Copyright 2010 Elsevier.

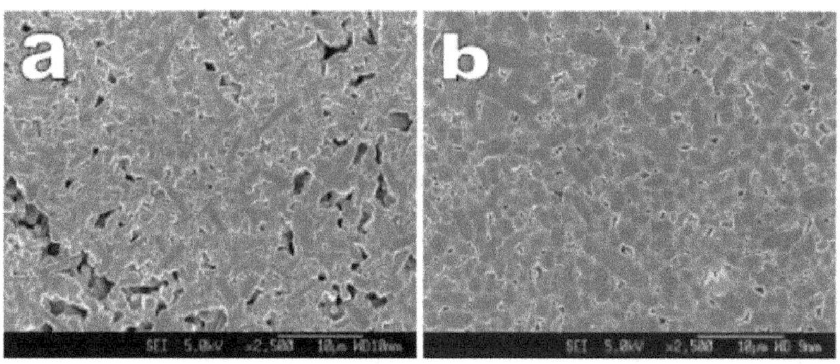

Figure 8.15. SEM images of selected samples derived from the activated powders sintered at 1500 °C for 5 h with different contents of TiO_2: (a) 3 wt% and (b) 5 wt%. Reproduced with permission from [39]. Copyright 2010 Elsevier.

the TiO_2 concentration range of 3–10 wt%. This trend was in good agreement with the SEM results and the relative densities, as discussed above.

Generally, the formation of mullite phase from the mixture of Al_2O_3 and SiO_2 requires very high temperatures when using the conventional ceramic process, owing mainly to the low reactivity of the precursor. With high-energy mechanochemical

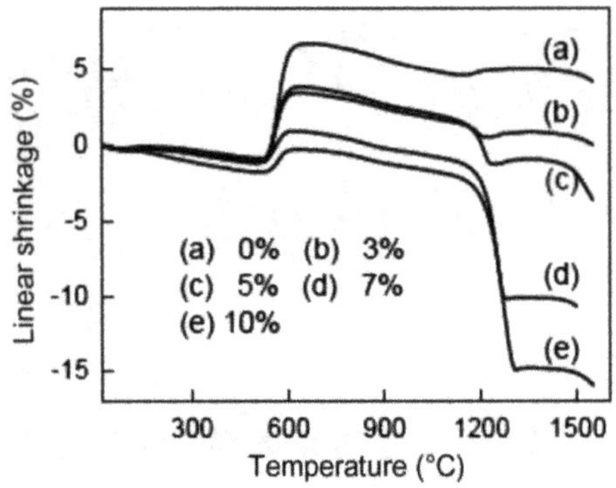

Figure 8.16. Linear shrinkage curves of the activated samples. The sample with 1 wt% TiO_2 was excluded due to overlapping with the undoped sample. Reproduced with permission from [39]. Copyright 2010 Elsevier.

activation, the particles of the precursor oxides were highly refined, so that their reactivity was significantly enhanced. It was found that the phase formation temperature from the mixture of Al_2O_3 and SiO_2 after mechanochemical activation could be decreased by about 200 °C – 300 °C, depending on the properties and types of the starting oxides. In addition, mullite whiskers were formed, because of the anisotropic grain growth characteristics of mullite.

Literature data indicate that Ti atoms can be incorporated into amorphous SiO_2 when precursors were prepared using the sol–gel process [40–42]. In the gels of TiO_2–SiO_2, the tetrahedrally coordinated Ti^{4+} would take the site of Si^{4+} with the same coordination environment, with a high concentration of 11.5 wt% TiO_2. The introduction of Ti^{4+} would cause the reduction in viscosity of the liquid siliceous phase, so that the diffusion kinetics were enhanced. In addition, alumina was also likely to be incorporated into the siliceous matrix, so that the dissolution of alumina was enhanced. In addition, the TiO_2 particles could act as the sites for heterogeneous nucleation of mullite. Therefore, the phase formation of mullite was accelerated. However, if the content of TiO_2 was too high the effect on phase formation became negative, as discussed above. This explained well the optimal concentrations of 5 wt% and 1 wt% TiO_2 for the unmilled and activated mixtures, respectively.

The effects of V_2O_5, Nb_2O_5 and Ta_2O_5 on the phase formation, microstructure and densification of mullite from an oxide mixture of Al_2O_3 and SiO_2, mixed through conventional ball milling and mechanochemically activated with high-energy ball milling with different milling media, were systematically studied [43]. The starting oxide powders included SiO_2 (Laboratory reagent), Al_2O_3 (99+% purity), V_2O_5 (99+% purity), Nb_2O_5 (99+% purity) and Ta_2O_5 (99+% purity).

The powders were mixed according to the nominal compositions of $(3Al_2O_3 \cdot 2SiO_2)_{1-x}(M_2O_5)_x$, with M = V, Nb and Ta; x = 0.05, 0.10 and 0.20. The oxide powders were thoroughly blended with a conventional ball milling process using ZrO_2 as the milling medium. The mixed samples were then subjected to high-energy

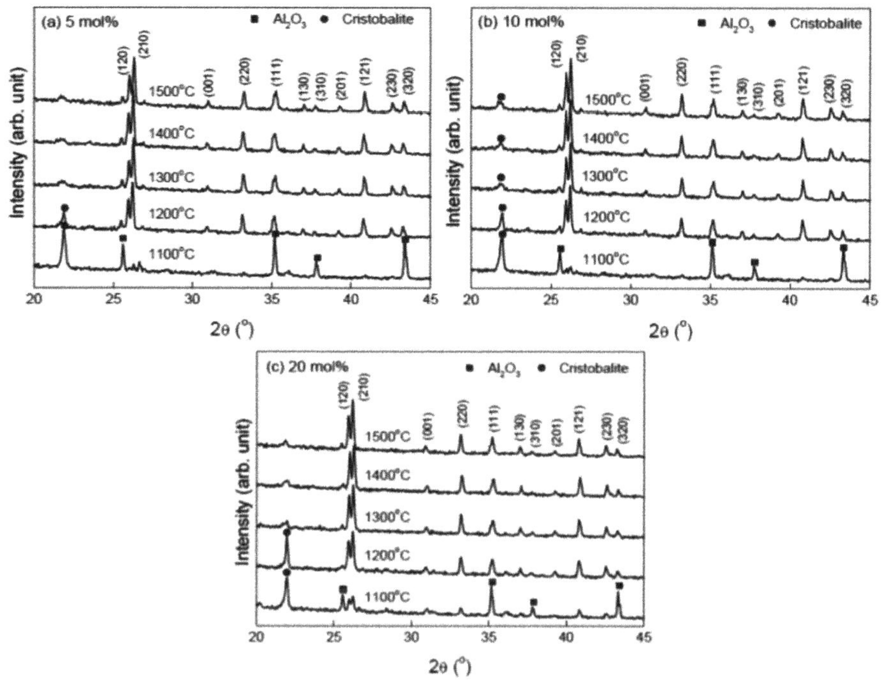

Figure 8.17. XRD patterns of the samples doped with V_2O_5 after sintering at different temperatures: (a) 5 mol%, (b) 10 mol% and (c) 20 mol%. Reproduced with permission from [43]. Copyright 2003 Elsevier.

mechanochemical activation, with stainless steel and WC milling media. The vials had a volume of 250 ml, while 100 balls with a diameter of 1 cm were used. Therefore, the ball-to-powder weight ratios were about 20:1 and 40:1 for the stainless steel and WC media, respectively. This is simply because WC has a higher density than stainless steel. The three groups of powders were compacted and then sintered in air for 4 h at temperatures of 1000 °C–1500 °C, with both the heating and cooling rates being 10 °C/min.

XRD patterns of the samples doped with V_2O_5, Nb_2O_5 and Ta_2O_5 are illustrated in figures 8.17–8.19. As shown in figure 8.17, the samples doped with V_2O_5 after sintering at 1100 °C contained Al_2O_3 and cristobalite SiO_2 as the major phases, indicating that the amorphous silica was converted to crystalline cristoalite. However, weak peaks of mullite phase were present in the samples doped with 10 mol% and 20 mol% V_2O_5 after sintering at this temperature, in particular for the sample with 20 mol% V_2O_5 as observed in figure 8.17(c). In other words, the higher the content of V_2O_5, the more pronounced the diffraction peaks of mullite would be. Therefore, it was concluded that the addition of V_2O_5 accelerated the phase formation of mullite. After sintering at 1200 °C the phase formation of mullite was nearly complete for the three groups of samples. A further increase in sintering temperature resulted in almost no change in the XRD patterns. This observation

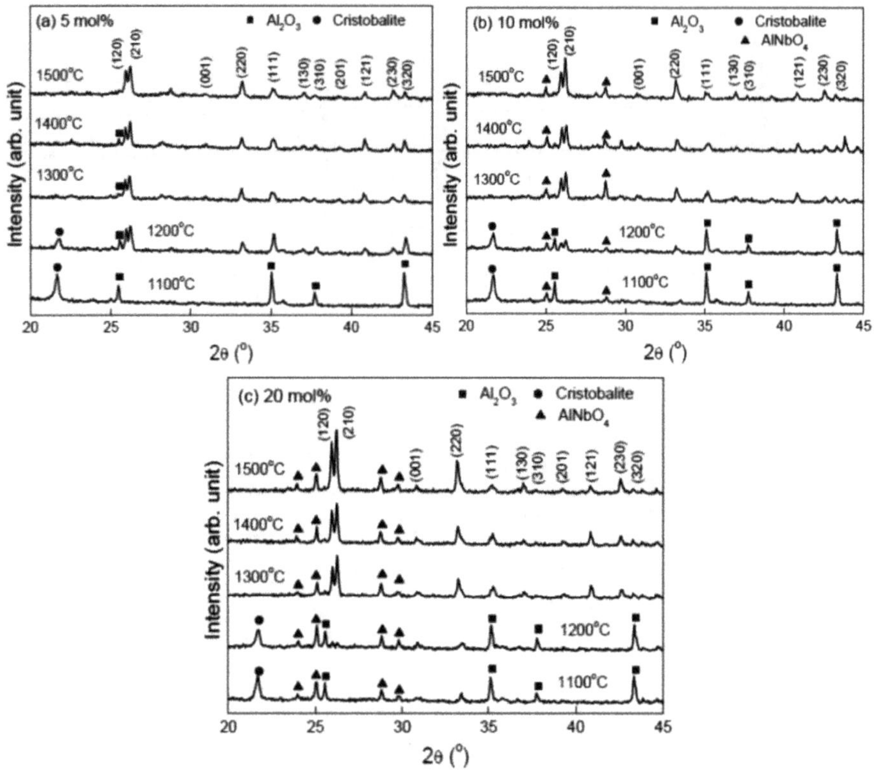

Figure 8.18. XRD patterns of the samples doped with Nb_2O_5 after sintering at different temperatures: (a) 5 mol%, (b) 10 mol% and (c) 20 mol%. Reproduced with permission from [43]. Copyright 2003 Elsevier.

indicated that the mullitization temperature was reduced by 100 °C compared to the mixture without the presence of dopants.

It was found that both Al_2O_3 and cristobalite were always present in all the samples regardless of the content of V_2O_5. This could be attributed to the fact that the SiO_2-rich liquid phase might be partially lost during the sintering at high temperatures, since the samples were directly put on alumina sample holders. In addition, V_2O_5 was not detected in all the samples, nor were any other compounds consisting of V and Al or Si. In other words, no reactions took place between V_2O_5 and Al_2O_3 or SiO_2. V_2O_5 was not present in the XRD patterns, mainly due to its refined particles and/or low degree of crystallinity. The positive effect of V_2O_5 on the mullite phase formation could be explained in terms of the mechanism of mullitization.

Previous studies indicated that the formation of mullite in diphasic aluminosilicate gels or in reaction sintering couples of quartz (SiO_2) and Al_2O_3 was governed by the reactions through dissolution–precipitation. In this case, Al_2O_3 was dissolved in the SiO_2-rich liquid with the content of Al_2O_3 finally reaching a critical level [44]. At high concentrations of Al_2O_3 a random mullite nucleation was induced in the SiO_2-rich liquid phase. The dissolution rate of Al_2O_3 into the SiO_2-rich liquid was the

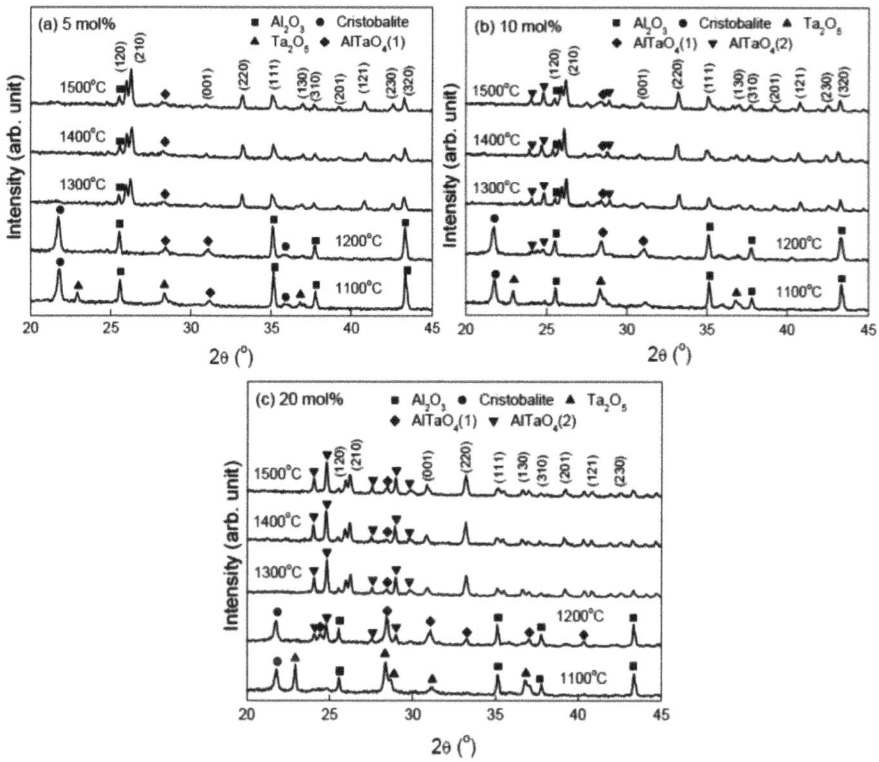

Figure 8.19. XRD patterns of the samples doped with Ta_2O_5 after sintering at different temperatures: (a) 5 mol%, (b) 10 mol% and (c) 20 mol%. Reproduced with permission from [43]. Copyright 2003 Elsevier.

rate-limiting step for the nucleation and crystal growth of mullite. Both the formation temperature of the SiO_2-rich liquid phase and the dissolution rate of Al_2O_3 into the liquid phase were highly dependent on the properties of SiO_2 and Al_2O_3. Therefore, V_2O_5 promoted liquid phase formation due to its relatively low melting point of 690 °C.

The influence of Nb_2O_5 on the mullite phase formation of the oxide mixture was not exactly the same as that of V_2O_5. As revealed in figure 8.18(a), with the presence of 5 mol% Nb_2O_5 the mullite phase formation had been nearly complete after sintering at 1200 °C, while new phases were not present according to the XRD patterns. When the content of Nb_2O_5 was increased to 10 mol%, $AlNbO_4$ was present as a secondary phase after the sample was sintered at 1100 °C. In addition, mullite phase formation at 1200 °C was suppressed at this level of Nb_2O_5. The suppression of mullite phase formation was accelerated as the content of Nb_2O_5 was further increased, as seen in figure 8.18(c). In this sample, the peaks of mullite were absent. The phase formation of mullite was suppressed, mainly because of the presence of the new phase $AlNbO_4$. During the formation of $AlNbO_4$, the dissolution of Al_2O_3 into the SiO_2-rich liquid phase was affected, while Al_2O_3 was partially consumed and thus less Al_2O_3 participated the mullitization. In addition, the presence of $AlNbO_4$ could also increase the viscosity of the liquid phase.

The phase formation due to the addition of Ta_2O_5 was much more complicated. In comparison, Ta_2O_5 was detected in all the three groups of samples after sintering at 1100 °C. At the same time, two types of $AlTaO_4$ were formed, which were $AlTaO_4$(1) (JCPDS 36–440) and $AlTaO_4$(2) (JCPDS 25–1490). In the sample doped with 5 mol% Ta_2O_5 the secondary phase was $AlTaO_4$(1) present at 1200 °C, which was stable thereafter. As the content of Ta_2O_5 was increased to 10 mol%, $AlTaO_4$(2) was formed at the same time together with $AlTaO_4$(1) after sintering at 1200 °C, although the diffraction peaks of $AlTaO_4$(1) were stronger than those of $AlTaO_4$(2).

However, as the sintering temperature was increased to 1300 °C, $AlTaO_4$(2) exhibited stronger peaks than $AlTaO_4$(1), suggesting that there was most likely a phase transformation from $AlTaO_4$(1) to $AlTaO_4$(2) at 1300 °C. Both phases were stable up to 1500 °C, as observed in figure 8.19(b). The sample doped with 20 mol% Ta_2O_5 exhibited a similar reaction sequence. In addition, mullite was formed at 1300 °C, indicating that Ta_2O_5 retarded the phase formation of mullite. Therefore, the addition of V_2O_5 enhanced the mullitization process, while the presence of Ta_2O_5 deaccelerated the mullite phase formation. The effect of Nb_2O_5 was between that of Ta_2O_5 and V_2O_5

SEM images of the samples doped with V_2O_5 are depicted in figures 8.20–8.22. Obviously, all the samples exhibited anisotropic grain growth behavior. For the sample doped with 5 mol% V_2O_5, equiaxed grains were present with an average grain size of about 2 μm, after sintering at 1100 °C for 4 h. This was simply because the mullite phase was still not formed at 1100 °C. As the sintering temperature was increased to 1200 °C, anisotropic grain growth occurred. As demonstrated in figure 8.17(a), the mullite phase was present after sintering at this temperature. The mullite grains possessed a length of about 20 μm and a thickness of about 1.5 μm. The dimensions of the mullite grains was not significantly increased as the sintering temperature was gradually increased.

After sintering at 1100 °C, the samples doped with 10 mol% and 20 mol% V_2O_5 consisted of an equiaxed grain matrix, in which large anisotropic grains were embedded. In addition, the number of the anisotropic grains was increased with increasing concentration of V_2O_5. After sintering at 1200 °C, the anisotropic grains in the samples doped with 10 mol% and 20 mol% V_2O_5 had lengths of 40 μm and 50 μm and widths of 2.5 μm and 5 μm, respectively. However, the dimensions of the anisotropic mullite grains were not obviously varied with increasing sintering temperature.

Representative SEM images of the samples doped with Nb_2O_5 and Ta_2O_5, after sintering at different temperatures, are illustrated in figures 8.23–8.25 and figures 8.26–8.27, respectively. These two groups of samples exhibited a different grain growth behavior compared to those doped with V_2O_5. For the samples doped with 5 mol% and 10 mol% Nb_2O_5, all the grains had an equiaxed characteristic. In contrast, weak anisotropic grain growth was observed in the samples doped with 20 mol% Nb_2O_5 that were sintered at high temperatures. In addition, the addition of Ta_2O_5 resulted in no anisotropic grain growth. The effect of the three oxides on

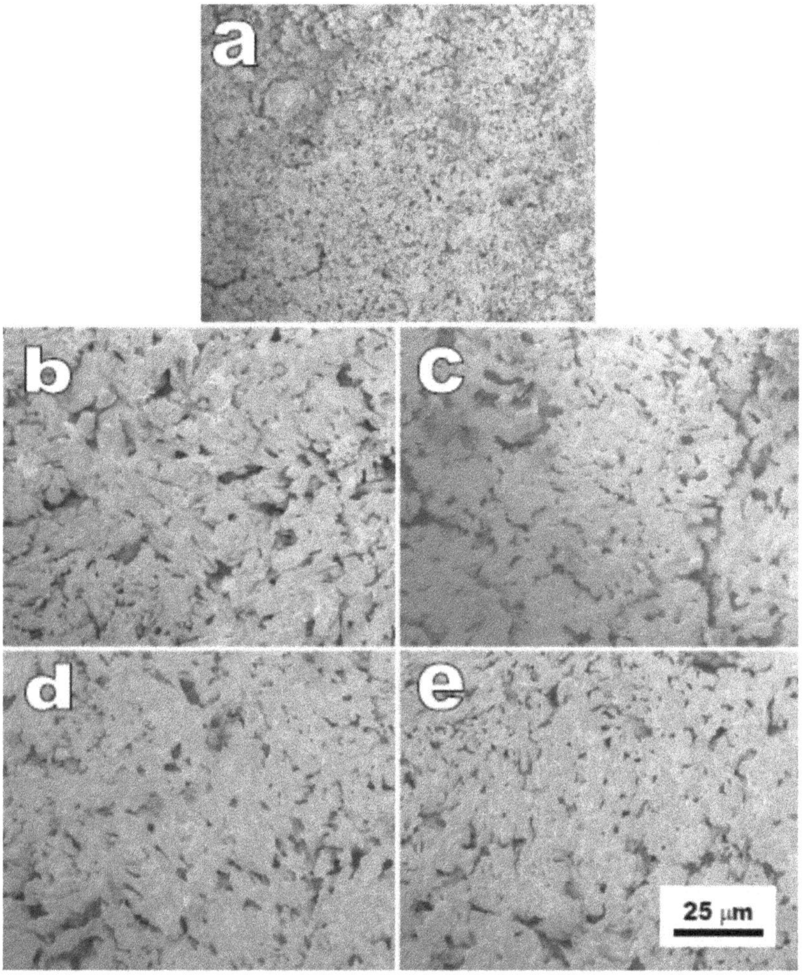

Figure 8.20. SEM images of the samples doped with 5 mol% V_2O_5 after sintering for 4 h at different temperatures: (a) 1100 °C, (b) 1200 °C, (c) 1300 °C, (d) 1400 °C and (e) 1500 °C. Reproduced with permission from [43]. Copyright 2003 Elsevier.

anisotropic grain growth was consistent with their effect on mullite phase formation, as discussed above.

Figure 8.28 shows the densities of the three groups of samples with different doping levels as a function of sintering temperature. The density of the samples doped with V_2O_5 was decreased almost monotonically with increasing doping content and sintering temperature. The densities of all the samples were much lower than the theoretical density of mullite, which was readily ascribed to the anisotropic grain growth in this group of samples. For the samples doped with Nb_2O_5 and Ta_2O_5, their densification behaviors were similar to that of the mixture of Al_2O_3 and SiO_2 without the addition of any dopant [33]. The densities of some

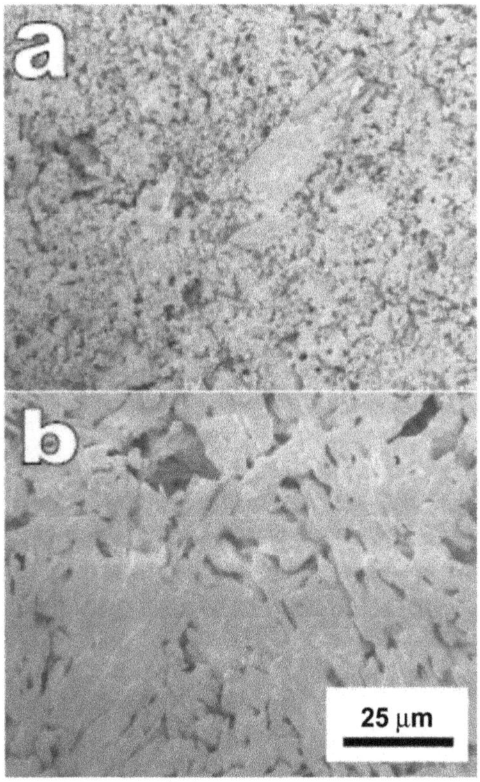

Figure 8.21. SEM images of the samples doped with 10 mol% V_2O_5 after sintering for 4 h at different temperatures: (a) 1100 °C and (b) 1400 °C. Reproduced with permission from [43]. Copyright 2003 Elsevier.

samples doped with Nb_2O_5 and Ta_2O_5 were higher than the theoretical density of mullite, owing to the higher densities of Nb_2O_5 and Ta_2O_5.

The mullite phase formation, microstructural development and anisotropic grain growth of the oxide mixture of Al_2O_3 and SiO_2 doped with different concentrations of V_2O_5, Nb_2O_5 and Ta_2O_5, mechanochemically activated with stainless steel media and sintered at different temperatures has been studied. Figure 8.29 shows the XRD patterns of the samples doped with different contents of V_2O_5 after sintering at different temperatures for 4 h. For the samples with 5 mol% V_2O_5 mullite phase formation was observed after sintering at 1100 °C. This phase formation temperature was lower than that of the sample under the same conditions without mechanochemical activation by 100 °C. The reduction in mullite phase formation was readily attributed to the refinement in particles and thus the enhancement in reactivity of the starting oxides, as a result of the mechanochemical activation. In addition, after sintering at 1100 °C, the samples with higher concentrations of V_2O_5, as illustrated in figures 8.29(b) and (c), confirmed that the presence of V_2O_5 had a positive effect on mullite phase formation, as discussed earlier.

Figure 8.30 shows the XRD patterns of the samples doped with different contents of Nb_2O_5, after sintering at different temperatures. It was found that the mullite

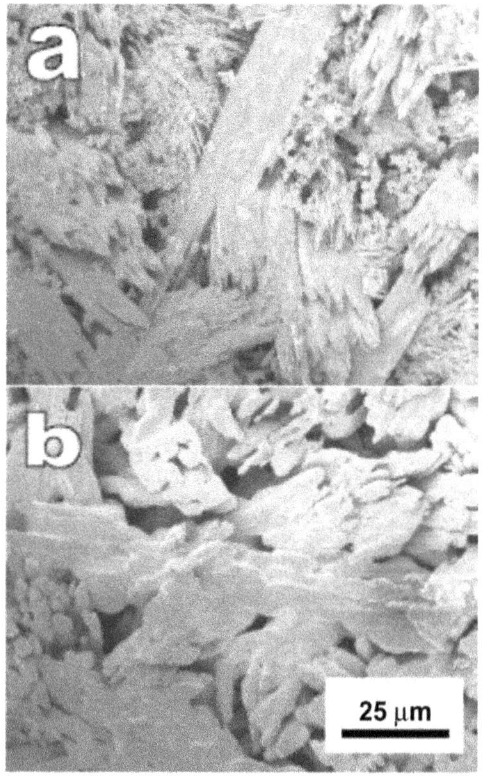

Figure 8.22. SEM images of the sample doped with 20 mol% V_2O_5 after sintering for 4 h at different temperatures: (a) 1100 °C and (b) 1400 °C. Reproduced with permission from [43]. Copyright 2003 Elsevier.

phase was formed at 1200 °C for the sample doped with 5 mol% Nb_2O_5, as seen in figure 8.30(a). In this regard, the effect of Nb_2O_5 on mullite phase formation was negative, compared to that of V_2O_5. Although the mullite phase was also observed in the samples doped with 10 mol% and 20 mol% Nb_2O_5, the intensities of the mullite diffraction peaks were relatively decreased, as displayed in figures 8.30(b) and (c), further confirming that the presence of Nb_2O_5 delayed the phase formation of mullite from the oxide mixture. In addition, a new phase with a composition of $AlNbO_4$ was present in the samples after sintering at 1100 °C. Moreover, this phase was stable up to 1500 °C.

This observation is similar to that of the samples doped with Nb_2O_5 without the application of mechanochemical activation. The reason behind the retarded mullitization due to the addition of Nb_2O_5 has been qualitatively discussed previously. The negative effect has also been evidenced using other measurement techniques. Figure 8.31 shows DTA curves of the mixtures doped with Nb_2O_5 that were mechanochemically activated with stainless steel media. The exothermic peak in the DAT curve was due to the mullitization from the mixture. The peak temperatures were 1293 °C, 1301 °C and 1312 °C for the mixtures with 5 mol%,

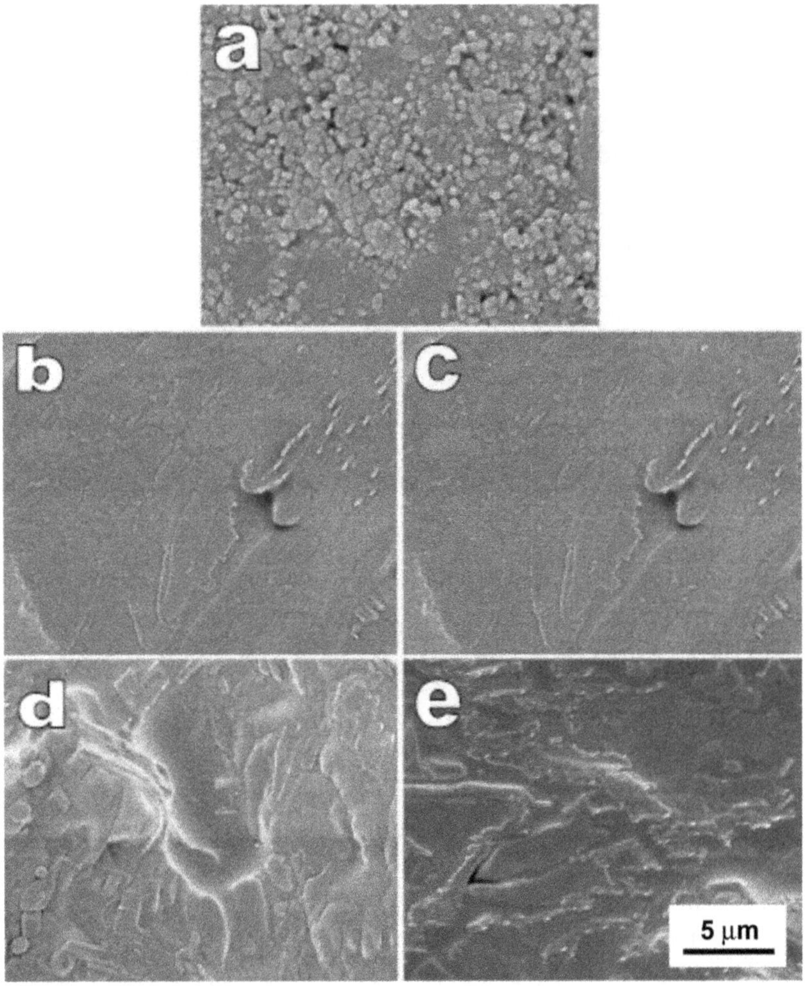

Figure 8.23. SEM images of the samples doped with 5 mol% Nb_2O_5 after sintering for 4 h at different temperatures: (a) 1100 °C, (b) 1200 °C, (c) 1300 °C, (d) 1400 °C and (e) 1500 °C. Reproduced with permission from [43]. Copyright 2003 Elsevier.

10 mol% and 20 mol%, respectively. In other words, the mullite phase formation temperature was increased with increasing content of Nb_2O_5.

Figure 8.32 depicts the XRD patterns of the samples doped with different contents of Ta_2O_5 after sintering at different temperatures. With the addition of 5 mol% Ta_2O_5, the mullite phase was detected in the samples sintered at 1200 °C. However, the relative intensity of the mullite diffraction peaks was quite low. Therefore, the negative effect of Ta_2O_5 on the mullite phase formation was stronger than that of Nb_2O_5. It was interesting to note that the new phase present in this group of samples also had a composition of $AlTaO_4$, while it had a different crystal structure from the two phases of $AlTaO_4$. For clarity, the new phase was labeled as $AlTaO_4(3)$, as illustrated in figure 8.32(a). The secondary phase $AlTaO_4(3)$ was

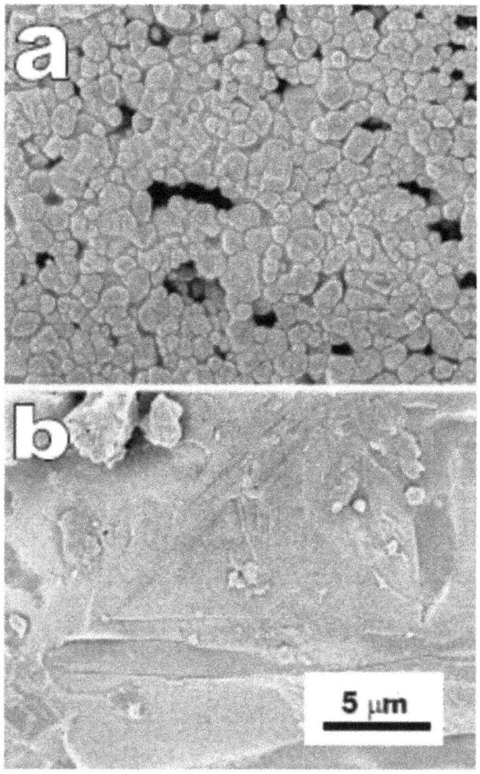

Figure 8.24. SEM images of samples doped with 10 mol% Nb_2O_5 after sintering for 4 h at different temperatures: (a) 1200 °C and (b) 1400 °C. Reproduced with permission from [43]. Copyright 2003 Elsevier.

stable up to 1500 °C. This interesting occurrence could be attributed to the refined particles of the oxides, while the exact mechanism should be clarified through further investigation.

As the content of Ta_2O_5 was increased to 10 mol% the mullite phase was also detected in the sample sintered at 1200 °C, but the peak intensity was lower than that with 5 mol% Ta_2O_5, as revealed in figure 8.32(b). This once again indicated that Ta_2O_5 had a negative effect on mullite phase formation. In addition to the new phase $AlTaO_4$(3), another secondary phase $FeTaO_4$ was present in this group of samples, which was mainly ascribed to the contamination of the stainless steel milling media. Additionally, the peak intensity of $AlTaO_4$(3) was gradually increased with increasing concentration of Ta_2O_5.

When the concentration of Ta_2O_5 was increased to 20 mol%, the mullite phase formation temperature was increased to 1300 °C, as observed in figure 8.32(c). At the same time, Ta_2O_5 was detected in the sample sintered at 1000 °C. In addition, besides $AlTaO_4$(3), $AlTaO_4$(1) was present, as observed in the case without mechanochemical activation. However, $AlTaO_4$(1) was absent after sintering at 1500 °C. The reason for this observation has not yet been clarified. Nevertheless, the negative effect of Ta_2O_5 on mullite phase formation was closely related to the

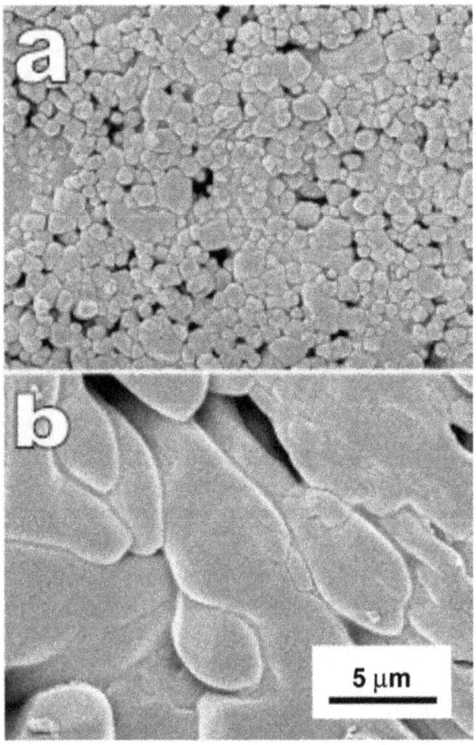

Figure 8.25. SEM images of the samples doped with 20 mol% Nb_2O_5 after sintering for 4 h at different temperatures: (a) 1200 °C and (b) 1400 °C. Reproduced with permission from [43]. Copyright 2003 Elsevier.

presence of the new secondary phases, which strongly compete with the dissolution of Al_2O_3 into the SiO_2-rich liquid phase, as discussed previously.

SEM images of the samples doped with 5 mol% V_2O_5, after sintering at different temperatures, are illustrated in figure 8.33. After sintering at 1000 °C, the sample consisted of only small particles with spherical morphology. At this temperature the mullite phase was not formed, as shown in figure 8.29(a). However, the sample sintered at 1100 °C contained a large number of whiskers, which had a bimodal size distribution. The larger whiskers had lengths >20 μm and thicknesses of up to 5 μm, while there were also many very thin whiskers. According to the XRD patterns, the mullite phase had been already formed after sintering at this temperature, as seen in figure 8.29(a). In comparison, the whiskers in the samples sintered at temperatures ⩾1200 °C exhibited a uniform size distribution.

This observation could be understood from the nucleation and grain growth point of view. Once a critical temperature was reached, nucleation began to occur due to the reaction of the reactants, forming mullite. As the number of nuclei reached a critical point, grain growth would take place. Because the grain growth involved mass transportation at the expense of the nuclei, it was a kinetic process. When the sample was sintered at 1100 °C, the nucleation was continued, while grain growth was possible at the same time. As a result, grains with relatively large dimensions

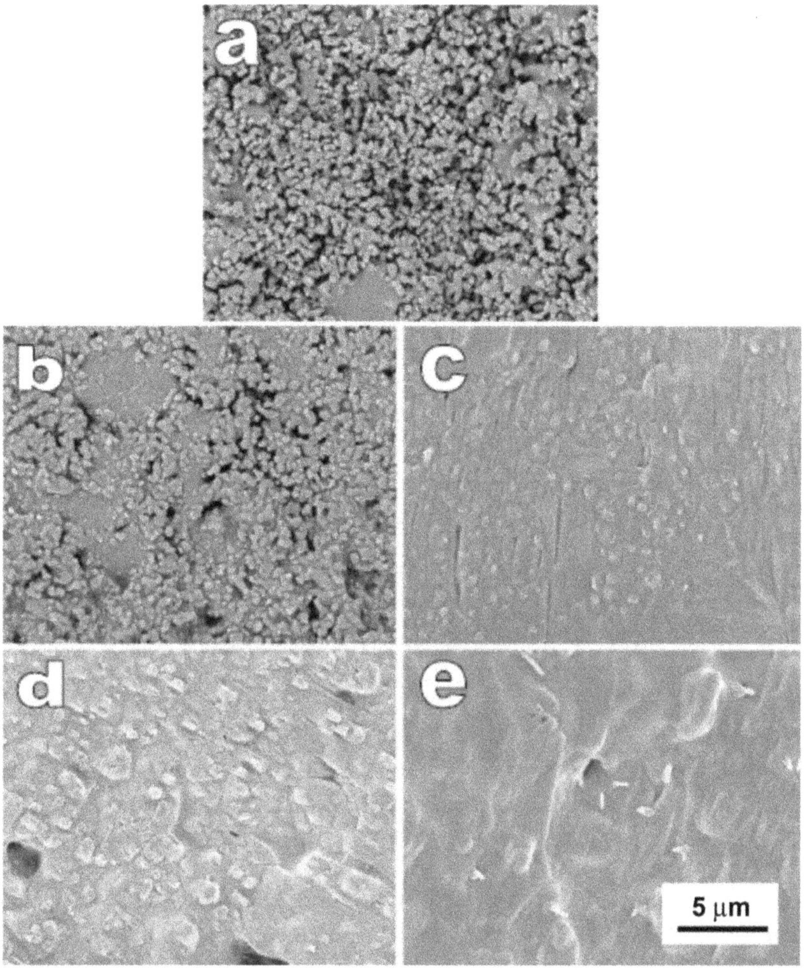

Figure 8.26. Ten SEM images of the samples doped with 5 mol% Ta_2O_5 after sintering for 4 h at different temperatures: (a) 1100 °C, (b) 1200 °C, (c) 1300 °C, (d) 1400 °C and (e) 1500 °C. Reproduced with permission from [43]. Copyright 2003 Elsevier.

were formed. It is expected that if the sintering time was prolonged, the size of the mullite grains would be further increased. In addition, since the mullitization occurred before densification, anisotropic grain growth occurred in this case.

For the sample sintered at 1200 °C the scenario was different, where the grain growth started after the nucleation was complete. As a consequence, the grain size distribution was relatively narrow, compared to that of the sample sintered at 1100 °C. The nucleation and grain growth behaviors of the samples sintered at higher temperatures were similar to those of the sample sintered at 1200 °C. In other words, the nucleation process was finished at a temperature between 1100 °C and 1200 °C. In this regard, there is a need to examine the effect of sintering temperature and sintering time

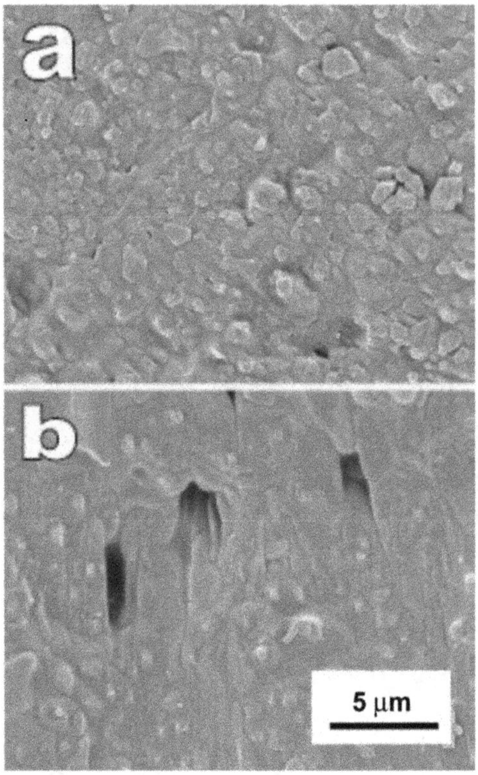

Figure 8.27. SEM images of the samples doped with different contents of Ta_2O_5 after sintering at 1300 °C: (a) 10 mol% and (b) 20 mol%. Reproduced with permission from [43]. Copyright 2003 Elsevier.

on the phase formation and grain growth in a more detailed way, so as to clarify the consequence of nucleation and grain growth in the material systems.

SEM images of the samples doped with 10 mol% and 20 mol% V_2O_5 after sintering at different temperatures are depicted in figures 8.34 and 8.35, respectively. It was found that the development of the microstructure was similar to that of the sample doped with 5 mol% V_2O_5 as a function of sintering temperature, as demonstrated in figures 8.33(b) and (c). However, it was slightly different in that the grains in the latter two samples were much larger than those in the 5 mol% sample. Also, the grain dimensions were increased with increasing concentration of V_2O_5. The same was observed for the samples sintered at higher temperatures. In other words, with an increasing concentration of V_2O_5, the nucleation temperature would be decreased or the nucleation process completed at a low temperature. Therefore, the grain growth started gradually at a low temperature. In addition, the amount of liquid phase increased with increasing content of V_2O_5, owing to its low melting point, which in turn promoted the nucleation and grain growth. It was also found that the microstructure of the samples remained nearly unchanged within a sintering temperature range 1200 °C–1500 °C.

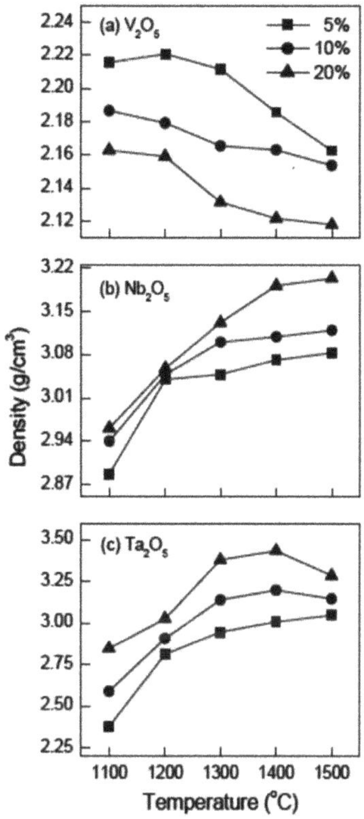

Figure 8.28. Densities of the sintered samples doped with different concentrations of three oxides as function of sintering temperature: (a) V_2O_5, (b) Nb_2O_5 and (c) Ta_2O_5. Reproduced with permission from [43]. Copyright 2003 Elsevier.

SEM images of the samples doped with Nb_2O_5 after sintering at different temperatures are displayed in figures 8.36–8.38. First, the anisotropic grain growth of this group of samples was different from that of those doped with V_2O_5. Unlike the spindle-shaped grains, the mullite grains in this group were shaped more like potatoes. The differences in grain morphology between the two groups of samples could be related to the differences in the environments in which the grains grew. Due to the negative effect of Nb_2O_5 on mullite phase formation, the samples sintered at 1100 °C were composed of spherical particles, because the mullite phase still had not formed. After sintering at 1200 °C anisotropic grain growth was observed, although the grain size distribution was quite wide. For the samples doped with 5 mol% Nb_2O_5, the dimensions of the mullite whiskers were increased gradually with increasing sintering temperature in the range 1200 °C–1500 °C. At a given sintering temperature, the size of the mullite whiskers decreased with increasing sintering temperature. In comparison, the samples doped with 20 mol% Nb_2O_5 exhibited some grains with regular shapes, which could be the secondary phases, although the observation has not been clarified.

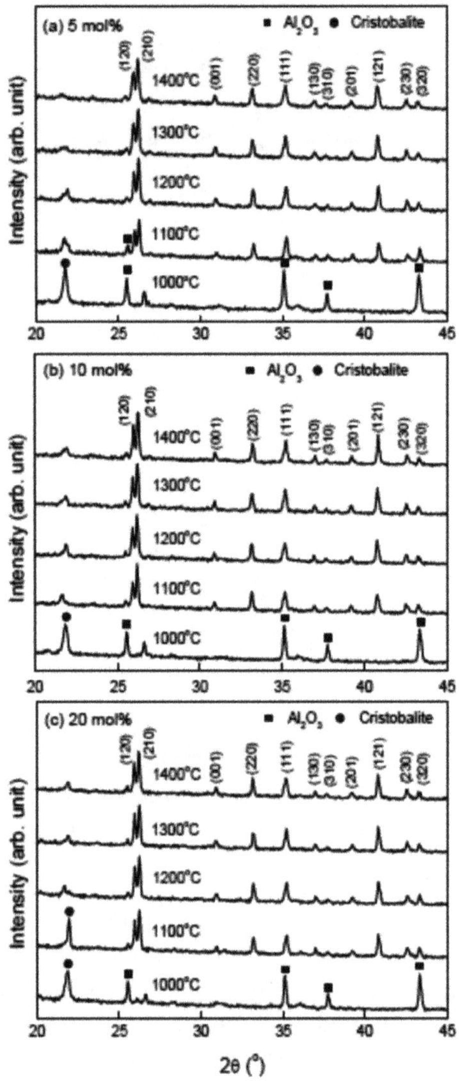

Figure 8.29. XRD patterns of the samples doped with V_2O_5 mechanochemically activated with stainless steel media after sintering at different temperatures.

SEM images of the samples doped with Ta_2O_5 after sintering at temperatures in the range 1100 °C–1500 °C for 4 h are shown in figures 8.39–8.41. The samples sintered at 1100 °C and 1200 °C possessed spherical grains, although the mullite phase was detected by XRD in those samples doped with 5 mol% and 10 mol% Ta_2O_5 after sintering at 1200 °C. This observation was slightly different from those discussed previously. After sintering at 1300 °C whisker-like grains were formed, together the presence of some spherical grains, for the samples doped with 5 mol% Ta_2O_5, as observed in figure 8.39(c). In addition, the sample exhibited a pretty dense microstructure, with the whisker-like grains being embedded in a matrix. As the

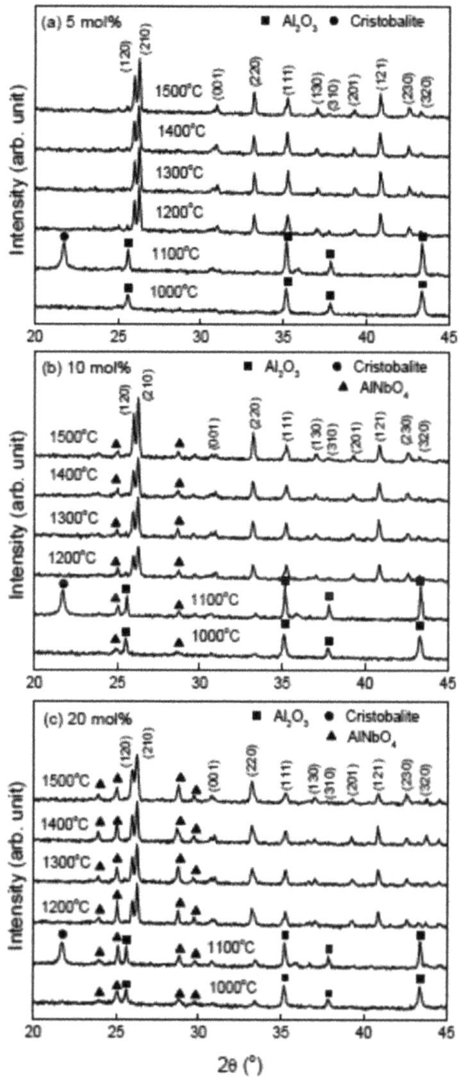

Figure 8.30. XRD patterns of the samples doped with Nb_2O_5 and mechanochemically activated with stainless steel media after sintering at different temperatures.

sintering temperature was increased, the mullite grains grew gradually and became less anisotropic. At the same time, the microstructure of the samples became less and less dense, because of the continuing growth of the grains.

The samples doped with 10 mol% Ta_2O_5 experienced a similar trend in development, but the grain sizes were much larger than those of the samples doped with 5 mol% Ta_2O_5, as seen in figure 8.40. As for the group doped with 20 mol% Ta_2O_5, the anisotropic grain growth was much less pronounced, while a very dense microstructure was observed, in particular for those sintered at higher temperatures, as revealed in figures 8.41(d) and (e). These results once again supported our

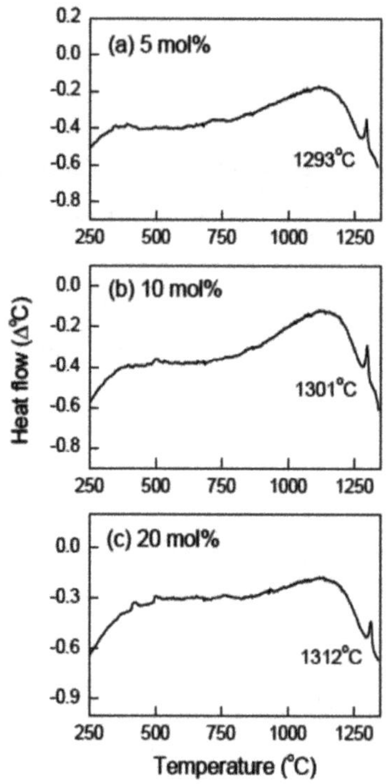

Figure 8.31. DTA curves of the samples doped with Nb_2O_5, mechanochemically activated with stainless steel media for 5 h.

statement that densification and anisotropic grain growth compete with each other as the mullite phase is formed from the mixture of oxides.

Figure 8.42 shows XRD patterns of the samples from the mixtures doped with different concentrations of V_2O_5, mechanochemically activated with WC milling media for 5 h, after sintering at different temperatures. Regardless of the concentrations of V_2O_5, mullitization of the all the samples was nearly complete after sintering at 1100 °C. This observation is very similar to that of the samples from the powders mechanochemically activated with stainless steel milling media. However, careful inspection revealed that the mullite phase formation capability of the mixture activated with WC media was slightly stronger than that treated with stainless steel media, in particular for the samples doped with 5 mol% V_2O_5. As shown clearly in figure 8.29(a), the diffraction peaks of one of the reactants (i.e. Al_2O_3) was quite strong, while Al_2O_3 is nearly invisible in the XRD pattern of figure 8.42(b). This is simply because the activation with WC media generated higher energy than that with stainless steel, due to the higher density of WC than stainless steel. In addition, the XRD patterns of all the samples remained almost unchanged after sintering in the temperature range 1200 °C–1500 °C, which was similar to those of the stainless steel activated samples.

Figure 8.43 shows the XRD patterns of the samples doped with Nb_2O_5 from the mixtures mechanochemically activated with WC milling media after sintering at

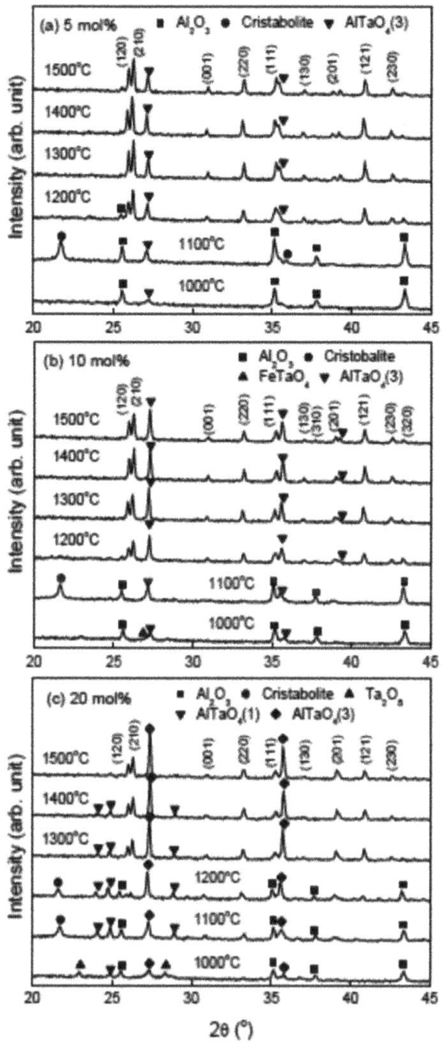

Figure 8.32. XRD patterns of the samples doped with Ta_2O_5 mechanochemically activated with stainless steel media after sintering at different temperatures.

different temperatures. In terms of mullite phase formation, the mixtures activated with WC media were almost the same as those activated with stainless steel, as seen in figure 8.30. The only difference was that the secondary phase $AlNbO_4$ had been formed in the samples doped with 5 mol% Nb_2O_5 in this case, whereas such a phase was only detected in the samples doped with 10 mol% Nb_2O_5 when using stainless steel milling media. These results once again confirmed that the WC media were stronger than the stainless steel media. Owing to the higher energy with the WC media, the oxides were refined to a higher degree, thus having higher reactivity. As a result, the reaction between Al_2O_3 and Nb_2O_5 became much easier. This small difference in reactivities of the mixtures activated with the two different milling

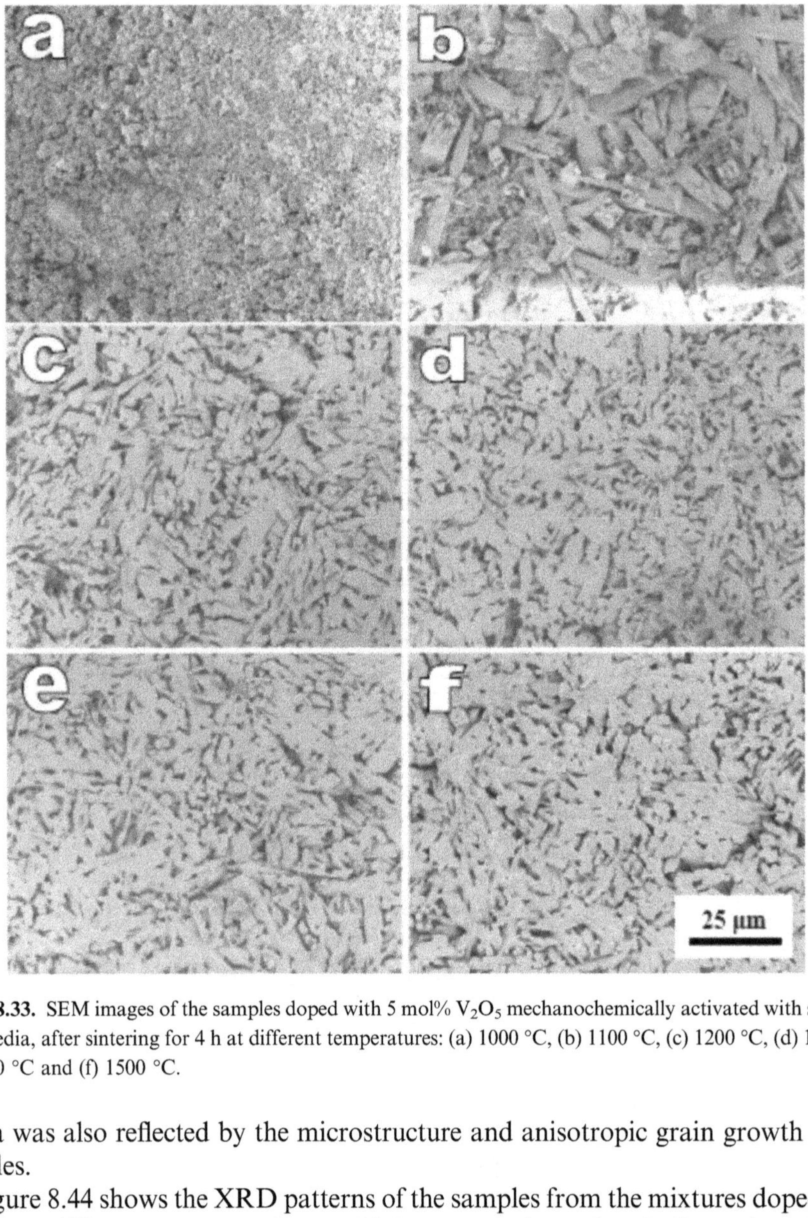

Figure 8.33. SEM images of the samples doped with 5 mol% V_2O_5 mechanochemically activated with stainless steel media, after sintering for 4 h at different temperatures: (a) 1000 °C, (b) 1100 °C, (c) 1200 °C, (d) 1300 °C, (e) 1400 °C and (f) 1500 °C.

media was also reflected by the microstructure and anisotropic grain growth of the samples.

Figure 8.44 shows the XRD patterns of the samples from the mixtures doped with Ta_2O_5 at different concentrations, which were mechanochemically activated with WC milling media after sintering at different temperatures. Compared to the results of the samples from the mixtures treated with stainless steel media, the negative effect of Ta_2O_5 on the mullite phase formation was confirmed. It was interesting to note that only $AlTaO_4$(1) was detected in the three doping groups, without the presence of $AlTaO_4$(2) and $AlTaO_4$(3), which were observed in the samples mechanochemically activated with stainless steel media and those without mechanochemical activation, as demonstrated in figures 8.32 and 8.19. The reason for this

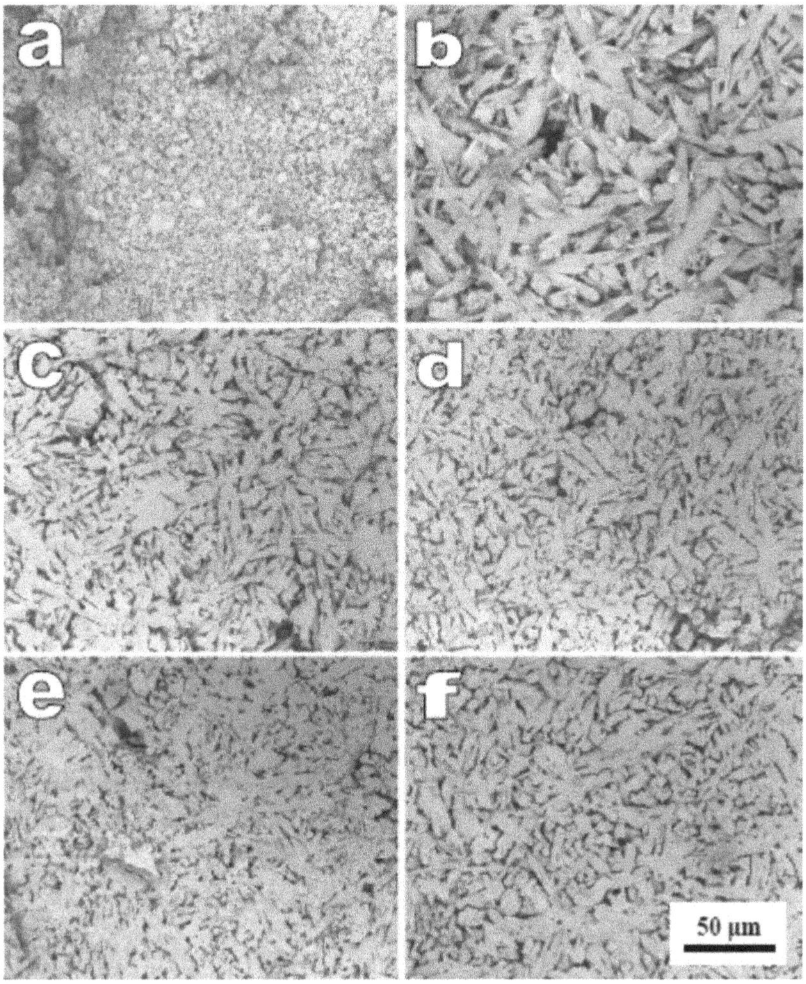

Figure 8.34. SEM images of the samples doped with 10 mol% V_2O_5 mechanochemically activated with stainless steel media, after sintering for 4 h at different temperatures: (a) 1000 °C, (b) 1100 °C, (c) 1200 °C, (d) 1300 °C, (e) 1400 °C and (f) 1500 °C.

difference had not yet been clarified, which could be an open question for future study. In addition to $AlTaO_4$(1), $AlWO_4$ was detected as another secondary phase, which was readily ascribed to the contamination by WC from the milling media.

For the samples doped with 5 mol% and 10 mol% Ta_2O_5, diffraction peaks of mullite with a weak intensity were observed after sintering at 1100 °C, while the mullite phase was visible for the sample doped with 20 mol% Ta_2O_5 at the sintering temperature of 1200 °C. These temperatures were obviously lower than those of the samples from the mixtures activated with stainless steel media, which was also a reflection of the stronger activation effect of the WC milling media. Similarly, the diffraction peaks of the secondary phase were gradually increased with increasing

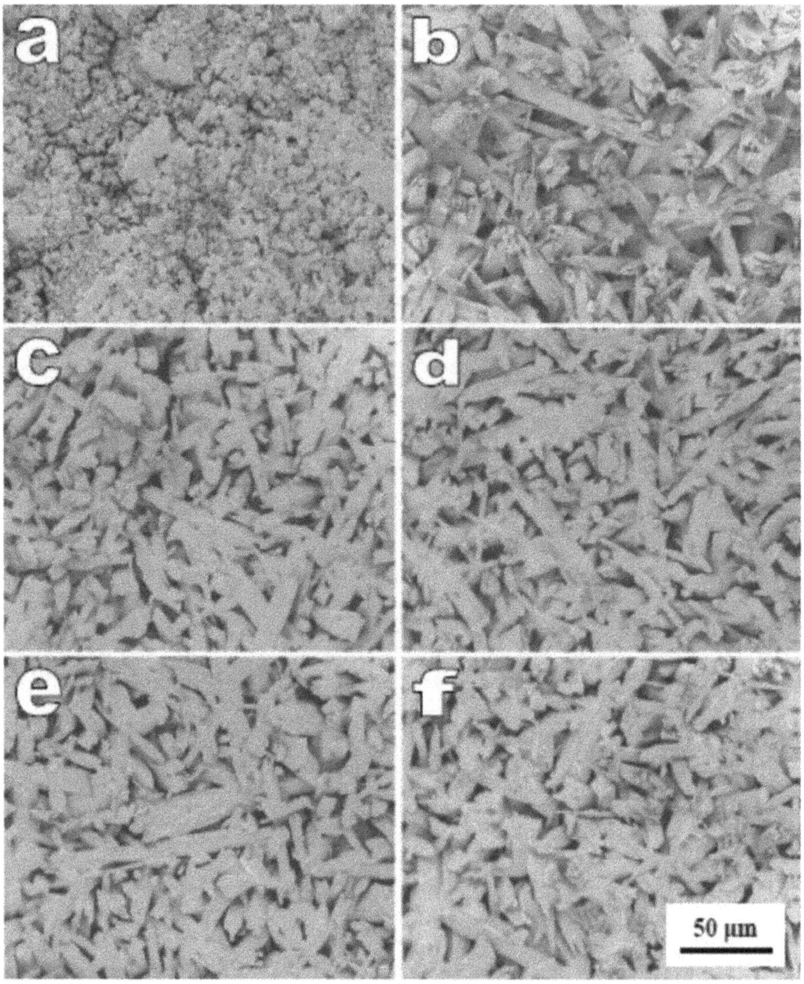

Figure 8.35. SEM images of the samples doped with 20 mol% V_2O_5 mechanochemically activated with stainless steel media, after sintering for 4 h at different temperatures: (a) 1000 °C, (b) 1100 °C, (c) 1200 °C, (d) 1300 °C, (e) 1400 °C and (f) 1500 °C.

concentration of the dopant, while they passed greatly over those of mullite, in particular for the samples with 20 mol% Ta_2O_5.

SEM images of the samples doped with different contents of V_2O_5 from the mixtures mechanochemically activated with WC milling media after sintering at different temperatures are depicted in figures 8.45–8.47. Interestingly, it was found that the samples from the mixtures activated with WC and stainless steel media shared a similar microstructure and grain morphology after sintering at 1100 °C for 4 h. They were all characterized by the grains with an anisotropic growth profile and a morphology more or less like spindles, although the two ends were branched. As the sintering was increased to 1200 °C the mullite whiskers became smaller, while their morphology became more regular. In addition, for a given concentration of

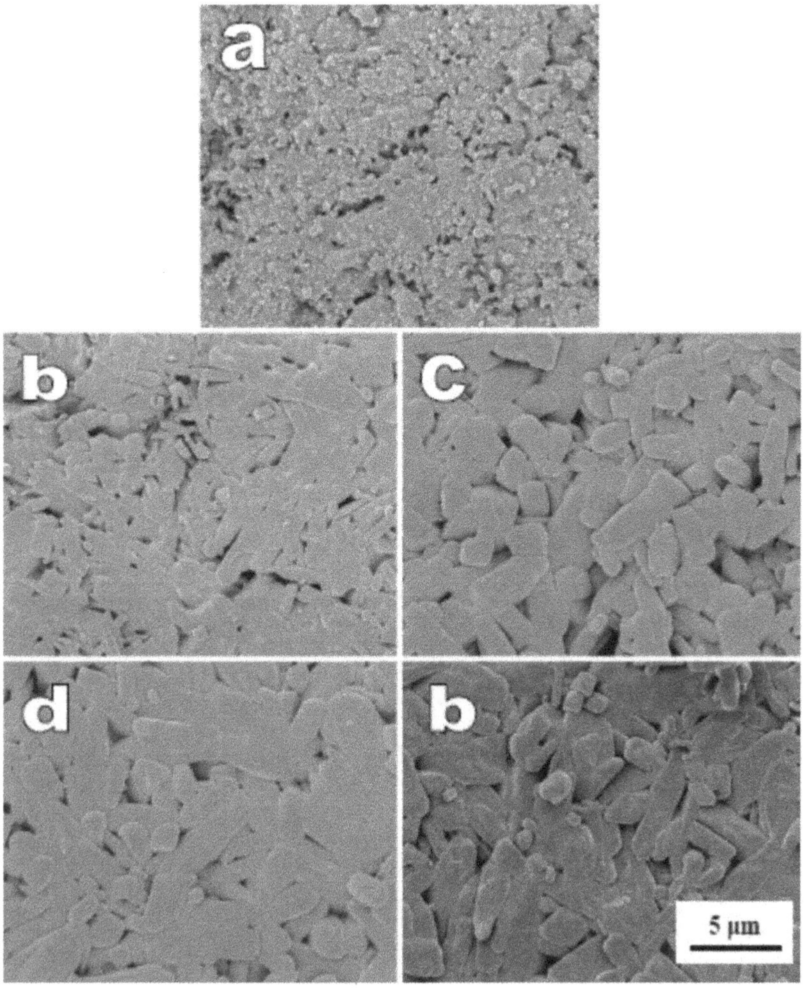

Figure 8.36. SEM images of the samples doped with 5 mol% Nb_2O_5 mechanochemically activated with stainless steel media, after sintering for 4 h at different temperatures: (a) 1000 °C, (b) 1100 °C, (c) 1200 °C, (d) 1300 °C and (e) 1400 °C.

V_2O_5 the size of the whiskers decreased slightly with increasing sintering temperature. At a given sintering temperature, the size of the mullite whiskers increased with increasing content of V_2O_5.

Figure 8.48 shows SEM images of the samples doped with 5 mol% Nb_2O_5 from the mixtures mechanochemically activated with WC milling media, after sintering at temperatures in the range 1100 °C – 1500 °C for 4 h. The sample sintered at 1100 °C possessed spherical particles/grains and the mullite phase was not form at this temperature. As the temperature was raised to 1200 °C, mullite whiskers with relatively regular shapes were present. When further increasing the sintering temperature to 1300 °C the whiskers became shorter and thicker with a square-shaped cross-section. As the temperature was increased to 1400 °C, the dimensions

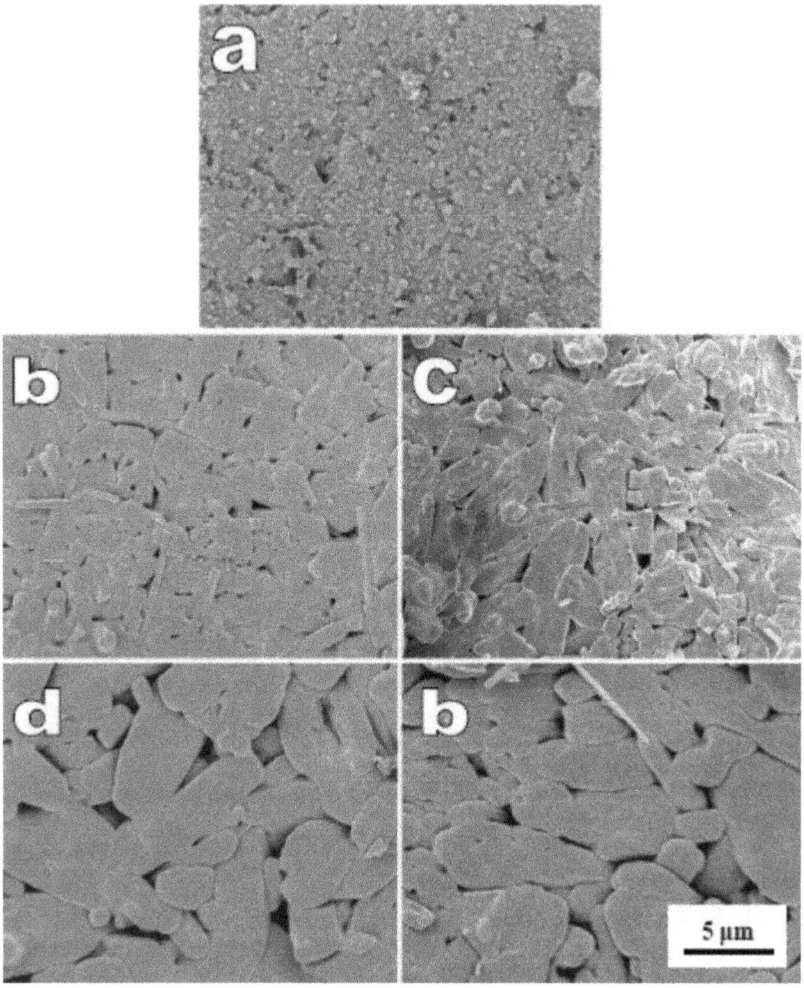

Figure 8.37. SEM images of the samples doped with 10 mol% Nb_2O_5 mechanochemically activated with stainless steel media, after sintering for 4 h at different temperatures: (a) 1000 °C, (b) 1100 °C, (c) 1200 °C, (d) 1300 °C and (e) 1400 °C.

of the whiskers were further enlarged, while a small number of them lost their regular shape, as observed in figure 8.48(d). Finally, after sintering at 1500 °C, almost all the mullite whiskers became irregular in shape, as seen in figure 8.48(e). However, this variation in the morphology of the mullite whiskers could not be attributed to the presence of the secondary phase, because it already formed after sintering at 1200 °C, which deserves to be further studied.

A similar result was observed in terms of microstructural evolution and grain growth of the samples doped with 10 mol% and 20 mol% Nb_2O_5, as illustrated in figures 8.49 and 8.50. However, with higher contents of Nb_2O_5 it was more apparent that two types of morphologies could be identified from the grains of the samples. Those with a different morphology from the typical mullite whiskers were most

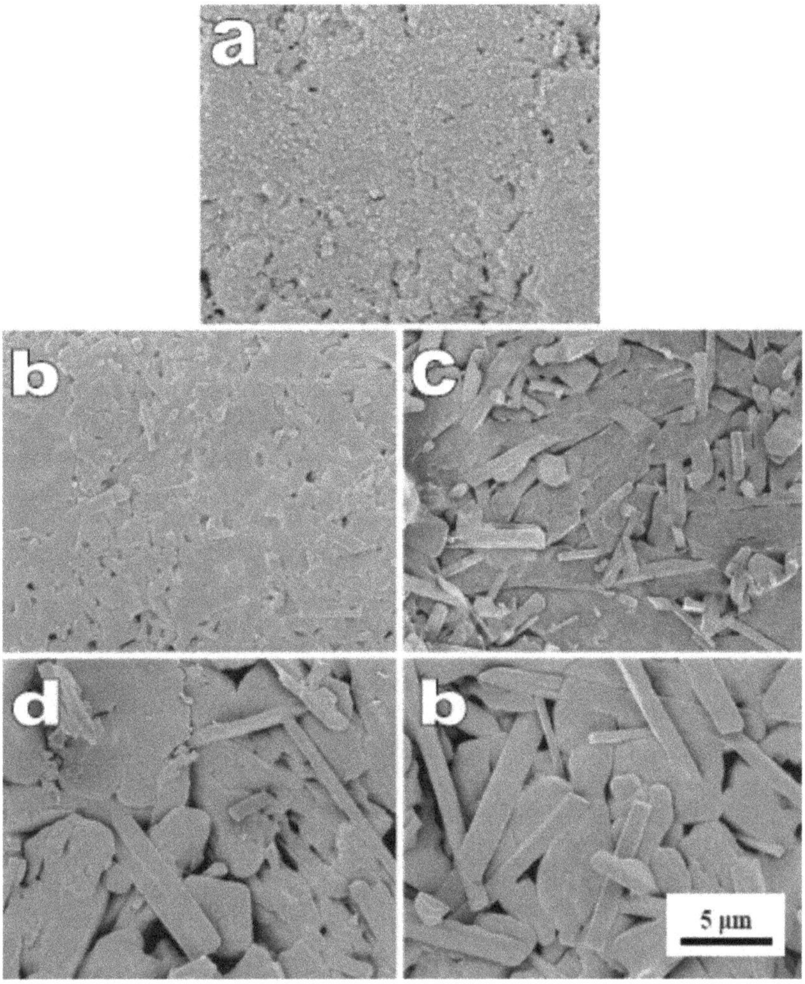

Figure 8.38. SEM images of the samples doped with 20 mol% Nb_2O_5 mechanochemically activated with stainless steel media, after sintering for 4 h at different temperatures: (a) 1000 °C, (b) 1100 °C, (c) 1200 °C, (d) 1300 °C and (e) 1400 °C.

likely the secondary phases. To clearly identify them, it is necessary to focus more research attention on this aspect by using more advanced and sophisticated analysis techniques, such as elemental analysis, TEM and so on.

Figure 8.51 shows SEM images of the samples doped with 5 mol% Ta_2O_5 from the mixtures mechanochemically activated with WC milling media, after sintering at different temperatures. As indicated by the XRD results, the presence of Ta_2O_5 had a negative effect on mullite phase formation. For example, the mullite phase was formed from the untreated mixture at 1300 °C, while the mullitization process in the mixture mechanochemically activated with stainless steel media was still incomplete after sintering at 1200 °C. Also, the mullite phase formation temperature was gradually increased with increasing content of Ta_2O_5. However, the high-energy

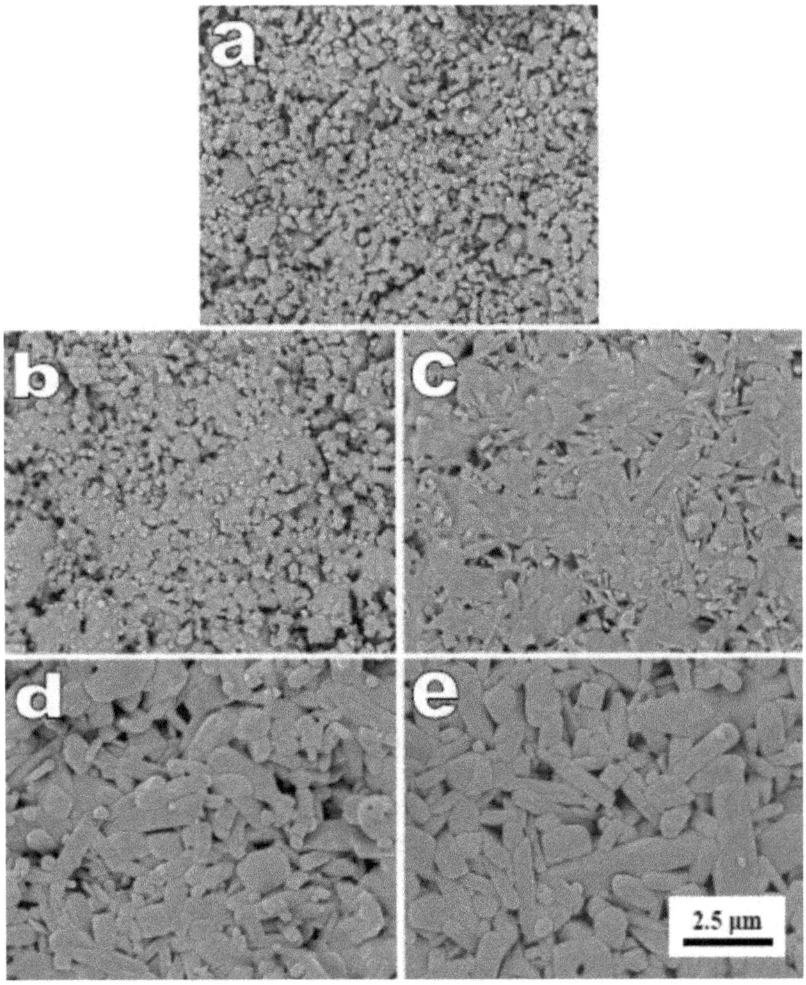

Figure 8.39. SEM images of the samples doped with 5 mol% Ta_2O_5 mechanochemically activated with stainless steel media, after sintering for 4 h at different temperatures: (a) 1000 °C, (b) 1100 °C, (c) 1200 °C, (d) 1300 °C and (e) 1400 °C.

mechanochemical activation with WC media could significantly promote the phase formation of mullite. After sintering at 1100 °C, many thin whiskers could be observed on the surface of the sample, which was similar to that of the mixture of Al_2O_3 and SiO_2 without the addition of dopant, as discussed in the last subsection. This result was in good agreement with the XRD pattern, as shown in figure 8.43(a) where the mullite phase formed as a minority in that case. As the sintering temperature was increased, the whisker grains were present inside the whole sample, consistent with the nearly full mullitization suggested by the XRD pattern. With increasing sintering temperature, the dimensions of the mullite whiskers were monotonically increased, while the relatively regular morphology was retained well.

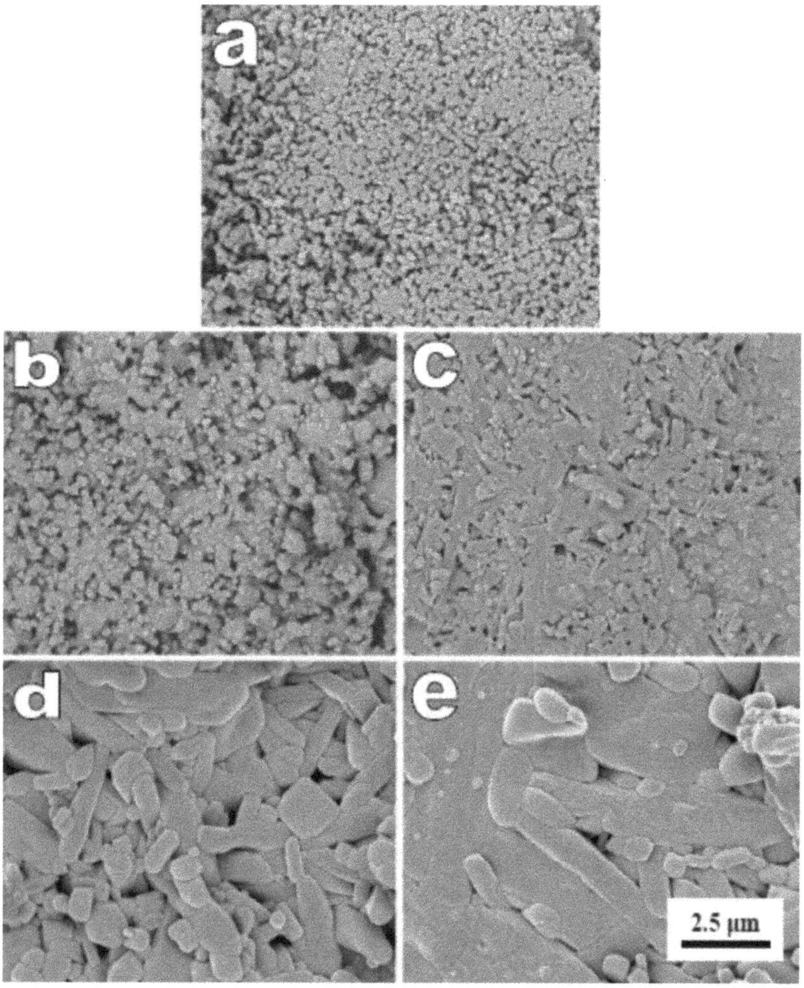

Figure 8.40. SEM images of the samples doped with 10 mol% Ta_2O_5 mechanochemically activated with stainless steel media, after sintering for 4 h at different temperatures: (a) 1000 °C, (b) 1100 °C, (c) 1200 °C, (d) 1300 °C and (e) 1400 °C.

The samples doped with 10 mol% Ta_2O_5 followed a similar variation trend in microstructure development and grain growth to that with 5 mol% Ta_2O_5, as observed in figure 8.52. Both the dimensions and aspect ratio of the mullite whiskers were reduced, while the whiskers were still regular in shape, as demonstrated in figure 8.53. However, as the content of Ta_2O_5 was increased to 20 mol%, the grain morphology was changed to be less regular, in particular for those sintered at higher temperatures, as revealed in figures 8.53(e) and (f). Similar to that observed in the samples doped with Nb_2O_5, a bimodal grain size distribution was presence in this group of samples, due to the formation of the secondary phase.

In this subsection, the effects of Cr_2O_3 and the milling conditions on mullite phase formation and microstructure will be presented and discussed. This work has been

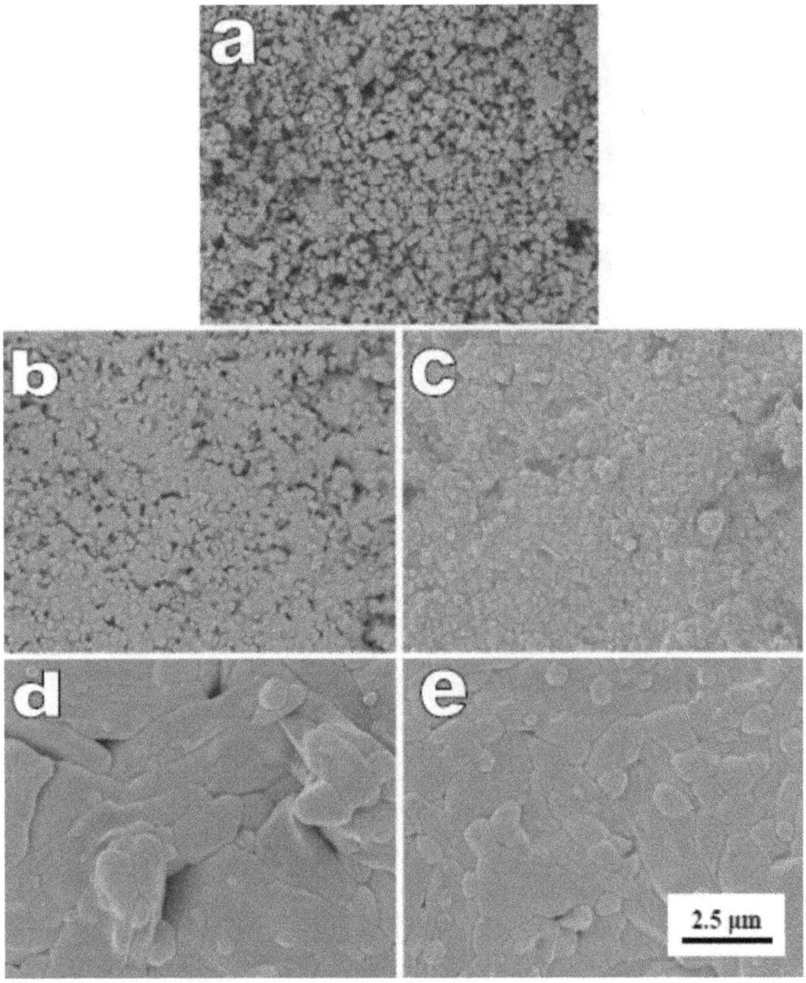

Figure 8.41. SEM images of the samples doped with 20 mol% Ta_2O_5 mechanochemically activated with stainless steel media, after sintering for 4 h at different temperatures: (a) 1000 °C, (b) 1100 °C, (c) 1200 °C, (d) 1300 °C and (e) 1400 °C.

partly published [45]. The starting powders were Al_2O_3 and SiO_2 (quartz). Mechanochemical activation was conducted in a similar way as stated above. The XRD patterns of the samples doped with Cr_2O_3 at different concentrations, from the mixtures without the treatment of mechanochemical activation, are shown in figures 8.54 and 8.55. After sintering at 1100 °C for 4 h, the major phases include Al_2O_3, quartz and cristobalite, which meant that SiO_2 experienced a phase transition from quartz to cristobalite. The peak intensity of cristobalite was higher than that of quartz in the sample sintered at 1200 °C, while mullitization had not yet started. The mullite phase was formed in the sample doped with 5 mol% Cr_2O_3 after sintering at 1300 °C for 4 h.

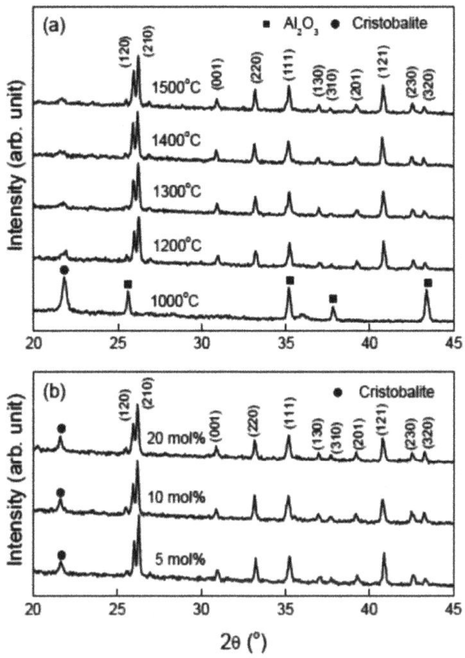

Figure 8.42. XRD patterns of the samples doped with V_2O_5 mechanochemically activated with WC milling media after sintering at different temperatures: (a) 5 mol% V_2O_5 and (b) 1100 °C.

In addition, the phase transition from quartz to cristobalite was complete. With increasing sintering temperature thereafter, the relative peak intensities of mullite were gradually increased, but both Al_2O_3 and cristobalite were still visible up to 1500 °C. After sintering at 1400 °C, the sample with 10 mol% Cr_2O_3 had an XRD pattern similar to that of the sample with 5 mol% Cr_2O_3. However, no mullite was detected in the sample doped with 20 mol% Cr_2O_3, as seen in figure 8.55(b). Therefore, it was concluded that the doping of Cr_2O_3 delayed the mullitizaion process of the oxide mixture of Al_2O_3 and SiO_2.

With the application of the mechanochemical activation with stainless steel milling media, the reactivity of the oxides was increased to a certain degree, as reflected by the XRD results of the samples, as illustrated in figures 8.56 and 8.57. The mullite phase evolution of the samples with mechanochemical activation was similar to that of the samples without the treatment of mechanochemical activation. For example, mullite was detected in the sample with 5 mol% Cr_2O_3 after sintering at 1300 °C, while no mullite was formed at temperatures ⩽1200 °C. The enhanced reactivity of the mechanochemical activated mixtures was evidenced by the reduction in the temperatures of mullitization and the quartz–cristobalite phase transition. As shown in figures 8.54 and 8.56, the peak intensity of cristobalite was lower than that of quartz in the sample without activation, while the relative peak intensity was reversed in the sample with activation, after sintering at 1200 °C. With

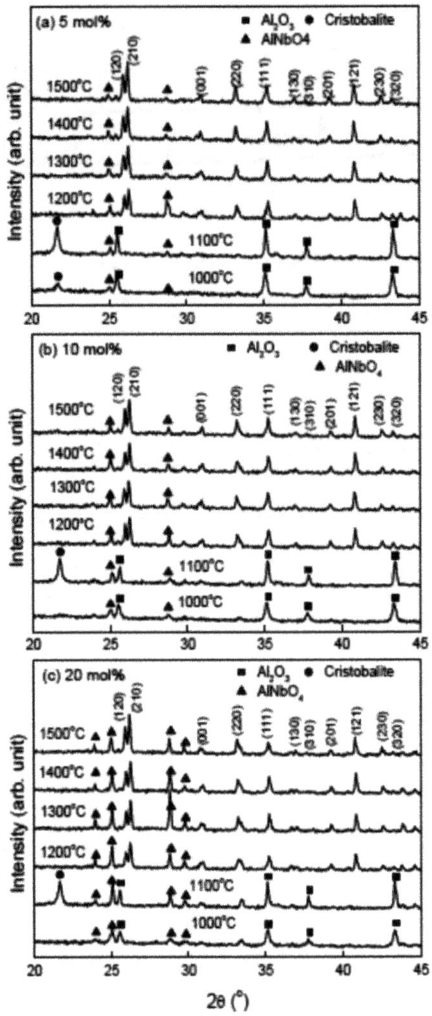

Figure 8.43. XRD patterns of the samples doped with Nb_2O_5 mechanochemically activated with WC milling media after sintering at different temperatures.

the application of mechanochemical activation, mullite was formed in the sample doped with 20 mol% Cr_2O_3 after sintering at 1400 °C.

The enhancement in mullite phase formation from the mixtures mechanochemically activated with WC milling media was more pronounced, as revealed in figures 8.58 and 8.59. As demonstrated in figure 8.58, for the samples doped with 5 mol% Cr_2O_3 the mullite phase was detectable after sintering at 1100 °C, although it was minor phase, while the mullitization process was almost complete at 1200 °C. In comparison, the mullite phase formation temperature of the mixture activated with stainless steel media was 1300 °C, as observed in figure 8.56, which was higher by 100 °C. Figure 8.59 indicated that the increase in the concentration of Cr_2O_3 corresponded to a decrease in the degree of mullitization, confirming the negative

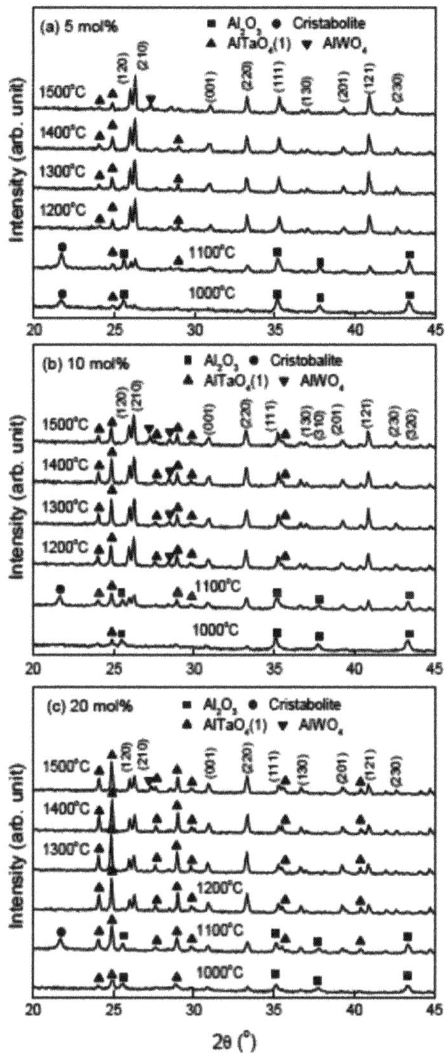

Figure 8.44. XRD patterns of the samples doped with Ta_2O_5 mechanochemically activated with WC milling media after sintering at different temperatures.

effect of Cr_2O_3 on the mullite phase formation from the mixtures. After sintering at 1200 °C, the sample doped with 10 mol% Cr_2O_3 contained only a trace of Al_2O_3, while that with 20 mol% Cr_2O_3 possessed quite a lot of Al_2O_3. In summary, the addition of Cr_2O_3 deaccelerated the formation of mullite phase from the oxide mixtures, while mechanochemical activation could promote the mullitization process.

The microstructure development and grain growth behavior of the samples as a function of Cr_2O_3 doping concentration and sintering temperature have been systematically studied. Figure 8.60 shows SEM images of the samples doped with

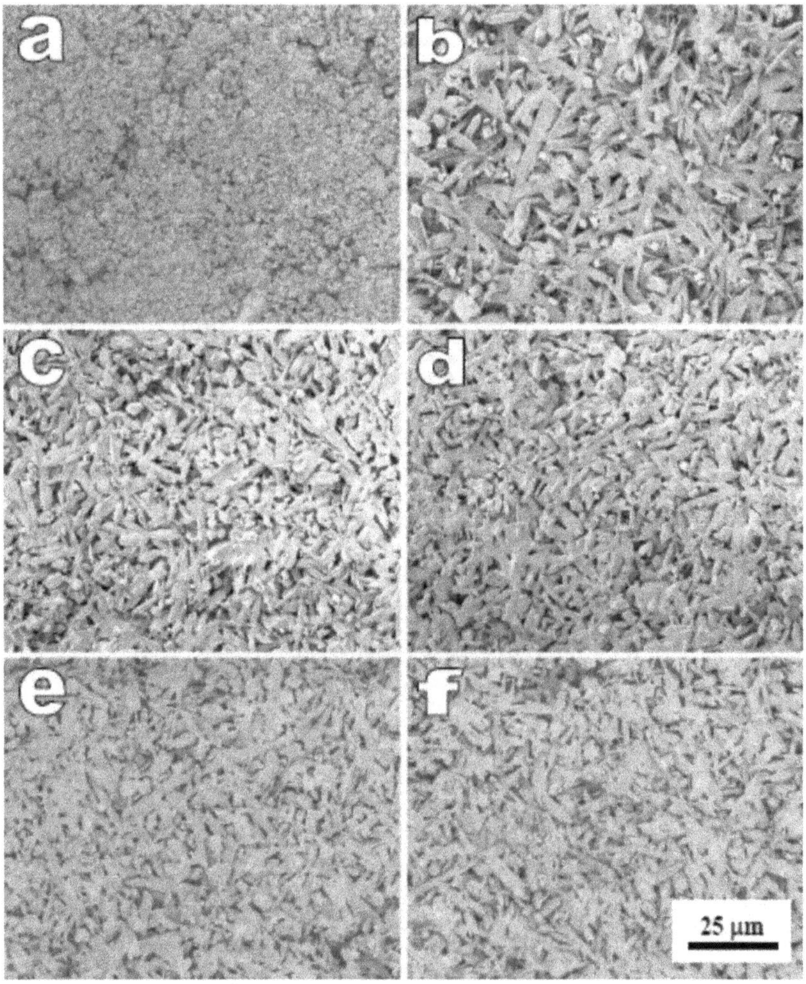

Figure 8.45. SEM images of the samples doped with 5 mol% V_2O_5 derived from the mixture mechanochemically activated with WC milling media, after sintering for 4 h at different temperatures: (a) 1000 °C, (b) 1100 °C, (c) 1200 °C, (d) 1300 °C, (e) 1400 °C and (f) 1500 °C.

5 mol% Cr_2O_3, after sintering at temperatures in the range 1100 °C – 1150 °C for 4 h. The samples mainly consisted of pseudo-spherical grains with a slightly fused microstructure. The degree of the fusion was more pronounced in the samples sintered at higher temperatures. Similar microstructure was observed for the samples from the mixtures doped with higher contents of Cr_2O_3. Figure 8.61 shows SEM images of the samples doped with 10 mol% and 20 mol% Cr_2O_3, after sintering at 1500 °C. In this case, anisotropic grain growth was suppressed because of the low mullitization reactivity of the precursor oxides.

After treatment with mechanochemical activation, the reactivity was enhanced, owing to the refinement of the particles of the oxides. As a result, the mullite phase formation temperature was reduced. However, anisotropic grain growth was still not

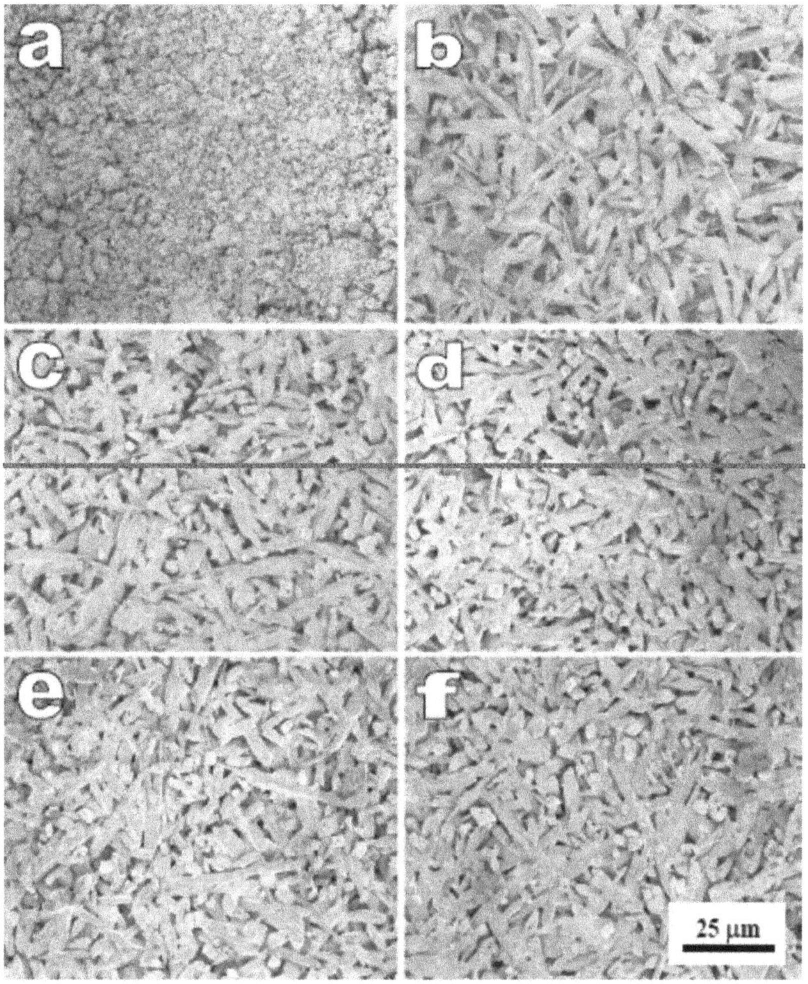

Figure 8.46. SEM images of the samples doped with 10 mol% V_2O_5 derived from the mixture mechanochemically activated with WC milling media, after sintering for 4 h at different temperatures: (a) 1000 °C, (b) 1100 °C, (c) 1200 °C, (d) 1300 °C, (e) 1400 °C and (f) 1500 °C.

obvious, as illustrated in figures 8.62 and 8.63. Although the samples doped with 5 mol% Cr_2O_3 and sintered at temperatures ⩾1300 °C contained whisker-like grains, they were highly fused in the form of an agglomerate, as seen in figures 8.62(d)–(f). The tendency of anisotropic grain growth was further prohibited as the concentration of Cr_2O_3 was increased, as revealed by the representative SEM images in figure 8.63.

SEM images of the samples doped with different contents of Cr_2O_3 from the mixtures mechanochemically treated with WC milling media, after sintering at different temperatures, are shown in figures 8.64–8.66. Compared to the previous two groups of samples, the samples in this group all exhibited well-developed whisker morphology. This was obviously attributed to the much finer particles and

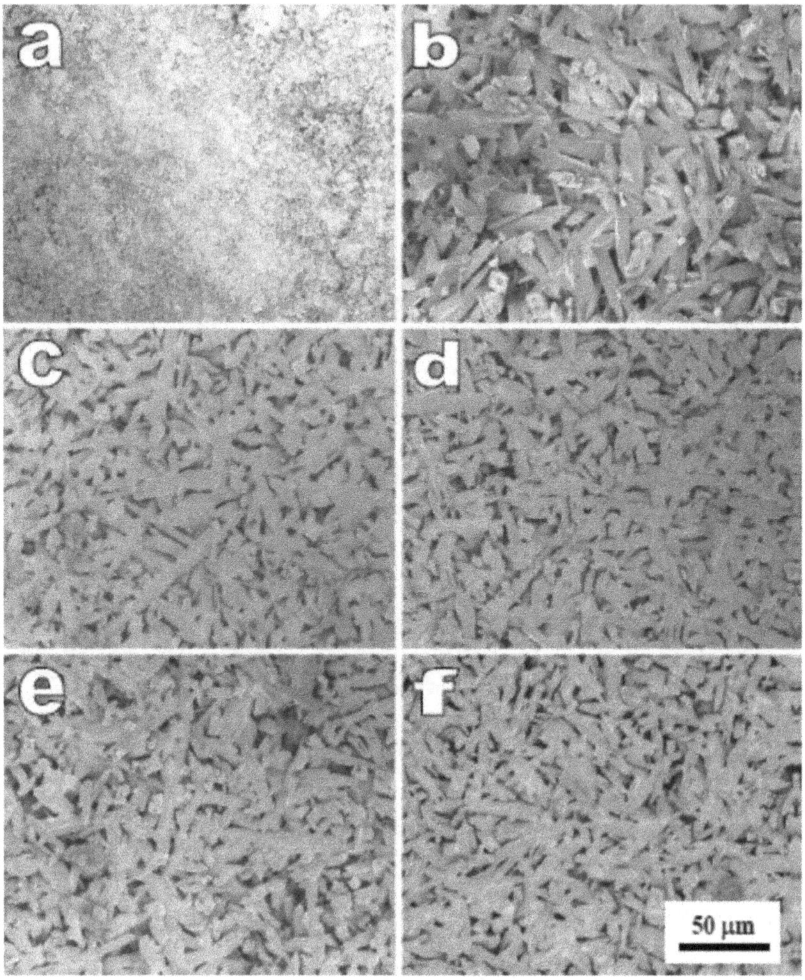

Figure 8.47. SEM images of the samples doped with 20 mol% V_2O_5 derived from the mixture mechanochemically activated with WC milling media, after sintering for 4 h at different temperatures: (a) 1000 °C, (b) 1100 °C, (c) 1200 °C, (d) 1300 °C, (e) 1400 °C and (f) 1500 °C.

thus higher reactivity of the precursor oxides, due to the higher efficiency of the WC milling media. However, the mullite whiskers in this case had a much less regular shape compared to those presented in the last section, as seen in figure 8.5. Two main reasons could be responsible for this observation. First, the SiO_2 powder was ultrafine amorphous silica with high reactivity, while that used in this case was quartz with coarse grains/particles. Second, the presence of the dopant Cr_2O_3 could have played a negative role in the development of the mullite whiskers.

It was found that the addition of WO_3 could promote mullite phase formation from the oxide mixtures [46]. Figure 8.67 shows XRD patterns of the samples doped with WO_3 at different concentrations without mechanochemical activation, after sintering at different temperatures. As observed in figure 8.67(a), for the sample with

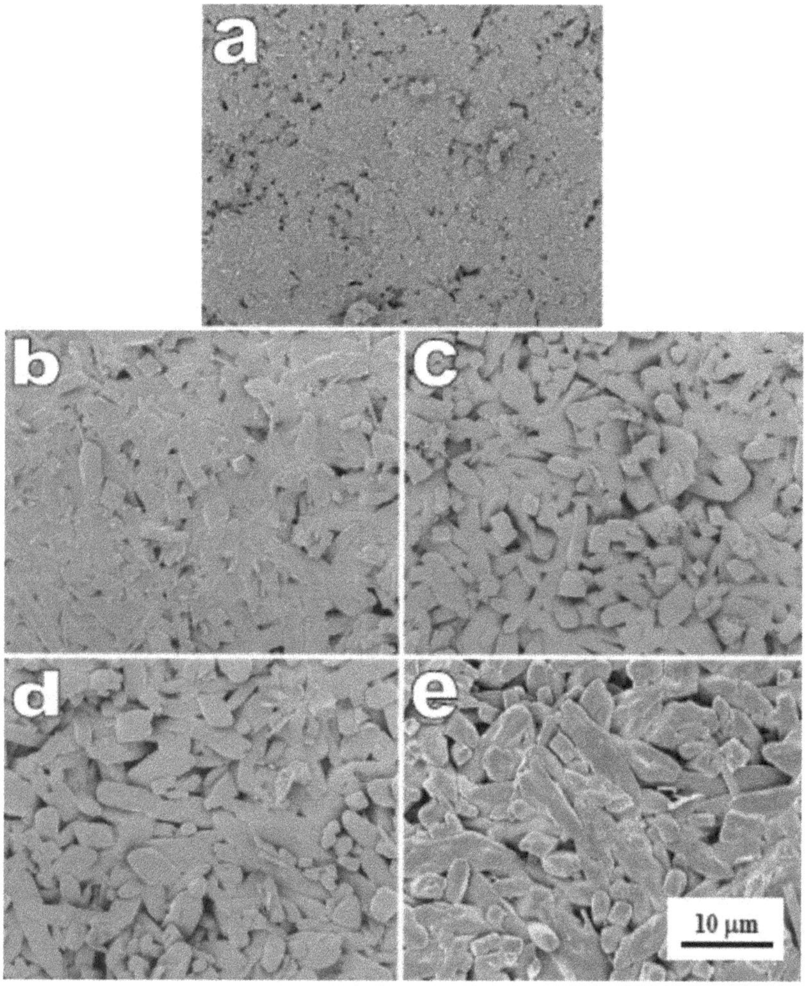

Figure 8.48. SEM images of the samples doped with 5 mol% Nb_2O_5 derived from the mixture mechanochemically activated with WC milling media, after sintering for 4 h at different temperatures: (a) 1000 °C, (b) 1100 °C, (c) 1200 °C, (d) 1300 °C and (e) 1400 °C for 4 h.

5 mol% WO_3 the mullite phase was formed after sintering at 1300 °C, which was the same as that of the oxide mixture without the presence of any dopant. In comparison, for the sample with 10 mol% WO_3, diffraction peaks of mullite were present after sintering at 1200 °C, while the mullite phase formation was nearly complete for the sample doped with 20 mol% WO_3, as revealed in figures 8.67(b) and (c). In other words, the addition of 20 mol% WO_3 induced a reduction in the mullite phase formation temperature by 100 °C. The positive effect of WO_3 on the phase formation of mullite from oxide mixtures has not yet been elucidated.

Although the mullite phase formation was promoted by the presence of WO_3, the reduction in mullitization temperature was not sufficiently large to ensure that the phase formation occurred before densification. As a result, anisotropic grain growth

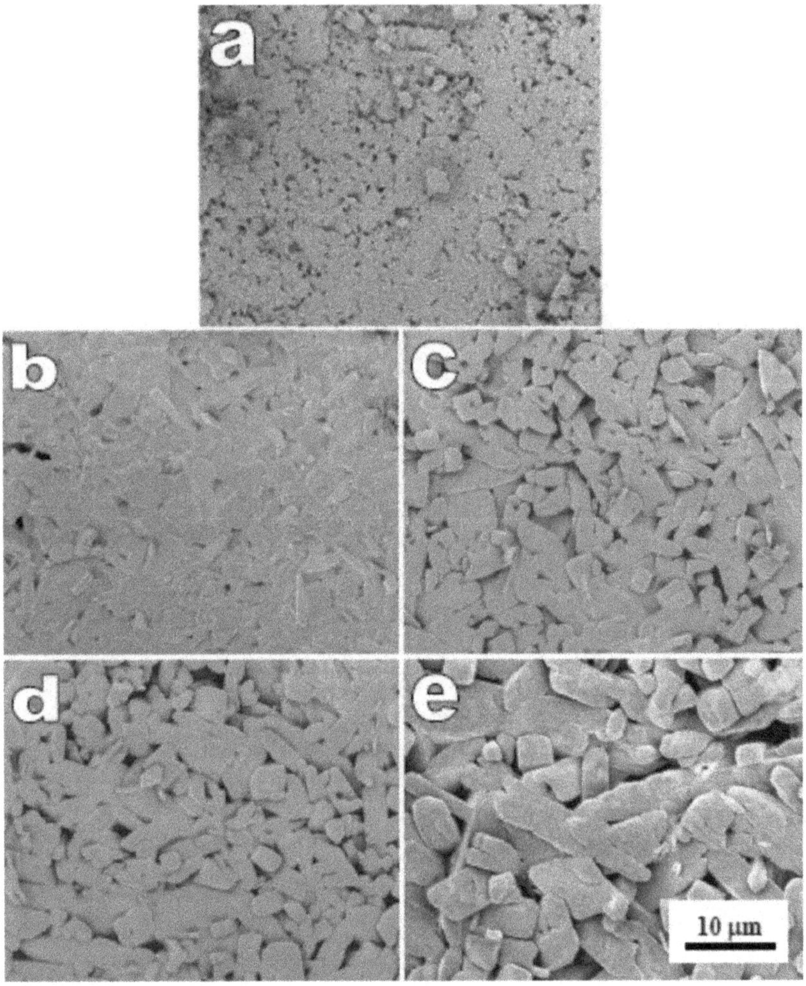

Figure 8.49. SEM images of the samples doped with 10 mol% Nb_2O_5 derived from the mixture mechanochemically activated with WC milling media, after sintering for 4 h at different temperatures: (a) 1000 °C, (b) 1100 °C, (c) 1200 °C, (d) 1300 °C and (e) 1400 °C for 4 h.

was not observed, i.e. no mullite whiskers were present in the samples. Representative SEM images of the samples sintered at different temperatures with different contents of WO_3 are depicted in figures 8.68–8.70. As expected, the sample with 5 mol% WO_3 after sintering at 1200 °C for 4 h exhibited a pretty dense microstructure, as seen in figure 8.68(b). At this temperature, the mullite phase was not detected in the XRD pattern, as illustrated in figure 8.67(a). After sintering at 1300 °C, the sample was further densified, without the observation of any pores. Even though whisker-like grains seemed to be visible, they were tightly embedded in a continuous matrix.

Figure 8.69 shows SEM images of the samples doped with 10 mol% and 20 mol% WO_3, after sintering at 1200 °C for 4 h. Similarly, they were characterized by a dense

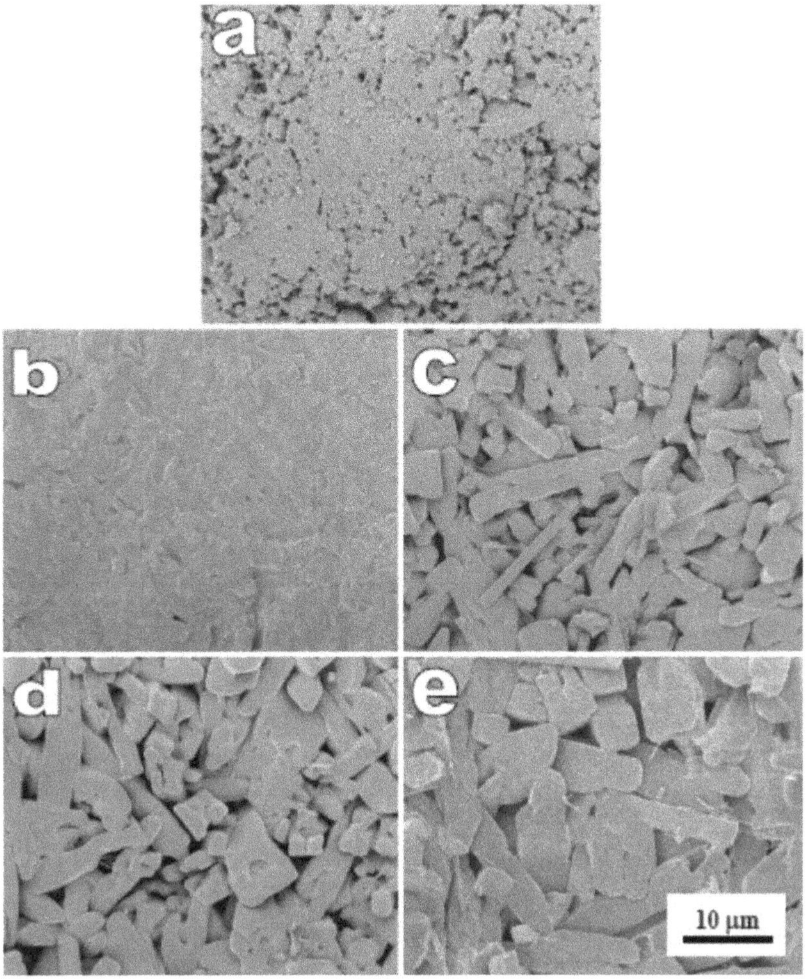

Figure 8.50. SEM images of the samples doped with 20 mol% Nb_2O_5 derived from the mixture mechanochemically activated with WC milling media, after sintering for 4 h at different temperatures: (a) 1000 °C, (b) 1100 °C, (c) 1200 °C, (d) 1300 °C and (e) 1400 °C for 4 h.

microstructure and spherical grains. In addition, with increasing concentration of WO_3, more and more grains with sharp facets were present in the samples, which could be attributed to the effect of WO_3 but has not been clearly clarified. For the sample doped with 20 mol% WO_3, mullite phase formation was nearly complete, while no anisotropic grain growth took place. In this case, the densification and the mullitization might happen simultaneously. However, after sintering at higher temperatures, there was more or less a tendency for the mullite grains to grow anisotropically, as revealed in figure 8.70. Nevertheless, all the samples were nearly fully densified.

With the aid of mechanochemical activation, the mullite phase formation behavior was enhanced. For the samples mechanochemically treated with stainless

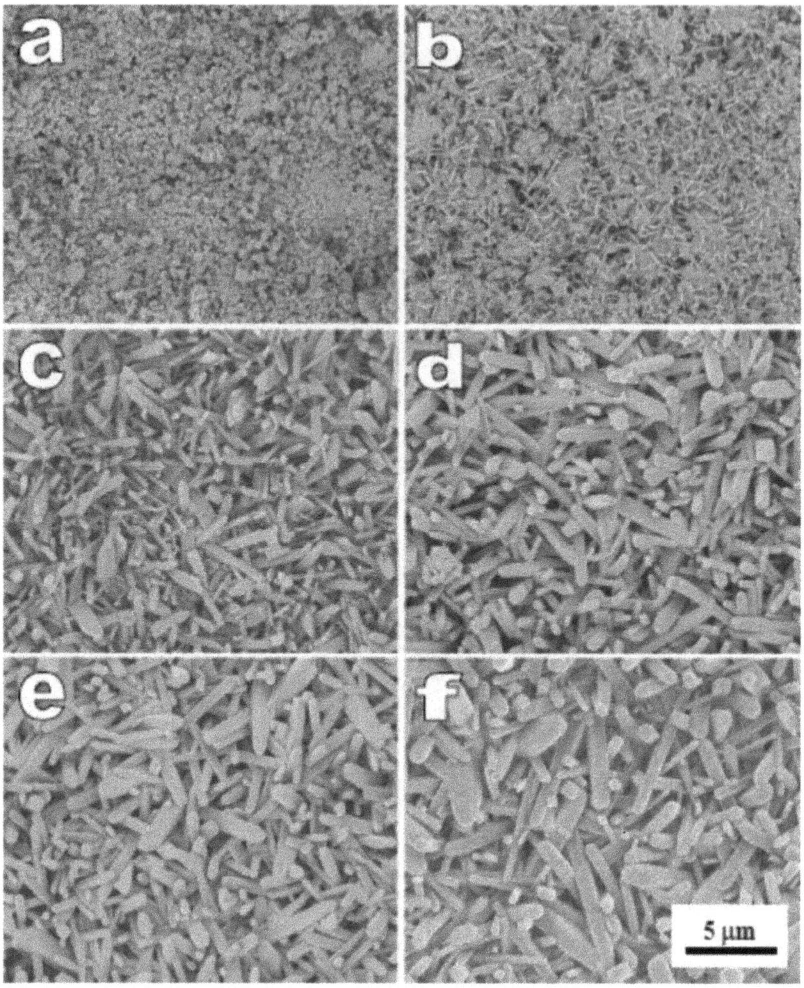

Figure 8.51. SEM images of the samples doped with 5 mol% Ta_2O_5 derived from the mixture mechanochemically activated with WC media, after sintering for 4 h at different temperatures: (a) 1000 °C, (b) 1100 °C, (c) 1200 °C, (d) 1300 °C, (e) 1400 °C and (f) 1500 °C.

steel milling media, the mullitization temperature was reduced by about 100 °C. All the samples were nearly entirely mullitized after sintering at 1200 °C. As a consequence, anisotropic grain growth occurred. Figure 8.71 shows SEM images of the samples doped with WO_3 from the mixture mechanochemically activated with stainless steel media after sintering at different temperatures. It was found that the sample sintered at 1100 °C already contained a small number of whiskers, although most of them had a very thin diameter. After sintering at 1200 °C for 4 h, mullite whiskers with much larger sizes were obtained, although their morphology was not regular. The observation was similar to that encountered by the mixture without the addition of WO_3, as discussed earlier.

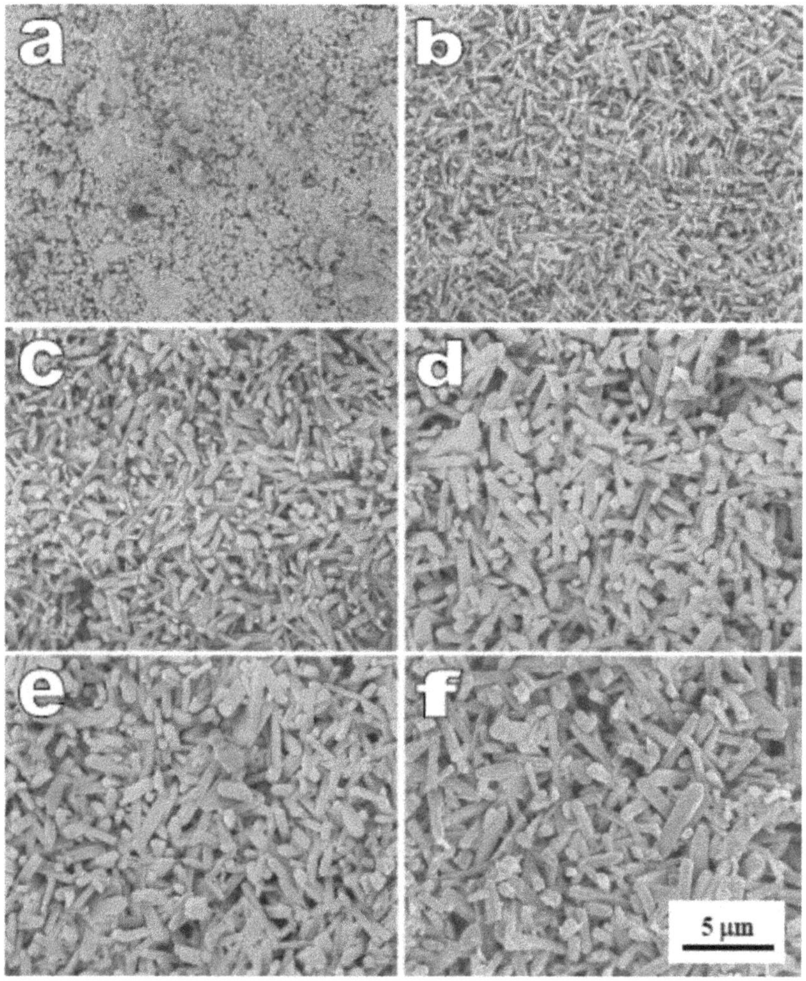

Figure 8.52. SEM images of the samples doped with 5 mol% Ta_2O_5 derived from the mixture mechanochemically activated with WC media, after sintering for 4 h at different temperatures: (a) 1000 °C, (b) 1100 °C, (c) 1200 °C, (d) 1300 °C, (e) 1400 °C and (f) 1500 °C.

According to the phase formation behavior and microstructure evolution of the samples doped with higher concentrations of WO_3, the positive effect of WO_3 on mullitization was not affected by the mechanochemical activation. Figure 8.72 shows SEM images of the samples doped with 10 mol% and 20 mol% WO_3, after sintering at 1100 °C. Obviously, the anisotropic grain growth was enhanced as the doping concentration of WO_3 was increased to 10 mol%. In this sample, the number of large whiskers was increased, while there were still whiskers with thin diameters. At the highest content of 20 mol%, fewer thick whiskers were observed, while more thin whiskers were present. Therefore, it could be deduced that too a high concentration of WO_3 delayed the grain growth of mullite, while the nucleation was not affected. As a result, the dimensions of the mullite whiskers were decreased

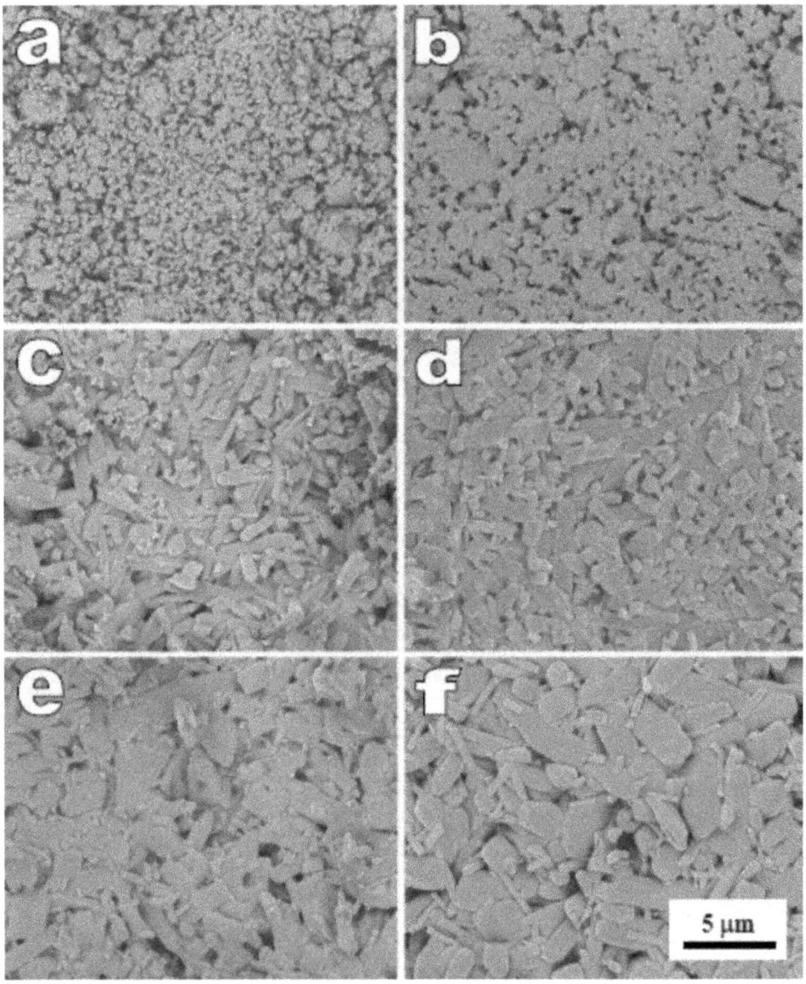

Figure 8.53. SEM images of the samples doped with 5 mol% Ta_2O_5 derived from the mixture mechanochemically activated with WC media, after sintering for 4 h at different temperatures: (a) 1000 °C, (b) 1100 °C, (c) 1200 °C, (d) 1300 °C, (e) 1400 °C and (f) 1500 °C.

with increasing concentration of WO_3, as observed in figure 8.73 where the samples were sintered at 1300 °C. The samples from the mixtures activated with WC media experienced a similar variation trend in the dimensions of whiskers, as elaborated in the following [46].

According to the XRD patterns of the as-activated mixtures of Al_2O_3 and SiO_2 with WC milling media, doped with different concentrations of WO_3, only Al_2O_3 was clearly visible, but with decreased and widened diffraction peaks, due to the significant refinement of the particles. SiO_2 was amorphous, thus having no diffraction peaks. Similarly to the results of figure 8.67, WO_3 was not detected, probably because the intensity of its diffraction peaks was too low for the XRD to

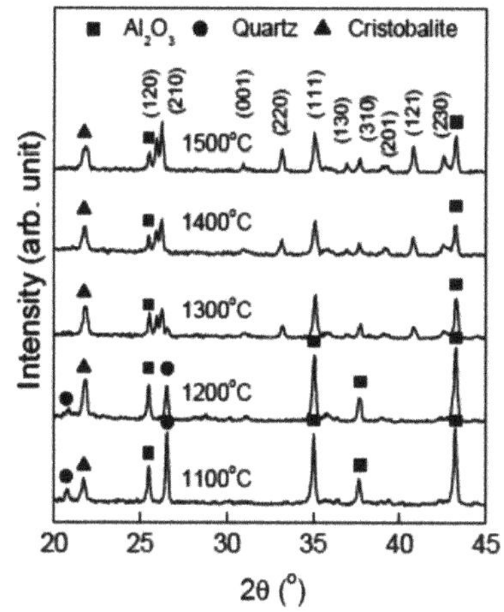

Figure 8.54. XRD patterns of the unactivated mixture doped with 5 mol% Cr_2O_3 sintered at different temperatures for 4 h. Reproduced with permission from [45]. Copyright 2018 Elsevier.

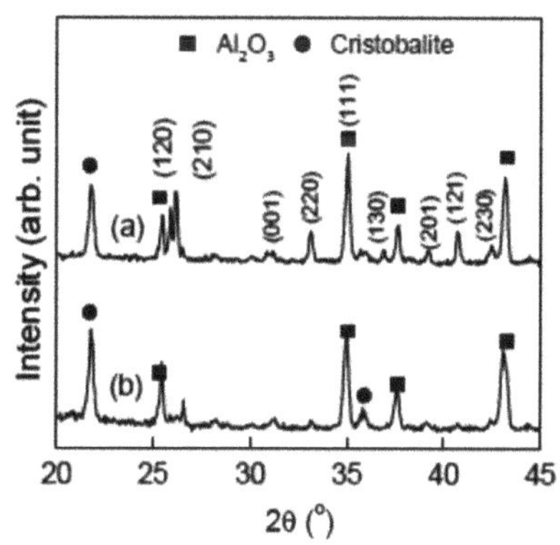

Figure 8.55. XRD patterns of the unactivated mixtures doped with different concentrations of Cr_2O_3 sintered at 1400 °C for 4 h: (a) 10 mol% and (b) 20 mol%. Reproduced with permission from [45]. Copyright 2018 Elsevier.

detect. However, a trace of WC was present in the as-activated mixtures due to contamination, as mentioned previously.

DTA analytic results of the as-activated mixtures indicated that the crystallization temperatures of amorphous silica to cristobalite were 853 °C, 854 °C and 855 °C, while mullitization occurred at the temperatures of 951 °C, 912 °C and 835 °C, corresponding to the WO_3 concentrations of 5 mol%, 10 mol% and 20 mol%, respectively (figure 8.74). In other words, the presence of WO_3 had nearly no effect on the crystallization of the amorphous silica, but strongly boosted the phase formation of mullite.

For the samples doped with 5 mol% WO_3, the mullite phase was present after sintering at 1000 °C for 4 h, while it became a major phase after firing at 1100 °C. Full mullitization was achieved as the sintering temperature was raised to 1200 °C.

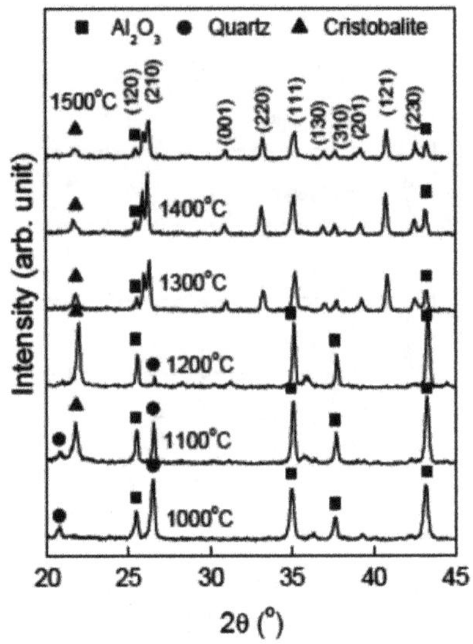

Figure 8.56. XRD patterns of the mixture with 5 mol% Cr_2O_3 activated mechanochemically with stainless steel milling media, after sintering at different temperatures. Reproduced with permission from [45]. Copyright 2018 Elsevier.

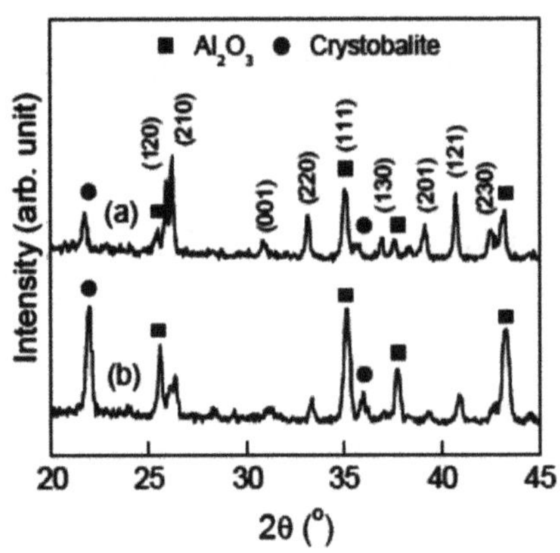

Figure 8.57. XRD patterns of mixtures with different contents of Cr_2O_3 mechanochemically activated with stainless steel milling media, after sintering at 1400 °C: (a) 10 mol% and (b) 20 mol%. Reproduced with permission from [45]. Copyright 2018 Elsevier.

For the samples doped with 10 mol% WO_3, the mullite phase formation process was almost complete at 1000 °C, so that well-developed mullite whiskers were obtained in this sample. Further increase in the content of WO_3 to 20 mol% resulted in no obvious change in the mullite phase formation according to the XRD results. However, the dimensions of the mullite whiskers were reduced, while the morphology of the whiskers became more imperfect. Therefore, the optimal concentration of WO_3 was 10 mol%, in terms of maintain strong mullization behavior and perfect mullite whiskers, as demonstrated by the SEM images of the samples sintered at

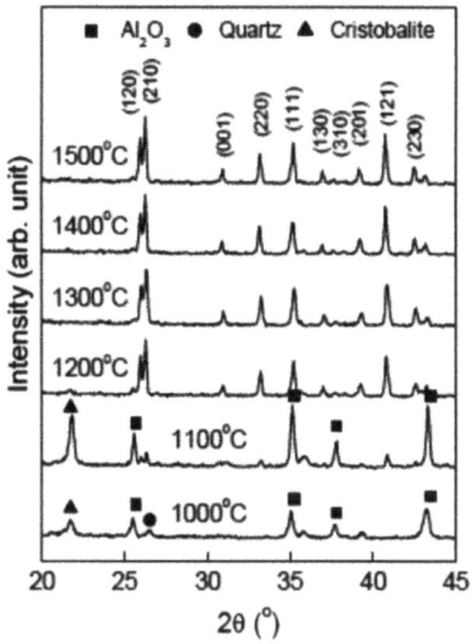

Figure 8.58. XRD patterns of the mixture with 5 mol% Cr_2O_3 mechanochemically activated with WC milling media, after sintering at different temperatures. Reproduced with permission from [45]. Copyright 2018 Elsevier.

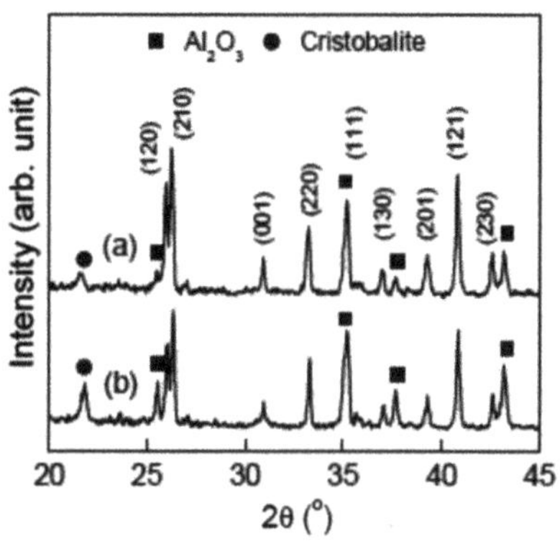

Figure 8.59. XRD patterns of the mixtures doped with different concentrations of Cr_2O_3 mechanochemically activated with WC milling media, after sintering at 1200 °C for 4 h: (a) 10 mol% and (b) 20 mol%. Reproduced with permission from [45]. Copyright 2018 Elsevier.

1300 °C, with different contents of WO_3. Cross-sectional SEM images revealed that the mullite whiskers grew throughout the bulk of the samples. It is strongly believed that the optimal concentration could be further tuned through more careful studies.

The effects of MnO on mullite phase formation and the microstructure of the samples derived from oxide mixtures, with and without mechanochemical activation, are discussed in this subsection. Figure 8.75 shows XRD patterns of the samples doped with 5 mol% MnO after sintering at different temperatures. The mullite phase formed at 1200 °C, although Al_2O_3 was visible up to 1500 °C. The incomplete mullization from the mixture without mechanochemical activation

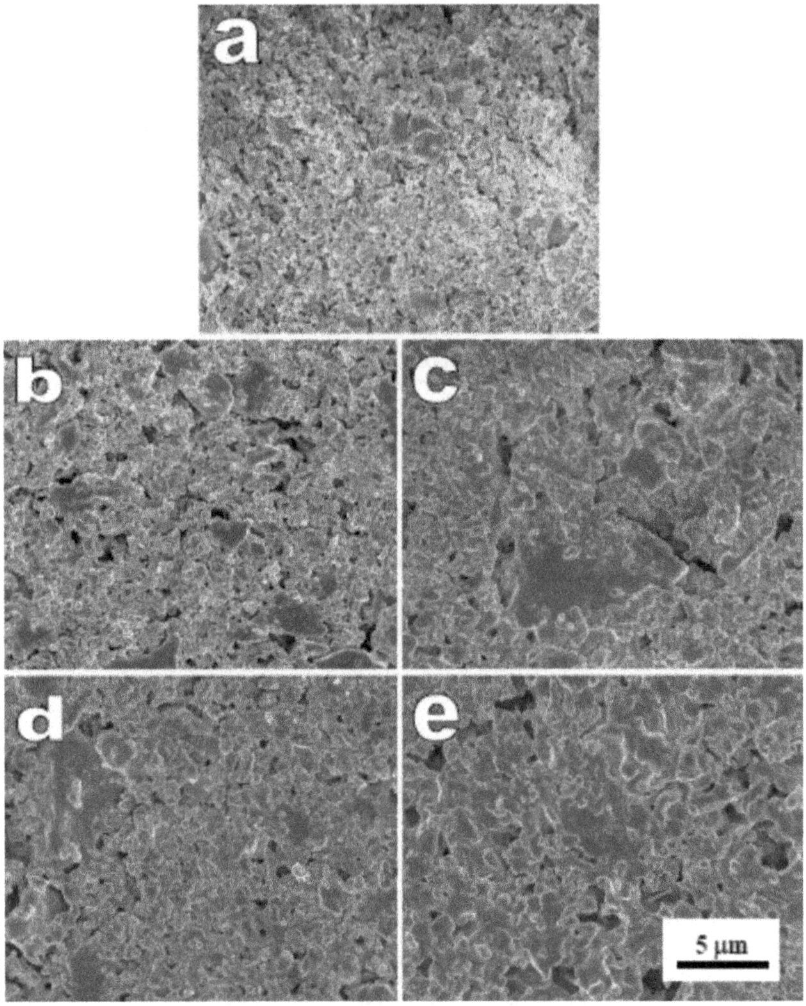

Figure 8.60. SEM images of the samples doped with 5 mol% Cr_2O_3 from the mixture without mechanochemical activation, after sintering at different temperatures: (a) 1100 °C, (b) 1200 °C, (c) 1300 °C, (d) 1400 °C and (e) 1500 °C.

could be attributed to the inhomogeneity of the reactants mixed using the conventional ball milling process, mainly due to their large particle sizes. As seen earlier, the mullization temperature from the oxide mixture without mechanochemical activation was most likely 1300 °C. Therefore, the presence of MnO enhanced the mullite phase from the oxide mixture of Al_2O_3 and quartz. The positive effect of MnO was further evidenced by the XRD results of the samples with higher contents of MnO after sintering at 1200 °C, as observed in figure 8.76. However, the effect was not strongly dependent on the concentration of MnO.

XRD patterns of the samples doped with 5 mol% MnO and those with different contents of MnO, from the mixtures mechanochemically activated with stainless

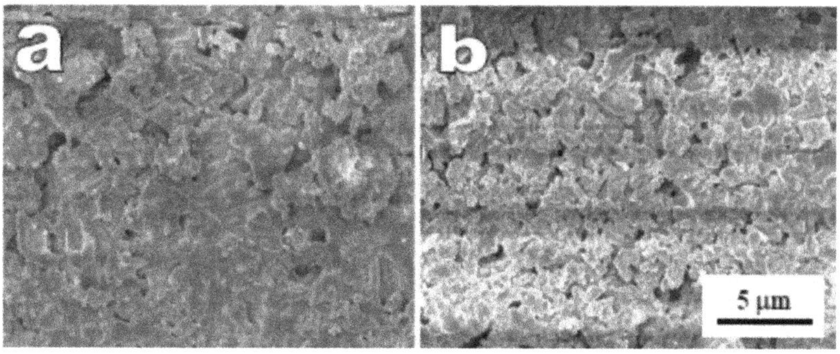

Figure 8.61. SEM images of the samples with different contents of Cr_2O_3 without mechanochemical activation, after sintering at 1500 °C: (a) 10 mol% and (b) 20 mol%.

steel milling media, are shown in figures 8.77 and 8.78. Careful inspection revealed that the mullite phase was detectable in the sample sintered at 1100 °C. This implied that the reactivity of the starting oxides was slightly increased due to the mechanochemical activation process. However, Al_2O_3 was still present in the sample sintered at 1500 °C, which was similar to that observed in the samples without mechanochemical activation.

Figure 8.79 shows the XRD patterns of the samples doped with different concentrations of MnO after sintering at 1100 °C. The three samples all exhibited a nearly complete mullitization at this temperature, meaning that the mullite phase formation temperature was decreased by 100 °C, compared to that for the mixtures activated with stainless steel media. At the same time, two new phases, Mn_3O_4 and $MnSiO_3$, formed as the secondary phase when the sintering temperature was below 1300 °C. After sintering at higher temperatures, they disappeared due to consumption by the reaction to form the mullite phase. These two phases were not observed in the samples that were not mechanochemically activated and those activated with stainless steel media. Therefore, activation with WC media has a higher efficiency in enhancing the reactivity of the starting materials.

Figure 8.80 shows SEM images of the samples doped with 5 mol% MnO from the mixture without the application of activation after sintering at different temperatures. As expected, no anisotropic grain growth was observed. The sample sintered at 1100 °C was relatively porous, with small grain sizes. After sintering at 1200 °C, the sample was obviously densified while grain growth was obviously observed. Above this temperature, all the samples exhibited a similar dense microstructure, consisting of pseudo-spherical grains that were fused together to a certain degree, which is typical for the case of liquid phase sintering. The samples with 10 mol% and 20 mol% MnO displayed a similar variation trend in microstructure.

Figure 8.81 shows SEM images of the samples doped with 5 mol% MnO from the mixture mechanochemically activated with stainless steel milling media, after being sintered at different temperatures. In this case, the microstructural development was similar to that of the mixtures without mechanochemical activation. After sintering at temperatures $\leqslant$1100 °C, the samples were of relatively high porosity, while no

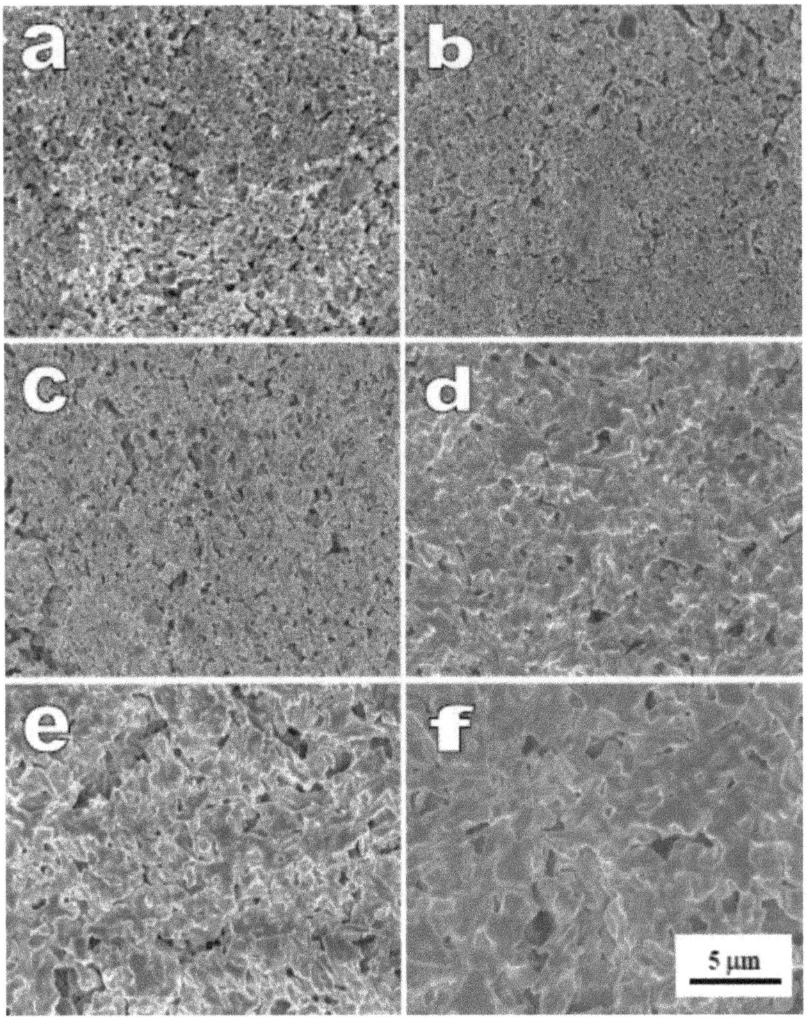

Figure 8.62. SEM images of the samples doped with 5 mol% Cr_2O_3 from the mixture mechanochemically activated with stainless steel milling media, after sintering at different temperatures for 4 h: (a) 1000 °C, (b) 1100 °C, (c) 1200 °C, (d) 1300 °C (e) 1400 °C and (f) 1500 °C. Reproduced with permission from [45]. Copyright 2018 Elsevier.

obvious grain growth occurred. In the sample sintered at 1200 °C both densification and grain growth took place. However, no mullite whiskers were formed in this group of samples. The concentration of MnO had no significant effect on the microstructure and grain growth of the mullite ceramics.

Figure 8.82 shows SEM images of the samples doped with 5 mol% MnO from the mixture, which was mechanochemically activated with WC milling media, after sintering at different temperatures. The sample sintered at 1000 °C was porous and displayed no grain growth. However, after sintering at 1100 °C a layer of thin whiskers was present on the surface of the sample. At the same time, thick whiskers

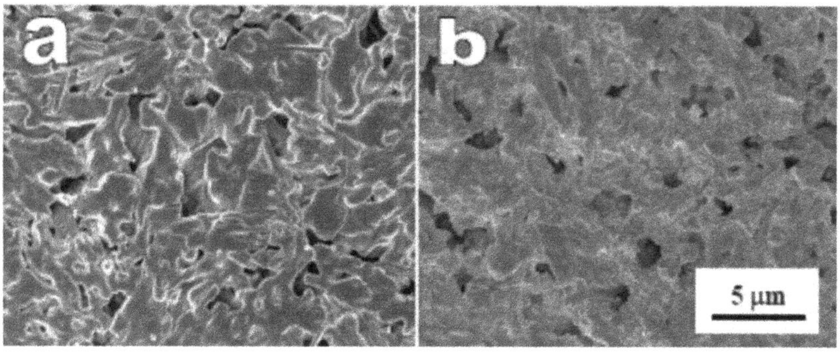

Figure 8.63. SEM images of the samples with different contents of Cr_2O_3 from the mixtures mechanochemically activated with stainless steel milling media, after sintering at 1500 °C: (a) 10 mol% and (b) 20 mol%. Reproduced with permission from [45]. Copyright 2018 Elsevier.

could be identified beneath the thin whisker layer. The XRD patterns indicated that the mullite phase was formed at this temperature. After sintering at 1200 °C, well-developed mullite whiskers with a uniform size distribution were obtained. With increasing sintering temperature, the dimensions and morphology of the whiskers remained almost unchanged.

The sample doped with 10 mol% MnO shared a similar microstructure to that with 5 mol% MnO, after they were sintered at 1100 °C, as can be seen in figures 8.82(b) and 8.83(a). However, the density of the thin whiskers was much higher in the latter than in the former, which could be related to the positive effect of MnO on the mullite phase formation. This statement was further supported by the SEM image of the sample doped with 20 mol%, as revealed in figure 8.83(b), in which a small number of thin whiskers were distributed in the matrix of thick whiskers. However, too high a content of MnO was not favorable to the morphology of the mullite whiskers. As demonstrated in figure 8.84, the whiskers in the sample with 10 mol% MnO could be clearly identified, while those in the samples with 20 mol% MnO were highly imperfect in terms of whisker morphology. The reduction in the perfection of the mullite whiskers was believed to be related to the presence of MnO. On the other hand, the results suggested that MnO was homogeneously distributed in the samples.

Fe_2O_3, CoO and NiO were examined for their influence on the phase formation and anisotropic grain growth from the mixtures mechanochemically activated with stainless steel and WC milling media [47, 48]. Commercially precipitated silica (SiO_2, Laboratory reagent), Al_2O_2 (99+% purity), Fe_2O_3 (>99% purity), CoO (99.99% purity) and NiO (99+% purity) powders were used as the starting materials. The mullite composition was $3Al_2O_3 \cdot 2SiO_2$. The transition oxides were expressed as $FeO_{1.5}$, CoO and NiO. The concentrations of the transition oxides were measured according to the formulation of $(3Al_2O_3 \cdot 2SiO_2)_{0.9}(MO_y)_{0.1}$ (M = Fe, Co and Ni). The weigh percentage of the transition oxides in the final mullite samples was about 2 wt%.

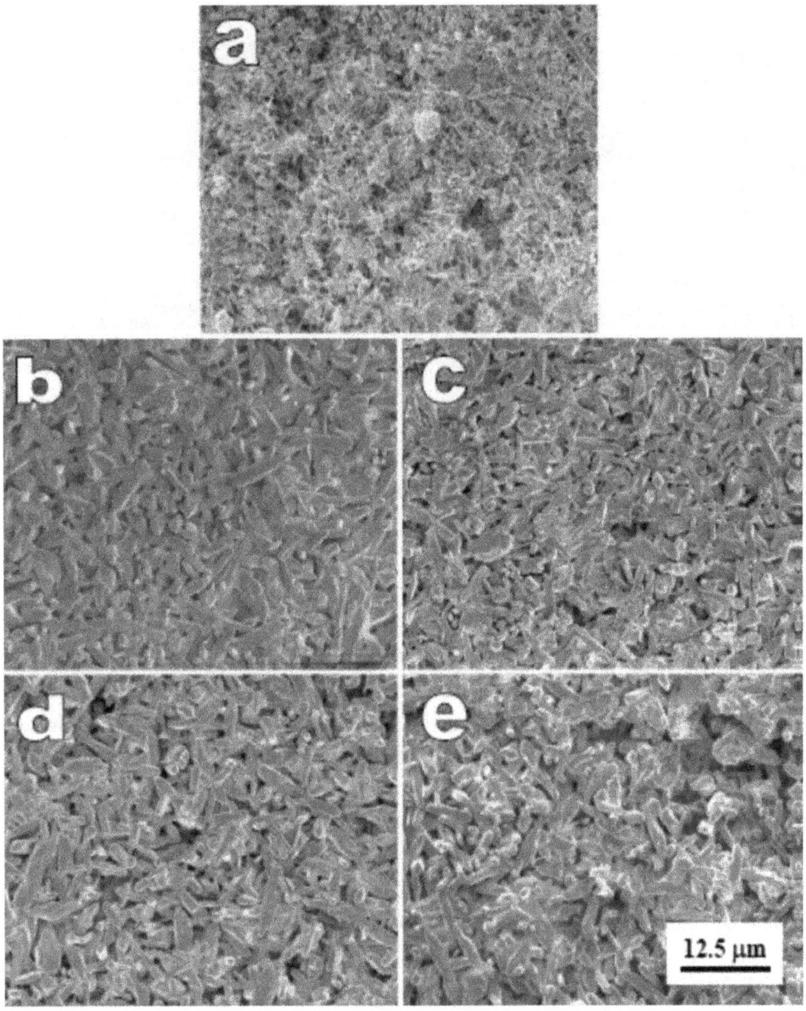

Figure 8.64. SEM images of the samples doped with 5 mol% Cr_2O_3 from the mixture mechanochemically activated with WC milling media, after sintering at different temperatures for 4 h: (a) 1100 °C, (b) 1200 °C, (c) 1300 °C, (d) 1400 °C and (e) 1500 °C. Reproduced with permission from [45]. Copyright 2018 Elsevier.

It was found that the effects of the three transition oxides on mullite phase formation in the as-mixed oxide powders were slightly different [47, 48]. Although nearly complete mullitization was realized in the three samples after sintering at 1300 °C, the mullite phase was detected in the XRD pattern of the sample doped with CoO after sintering at 1200 °C. Therefore, the effect of CoO was slightly stronger than that of $FeO_{1.5}$ and NiO. In addition, the $CoAl_2O_4$ and $NiAl_2O_4$ spinel phases were present in the two groups of samples sintered at temperatures $\leqslant$1200 °C, both of which disappeared at higher sintering temperatures.

In comparison, after the application of mechanochemical activation with stainless steel media, the three groups of mixtures exhibited similar mullitization behavior.

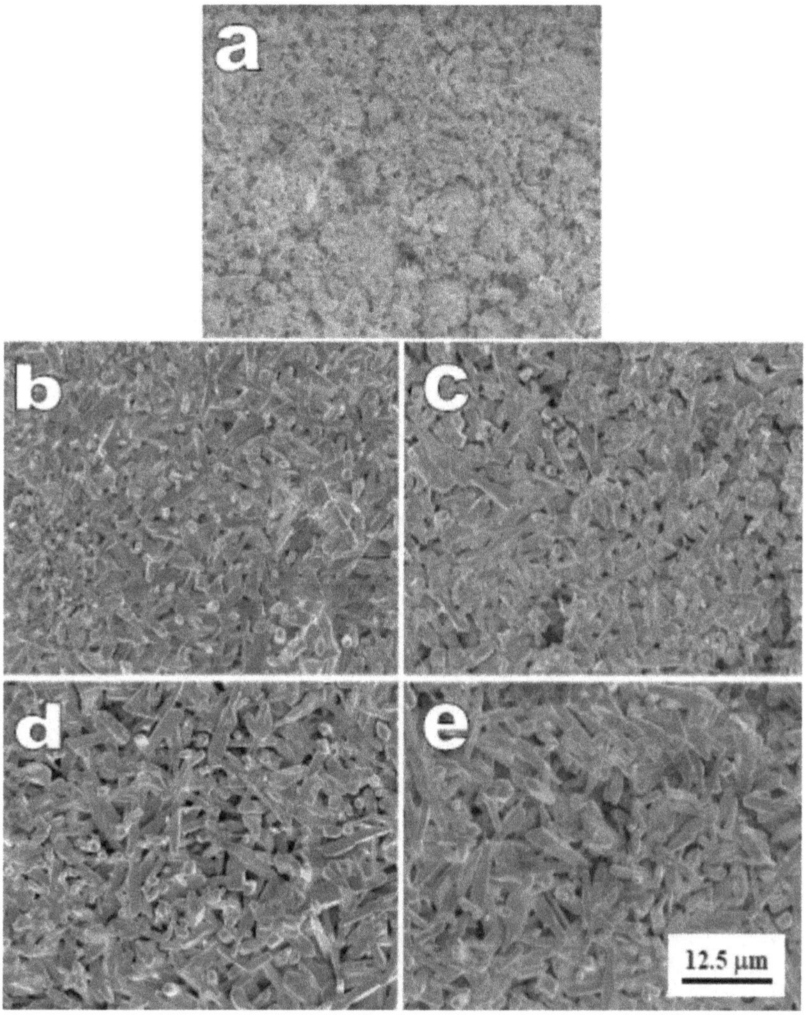

Figure 8.65. SEM images of the samples doped with 10 mol% Cr_2O_3 from the mixture mechanochemically activated with WC milling media, after sintering at different temperatures for 4 h: (a) 1100 °C, (b) 1200 °C, (c) 1300 °C, (d) 1400 °C and (e) 1500 °C. Reproduced with permission from [45]. Copyright 2018 Elsevier.

Mullite phase formation was almost complete after sintering at 1200 °C. In other words, the activation with stainless steel media resulted in a reduction in mullite phase formation temperature by 100 °C, similar to those discussed previously. In addition, Fe and Fe_2O_3 were detected in the samples sintered at low temperatures, due to the contamination from the stainless steel vials and balls. However, the three transition oxides were not observed in the as-activated mixtures, because their concentrations were below the detection limit of XRD.

When WC milling media were used to conduct the mechanochemical activation, the difference in their effect on mullite phase formation was increased, with an order of $FeO_{1.5}$ > CoO > NiO. Figure 8.85 shows the XRD patterns of the samples doped

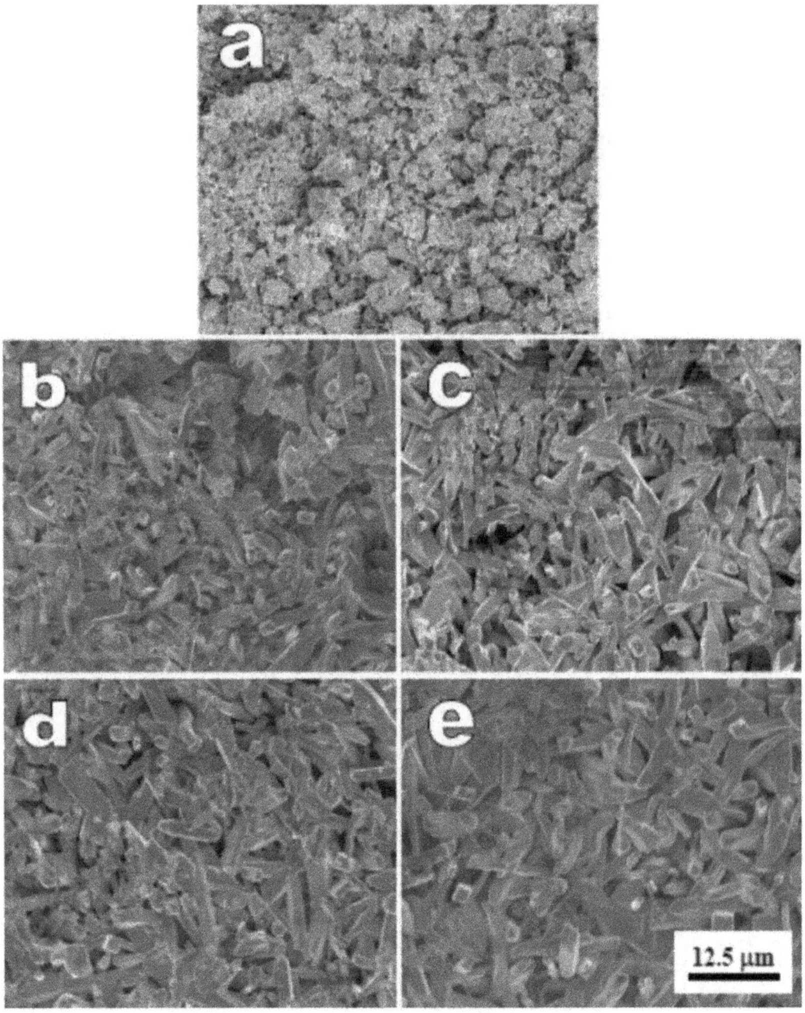

Figure 8.66. SEM images of the samples doped with 20 mol% Cr_2O_3 from the mixture mechanochemically activated with WC milling media, after sintering at different temperatures for 4 h: (a) 1100 °C, (b) 1200 °C, (c) 1300 °C, (d) 1400 °C and (e) 1500 °C. Reproduced with permission from [45]. Copyright 2018 Elsevier.

with the three transition oxides from the mixtures mechanochemically activated with WC media, after sintering at different temperatures for 4 h [48]. Mullite was present as a minor phase in the sample doped with $FeO_{1.5}$ after sintering at 1000 °C. In addition, the amorphous silica was crystallized to cristobalite. As the sintering temperature was increased to 1100 °C, the mullite phase was dominant, with Al_2O_3 and cristabolite being minor phases. Complete mullitization was achieved after sintering at 1200 °C.

No mullite phase was detected in the sample doped with CoO after sintering at 1000 °C, while XRD patterns sintering at higher temperatures were similar to those with $FeO_{1.5}$. In the sample that was doped with CoO, mullite was not present after it

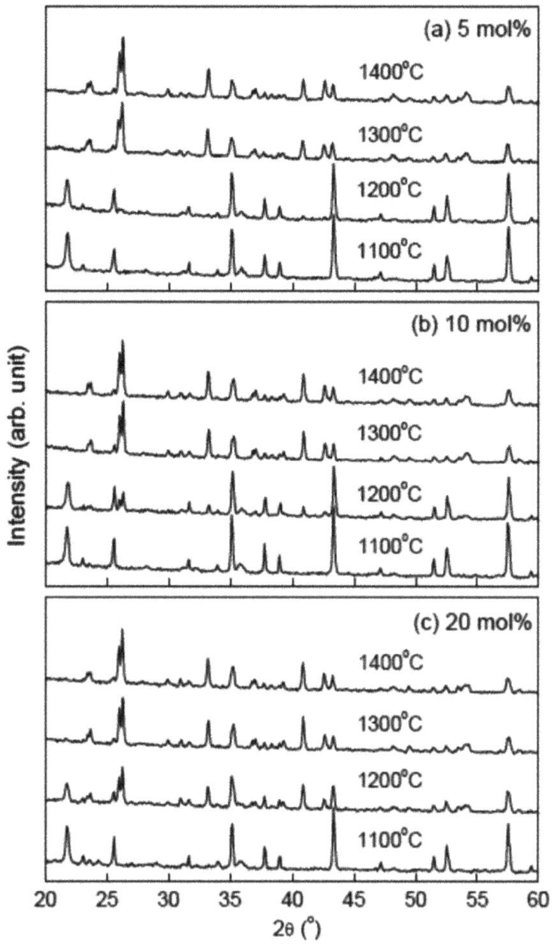

Figure 8.67. XRD patterns of the samples doped with different concentrations of WO_3, after sintering at different temperatures for 4 h.

was sintered at 1000 °C, suggesting that mullite formation is more difficult in this case than in the case of doping with iron oxide. The peak intensity of the cristabolite is much weaker than that of the one in the $FeO_{1.5}$-doped sample at this sintering temperature. Mullite is detected after sintering at 1100 °C, with an XRD pattern similar to that of the $FeO_{1.5}$-doped sample. After sintering at 1200 °C mullitization was complete. For the samples doped with NiO, mullitization occurred only at 1200 °C.

In summary, the effects of the three transition oxides on the phase formation of mullite from the oxide mixture of Al_2O_3 and silica were dependent on the states of the mixtures. For the mixtures obtained using the conventional ball milling process, the effect of CoO was the strongest in terms of promoting mullite phase formation. The possible reason for this observation was that the CoO powder had relatively small particles, so CoO was easily engaged in the reaction of Al_2O_3 and SiO_2. After

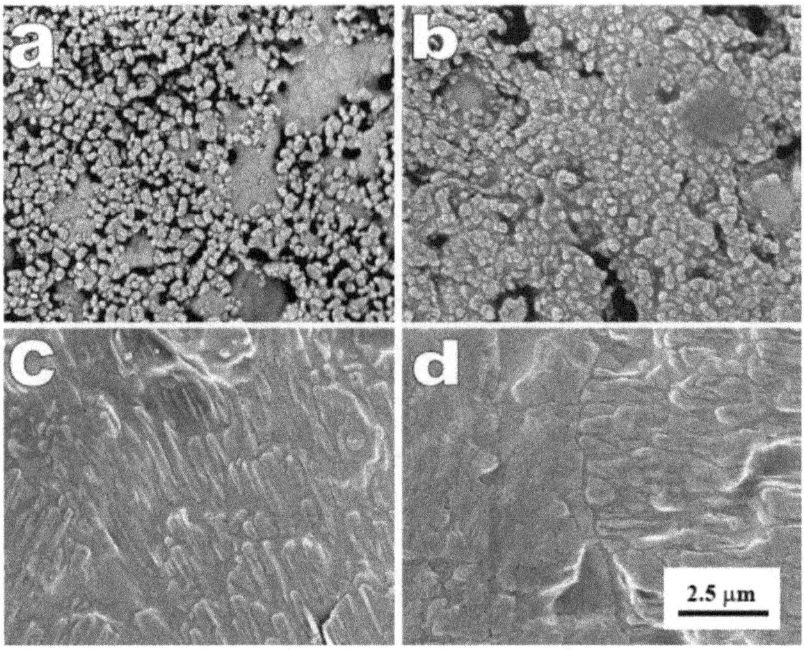

Figure 8.68. SEM images of the samples doped with 5 mol% WO_3, after sintering at different temperatures for 4 h: (a) 1100 °C, (b) 1200 °C, (c) 1300 °C and (d) 1400 °C.

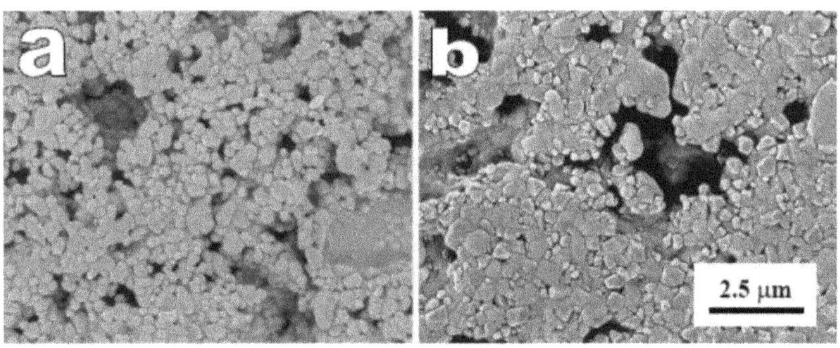

Figure 8.69. SEM images of the samples doped with higher contents of WO_3, after sintering at 1200 °C for 4 h: (a) 10 mol% and (b) 20 mol%.

being mechanochemically activated with stainless steel media, the reactants were refined and thus the reaction between Al_2O_3 and silica became the limiting step.

Therefore, the effects of the transition oxides were all screened out and the three groups of mixtures exhibited similar mullitization behaviors. As for the mechanochemical activation with WC media, much higher energy was applied to the mixtures. As a result, both the reactants and the dopants were significantly activated. In this case, the effects of the dopants became pronounced. Probably because the activation efficiency of NiO was lower than those of $FeO_{1.5}$ and CoO, the samples

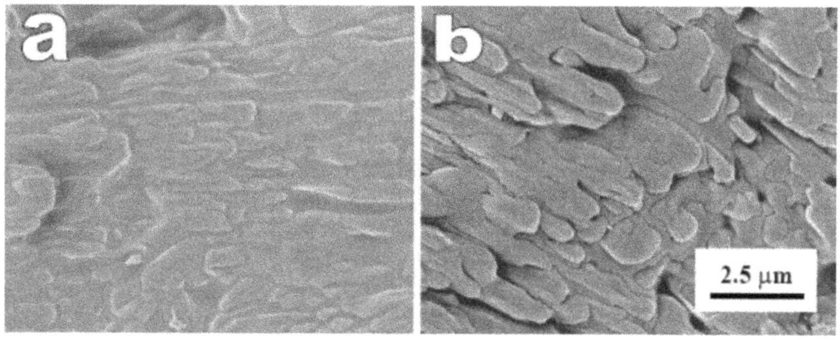

Figure 8.70. SEM images of the samples doped with higher contents of WO_3, after sintering at 1300 °C for 4 h: (a) 10 mol% and (b) 20 mol%.

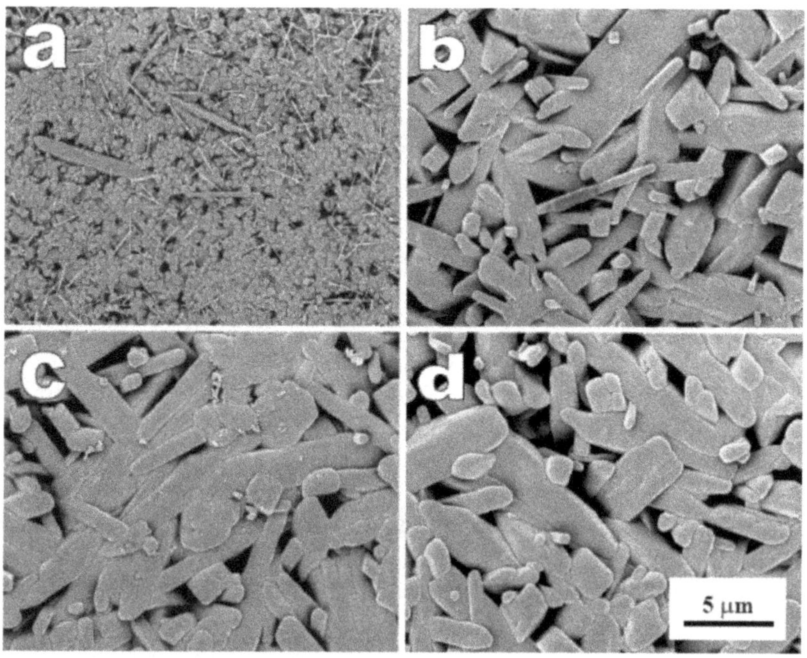

Figure 8.71. SEM images of the samples doped with 5 mol% WO_3 from the mixture mechanochemically activated with stainless steel media, after sintering at different temperatures for 4 h: (a) 1100 °C, (b) 1200 °C, (c) 1300 °C and (d) 1400 °C.

doped with NiO displayed the highest mullitization temperature. This could be an interesting topic for future study.

SEM images of the samples doped with the transition oxides from the mixtures mechanochemically activated with stainless steel media, after sintering at different temperatures, are shown in figures 8.86–8.88. Anisotropic grain growth took place in all three groups. Figure 8.85 indicates that the samples doped with $FeO_{1.5}$ and sintered at 1000 °C and 1100 °C consisted of pseudo-spherical grains. At these temperatures, the mullite phase was not formed. After sintering at 1200 °C,

Figure 8.72. SEM images of the samples doped with higher contents of WO_3 from the mixtures mechanochemically activated with stainless steel media, after sintering at 1100 °C for 4 h: (a) 10 mol% and (b) 20 mol%.

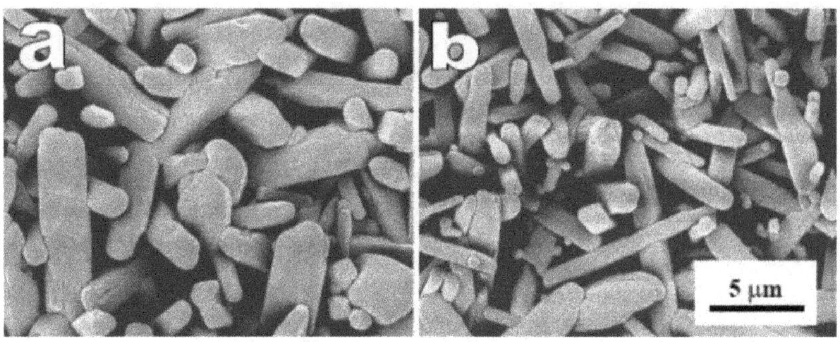

Figure 8.73. SEM images of the samples doped with higher contents of WO_3 from the mixtures mechanochemically activated with stainless steel media, after sintering at 1300 °C for 4 h: (a) 10 mol% and (b) 20 mol%.

anisotropic grains were obtained, together with a small quantity of spherical particles. As the sintering temperature was further increased, the dimensions of the anisotropic grains were increased gradually. The anisotropic grain growth and mullite phase formation occurred simultaneously, as discussed before. A similar microstructure development was observed for the samples doped with CoO and NiO.

SEM images of the three groups of samples from the mixtures mechanochemically activated with WC milling media, after sintering at different temperatures, are shown in figures 8.89–8.91. By comparing the XRD patterns shown in figure 8.85, it was quickly found that the mullite phase formation was accompanied by anisotropic grain growth. The samples sintered at temperatures of 1200 °C – 1500 °C shared a similar microstructure in terms of whisker morphology and dimensions. The most distinct difference was observed among the three samples sintered at 1100 °C. For the sample doped with $FeO_{1.5}$, well-developed mullite whiskers with a uniform size distribution were obtained. In comparison, the sample doped with CoO had a bimodal size distribution, indicating that it was in a state with the presence of both nucleation and grain growth. However, anisotropic grain growth was not observed

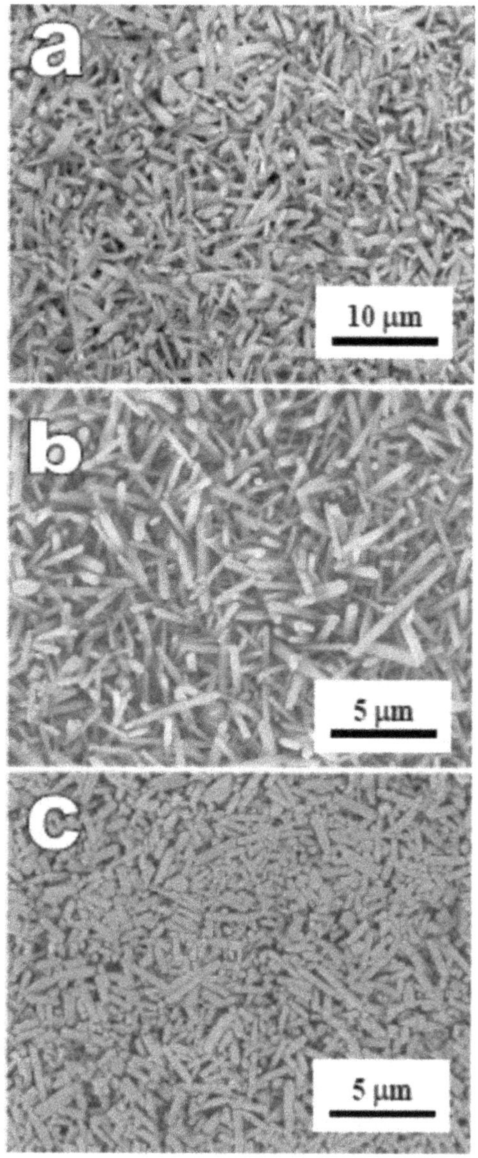

Figure 8.74. SEM images of the samples doped with different contents of WO_3 from the mixtures mechanochemically activated with WC milling media, after sintering at 1300 °C for 4 h: (a) 5 mol%, (b) 10 mol% and (c) 20 mol%. Reproduced with permission from [46]. Copyright 2003 Elsevier.

in the sample doped with NiO. The observation was in good agreement with the XRD results.

It is important to note that all the samples from the mixtures mechanochemically activated with stainless steel media possessed a very dense microstructure, although they experienced anisotropic grain growth. Ceramics with such a microstructure are expected to have high mechanical strengths due to the interlocking anisotropic

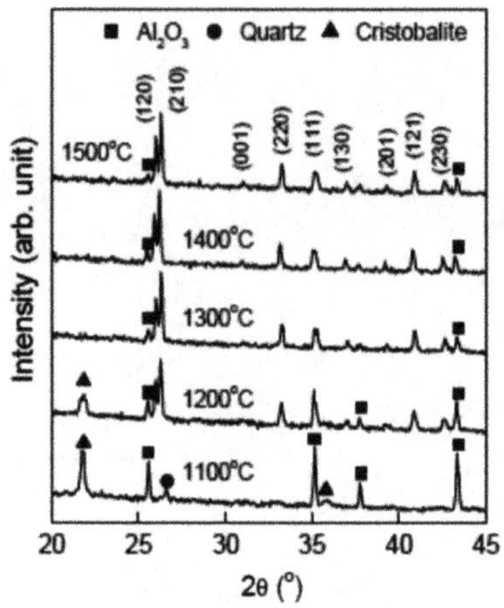

Figure 8.75. XRD patterns of the samples doped with 5 mol% MnO without mechanochemical activation, after sintering at different temperatures.

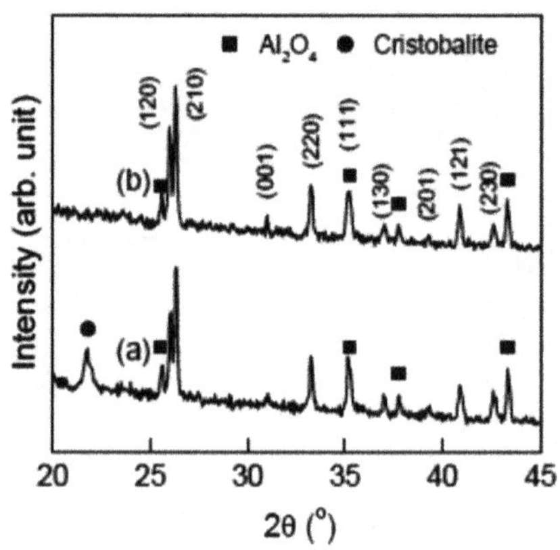

Figure 8.76. XRD patterns of the samples doped with different concentrations of MnO, after sintering at 1200 °C for 4 h: (a) 10 mol% and (b) 20 mol%.

grains. In addition, the microstructure could be controlled by selecting different transition oxides as the dopants. Moreover, it is strongly believed that the microstructure could be further tuned or tailored by adjusting the concentration of the dopants and/or using a combination of them. In comparison, the samples from the mixtures mechanochemically treated with WC media exhibited a porous microstructure due to the presence of the regular mullite whiskers. Similarly, the dimensions and porosity of the mullite ceramics could be controlled by selecting different dopants and sintering at different temperatures.

Mullite phase formation and grain growth from the mixture of Al_2O_3 and silica doped with CuO have been reported [49]. The composition was $(3Al_2O_3 \cdot 2SiO_2)_{0.9}(CuO)_{0.1}$, corresponding to the weight percent of CuO being about 2%. The oxide mixture was mechanochemically activated with both stainless steel and

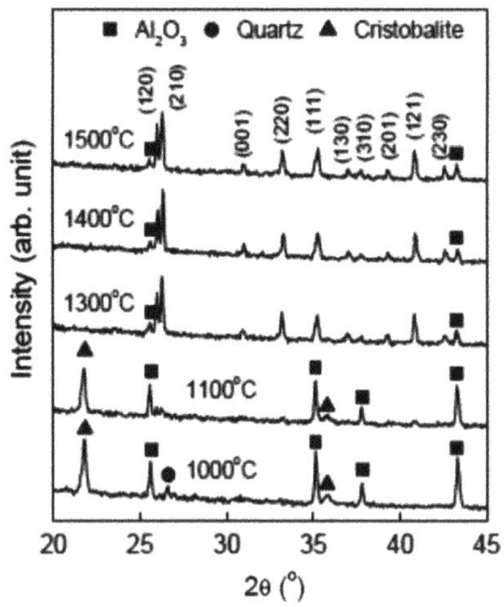

Figure 8.77. XRD patterns of the samples doped with 5 mol% MnO from the mixture activated with stainless steel media, after sintering at different temperatures.

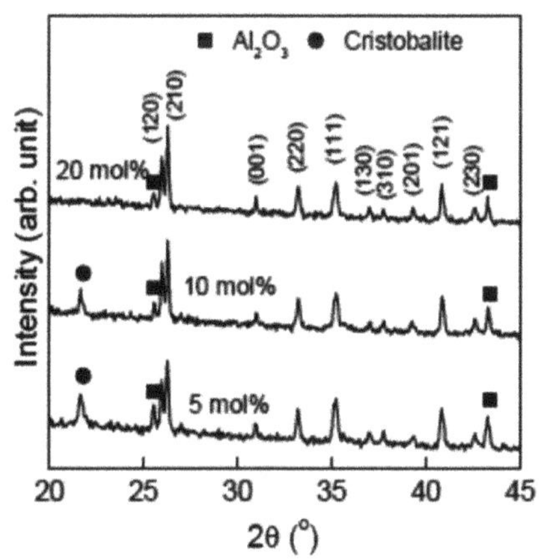

Figure 8.78. XRD patterns of the samples doped with different concentrations of MnO from the mixtures mechanochemically activated with stainless steel milling media, after sintering at 1200 °C for 4 h.

WC milling media. The activated mixtures showed much lower mullite formation temperatures. In addition, anisotropic grain growth was observed in the mixture activated with stainless steel media, but the grains had a huge size and irregular morphology. In comparison, well-shaped mullite whiskers were obtained from the mixture treated with WC vials and balls. The reduced mullite phase formation temperature and the anisotropic grain growth were ascribed to the refined particles of the mixtures, due to the high-energy activation.

To further demonstrate the effect of CuO, a series of samples have been studied, with CuO weight percentages of 1%, 3%, 5% and 7%. Commercial oxide powders of Al_2O_3, quartz (SiO_2) and CuO were thoroughly mixed using the conventional ball milling process. The mixtures were then subjected to high-energy mechanochemical

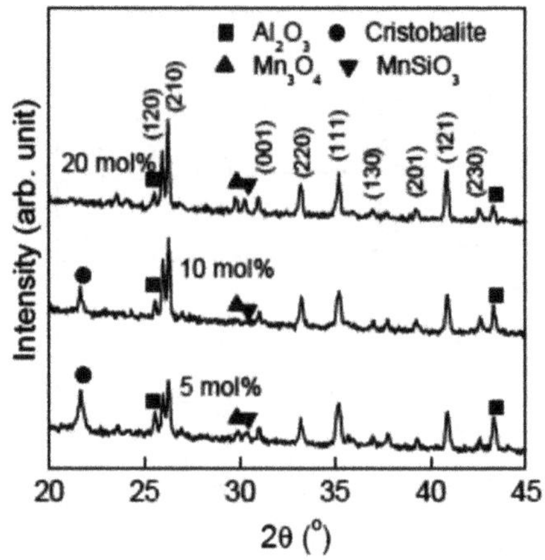

Figure 8.79. XRD patterns of the samples doped with different concentrations of MnO from the mixtures mechanochemically activated with WC milling media, after sintering at 1100 °C for 4 h.

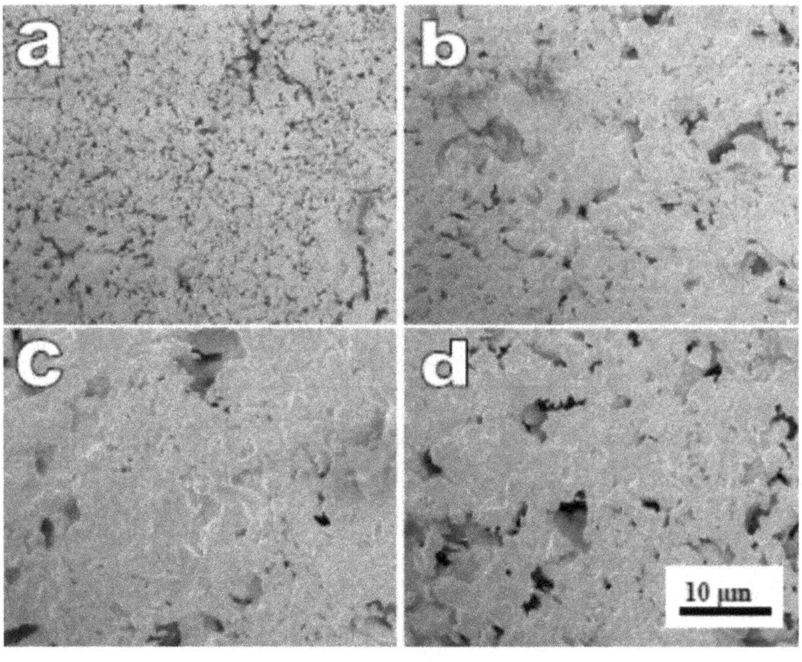

Figure 8.80. SEM images of the samples doped with 5 mol% MnO without activation, after sintering at different temperatures for 4 h: (a) 1100 °C, (b) 1200 °C, (c) 1300 °C and (d) 1400 °C.

activation with WC milling media. The milling was carried out for 5 h, with experimental parameters similar to those stated before. The activated mixtures were compacted and sintered at different temperatures for 4 h.

Figure 8.92 shows the XRD patterns of the samples doped with 1% CuO after sintering at different temperatures. The phase transition of quartz to cristobalite

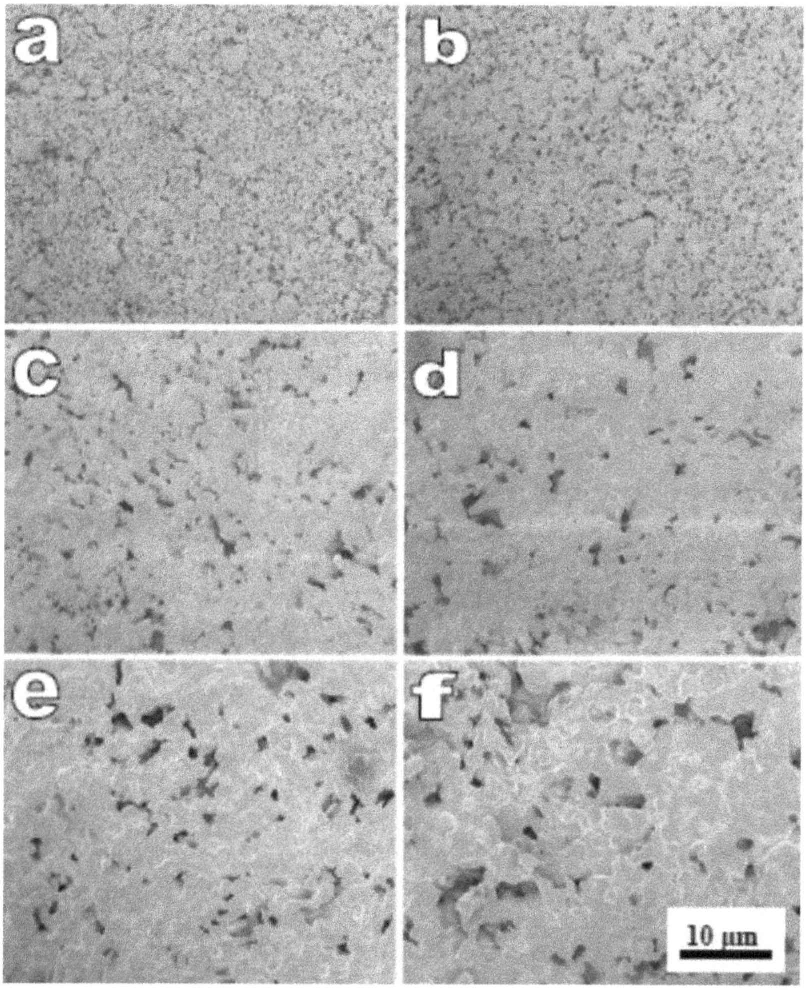

Figure 8.81. SEM images of the samples doped with 5 mol% MnO from the mixture activated with stainless steel media, after sintering at different temperatures: (a) 1000 °C, (b) 1100 °C, (c) 1200 °C, (d) 1300 °C, (e) 1400 °C and (f) 1500 °C.

occurred at 1000 °C. Mullite phase formation was almost finished at 1150 °C. No phase change happened at a higher sintering temperature. XRD patterns of the samples doped with different contents of CuO sintered at 1100 °C are depicted in figure 8.93. The mullite phase was not formed in all four samples. However, it was found that the phase transition of quartz to cristobalite was complete in the samples with CuO ⩾3%, while quartz was still visible in the sample with 1% CuO. In other words, the presence of CuO was also favorable for the quartz–cristobalite phase transition. As the sintering temperature was slightly increased to 1125 °C, mullitization was complete in all four samples, as illustrated in figure 8.94.

Representative SEM images of the samples with different contents of CuO sintered at different temperatures are shown in figures 8.95–8.97. As revealed in

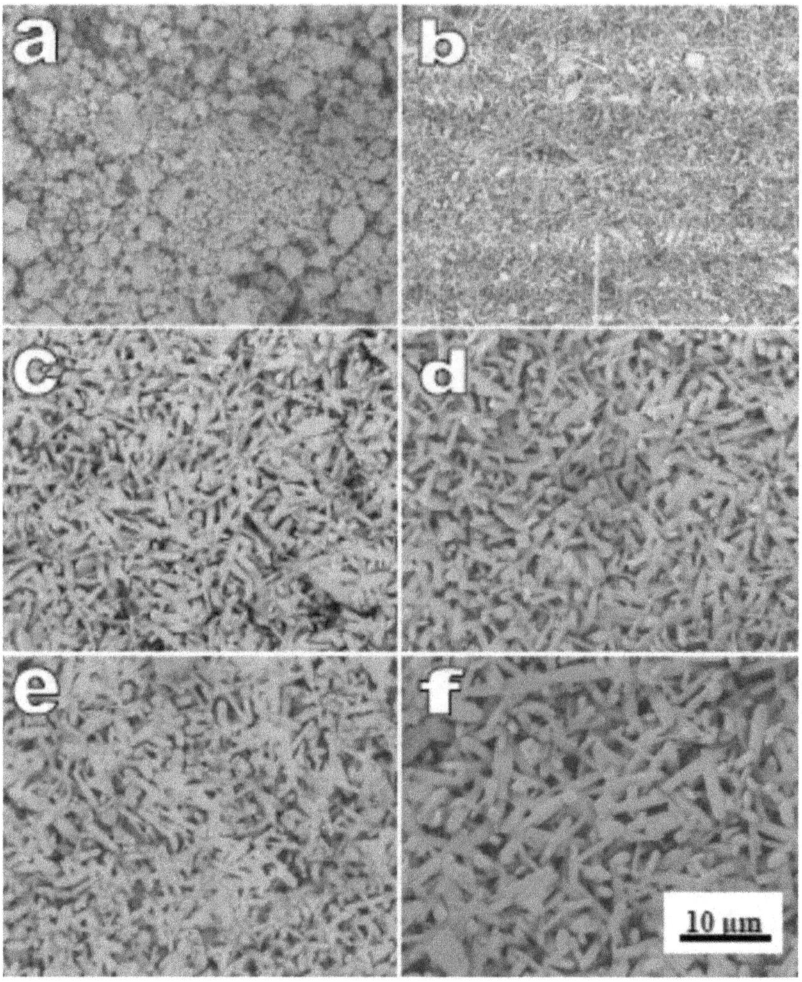

Figure 8.82. SEM images of the samples doped with 5 mol% MnO form the mixture mechanochemically activated with WC milling media, after sintering at different temperatures for 4 h: (a) 1000 °C, (b) 1100 °C, (c) 1200 °C, (d) 1300 °C, (e) 1400 °C and (f) 1500 °C.

figure 8.95, well-shaped mullite whiskers were formed in all four samples after sintering at 1250 °C, which was consistent with the XRD results. In other words, the mullitization process occurred before the densification, so that anisotropic grain growth was allowed. Careful inspection indicated that the dimensions of the mullite whiskers were gradually increased with increasing concentration of CuO. Therefore, the presence of CuO had a positive effect on mullite phase formation from the oxide mixture of Al_2O_3 and SiO_2. However, as the sintering temperature was further increased, the difference in the morphology and size of the mullite whiskers among the samples with different contents of CuO was gradually reduced, as observed in figure 8.96. In addition, for a given composition, the samples sintered at temperatures ⩾1200 °C displayed almost the same microstructure.

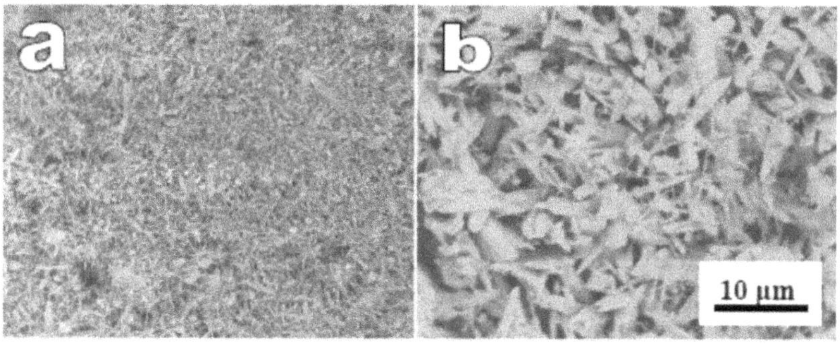

Figure 8.83. SEM images of the samples doped with different contents of MnO from the mixtures mechanochemically activated with WC milling media, after sintering at 1100 °C for 4 h: (a) 10 mol% and (b) 20 mol%.

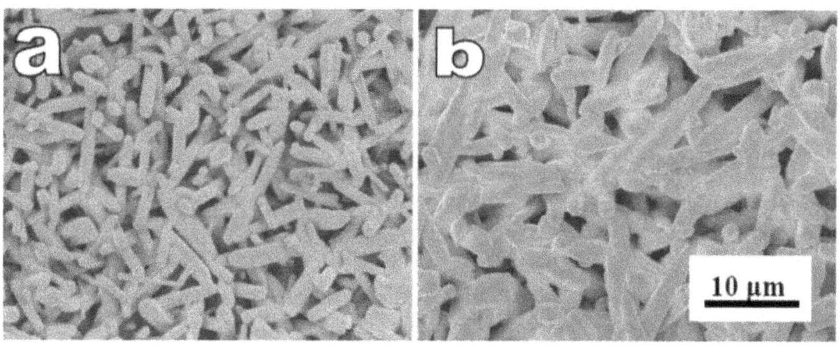

Figure 8.84. SEM images of the samples doped with different contents of MnO from the mixtures mechanochemically activated with WC milling media, after sintering at 1500 °C for 4 h: (a) 10 mol% and (b) 20 mol%.

8.4 Summary

A new method has been presented with the initial purpose of developing mullite ceramics with a low sintering temperature from mixtures of oxide precursors, which were mechanochemically activated with a high-energy milling process. It was found that the highly activated mixtures could not be fully densified due to the formation of mullite whiskers, which were formed through anisotropic grain growth. The anisotropic grain growth occurred before the densification process began. Once mullite whiskers were formed, their special structures would prohibit the procedure of densification.

The microstructure of the ceramics and the morphology of the mullite grains were highly dependent on: (i) the quality of the starting oxides, (ii) the mechanochemical activation milling media, (iii) the types of dopants and (iv) the sintering temperature/time. A wide range of morphologies has been obtained, most of which, however, have not been fully investigated and understood. Hopefully, the results presented in this chapter will inspire deep studies to reveal the underlying mechanisms that govern the formation of the special microstructure and grain morphology. It is also

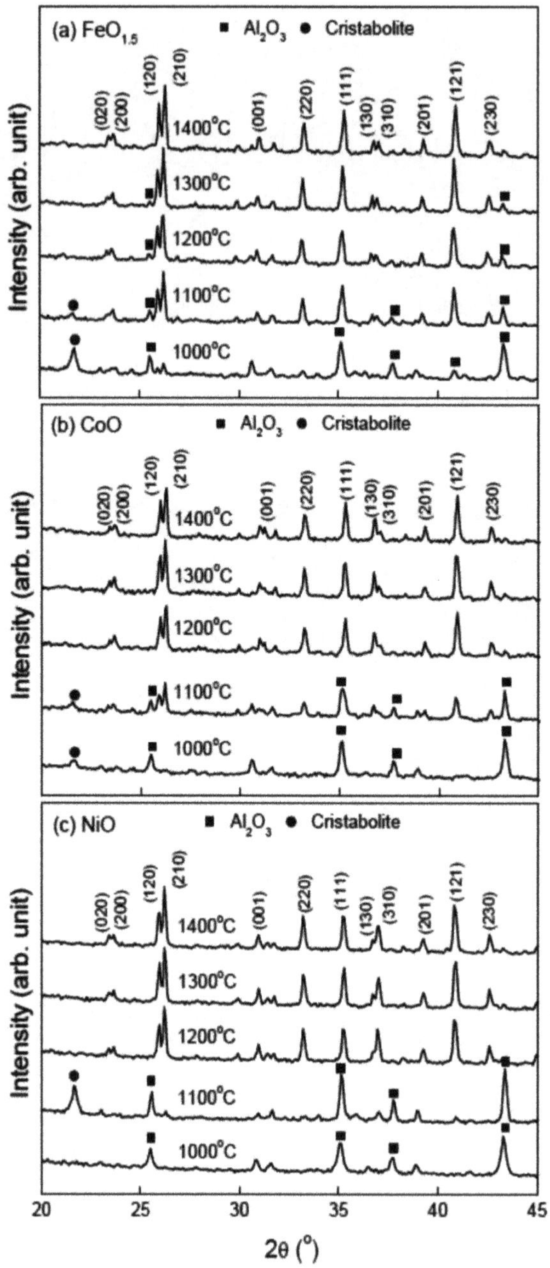

Figure 8.85. XRD patterns of the samples doped with the three groups of transition oxides from mixtures mechanochemically activated with WC milling media, after sintering at different temperatures. Reproduced with permission from [48]. Copyright 2003 Elsevier.

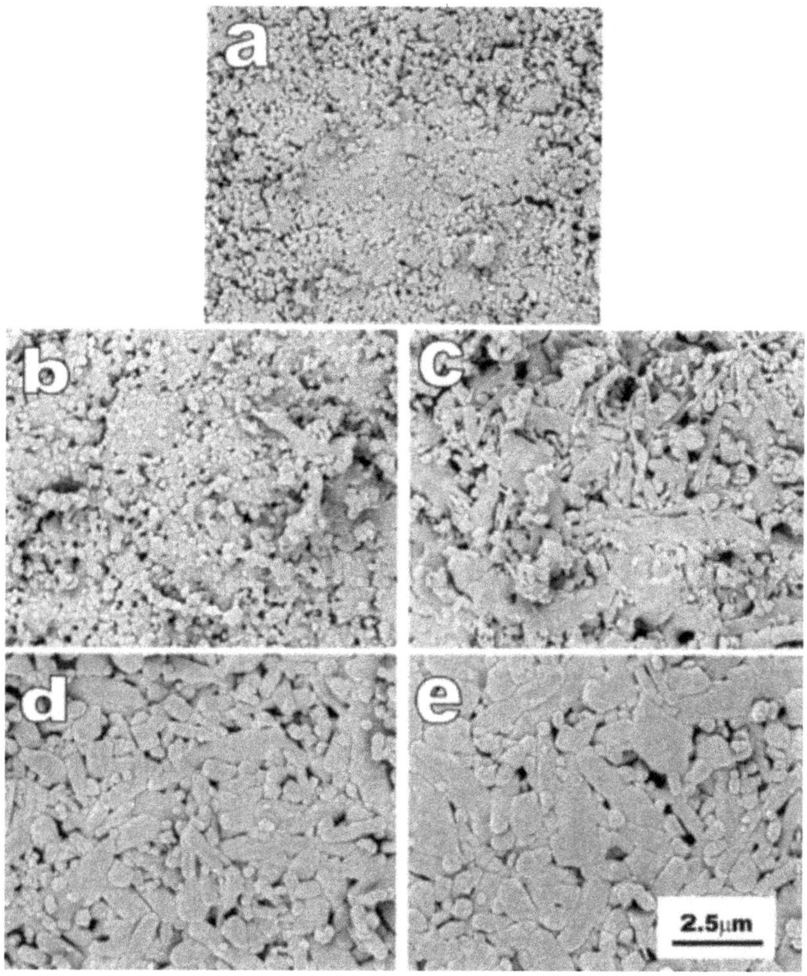

Figure 8.86. SEM images of the samples from the mixture doped with $FeO_{1.5}$ mechanochemically activated with stainless steel media, after sintering at different temperatures for 4 h: (a) 1100 °C, (b) 1200 °C, (c) 1300 °C, (d) 1400 °C and (e) 1500 °C. Reproduced with permission from [47]. Copyright 2003 Elsevier.

highly expected that a variety of mullite ceramics with special structures could be developed by making full use of this simple yet effective method, such as porous structures, composites, gradient structures, textures and so on.

Acknowledgments

Shenzhen Technology University (SZTU) is acknowledged for the financial support of a start-up grant (2018) and also the Natural Science Foundation of Top Talent of SZTU (grant no. 2019010801002).

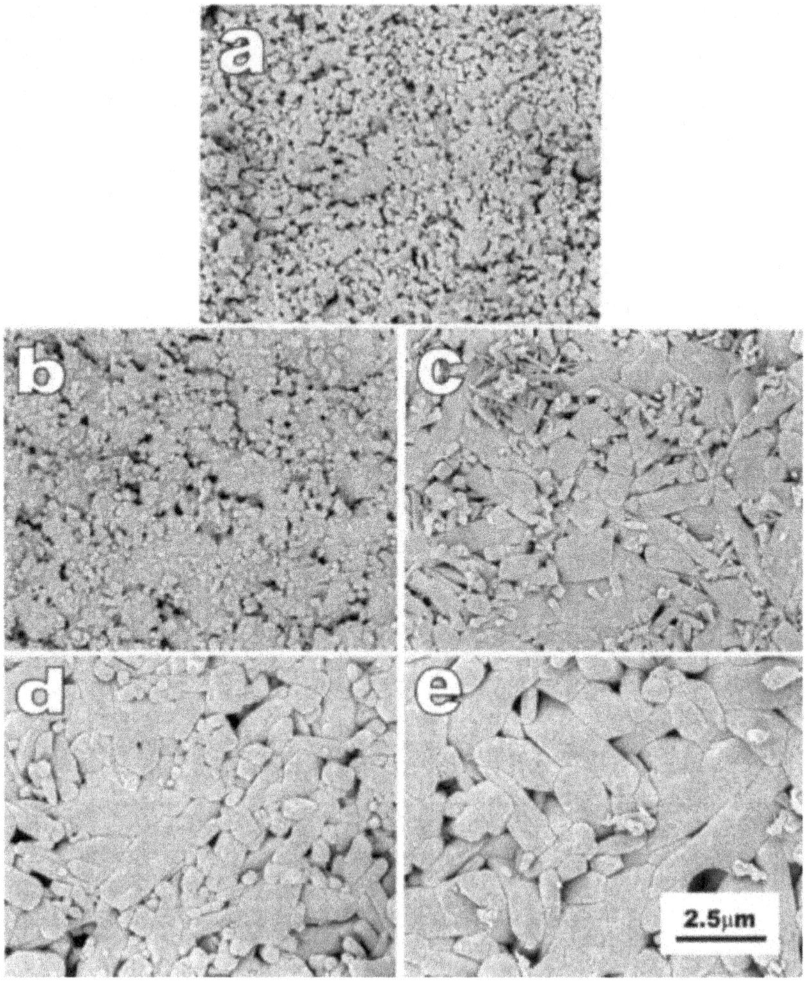

Figure 8.87. SEM images of the samples from the mixture doped with CoO mechanochemically activated with stainless steel media, after sintering at different temperatures for 4 h: (a) 1100 °C, (b) 1200 °C, (c) 1300 °C, (d) 1400 °C and (e) 1500 °C. Reproduced with permission from [47]. Copyright 2003 Elsevier.

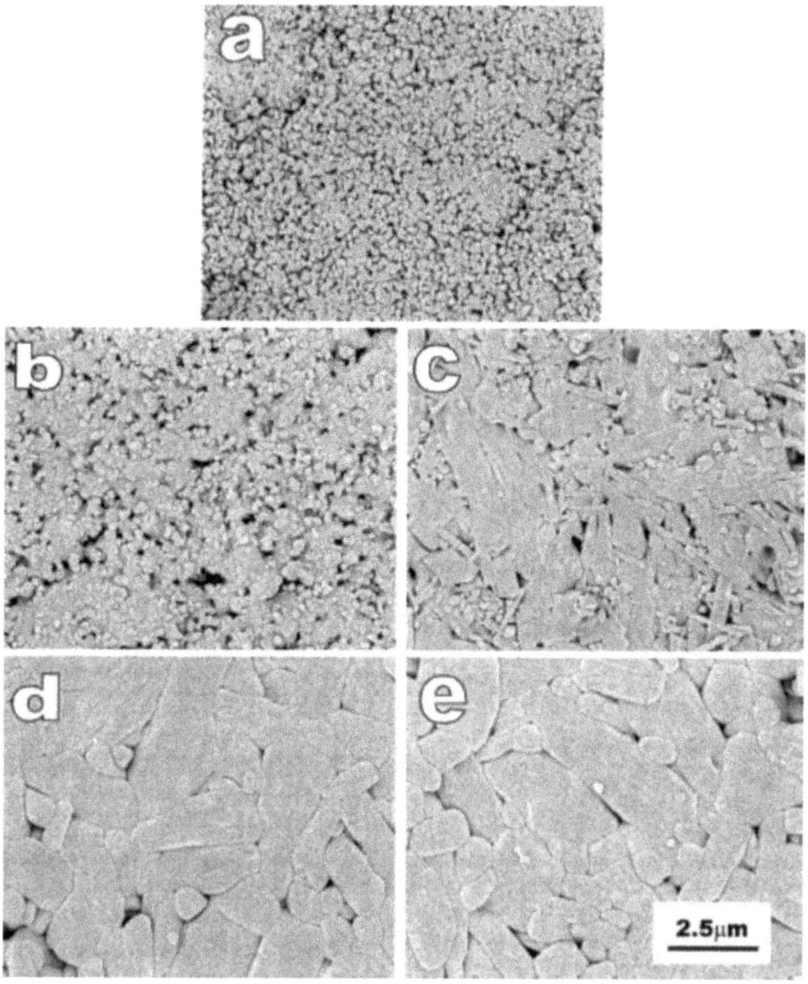

Figure 8.88. SEM images of the samples from the mixture doped with NiO mechanochemically activated with stainless steel media, after sintering at different temperatures for 4 h: (a) 1100 °C, (b) 1200 °C, (c) 1300 °C, (d) 1400 °C and (e) 1500 °C. Reproduced with permission from [47]. Copyright 2003 Elsevier.

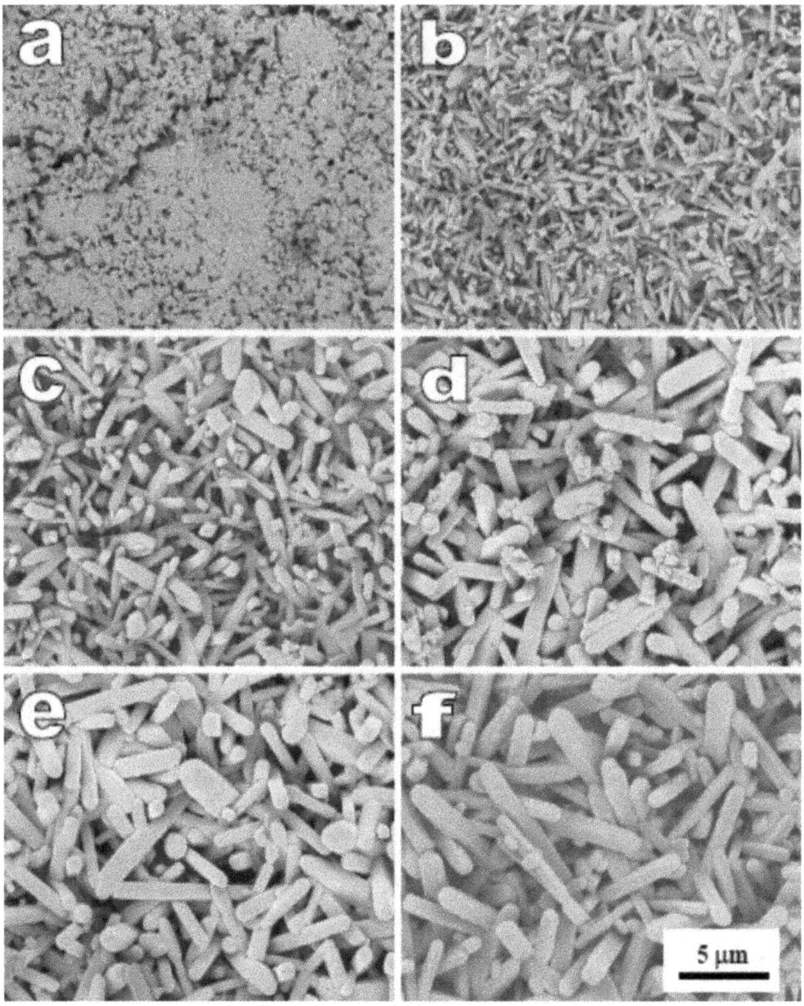

Figure 8.89. SEM images of the samples from the mixture doped with $FeO_{1.5}$ mechanochemically activated with WC media, after sintering at different temperatures for 4 h: (a) 1000 °C, (b) 1100 °C, (c) 1200 °C, (d) 1300 °C, (e) 1400 °C and (f) 1500 °C. Reproduced with permission from [48]. Copyright 2003 Elsevier.

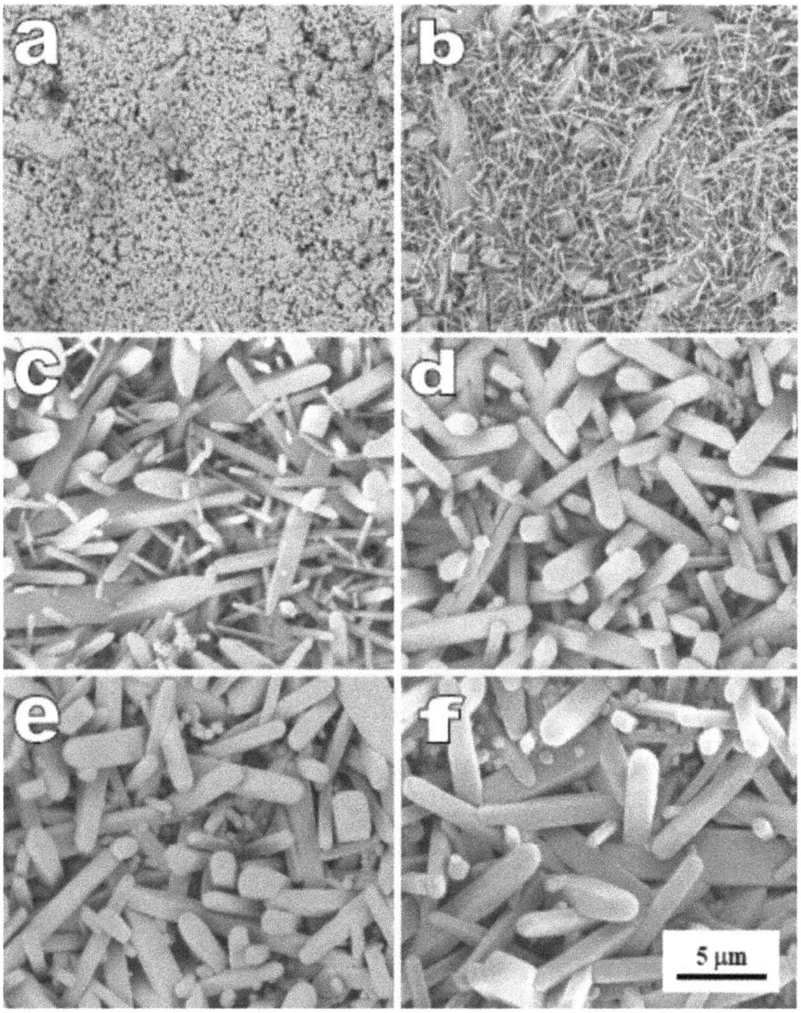

Figure 8.90. SEM images of the samples from the mixture doped with CoO mechanochemically activated with WC media, after sintering at different temperatures for 4 h: (a) 1000 °C, (b) 1100 °C, (c) 1200 °C, (d) 1300 °C, (e) 1400 °C and (f) 1500 °C. Reproduced with permission from [48]. Copyright 2003 Elsevier.

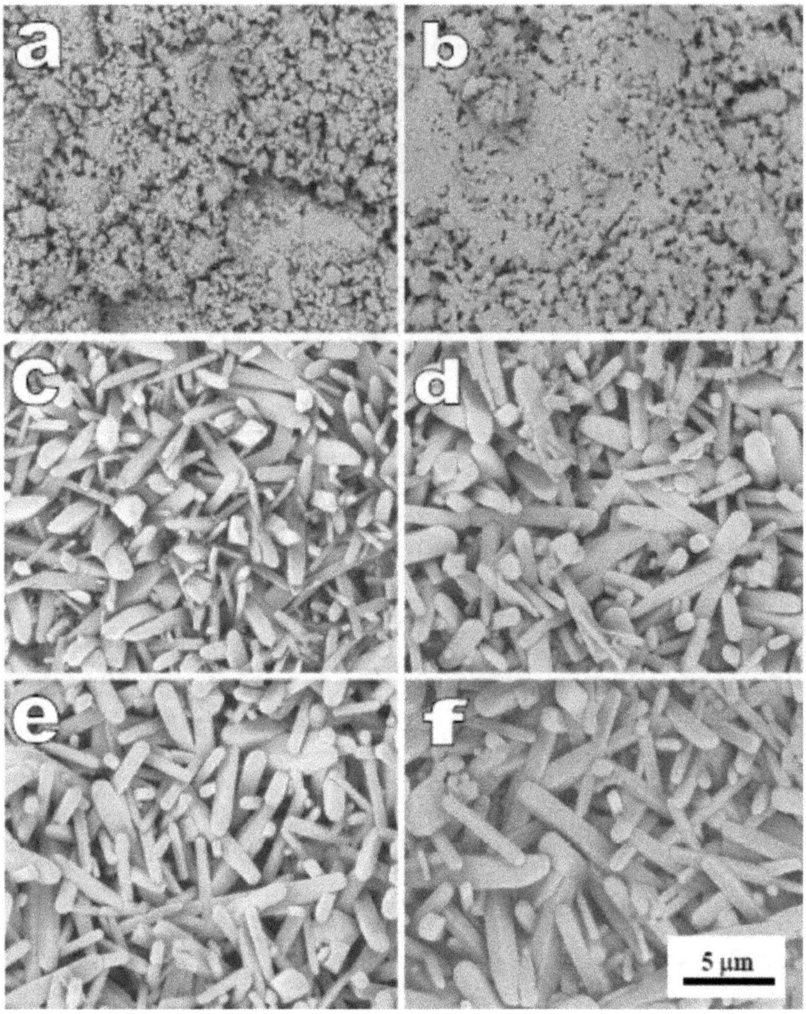

Figure 8.91. SEM images of the samples from the mixture doped with NiO mechanochemically activated with WC media, after sintering at different temperatures for 4 h: (a) 1000 °C, (b) 1100 °C, (c) 1200 °C, (d) 1300 °C, (e) 1400 °C and (f) 1500 °C. Reproduced with permission from [48]. Copyright 2003 Elsevier.

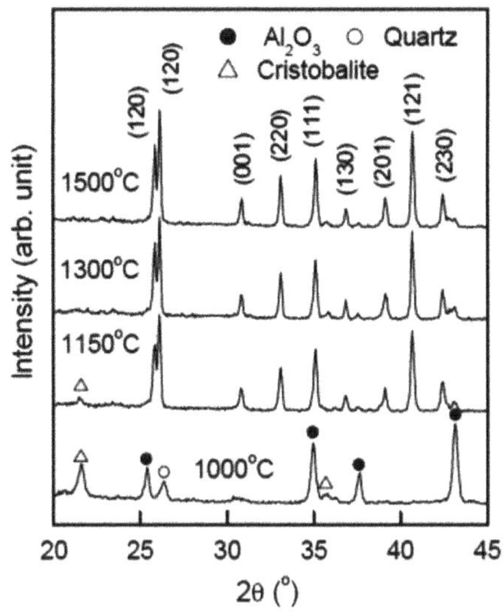

Figure 8.92. XRD patterns of the mixture with 1% CuO sintered at different temperatures for 4 h.

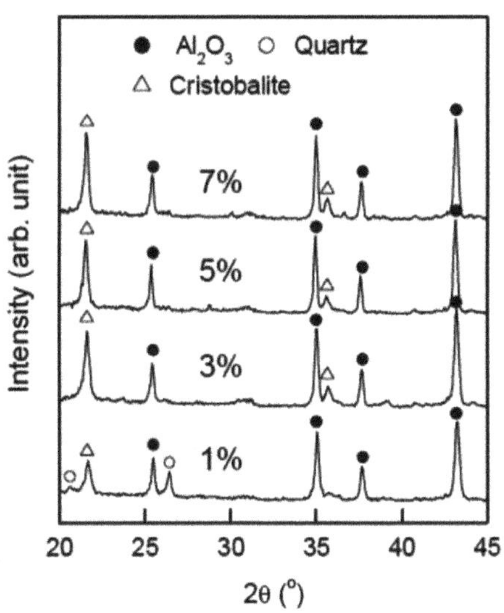

Figure 8.93. XRD patterns of the mixtures with different concentrations of CuO sintered at 1100 °C for 4 h.

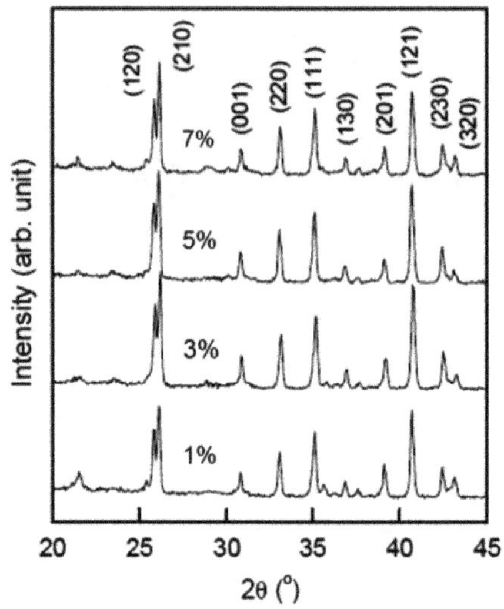

Figure 8.94. XRD patterns of the mixtures with different concentrations of CuO sintered at 1125 °C for 4 h.

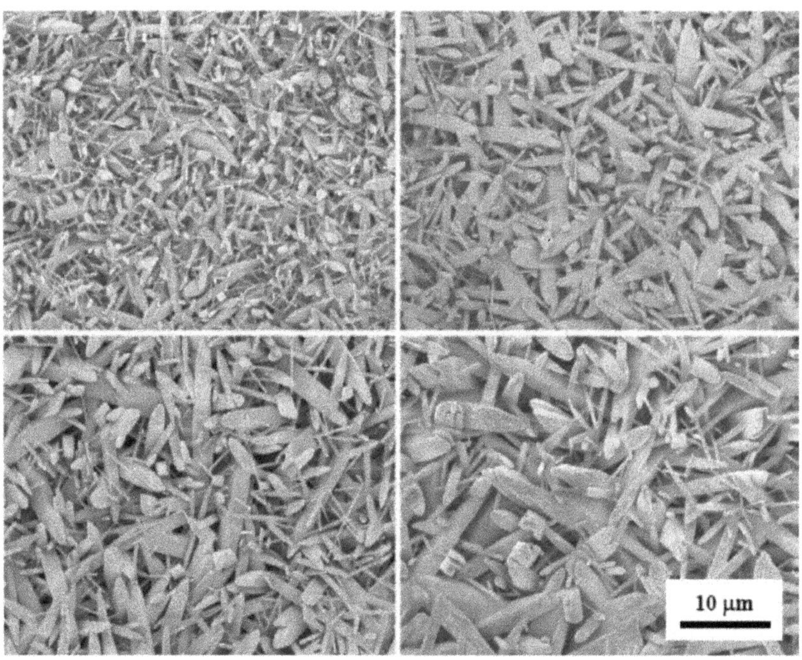

Figure 8.95. SEM images of the samples with different concentrations of CuO, after sintering at 1125 °C for 4 h: (a) 1%, (b) 3%, (c) 5% and (d) 7%.

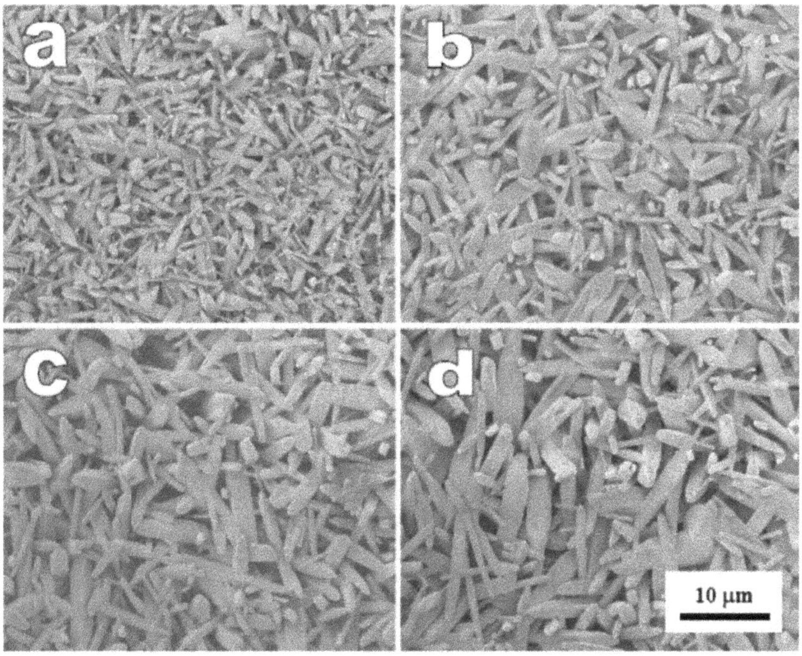

Figure 8.96. SEM images of the samples with different concentrations of CuO, after sintering at 1150 °C for 4 h: (a) 1%, (b) 3%, (c) 5% and (d) 7%.

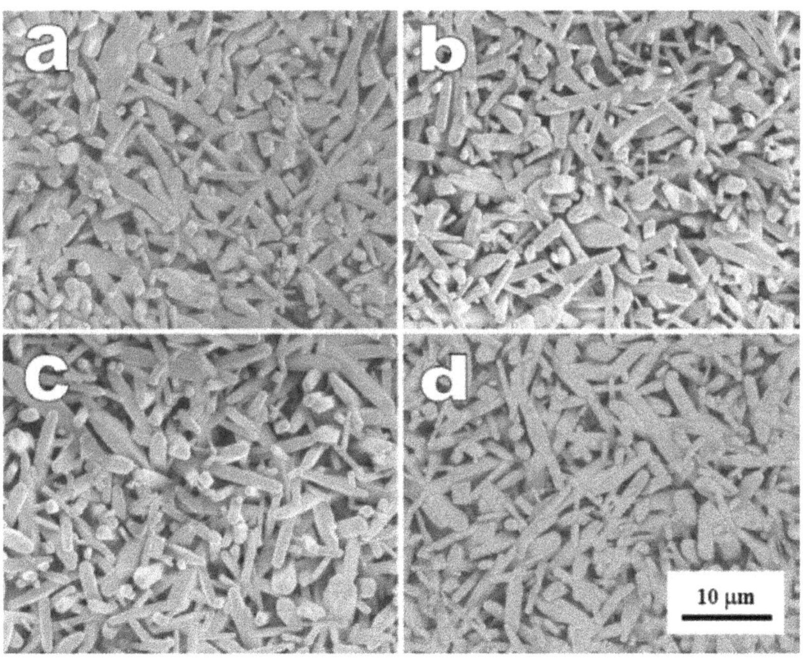

Figure 8.97. SEM images of the samples with 3% CuO sintered at different temperatures for 4 h: (a) 1200 °C, (b) 1300 °C, (c) 1400 °C and (d) 1500 °C.

References

[1] Aksay I A, Dabbs D M and Sarikaya M 1991 Mullite for structural, electronic, and optical applications *J. Am. Ceram. Soc.* **74** 2343–58

[2] Achari S and Satapathy L N 2003 Mullite-based refractories for molten-metal applications *Am. Ceram. Soc. Bull.* **82** 33–8

[3] Gao L M, Li J, Li Y and Zhang F Q 2011 Mechanical properties and microstructure of mullite modified zirconia ceramic for dental applications *J. Ceram. Proces. Res.* **12** 640–5

[4] Medvedovski E 2006 Alumina–mullite ceramics for structural applications *Ceram. Int.* **32** 369–75

[5] Torrecillas R *et al* 1999 Suitability of mullite for high temperature applications *J. Eur. Ceram. Soc.* **19** 2519–27

[6] Zaki Z I, Mostafa N Y and Ahmed Y M Z 2014 Synthesis of dense mullite/$MoSi_2$ composite for high temperature applications *Int. J. Refract. Met. Hard Mater.* **45** 23–30

[7] Wang Y G and Liu J L 2009 Aluminum phosphate–mullite composites for high-temperature radome applications *Int. J. Appl. Ceram. Technol.* **6** 190–94

[8] Jing Y N, Deng X Y, Li J B, Bai C Y and Jiang W K 2014 Fabrication and properties of SiC/mullite composite porous ceramics *Ceram. Int.* **40** 1329–34

[9] Ramakrishnan V, Goo E, Roldan J M and Giess E A 1992 Microstructure of mullite ceramics used for substrate and packaging applications *J. Mater. Sci.* **27** 6127–30

[10] Bourdais S, Mazel F, Fantozzi G and Slaoui A 1999 Silicon deposition on mullite ceramic substrates for thin-film solar cells *Prog. Photovol.* **7** 437–47

[11] Mazel F, Gonon M and Fantozzi G 2002 Manufacture of mullite substrates from andalusite for the development of thin film solar cells *J. Eur. Ceram. Soc.* **22** 453–61

[12] Ohashi M, Iida Y and Wada S 2000 Mullite-based substrates for polycrystalline silicon thin-film solar cells *J. Ceram. Soc. Jpn* **108** 105–7

[13] Slaoui A, Pihan E and Focsa A 2006 Thin-film silicon solar cells on mullite substrates *Sol. Ener. Mater. Sol. Cells* **90** 1542–52

[14] Snijkers F M M, van Hoolst J, Schoeters M, Alvarez E, Matamala J M and Luyten J J 2007 Composed mullite substrates for the development of crystalline silicon solar cells *Adv. Eng. Mater.* **9** 385–8

[15] Viswabaskaran V, Gnanam F D and Balasubramanian M 2004 Mullite from clay-reactive alumina for insulating substrate application *Appl. Clay Sci.* **25** 29–35

[16] Prochazka S and Klug F J 1983 Infrared-transparent mullite ceramic *J. Am. Ceram. Soc.* **66** 874–80

[17] Song X R, Liu H C, Jing W, Liu H Y and Zhang J D 2007 Study on TiO_2–mullite far-infrared ceramic *Rare Met. Mater. Eng.* **36** 471–3

[18] Sonuparlak B 1988 Sol–gel processing of infrared transparent mullite *Adv. Ceram. Mater.* **3** 263–7

[19] Bobkova N M, Kavrus I V, Radion E V and Popovskaya N F 1998 Formation of mullite obtained by coprecipitation *Glass Ceram.* **55** 186–8

[20] Satoshi S, Contreras C, Juarez H, Aguilera A and Serrato J 2001 Homogeneous precipitation and thermal phase transformation of mullite ceramic precursor *Int. J. Inorg. Mater.* **3** 625–32

[21] Sueyoshi S S and Soto C A C 1998 Fine pure mullite powder by homogeneous precipitation *J. Eur. Ceram. Soc.* **18** 1145–52

[22] Zhou M H, Ferreira J M F, Fonseca A T and Baptista J L 1996 Coprecipitation and processing of mullite precursor phases *J. Am. Ceram. Soc.* **79** 1756–60

[23] Boccaccini A R and Ponton C B 2000 Colloidal processing of mullite ceramics: are silica–alumina 'composite' precursor particles essential? *J. Mater. Sci. Lett.* **19** 1687–8
[24] Cho Y I, Kamiya H, Suzuki Y, Horio M and Suzuki H 1998 Processing of mullite ceramic from alkoxide-derived silica and colloidal alumina with ultra-high cold isostatic pressing *J. Eur. Ceram. Soc.* **18** 261–8
[25] Choi H J and Lee J G 2002 Synthesis of mullite whiskers *J. Am. Ceram. Soc.* **85** 481–3
[26] De Souza M F, Yamamoto J, Regiani I, Paiva-Santos C O and de Souza D P F 2000 Mullite whiskers grown from erbia-doped aluminum hydroxide–silica gel *J. Am. Ceram. Soc.* **83** 60–4
[27] Hashimoto S and Yamaguchi A 2004 Synthesis of mullite whiskers using Na_2SO_4 flux *J. Ceram. Soc. Jpn.* **112** 104–9
[28] Jiang X Q, Li J H, Ma H W, Zhou W C and Tong L X 2011 Temperature feature on synthesizing mullite whiskers from coal fly ash and andalusite–sericite phyllite *Rare Met.* **30** 379–82
[29] Katsuki H, Ichinose H and Furuta S 1996 Growth of mullite whiskers on alumina particle by thermal decomposition of clay minerals *J. Ceram. Soc. Jpn.* **104** 788–91
[30] Li X D, Zhu B Q and Zhang S W 2008 The effects of molten salts on the formation of mullite whiskers *Rare Met. Mater. Eng.* **37** 300–02
[31] Okada K and Otuska N 1991 Synthesis of mullite whiskers and their application in composites *J. Am. Ceram. Soc.* **74** 2414–18
[32] Zhang P Y *et al* 2010 Molten salt synthesis of mullite whiskers from various alumina precursors *J. Alloys Compd.* **491** 447–51
[33] Kong L B, Ma J and Huang H 2002 Mullite whiskers derived from an oxide mixture activated by a mechanochemical process *Adv. Eng. Mater.* **4** 490–94
[34] Jiang J Z, Poulson F W and Morup S 1999 Structure and thermal stability of nanostructured iron-doped zirconia prepared by high-energy ball milling *J. Mater. Res.* **14** 1343–2352
[35] Sacks M D, Bozkurt N and Scheiffele G W 1991 Fabrication of mullite and mullite–matrix composites by transient viscus sintering of composite powders *J. Am. Ceram. Soc.* **74** 2428–37
[36] Sacks M D, Lin Y J, Scheiffele G W, Wang K Y and Bozkurt N 1995 Effect of seeding on phase development, densification behavior, and microstructure evolution in mullite fabricated from microcomposite particles *J. Am. Ceram. Soc.* **78** 2897–906
[37] Sacks M D and Pask J A 1980 Intermediate-state sintering kinetics of mullite *Am. Ceram. Soc. Bull.* **59** 361
[38] Wang K Y and Sacks M D 1996 Mullite formation by endothermic reaction of alpha-alumina/silica microcomposite particles *J. Am. Ceram. Soc.* **79** 12–6
[39] Zhang T S, Kong L B, Du Z H, Ma J and Li S 2010 Effect of TiO_2 on anisotropic grain growth and densification of mullite ceramics via a high-energy ball milling *J. Alloys Compd.* **506** 777–83
[40] Hong S H, Lee N, Carim A H and Messing G L 1998 Interfacial precipitation in titania-doped diphasic mullite gels *J. Mater. Res.* **13** 974–8
[41] Hong S H and Messing G L 1997 Mullite transformation kinetics in P_2O_5-, TiO_2-, and B_2O_3-doped aluminosilicate gels *J. Am. Ceram. Soc.* **80** 1551–9
[42] Hong S H and Messing G L 1998 Anisotropic grain growth in diphasic-gel-derived titania-doped mullite *J. Am. Ceram. Soc.* **81** 1269–77

[43] Kong L B, Gan Y B, Ma J, Zhang T S, Boey F and Zhang R F 2003 Mullite phase formation and reaction sequences with the presence of pentoxides *J. Alloys Compd.* **351** 264–72

[44] Bartsch M, Saruhan B, Schmucker M and Schneider H 1999 Novel low-temperature processing route of dense mullite ceramics by reaction sintering of amorphous SiO_2-coated γ-Al_2O_3 particle nanocomposites *J. Am. Ceram. Soc.* **82** 1388–92

[45] Xiao Z H *et al* 2018 Phase formation and microstructure evolution of mullite ceramics from mechanochemical activated oxide powders doped with Cr_2O_3 *J. Phys. Chem. Solids* **123** 198–205

[46] Kong L B *et al* 2003 Growth of mullite whiskers in mechanochemically activated oxides doped with WO_3 *J. Eur. Ceram. Soc.* **23** 2257–64

[47] Kong L B, Zhang T S, Ma J and Boey F 2003 Anisotropic grain growth of mullite in high-energy ball milled powders doped with transition metal oxides *J. Eur. Ceram. Soc.* **23** 2247–56

[48] Kong L B *et al* 2003 Effect of transition metal oxides on mullite whisker formation from mechanochemically activated powders *Mater. Sci. Eng.* A **359** 75–81

[49] Kong L B, Ma J, Huang H T, Zhang T S and Boey F 2003 Anisotropic mullitization in CuO-doped oxide mixture activated by high-energy ball milling *Mater. Lett.* **57** 3660–66

Chapter 9

Mullite ceramics (II)

Ling Bing Kong, Zhuohao Xiao, Xiuying Li, Shijin Yu, Wenxiu Que, Yin Liu, Tianshu Zhang, Kun Zhou and Hongfang Zhang

9.1 The effects of rare-earth oxides

The effects of Y_2O_3, La_2O_3 and CeO_2 on the phase formation of mullite from oxide mixtures of Al_2O_3 and quartz (SiO_2) have been examined [1]. Commercial powders of precipitated SiO_2 (laboratory reagent), quartz (99+% purity), Y_2O_3 (99+% purity), La_2O_3 (99+% purity) and CeO_2 (99+% purity) were utilized as the starting materials, with the nominal composition of $(3Al_2O_3{\cdot}2SiO_2)_{1-x}(MO)_x$ (MO = Y_2O_3, La_2O_3 and CeO_2), with $x = 0.05$, 0.10 and 0.20. Their weight percentages were in the ranges 2.7–11.7%, 3.9%–16.1% and 2.1%–9.2%, respectively. Precipitated silica was used in the case of La_2O_3 doping, while quartz was used in the Y_2O_3 and CeO_2 doped samples. The oxides were mixed using the conventional ball milling process, with ZrO_2 milling media. The powder mixtures were then compacted and sintered at temperatures of 1100 °C–1500 °C for 4 h. The densification behaviors of the samples were monitored using a Setaram Setsys 16/18 type dilatometer at a heating rate of 10 °C/min in air.

Figure 9.1 shows the XRD patterns of the samples doped with 5 mol% Y_2O_3 after sintering at different temperatures. After sintering at 1100 °C, four phases were present according to the XRD, as seen in figure 9.1(a), which were Al_2O_3, quartz, cristobalite and $Y_2Si_2O_7$(1) (JCPDS 38–223). Therefore, at this temperature the phase transition from quartz to cristobalite took place, while Y_2O_3 and SiO_2 reacted to form a new phase with a composition of $Y_2Si_2O_7$(1). The $Y_2Si_2O_7$(1) has a triclinic crystal structure, with $a = 0.659$ nm, $b = 0.664$ nm and $c = 1.225$ nm, as well as $\alpha = 94°$, $\beta = 89.2°$ and $\gamma = 93.1°$. This phase will transform into another phase with monoclinic structure.

Figure 9.2 depicts the XRD patterns of the samples doped with different concentrations of Y_2O_3 after sintering at 1200 °C for 4 h. At this temperature mullite phase formation was almost complete. In addition, Al_2O_3, cristobalite and $Y_2Si_2O_7$(1) were still present as minor phases. In addition, quartz was completely

doi:10.1088/978-0-7503-2191-4ch9

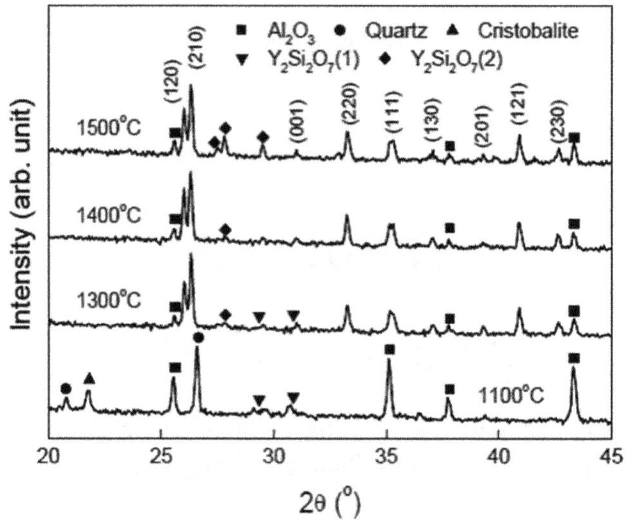

Figure 9.1. XRD patterns of the 5 mol% Y_2O_3-doped sample sintered at different temperatures. Reproduced with permission from [1]. Copyright 2004 Elsevier.

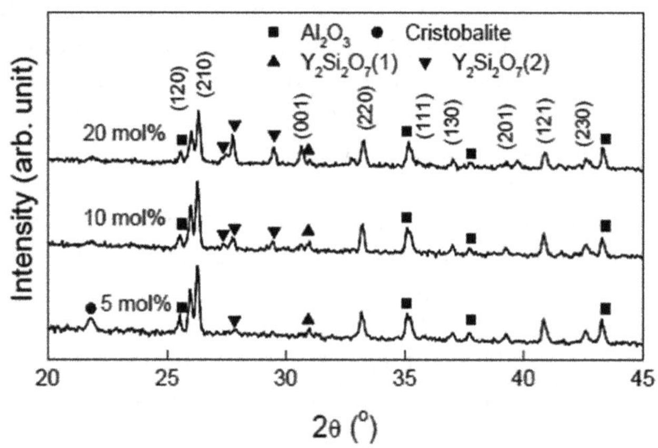

Figure 9.2. XRD patterns of the 1200 °C-sintered samples with different Y_2O_3 contents. Reproduced with permission from [1]. Copyright 2004 Elsevier.

transferred to cristobalite. After sintering at 1300 °C the cristobalite phase was absent. At the same time, another yttrium silicate, denoted as $Y_2Si_2O_7(2)$, was detected which has a monoclinic crystal structure with $a = 0.6875$ nm, $b = 0.897$ nm, $c = 0.4721$ nm and $\beta = 101.74°$ (JCPDS, 38–440). $Y_2Si_2O_7(1)$ was absent after the sample was sintered at 1400 °C, implying that $Y_2Si_2O_7(1)$ was stable at low temperatures and $Y_2Si_2O_7(2)$ was stable at high temperatures. Figure 9.2 also indicated that the peak intensity of $Y_2Si_2O_7(2)$ increased as the doping concentration of Y_2O_3 was increased. In addition, no cristobalite phase was present in the samples with 10 and 20 mol% Y_2O_3, suggesting the completion of the mullitization process.

Therefore, it could be concluded that the presence of Y_2O_3 had a positive effect on mullite phase formation from the oxide mixture.

Figure 9.3 shows the XRD patterns of the samples doped with 5 mol% La_2O_3 after sintering at different temperatures. The sample sintered at 1100 °C for 4 h consisted of only Al_2O_3 and cristobalite. After sintering at 1200 °C, in addition to Al_2O_3 and cristobalite, a trace of mullite phase was visible. The mullitization process was almost finished after sintering at 1300 °C. In other words, the mullite phase was formed at 1200 °C and the reaction was complete at 1300 °C.

The XRD patterns of the samples doped with 10 and 20 mol% La_2O_3 after sintering at 1200 °C are depicted in figure 9.4. Comparatively, the peak intensity of the mullite phase was significantly increased in the sample doped with 10 mol%

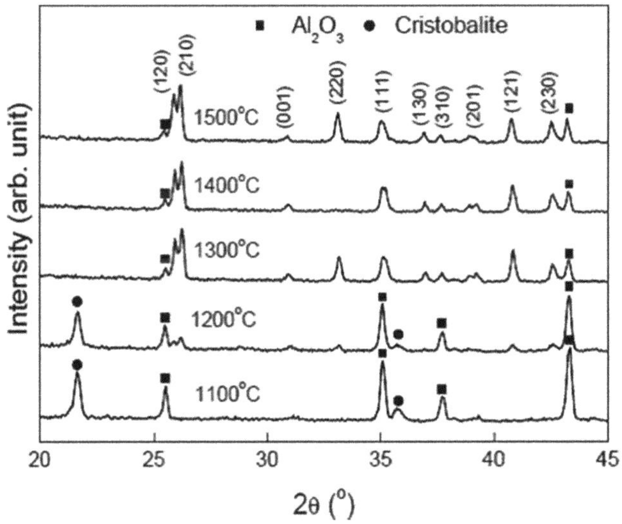

Figure 9.3. XRD patterns of the 5 mol% La_2O_3-doped sample sintered at different temperatures. Reproduced with permission from [1]. Copyright 2004 Elsevier.

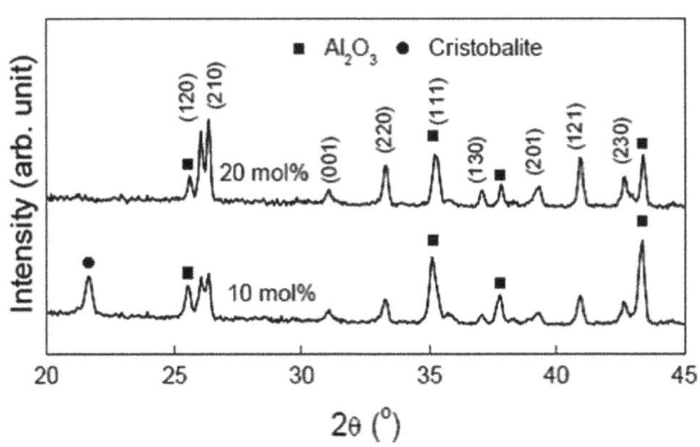

Figure 9.4. XRD patterns of the 1200 °C-sintered samples with 10 and 20 mol% Y_2O_3. Reproduced with permission from [1]. Copyright 2004 Elsevier.

La_2O_3, while mullitization was nearly complete in the sample with 20 mol% La_2O_3. Therefore, the addition of La_2O_3 also promoted the mullite phase formation from the mixture of Al_2O_3 and SiO_2. Comparing figure 9.2 and figure 9.4, it can be seen that the positive effect of La_2O_3 was slightly stronger than that of Y_2O_3. In addition, La_2O_3 was not detected according to the XRD results, nor was there a La_2O_3 related compound present in the samples.

Figure 9.5 shows the XRD patterns of the samples doped with 20 mol% CeO_2 after sintering at different temperatures, while those of the samples sintered at 1200 °C with different contents of CeO_2 are shown in figure 9.6. The XRD patterns of the samples doped with CeO_2 were almost the same as those of the samples doped with Y_2O_3. The small difference was that CeO_2 had no reactions with the starting oxides. Instead, CeO_2 was present in all the samples, with the peak intensity increasing with increasing content of CeO_2. In addition, CeO_2 also promoted mullite phase formation. Therefore, the three oxide dopants, Y_2O_3, La_2O_3 and CeO_2, all had a positive effect on the phase formation of mullite from the oxide mixture of Al_2O_3 and SiO_2. The promotion of mullitization was closely related to the mechanism of mullite phase formation, which is called the dissolution–precipitation reaction process [2–4].

According to the glass formation ability and phase diagram in the system of Y_2O_3–Al_2O_3–SiO_2, there was a ternary eutectic phase formed at the temperature of 1371 °C [5]. The formation of the eutectic liquid was the main reason for the positive effect of Y_2O_3 on mullite phase formation, as observed in figures 9.1 and 9.2. The promoted mullitization effect of Y_2O_3 was also observed in other studies when fabricating mullite ceramics through reaction-bonding technology [6–9]. The reduction in viscosity of the glass, due to the involvement of Y_2O_3, made a great contribution to the promotion of mullite phase formation. Similarly, in the ternary

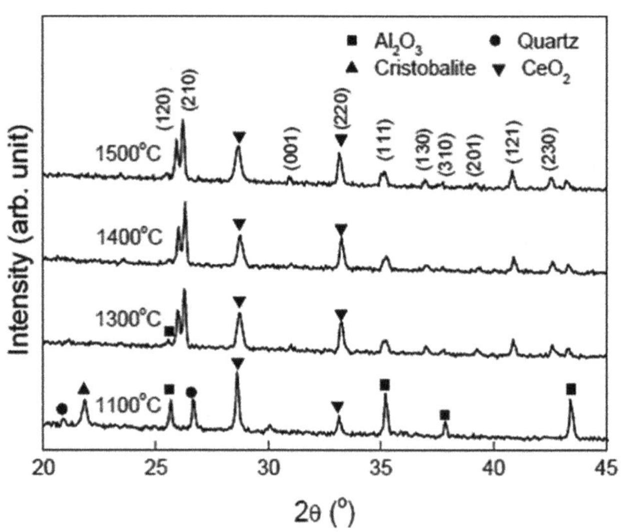

Figure 9.5. XRD patterns of the 20 mol% CeO_2-doped sample sintered at different temperatures. Reproduced with permission from [1]. Copyright 2004 Elsevier.

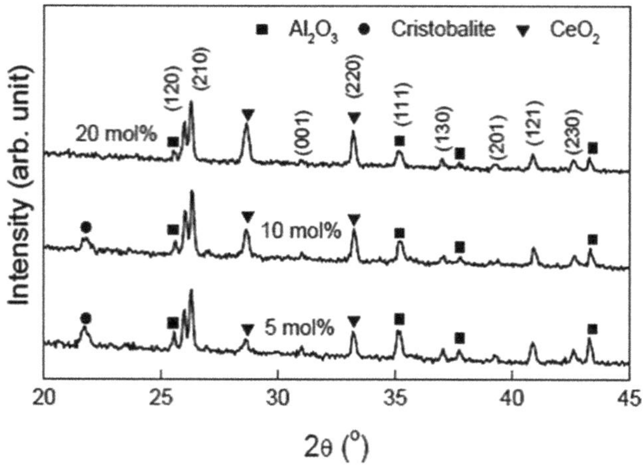

Figure 9.6. XRD patterns of the 1200 °C-sintered samples with different CeO_2 levels. Reproduced with permission from [1]. Copyright 2004 Elsevier.

system of La_2O_3–Al_2O_3–SiO_2 La_2O_3 there was a eutectic temperature of about 1240 °C in the composition with 3 mol% La_2O_3. As a result, the mullite phase was formed in the mixture of Al_2O_3 and SiO_2 close to this temperature.

Previous studies indicated that the presence of CeO_2 boosted the reaction bonding between Si and Al_2O_3 to develop mullite ceramics [10, 11]. The promoted reaction bonding was ascribed to the production of Ce–Al–Si–O liquid phase with low viscosity. As a result, the mass transportation was accelerated and thus the oxidation of Si and the mullitization process were enhanced. In fact, Ce^{4+} is a glass network modifier, so that the ionic diffusion rate was increased [12]. With the formation of mullite, CeO_2 was released from the glass phase as the SiO_2 was gradually consumed.

In addition, the three dopants resulted in different mullitization behaviors. In the group doped with Y_2O_3, Al_2O_3 was still present in the sample after sintering at 1500 °C, which was similar to that in the fabrication of mullite through the reaction-bonding process [7]. The key reason was the formation of yttrium silicates, which consumed SiO_2, thus leaving the Al_2O_3 in the sample. However, La_2O_3 was not involved in forming new compounds, but Al_2O_3 was also left in the samples. According to the mechanisms of mullite phase formation, mullite grains were precipitated from the siliceous liquid phase, where Al_2O_3 particles were wrapped by the newly formed mullite grains. Therefore, the Al_2O_3 particles that could not diffuse out of the mullite grains would be left, thus requiring a temperature as high as 1600 °C to remove [13]. In the case of La_2O_3–Al_2O_3–SiO_2, it was also likely that La_2O_3 stabilized the SiO_2-rich glass phase, so that the Al_2O_3 became excessive. To obtain a deeper understanding of these observations, further research should be conducted in future studies.

The densification behaviors of the samples doped with the three oxides are illustrated in figures 9.7–9.9. The mixtures doped with Y_2O_3 and CeO_2 shared a similar densification profile. In both cases, a continuous expansion was observed

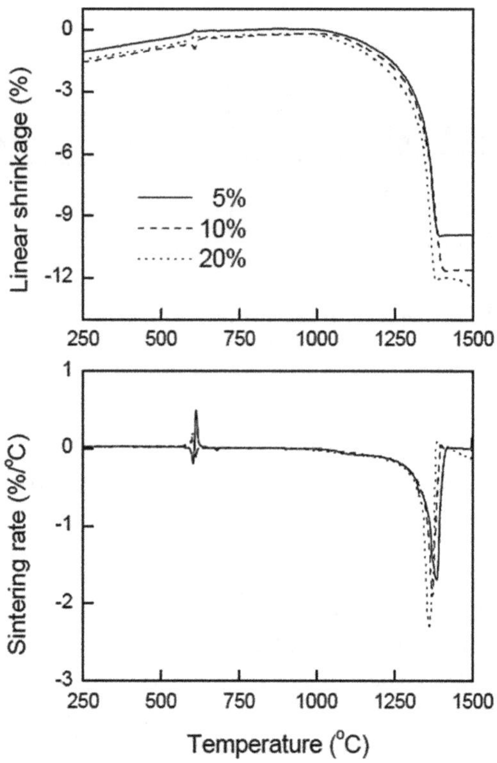

Figure 9.7. Sintering behavior of the Y_2O_3-doped samples. Reproduced with permission from [1]. Copyright 2004 Elsevier.

during the heating process, before the rapid densification began. The densification rates peaked at 1384 °C, 1371 °C and 1360 °C for the samples doped with Y_2O_3, while the temperatures were 1362 °C, 1341 °C and 1331 °C for the samples doped with CeO_2. The final linear shrinkages were 9.9%, 11.6% and 12.1% for the group with Y_2O_3, which were comparable to those reported in the available literature [5]. For the mixtures doped with CeO_2, the corresponding final shrinkage values were 7.7%, 10.5% and 13%. The densification rate peak is a reflection of the formation of the liquid phase. Therefore, the samples doped with CeO_2 had lower maximum densification temperatures than those doped with Y_2O_3 by 20 °C–30 °C. Additionally, another densification maxima was present at about 1178 °C for the group with CeO_2, which could be attributed to the transition of Ce^{4+} to Ce^{3+} [11].

Two peaks were observed in the densification rate curves of the samples doped with La_2O_3, with concentrations of 5, 10 and 20 mol%, at 1265 °C, 1263 °C and 1260 °C for the first, and 1416 °C, 1371 °C and 1308 °C for the second. Obviously, the effect of La_2O_3 on densification increased with increasing concentration and was much more pronounced compared to those for the two groups doped with Y_2O_3 and CeO_2. Moreover, the La_2O_3 doped mixtures exhibited maximum linear shrinkages of 23.1%, 23.8% and 27.5%, for the concentrations of 5 mol%, 10 mol% and 20 mol%,

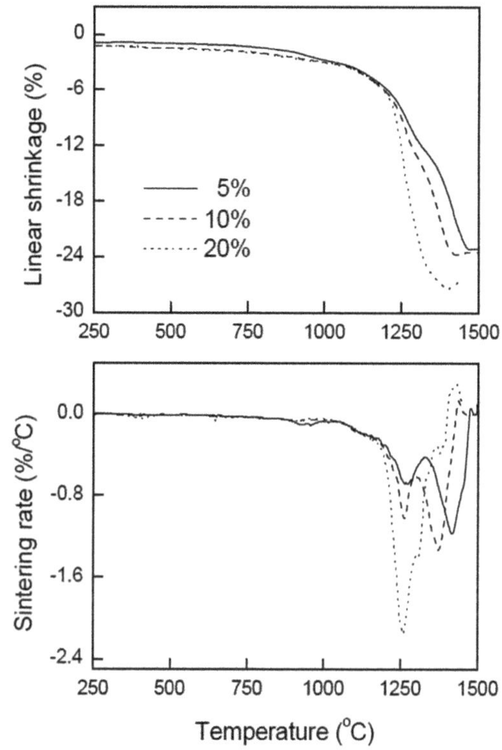

Figure 9.8. Sintering behavior of the La_2O_3-doped samples. Reproduced with permission from [1]. Copyright 2004 Elsevier.

respectively. In all three groups, the densification began before mullite phase formation. As a consequence, no anisotropic grain growth occurred.

Figure 9.10 shows the densities of the three groups of samples doped with different concentrations as a function of sintering temperature. Their densification behaviors were indirectly reflected by the density values. At a given sintering temperature, the density was increased with increasing content of dopants. For a given concentration of dopant, the density was monotonically increased with sintering temperature. The group of samples doped with La_2O_3 had the highest densities. The samples with high contents of La_2O_3 possessed densities that were even higher than the theoretical density of mullite. This could be attributed to two factors. First, the effect of La_2O_3 on the densification behavior was the strongest. Second, La_2O_3 has a much higher density of 6.51 g cm^{-3} than mullite. In other words, La_2O_3 was a favorable dopant to obtain fully dense mullite ceramics.

Figure 9.11 shows SEM images of the samples doped with 5 mol% Y_2O_3 after sintering at different temperatures for 4 h. The sample sintered at 1100 °C was still highly porous, and the sintering process had not started and mullite was not formed, as indicated in figure 9.1. After being sintered at 1200 °C, a glass-like amorphous layer was formed on the surface of the sample, while densification was clearly observed. A further increase in sintering temperature led to nearly no variation in

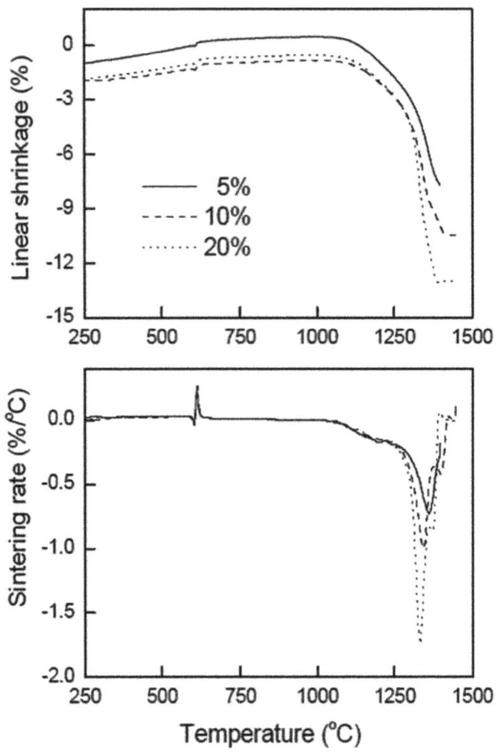

Figure 9.9. Sintering behavior of the CeO_2-doped samples. Reproduced with permission from [1]. Copyright 2004 Elsevier.

the microstructure of the samples. Even the grain size of the mullite phase remained almost unchanged as the sintering temperature was continuously raised.

In addition, the samples were characterized by the presence of pores with irregular shapes. For the samples doped with 10 mol% and 20 mol% Y_2O_3, their microstructures after sintering at low temperatures were similar to those of the samples doped with 5 mol% Y_2O_3, as revealed in figures 9.12(a) and (c). In comparison, sintering at high temperatures resulted in slight differences in their microstructures. Specifically, the sample doped with 20 mol% Y_2O_3 displayed visible mullite grains, as illustrated in figure 9.12(d).

The samples doped with La_2O_3 exhibited a much denser microstructure than those doped with Y_2O_3. Representative SEM images of the samples doped with La_2O_3 are presented in figures 9.13 and 9.14. After sintering at 1100 °C and 1200 °C, the samples doped with 5 mol% La_2O_3 were still porous, with particles being loosely compacted. As the sintering temperature was increased to 1300 °C, an almost fully dense microstructure was obtained. For the sample doped with 10 mol% Y_2O_3, the microstructure was also porous, while the sample with 20 mol% was already high densified, as demonstrated in figure 9.14, confirming an enhanced densification behavior due to the high concentration of La_2O_3.

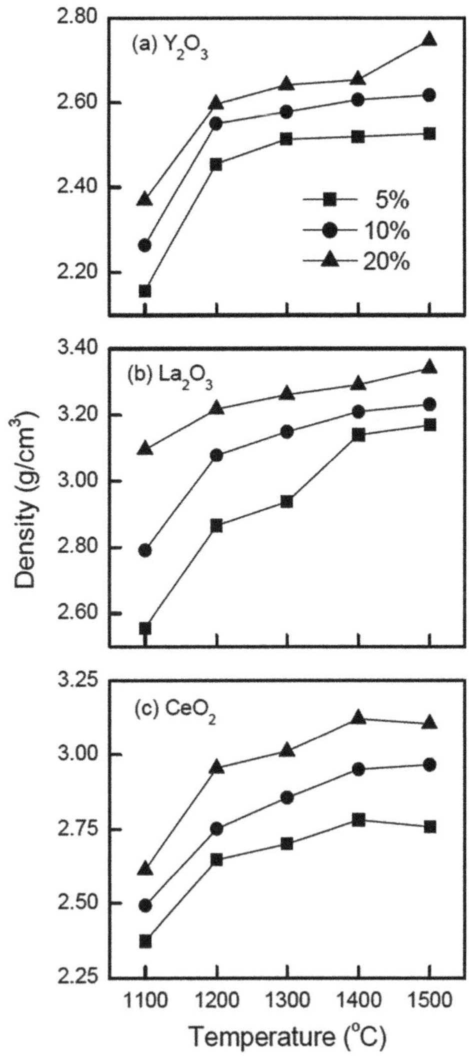

Figure 9.10. Densities of the three groups of mullite ceramics as a function of sintering temperature. Reproduced with permission from [1]. Copyright 2004 Elsevier.

The microstructure evolution of the samples doped with CeO_2 was similar to that of the samples doped with Y_2O_3 at low doping concentrations. Figure 9.15 shows the SEM images of the samples doped with 5 mol% CeO_2 after sintering at different temperatures for 4 h. The sample sintered at 1200 °C possessed an average mullite grain size of about 1 μm, which was only slightly increased with increasing sintering temperature thereafter. In addition, the sample with 5 mol% CeO_2 was porous even after sintering at 1500 °C. In contrast, the densification of the samples with higher concentrations of CeO_2 was largely enhanced. For example, the sample doped with 10 mol% CeO_2 sintered at 1200 °C was just slightly porous, while the sample sintered at 1400 °C was nearly fully densified without the observation of pores, as depicted in

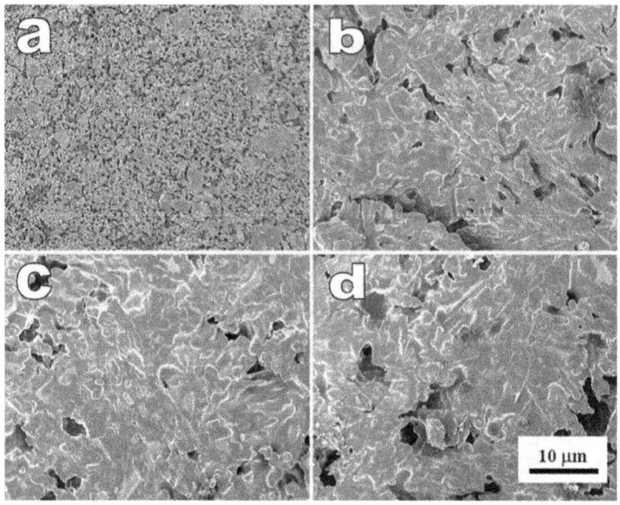

Figure 9.11. SEM images of the 5 mol% Y_2O_3-doped sample sintered at: (a) 1100 °C, (b) 1200 °C, (c) 1300 °C and (d) 1400 °C for 4 h. Reproduced with permission from [1]. Copyright 2004 Elsevier.

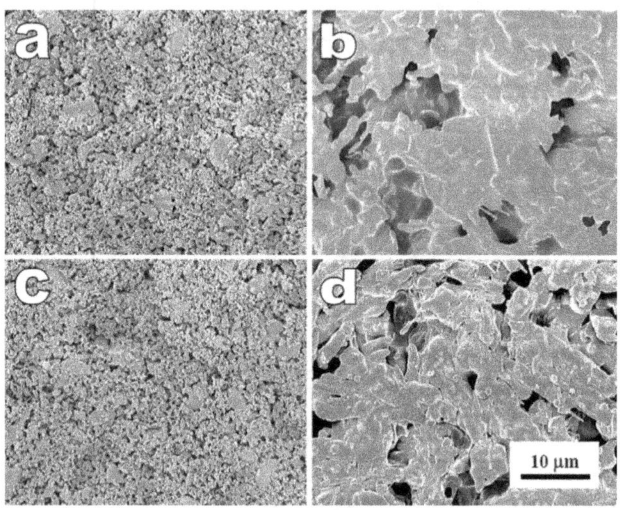

Figure 9.12. SEM images of the Y_2O_3-doped samples: (a) 10 mol%, 1100 °C, (b) 10 mol%, 1400 °C, (c) 20 mol%, 1100 °C and (d) 20 mol%, 1400 °C. Reproduced with permission from [1]. Copyright 2004 Elsevier.

figures 9.16(a) and (b). However, the samples doped with 20 mol% CeO_2 were fully densified after sintering at 1200 °C.

According the previous conclusions, the concentration of a dopant should not be over a critical value [10]. Otherwise, excessive glassy phase could be formed which could enclose the pores. It is very difficult for the enclosed pores to escape during sintering at high temperatures. As a consequence, the densification of the samples would be deaccelerated. However, this principle was only applicable to the doping of Y_2O_3. In contrast, when CeO_2 and La_2O_3 were used as the dopants, fully dense

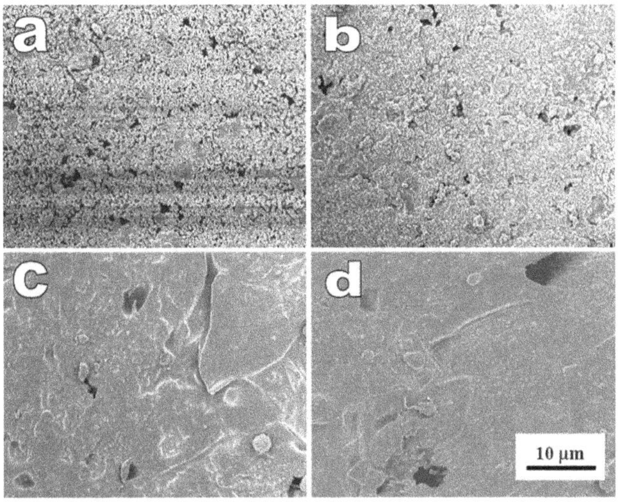

Figure 9.13. SEM images of the 5 mol% La_2O_3-doped sample sintered at: (a) 1100 °C, (b) 1200 °C, (c) 1300 °C and (d) 1400 °C for 4 h. Reproduced with permission from [1]. Copyright 2004 Elsevier.

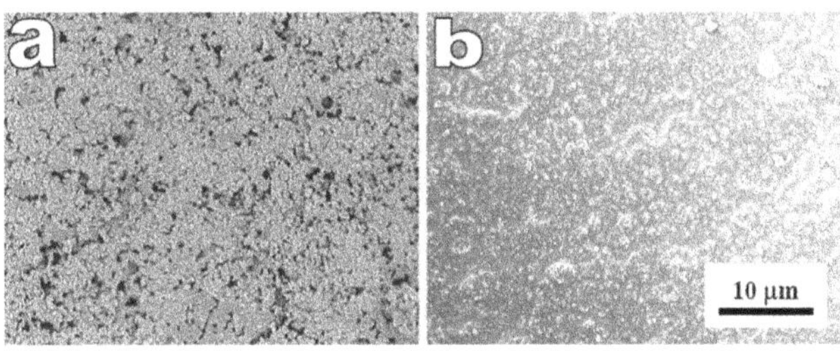

Figure 9.14. SEM images of the 1100 °C-sintered samples with La_2O_3 concentration: (a) 10 mol% and (b) 20 mol%. Reproduced with permission from [1]. Copyright 2004 Elsevier.

mullite ceramics were only achievable at high doping concentrations, as discussed above. The oxides showed different effects on the microstructure development of the mullite ceramics, which could be attributed to the difference in properties of the glassy phases due to their engagement.

The mixtures doped with the three oxides were also subject to mechanochemical activation with stainless steel and WC media. The effect of activation with stainless steel media was not as pronounced compared to that of the sample with WC media. The phase formation and microstructure evolution of the three groups of the samples from the mixtures activated with WC media will be discussed in this subsection. The consequent reactions of the samples were similar to those of the samples from the mixtures without mechanochemical activation, but the temperatures of mullite phase formation and phase transition were most likely reduced, owing to the significant refinement of the starting materials.

The XRD patterns of the samples doped with 5 mol% Y_2O_3 after sintering at different temperatures and those with different contents of Y_2O_3 after sintering at

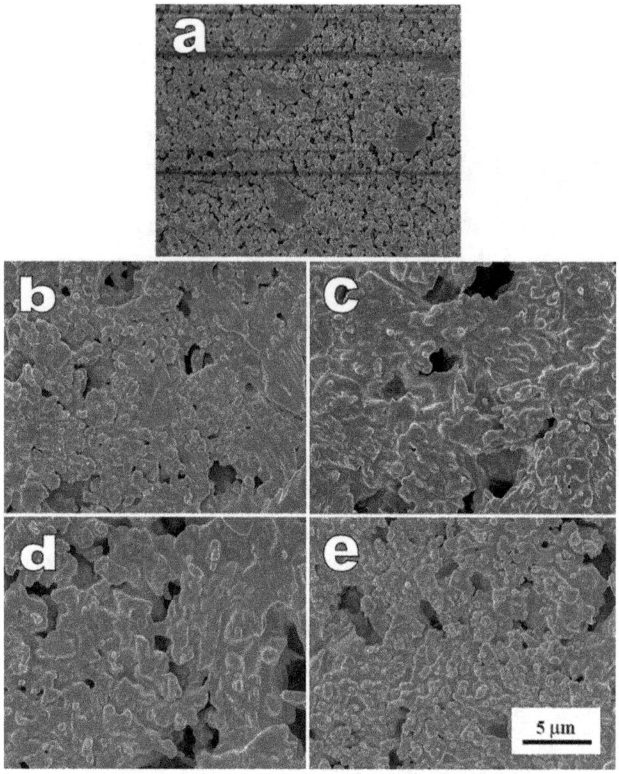

Figure 9.15. SEM images of the 5 mol% CeO_2 group sintered at: (a) 1100 °C, (b) 1200 °C, (c) 1300 °C, (d) 1400 °C and (e) 1500 °C for 4 h. Reproduced with permission from [1]. Copyright 2004 Elsevier.

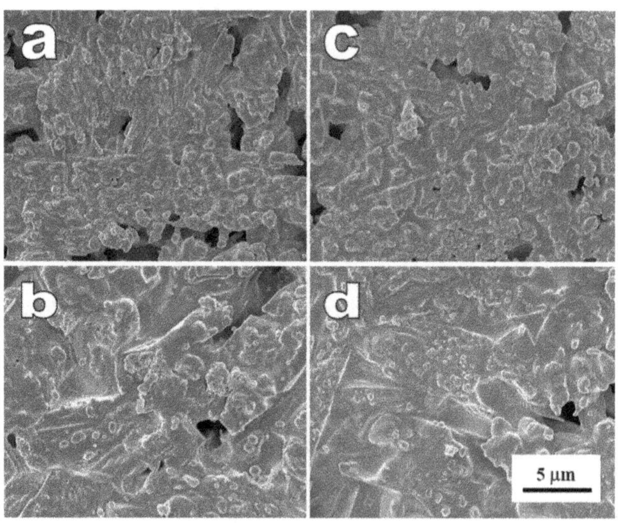

Figure 9.16. SEM images of the CeO_2-doped samples: (a) 10 mol%, 1200 °C, (b) 10 mol%, 1400 °C, (c) 20 mol%, 1200 °C and (d) 20 mol%, 1400 °C. Reproduced with permission from [1]. Copyright 2004 Elsevier.

1200 °C for 4 h are shown in figures 9.17 and 9.18, respectively. Similarly, the two secondary phases, $Y_2Si_2O_7$(1) and $Y_2Si_2O_7$(2), were formed in the sample doped with 5 mol% Y_2O_3 after sintering at 1100 °C. $Y_2Si_2O_7$(2) was stable up to 1500 °C, while $Y_2Si_2O_7$(1) disappeared after sintering at 1200 °C. This observation was similar to that for the samples from the unactivated mixtures. In addition, the mullite phase was formed at 1200 °C, which was the same as that for the unactivated samples. However, careful inspection revealed that there was a small difference in XRD patterns between the two samples with 5 mol% Y_2O_3 after sintering at 1200 °C. As observed in figures 9.2 and 9.18, the sample from the mixture without activation contained a trace of cristobalite, while no such phase was visible in the samples from

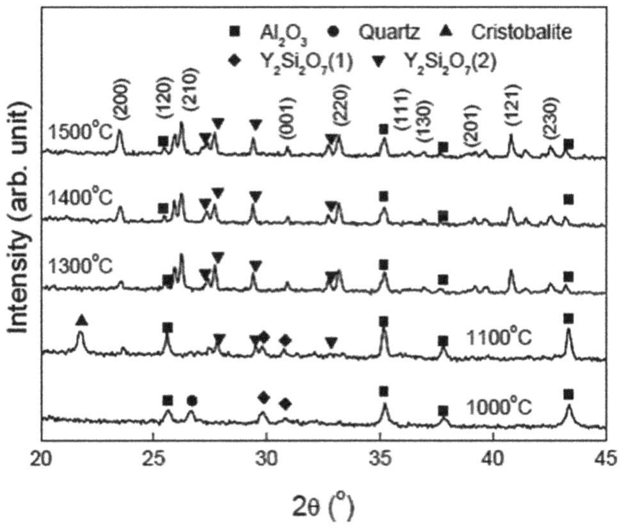

Figure 9.17. XRD patterns of the samples doped with 5 mol% Y_2O_3 from the mixture mechanochemically activated with WC milling media, after sintering at different temperatures for 4 h.

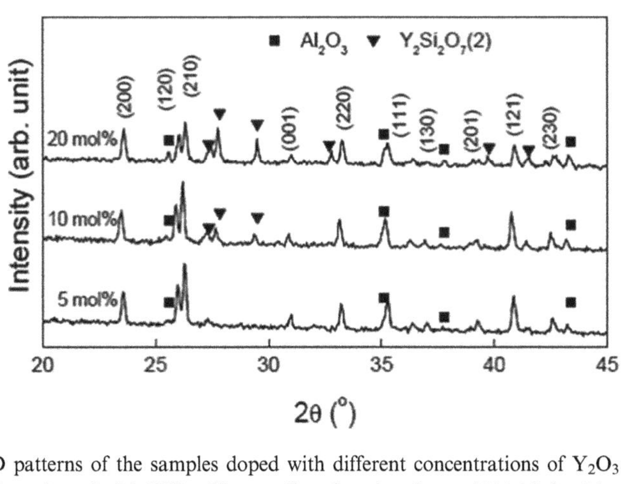

Figure 9.18. XRD patterns of the samples doped with different concentrations of Y_2O_3 from the mixtures mechanochemically activated with WC milling media, after sintering at 1200 °C for 4 h.

the mixture activated with WC media. Therefore, it could be concluded that the activation promoted the phase formation of mullite, but the effect was relatively weak compared to those observed in other material systems, as discussed previously. The underlying reason deserves to be clarified in future studies.

The effect of mechanochemical activation with WC media on the phase formation of the samples doped with La_2O_3 was similar. As shown in figures 9.19 and 9.20, the mullite phase was formed in the three samples after sintering at 1200 °C. In comparison, although mullitization also occurred in the unactivated mixture after sintering at 1200 °C, the mullite phase was in the minority. Although neither La_2O_3 nor La_2O_3 related new phases were detected in this group of samples, the presence of La_2O_3 was well reflected by the phase evolution and microstructure development, as discussed later. These results were similar to those for the samples from the mixtures without the application of mechanochemical activation.

The reduction in the temperature of mullite phase formation due to mechanochemical activation was the most significant in the samples doped with CeO_2. Figure 9.21 shows the XRD patterns of the samples doped with 20 mol% CeO_2 after sintering at different temperatures, while figure 9.22 depicts the XRD patterns of those with different concentrations of CeO_2 after sintering at 1200 °C. With mechanochemical activation, the sample doped with 20 mol% CeO_2 was already fully mullitized after sintering at 1100 °C, which was lower than that of the sample without the application of mechanochemical activation by 100 °C. Similarly, CeO_2 was clearly detected in this group of samples.

Densification curves of the three groups of samples are shown in figures 9.23–9.25. Compared to the samples without mechanochemical activation, the samples with mechanochemical activation exhibited poor densification behaviors, in

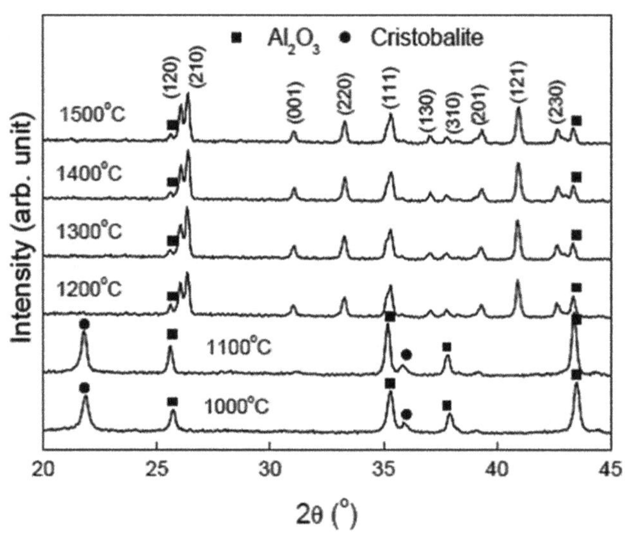

Figure 9.19. XRD patterns of the samples doped with 5 mol% La_2O_3 from the mixture mechanochemically activated with WC milling media after sintering at different temperatures for 4 h.

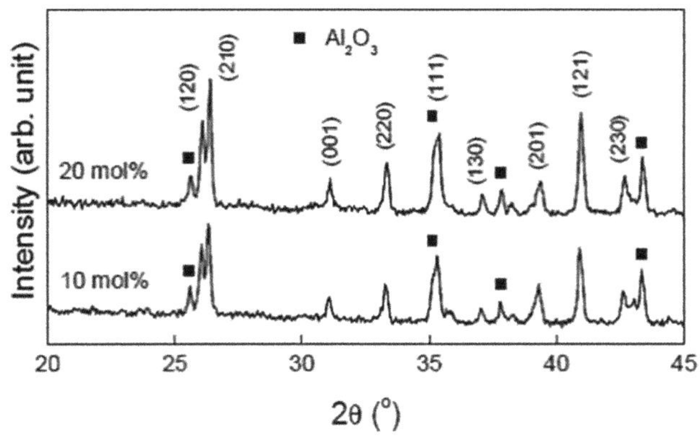

Figure 9.20. XRD patterns of the samples doped with 10 mol% and 20 mol% La_2O_3 from the mixtures mechanochemically activated with WC milling media after sintering at 1200 °C for 4 h.

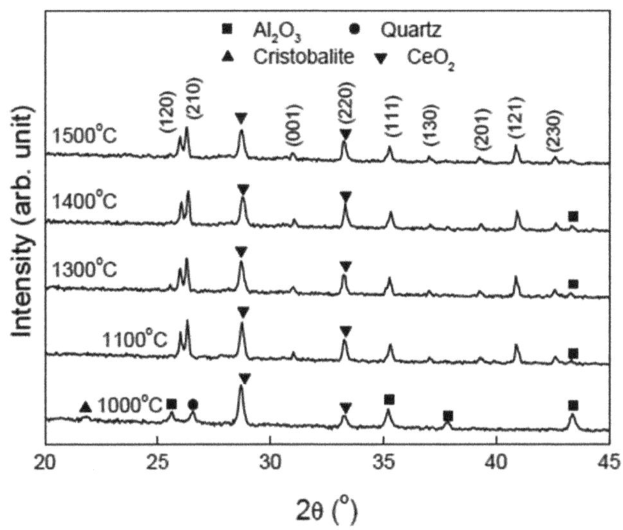

Figure 9.21. XRD patterns of the samples doped with 20 mol% CeO_2 from the mixture mechanochemically activated with WC media after sintering at different temperatures for 4 h.

particular for the two groups with Y_2O_3 and CeO_2. As shown in figures 9.23 and 9.7, the unactivated samples had final linear shrinkages of 10%–13%, the corresponding values of the activated sample were in the range 5%–7%. The final linear shrinkages of the samples doped with CeO_2 were less than 6%, as seen in figure 9.25. Although the maximum linear shrinkages of the samples doped with La_2O_3 were not greatly affected, the densification processes were complicated at high temperatures, as demonstrated in figure 9.24. The decreased densification rates and the variation in densification profiles of the activated mixtures were attributed to the occurrence of anisotropic grain growth in the samples, as discussed in the following section.

The microstructures of the samples from the activated mixtures were distinctly different from those that were derived from the mixtures without mechanochemical

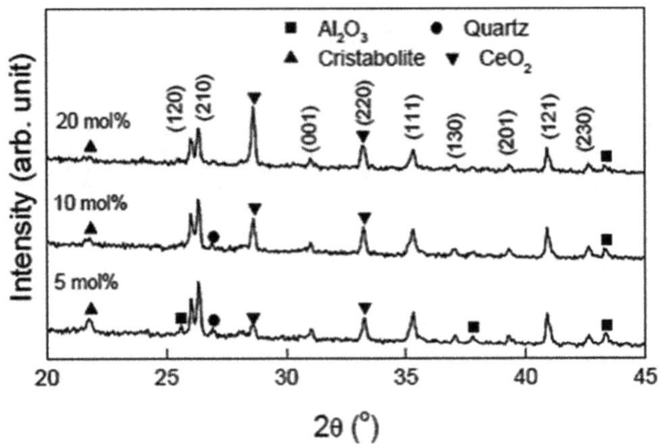

Figure 9.22. XRD patterns of the samples doped with different concentrations of CeO_2 from the mixtures mechanochemically activated with WC media after sintering at 1200 °C for 4 h.

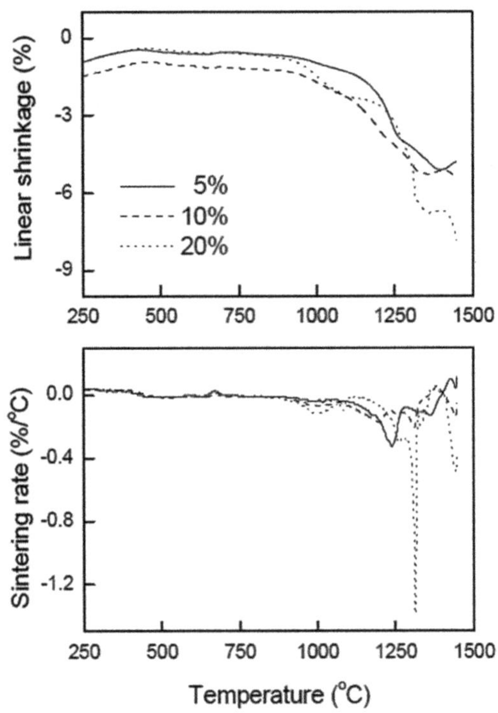

Figure 9.23. Densification behaviors of the mixtures doped with different contents of Y_2O_3 and mechanochemically activated with WC milling media.

activation. Figure 9.26 shows SEM images of the samples doped with 5 mol% Y_2O_3 from the mixture mechanochemically activated with WC milling media after sintering at different temperatures. The sample sintered at 1100 °C was porous with spherical particles/grains, in which the mullite phase had not formed. After

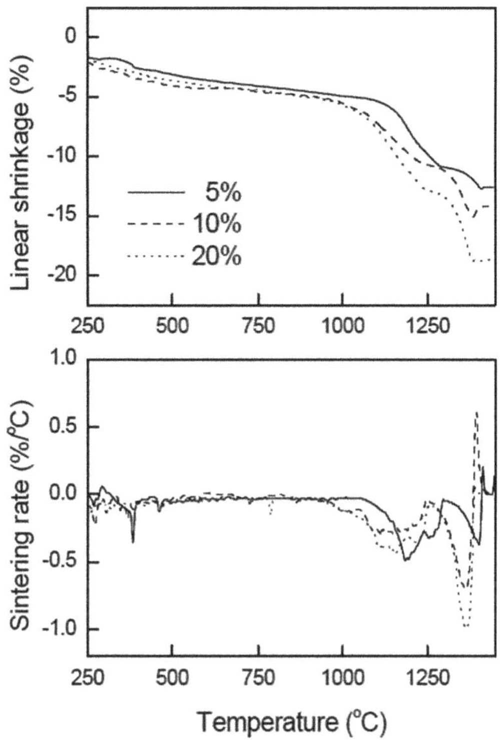

Figure 9.24. Densification behaviors of the mixtures doped with different contents of La_2O_3 and mechanochemically activated with WC milling media.

sintering at 1200 °C whisker-like grains appeared, while the sample was still porous. At this temperature the mullite phase was obtained, which meant that anisotropic grain growth took place, although the morphology of the whiskers was not very regular. Such anisotropic grains were not observed in the samples from the unactivated mixtures.

In addition, a bimodal size distribution was observed in this sample, where small whiskers with a relatively low quantity were mixed with rod-like grains with larger thicknesses. Due to the occurrence of the anisotropic grain growth, densification of the samples was retarded, which was responsible for the poor densification behavior, as illustrated in figure 9.23. As the sintering temperature was further increased to 1300 °C and 1400 °C, the small whiskers disappeared and the dimensions of the rod-like grains were slightly increased. At the same time, the samples became more densified in their microstructure.

The trend in microstructure development of the samples doped with 10 mol% Y_2O_3 was similar to that with 5 mol% Y_2O_3, as revealed in figure 9.27. However, after sintering at 1200 °C for 4 h, the bimodal size distribution of the sample was much less pronounced. Almost no thin whiskers were present in the SEM image of the sample, which means that the addition of Y_2O_3 promoted mullite grain growth. In addition, the samples sintered at higher temperatures showed more dense

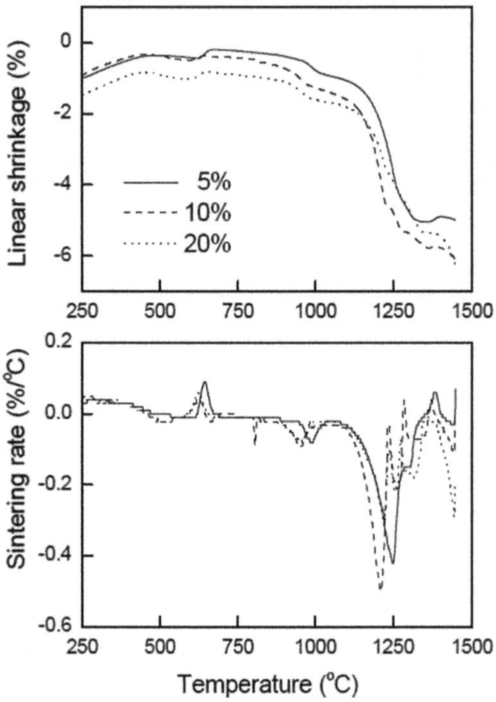

Figure 9.25. Densification behaviors of the mixtures doped with different contents of CeO_2 and mechanochemically activated with WC milling media.

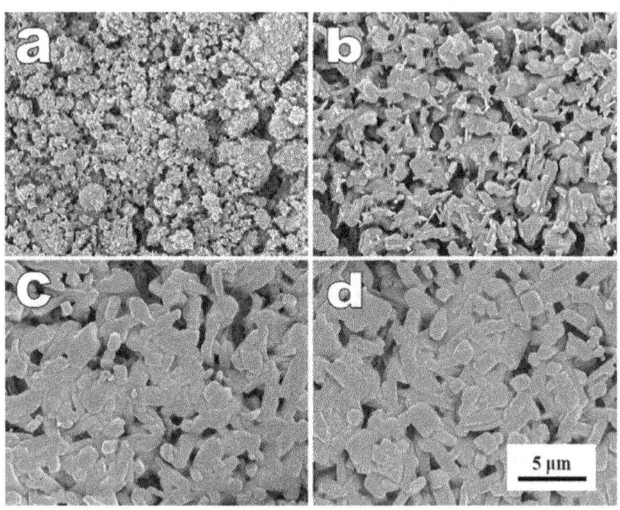

Figure 9.26. SEM images of the samples doped with 5 mol% Y_2O_3 from the mixture mechanochemically activated with WC milling media, after sintering at different temperatures for 4 h: (a) 1100 °C, (b) 1200 °C, (c) 1300 °C and (d) 1400 °C.

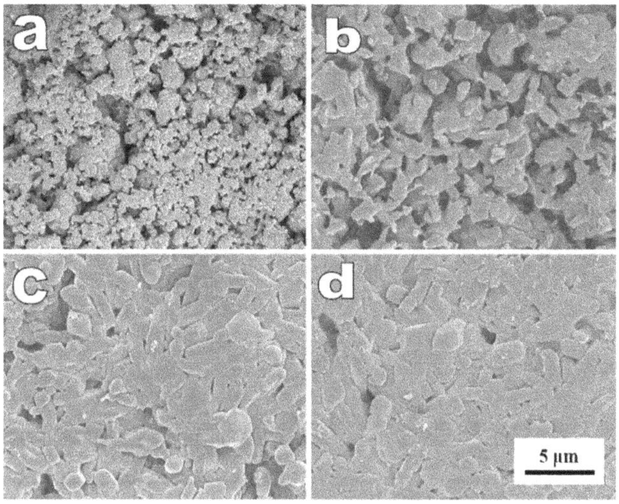

Figure 9.27. SEM images of the samples doped with 10 mol% Y_2O_3 from the mixture mechanochemically activated with WC milling media, after sintering at different temperatures for 4 h: (a) 1100 °C, (b) 1200 °C, (c) 1300 °C and (d) 1400 °C.

microstructures, as observed in figures 9.27(c) and (d). In addition, the mullite grains were approximately spherical.

However, the samples doped with 20 mol% Y_2O_3 exhibited different microstructures compared to the two groups of samples with lower contents of Y_2O_3, as shown in figure 9.29. The sample sintered at 1200 °C was also porous, but the mullite grains were not isolated, as illustrated in figure 9.28(b). It seemed that the mullite grains were fused with glassy phases. Higher sintering temperatures resulted in a microstructure consisting of pseudo-spherical particles with irregular morphologies. With increasing sintering temperature, the amount of the glassy phase was reduced, while the sample became more and more densified.

SEM images of the samples doped with different concentrations of La_2O_3 from the mixtures mechanochemically activated with WC milling media, after sintering at different temperatures, are shown in figures 9.29–9.31. In this group, the samples sintered at 1000 °C and 1200 °C exhibited a similar microstructure, with spherical grains/particles. The samples sintered at 1000 °C were highly porous, while those sintered at 1100 °C displayed a low degree of sintering. The samples doped with 5 mol% La_2O_3 after sintering at 1200 °C–1400 °C consisted of cuboid-like grains, with lengths of 5–10 μm and thicknesses of about 3 μm, as observed in figures 9.29(c)–(e). In this case, anisotropic grain growth was also achieved, but to a relatively lower degree. The cuboid grains were similar to rectangular mullite microtubes that were synthesized using a sol–gel reaction method [14]. However, the reaction conditions in this case were entirely different. Therefore, the formation mechanism of the mullite cuboid grains should be further studied.

In this temperature range, with increasing sintering temperature, the morphology of the cuboid grains gradually became irregular. After sintering at 1500 °C the

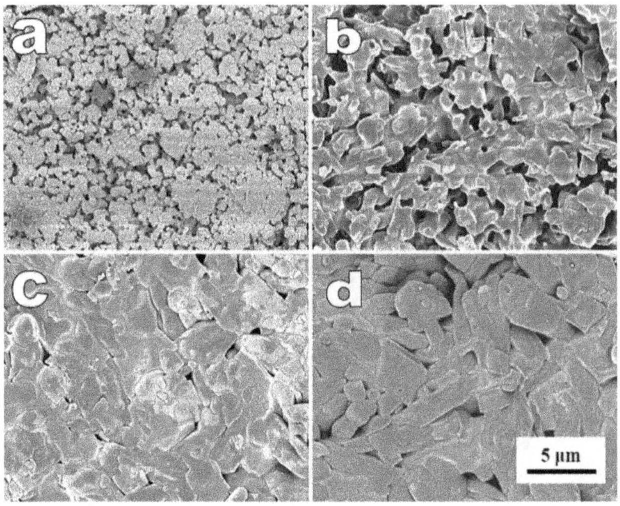

Figure 9.28. SEM images of the samples doped with 20 mol% Y_2O_3 from the mixture mechanochemically activated with WC milling media, after sintering at different temperatures for 4 h: (a) 1100 °C, (b) 1200 °C, (c) 1300 °C and (d) 1400 °C.

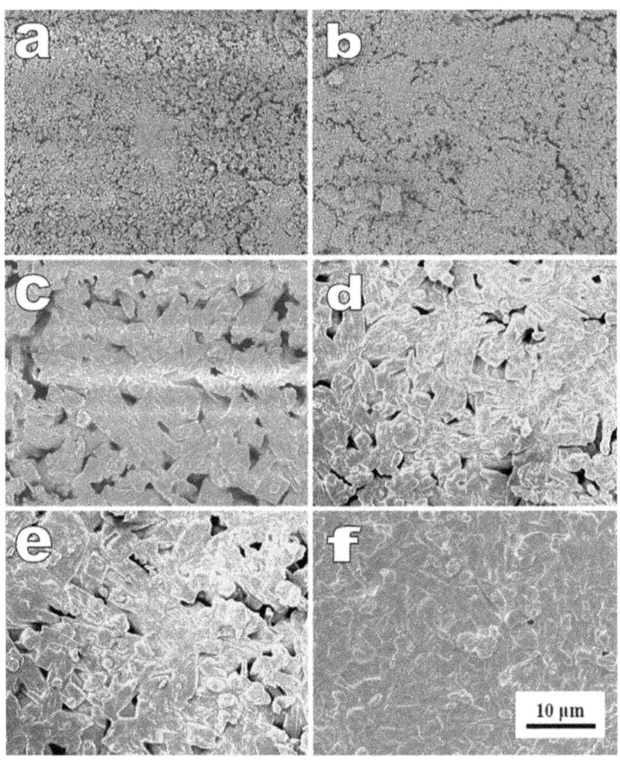

Figure 9.29. SEM images of the samples doped with 5 mol% La_2O_3 from the mixture mechanochemically activated with WC milling media, after sintering at different temperatures for 4 h: (a) 1000 °C, (b) 1100 °C, (c) 1200 °C (d) 1300 °C, (e) 1400 °C and (f) 1500 °C.

sample was nearly fully densified, as revealed in figure 9.29(f). However, such grain morphology variation was not observed in the samples derived from the unactivated mixtures. Although the cuboid grains were also present in the samples doped with 10 mol% and 20 mol% La_2O_3, they were all fully densified, without visible pores. Therefore, the development of the microstructures of the samples was in good agreement with their densification behaviors.

The samples doped with CeO_2 experienced typical anisotropic grain growth. SEM images of the sample doped with different concentrations of CeO_2 from the mixtures mechanochemically activated with WC milling media are shown in figures 9.32–9.34. After sintering at 1000 °C, the three samples were all porous and had spherical particles, whereas mullitization had not started. In the samples sintered at 1100 °C, a bimodal size distribution was observed, corresponding to mullite phase formation at this temperature. According to the XRD patterns shown in figure 9.22, the addition of CeO_2 accelerated the formation of mullite. However, a high concentration of CeO_2 corresponded to more irregular mullite whiskers. This observation was inconsistent with the previous statement, i.e. accelerated mullite phase formation would result in stronger anisotropic grain growth. Therefore, the competition between densification and anisotropic grain growth was also dependent on the content of secondary phases.

9.2 The effects of other oxides

The effects of alkaline-earth oxides, including MgO, CaO, SrO and BaO, on the formation of mullite phase and the morphological development of mullite ceramics from the mixtures of oxides, which were mechanochemically activated using high-energy milling process have been studied [15]. It was found that the mullite phase formation temperature was increased as the dopants were gradually varied from MgO to BaO. Well-developed mullite whiskers were formed in the samples doped with MgO, while spherical grains were formed in the samples doped with CaO, SrO and BaO. These interesting results could also be understood by considering the mullitization mechanism of dissolution–precipitation. Commercial powders of SiO_2 (quartz) and Al_2O_3 were used for mullite, while MgO, CaO, $SrCO_3$ and $BaCO_3$ powders were employed as the dopants. The nominal composition was $(3Al_2O_3{\cdot}2SiO_2)_{0.9}(MO)_{0.1}$ (M = Mg, Ca, Sr and Ba). WC vials and balls were used as the milling media, with milling parameters that have been described previously.

The XRD patterns of the samples doped with different alkaline-earth oxides after sintering at different temperatures are shown in figures 9.35–9.38. The four groups of samples had similar XRD patterns after sintering at 1000 °C, with Al_2O_3 and quartz being the major phases. For the groups with MgO and CaO, cristobalite was detected, implying that the phase transformation from quartz to cristobalite took place at this temperature. In comparison, cristobalite was not present in the samples doped with SrO and BaO. Therefore, the four oxides exhibited different effects on the quartz–cristobalite phase transition. After sintering at 1100 °C for 4 h, the mullite phase was formed in the sample doped with MgO, although the mullitization process was only half complete. In contrast, the mullite phase was not visible in the

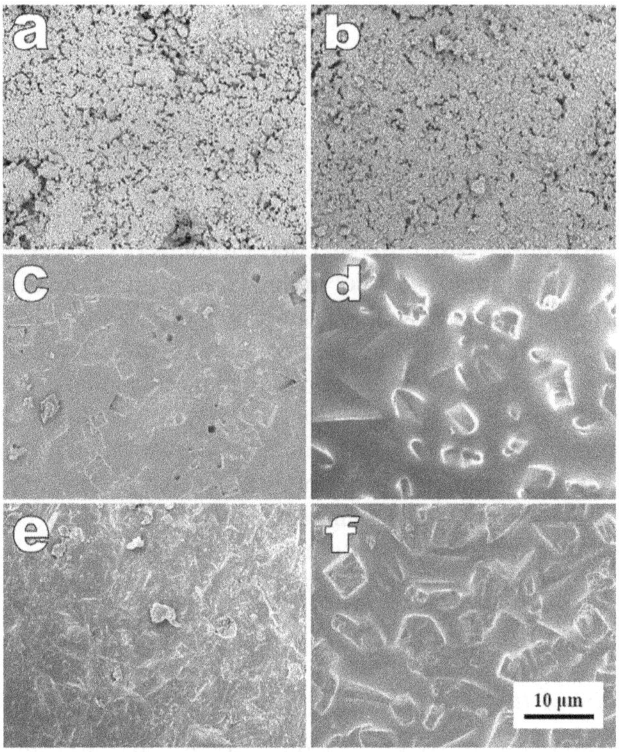

Figure 9.30. SEM images of the samples doped with 10 mol% La_2O_3 from the mixture mechanochemically activated with WC milling media, after sintering at different temperatures for 4 h: (a) 1000 °C, (b) 1100 °C, (c) 1200 °C (d) 1300 °C, (e) 1400 °C and (f) 1500 °C.

other three groups after sintering at this temperature. It is also noted that the phase transformation from quartz to cristobalite is already complete.

As the sintering temperature was increased to 1200 °C, mullitization was nearly finished in the sample with MgO. In the samples doped with CaO and SrO, mullite became the major phase, while the reactants were still the major phases in the sample doped with BaO. Therefore, CaO, SrO and BaO displayed a negative effect on the phase formation of mullite from the oxide mixtures. The effect was in the order CaO < SrO < BaO. At temperatures ⩾1300 °C, the XRD patterns of all the samples were almost the same and nearly unchanged with increasing temperature.

Figure 9.39 shows the densification behaviors of the activated mixtures doped with different alkaline-earth oxides. Obvious densification started at about 1200 °C, where the four samples demonstrated a similar densification character. However, significant differences were present between them as the temperature was over 1200 °C. For the sample doped with MgO, the final linear shrinkage was leveled off at about 6%. The maximum linear shrinkages of the samples doped with CaO, SrO and BaO were 10.5%, 15.9% and 18.5%, respectively. In addition, the happenings at high temperatures were different for the four groups of samples. For example, the sample with MgO showed an obvious fluctuation in the linear shrinkage curve, while no such

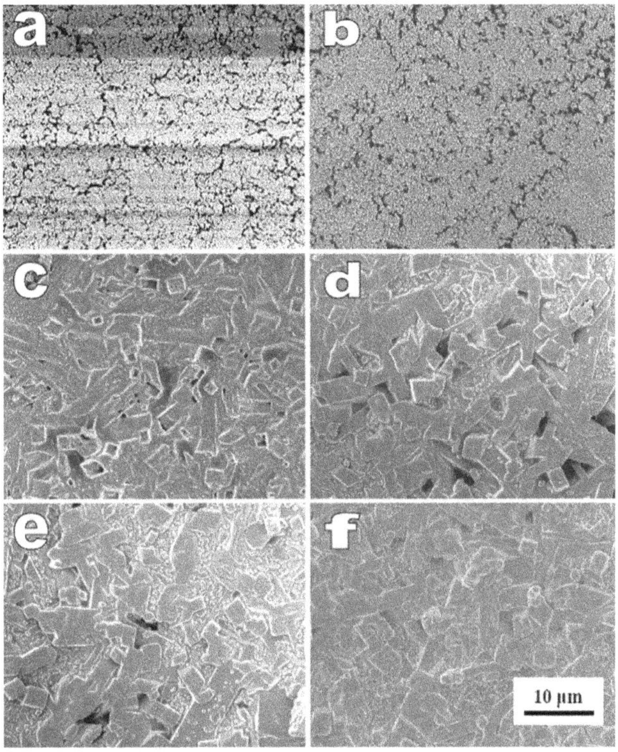

Figure 9.31. SEM images of the samples doped with 20 mol% La_2O_3 from the mixture mechanochemically activated with WC milling media, after sintering at different temperatures for 4 h: (a) 1000 °C, (b) 1100 °C, (c) 1200 °C (d) 1300 °C, (e) 1400 °C and (f) 1500 °C.

fluctuation was visible for the sample with BaO. The samples doped with CaO and SrO fell just in between. A clear trend of MgO, CaO, SrO and BaO was observed, which was in good agreement with the XRD results. The multiple peaks in the linear shrinkage rate curves reflected that the final densification processes of the mixtures were complicated, which could be governed by different mechanisms.

To further reveal the effects of the alkaline-earth oxides, samples with different concentrations of the dopants were studied. SEM images of the samples doped with different contents of the alkaline-earth oxides, after sintering at 1300 °C for 4 h, are depicted in figures 9.40–9.43. The samples doped with 1 wt% alkaline-earth oxides all consisted of mullite whiskers, but the whiskers were highly sintered in the sample with 1 wt% BaO. For a given alkaline-earth oxide, the morphology of the mullite whiskers became less and less regular, with increasing doping concentration. This result further confirmed the trend in the effect of the alkaline-earth oxides. Therefore, it could be concluded that the doping of alkaline-earth oxides had a negative effect on the formation of mullite whiskers.

According to the dissolution–precipitation mechanism, a silica-rich liquid-like layer was produced at relatively low temperatures in Al_2O_3–SiO_2 binary systems [16]. Al_2O_3 grains were dissolved in the liquid phase, followed by nucleation and

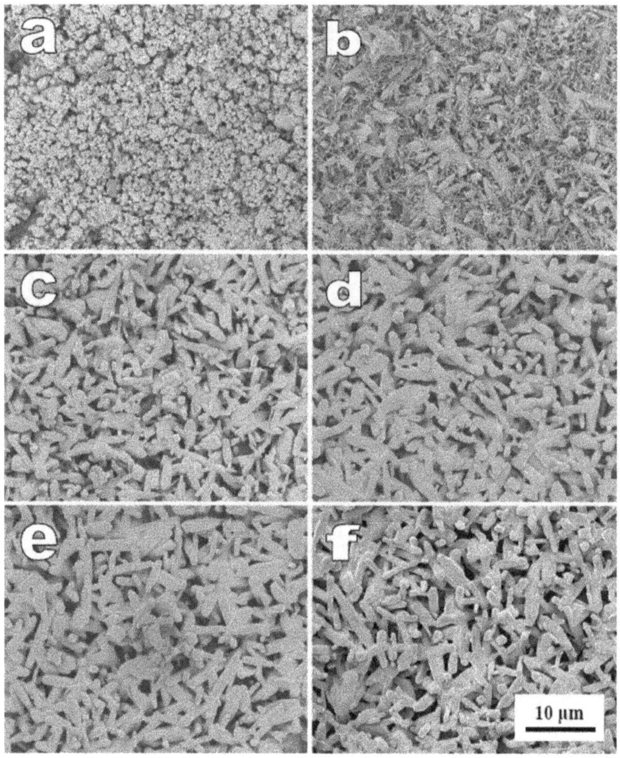

Figure 9.32. SEM images of the samples doped with 5 mol% CeO_2 from the mixture mechanochemically activated with WC milling media, after sintering at different temperatures for 4 h: (a) 1000 °C, (b) 1100 °C, (c) 1200 °C (d) 1300 °C, (e) 1400 °C and (f) 1500 °C.

grain growth of mullite, when the concentration of Al_2O_3 arrived at a critical value. In addition, this liquid-like layer also promoted the densification of the samples as a sintering aid, which was called viscous transient flow sintering. The competition between mullite phase formation and densification determined the microstructural development of the samples. If mullite phase formation was advanced over densification, anisotropic grain growth was allowed, so that mullite whiskers could be obtained. Otherwise, spherical grains were formed and the samples could be fully densified.

Between these two extreme cases, i.e. full densification with spherical grains and entirely anisotropic grain growth with perfect mullite whiskers, there have also been many transient states, as discussed previously. The mullite whiskers could adopt different morphologies. In this case, it was reasonable to suggest that the alkaline-earth oxides altered the properties of the SiO_2-rich liquid-like phase to different degrees, with MgO and BaO being the two end cases. The $MgO–Al_2O_3–SiO_2$ ternary system was close to the high-energy activated $Al_2O_3–SiO_2$ binary, while the $BaO–Al_2O_3–SiO_2$ system behaved in a similar way to the $Al_2O_3–SiO_2$ system without mechanochemical activation. In other words, the microstructure and grain morphology of the mullite

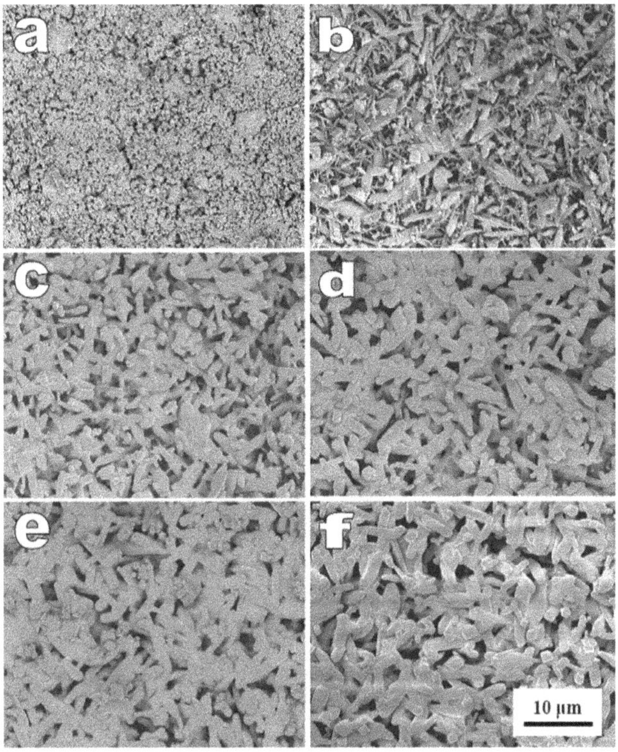

Figure 9.33. SEM images of the samples doped with 10 mol% CeO_2 from the mixture mechanochemically activated with WC milling media, after sintering at different temperatures for 4 h: (a) 1000 °C, (b) 1100 °C, (c) 1200 °C (d) 1300 °C, (e) 1400 °C and (f) 1500 °C.

ceramics could be finely tuned using different alkaline-earth oxides and adjusting their doping concentrations.

The effects of the three main group oxides, including SnO_2, Sb_2O_3 and Bi_2O_3, on the phase composition and microstructural development of mullite ceramics from the mixtures of Al_2O_3 and SiO_2 have been studied [17]. The addition of SnO_2 had a negligible effect on mullitization, while that of Sb_2O_3 and Bi_2O_3 boosted the formation of mullite, with Bi_2O_3 being more effective than Sb_2O_3 and SnO_2 having no reaction with the reactants, while Sb_2O_3 reacted with Al_2O_3 to form a new phase $AlSbO_4$ at 1100 °C, which disappeared at 1200 °C. Bi_2O_3 was not detectable in the XRD patterns. Bi_2O_3 had the strongest effect on densification among the three dopants. Commercial oxides of quartz, Al_2O_3, SnO_2, Sb_2O_3 and Bi_2O_3 powders, were mixed, with nominal compositions of $(3Al_2O_3{\cdot}2SiO_2)_{1-x}(MO)_x$ (MO = SnO_2, Sb_2O_3 and Bi_2O_3), with $x = 0.05$, 0.10 and 0.20.

Figure 9.44 shows the XRD patterns of the samples doped with different contents of SnO_2 after they were sintered at 1400 °C for 4 h. The mullite phase was detected in the sample doped with 5 mol% SnO_2 as a minor phase after sintering at 1300 °C. Cristobalite and Al_2O_3 were present after sintering at 1500 °C. Once the mullite phase was formed, there was no obvious difference in the XRD patterns, but the

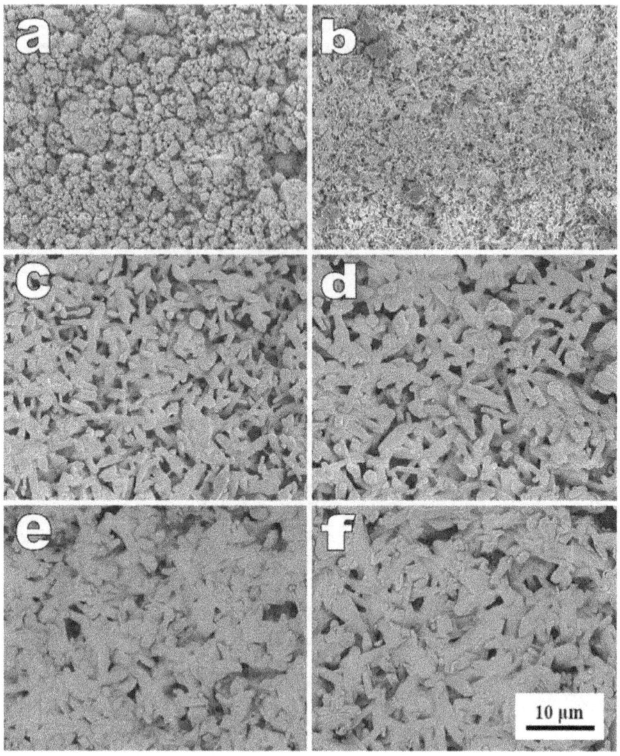

Figure 9.34. SEM images of the samples doped with 20 mol% CeO_2 from the mixture mechanochemically activated with WC milling media, after sintering at different temperatures for 4 h: (a) 1000 °C, (b) 1100 °C, (c) 1200 °C (d) 1300 °C, (e) 1400 °C and (f) 1500 °C.

peak intensity of SnO_2 was clearly increased. Therefore, SnO_2 was not engaged in the phase formation of mullite from Al_2O_3 and SiO_2.

The positive effect of Sb_2O_3 on mullite phase formation was confirmed by the XRD results. A new compound $AlSbO_4$ (JCPDS card no. 4–564) was detected in the XRD patterns, due to the oxidation of Sb^{3+} to Sb^{5+}. In addition, a trace of mullite was present, indicating that mullitization began at this temperature. Figure 9.45 shows the samples doped with different contents of Sb_2O_3 after sintering at 1300 °C. The mullitization process was enhanced as the content of Sb_2O_3 was increased, evidenced by the gradual reduction in the peak intensities of Al_2O_3 and the disappearance of the diffraction peaks of cristobalite in the sample with 20 mol% Sb_2O_3.

Figure 9.46 shows XRD patterns of the samples doped with different contents of Bi_2O_3, after sintering at 1200 °C for 4 h. Mullite was already formed as the dominant phase in the 5 mol% sample, while an enhanced mullitization process was observed in the 10 mol% sample. Almost single-phase mullite was obtained in the 20 mol% sample. In fact, mullitization was nearly complete in the sample doped with 20 mol% Bi_2O_3 after sintering at 1100 °C. Therefore, the effect of Bi_2O_3 on the mullitization was stronger than that of Sb_2O_3. Additionally, Bi_2O_3 was not visible in all the XRD patterns.

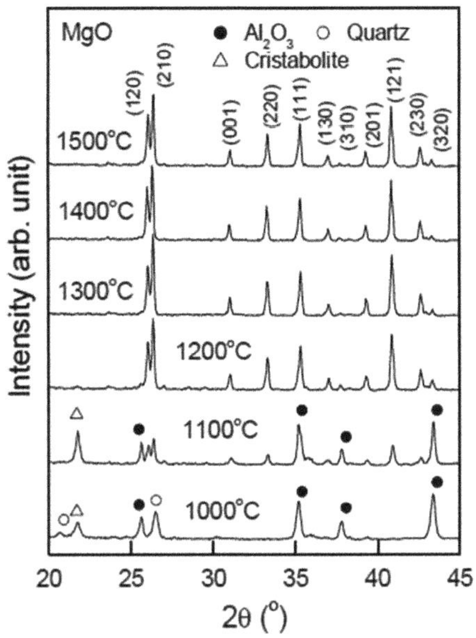

Figure 9.35. XRD patterns of the samples doped with MgO after sintering at different temperatures for 4 h. Reproduced with permission from [15]. Copyright 2004 Elsevier.

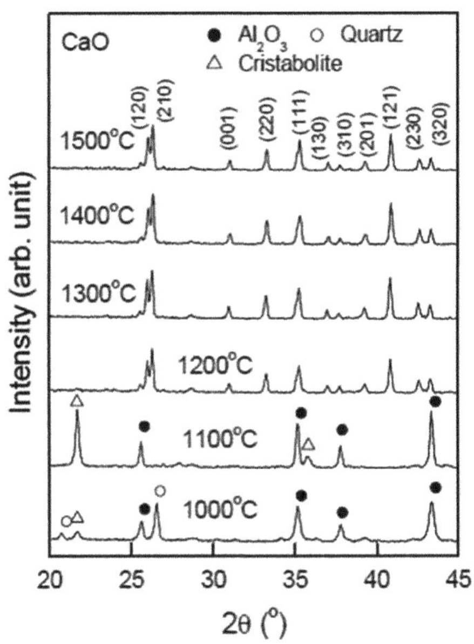

Figure 9.36. XRD patterns of the samples doped with CaO after sintering at different temperatures for 4 h. Reproduced with permission from [15]. Copyright 2004 Elsevier.

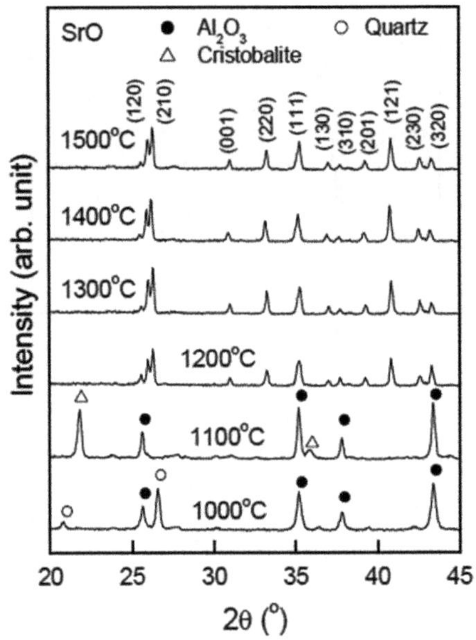

Figure 9.37. XRD patterns of the samples doped with SrO after sintering at different temperatures for 4 h. Reproduced with permission from [15]. Copyright 2004 Elsevier.

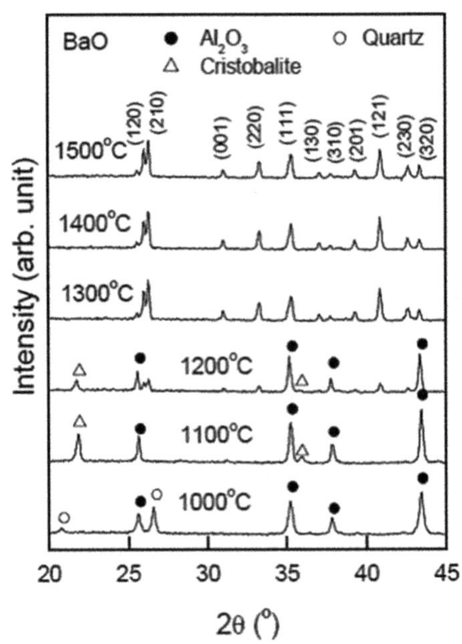

Figure 9.38. XRD patterns of the samples doped with BaO after sintering at different temperatures for 4 h. Reproduced with permission from [15]. Copyright 2004 Elsevier.

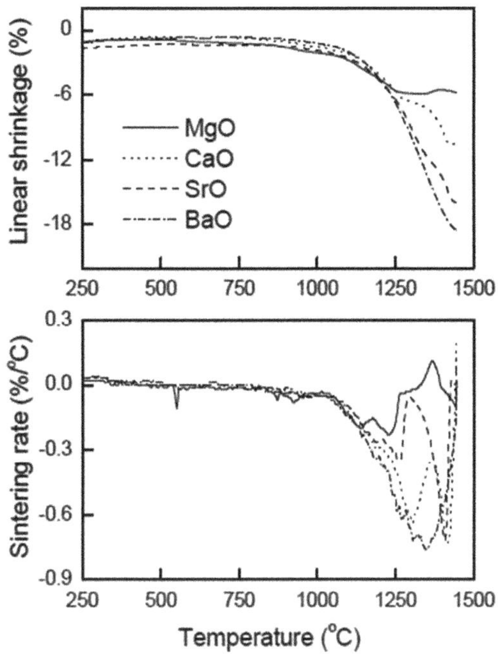

Figure 9.39. Densification behaviors of the mixtures doped with different alkaline-earth oxides after being subjected to mechanochemical activation with WC milling media for 5 h. Reproduced with permission from [15]. Copyright 2004 Elsevier.

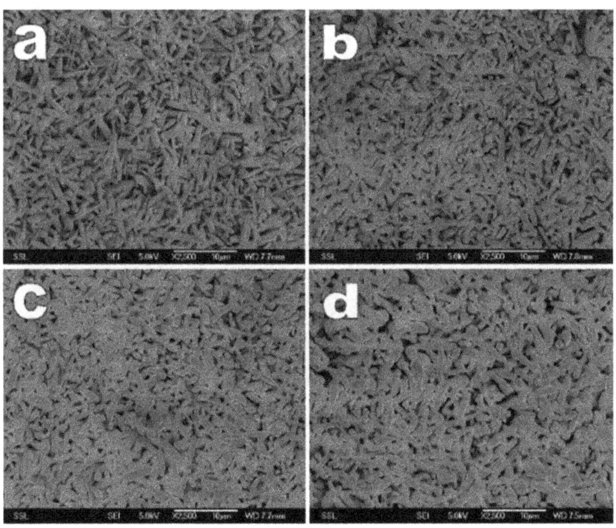

Figure 9.40. SEM images of the samples doped with different concentrations of MgO from the mixtures mechanochemically activated with WC milling media, after sintering at 1300 °C for 4 h: (a) 1 wt%, (b) 3 wt%, (c) 5 wt% and (d) 7 wt%.

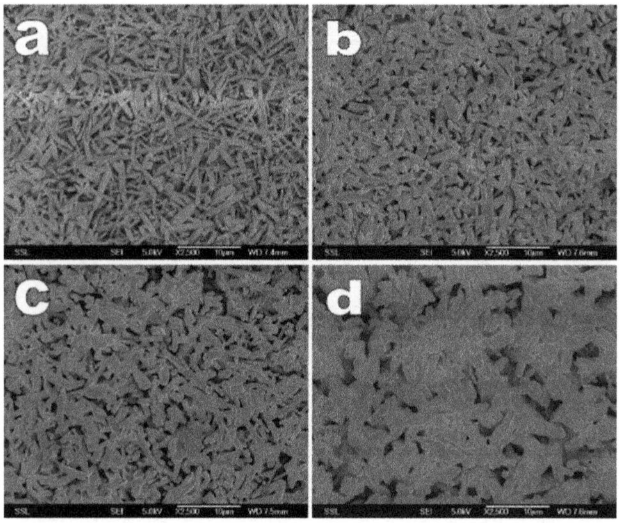

Figure 9.41. SEM images of the samples doped with different concentrations of CaO from the mixtures mechanochemically activated with WC milling media, after sintering at 1300 °C for 4 h: (a) 1 wt%, (b) 3 wt%, (c) 5 wt% and (d) 7 wt%.

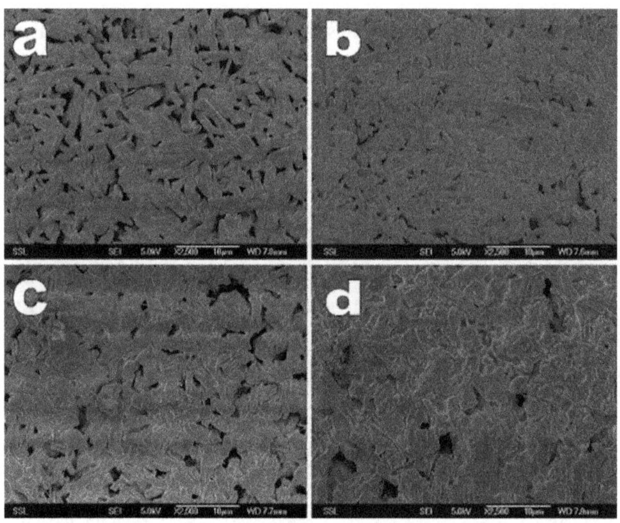

Figure 9.42. SEM images of the samples doped with different concentrations of SrO from the mixtures mechanochemically activated with WC milling media, after sintering at 1300 °C for 4 h: (a) 1 wt%, (b) 3 wt%, (c) 5 wt% and (d) 7 wt%.

It was reported that Sn^{4+} was able to enter the structure of the mullite, forming a solid solution with a mole fraction of tin of about 5% [18]. If this is true, the absence of SnO_2 in the XRD patterns of the samples doped with 5 mol% SnO_2 could be explained. Sb_2O_3 and Bi_2O_3 have relatively low melting points of 665 °C and 825 °C [19, 20], respectively. Both Sb_2O_3 and Bi_2O_3 are strongly capable of forming glasses

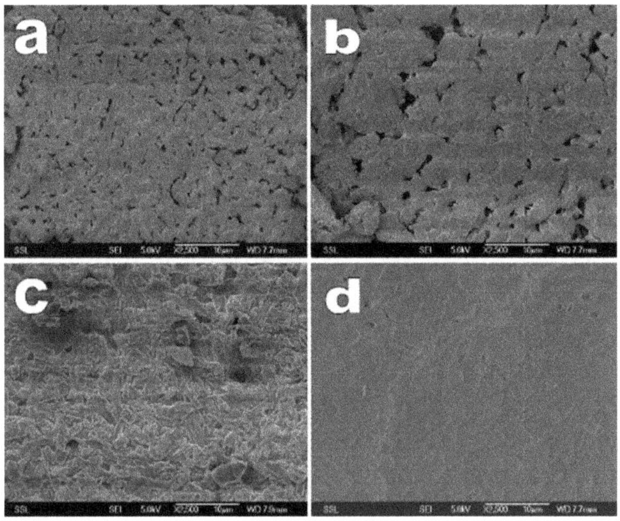

Figure 9.43. SEM images of the samples doped with different concentrations of BaO from the mixtures mechanochemically activated with WC milling media, after sintering at 1300 °C for 4 h: (a) 1 wt%, (b) 3 wt%, (c) 5 wt% and (d) 7 wt%.

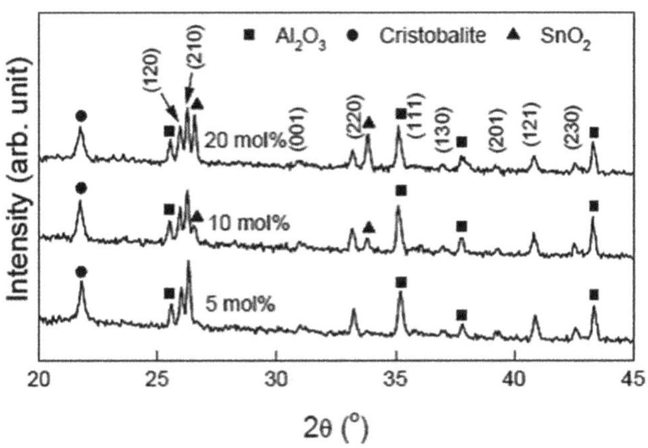

Figure 9.44. XRD patterns of the 1400 °C-sintered samples with different SnO_2 levels. Reproduced with permission from [17]. Copyright 2004 Elsevier.

[21]. Therefore, they could promote the formation of the liquid phase in mixtures, thus enhancing the phase formation of mullite.

Figure 9.47 shows SEM images of the samples doped with 5 mol% SnO_2 after sintering at 1100 °C–1500 °C for 4 h. The sample sintered at 1100 °C was porous, with large particles being embedded in the matrix of small grains, which could be from the starting materials. The large particles were quartz, which disappeared after sintering at 1200 °C due to the phase transition to cristobalite. After sintering at 1300 °C, the sample was covered by an amorphous-like layer in some surface areas, which could be related to the formation of a SiO_2-rich liquid phase. As the sintering

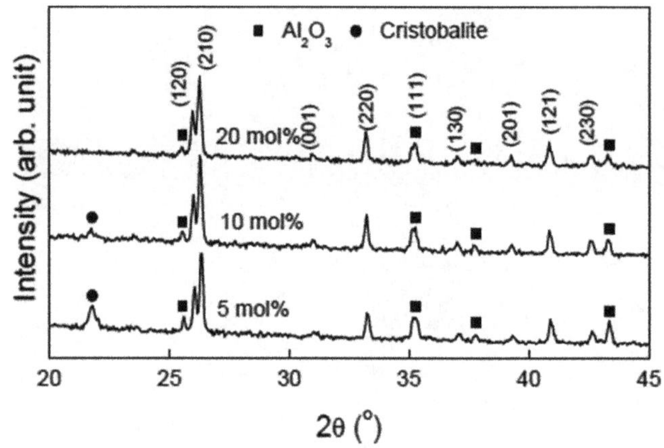

Figure 9.45. XRD patterns of the 1300 °C-sintered samples with different Sb_2O_3 contents. Reproduced with permission from [17]. Copyright 2004 Elsevier.

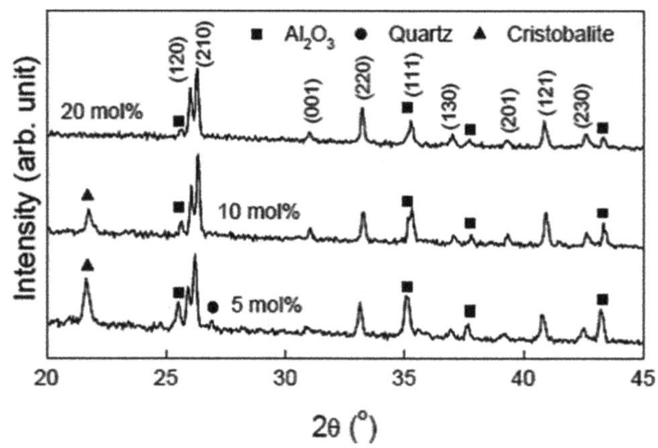

Figure 9.46. XRD patterns of the 1200 °C-sintered samples containing different amounts of Bi_2O_3. Reproduced with permission from [17]. Copyright 2004 Elsevier.

temperature was increased to 1400 °C, the grain size was slightly increased. The sample sintered at 1500 °C had an obvious variation in microstructure, where anisotropic grains were formed. The samples with 10 mol% and 20 mol% SnO_2 displayed a similar microstructure, as illustrated in figure 9.48.

Figure 9.49 shows SEM images of the samples doped with 5 mol% Sb_2O_3 after sintering at different temperatures for 4 h. The microstructural development of the samples doped with Sb_2O_3 was similar to that of the samples doped with SnO_2. However, the anisotropic-like structures were observed at lower temperatures, as shown in figures 9.49(b)–(d). At the same time, grain size was greatly increased due to the addition of Sb_2O_3, as revealed in figure 9.50. This was closely related to the low melting point of Sb_2O_3, as stated above.

As a representative example, figure 9.51 shows SEM images of the samples doped with 20 mol% Bi_2O_3 after they were sintered at different temperatures. The

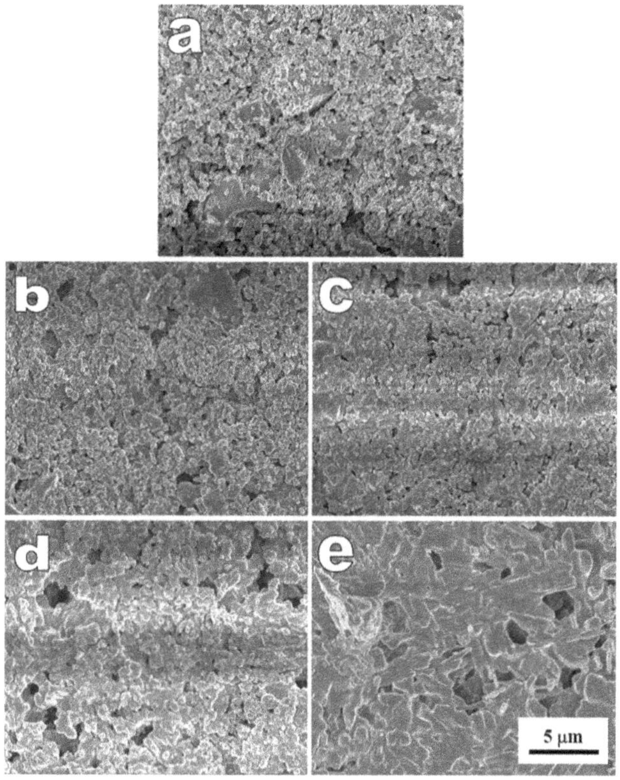

Figure 9.47. SEM images of the 5 mol%-SnO_2-doped samples sintered at: (a) 1100 °C, (b) 1200 °C, (c) 1300 °C, (d) 1400 °C and (e) 1500 °C for 4 h. Reproduced with permission from [17]. Copyright 2004 Elsevier.

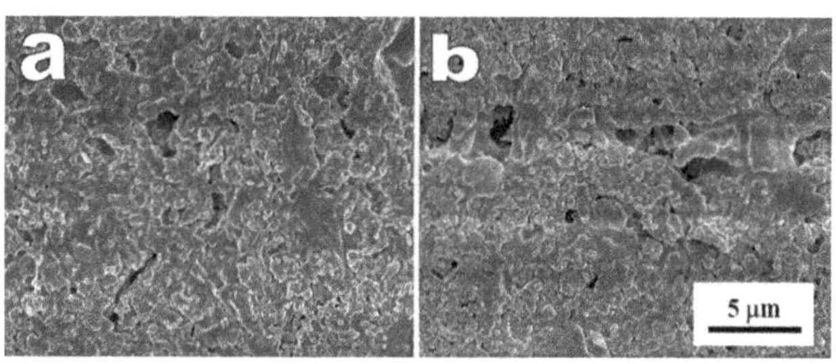

Figure 9.48. SEM images of the 1400 °C-sintered samples with SnO_2 of (a) 10 mol% and (b) 20 mol%. Reproduced with permission from [17]. Copyright 2004 Elsevier.

microstructure development of the samples doped with Bi_2O_3 was examined using SEM images. A significantly different microstructural characteristics was observed in this group of samples, in particular for those with high doping concentrations. The samples with 5 mol% Sb_2O_3 were all porous. However, as the concentration of Sb_2O_3 was increased to 10 mol%, the densification behavior of the samples was significantly enhanced. Sintering at 1200 °C resulted in almost fully densified ceramics. The sample with 20 mol% Bi_2O_3 could be fully densified at 1100 °C, as

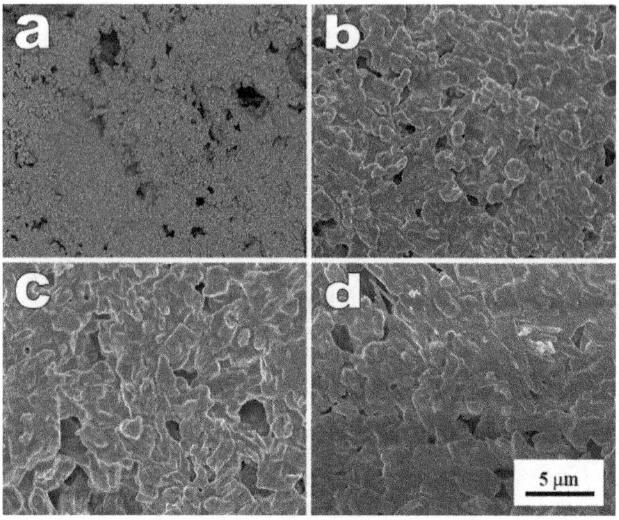

Figure 9.49. SEM images of the 5 mol%-Sb_2O_3-doped samples sintered at: (a) 1200 °C, (b) 1300 °C, (c) 1400 °C and (d) 1500 °C for 4 h. Reproduced with permission from [17]. Copyright 2004 Elsevier.

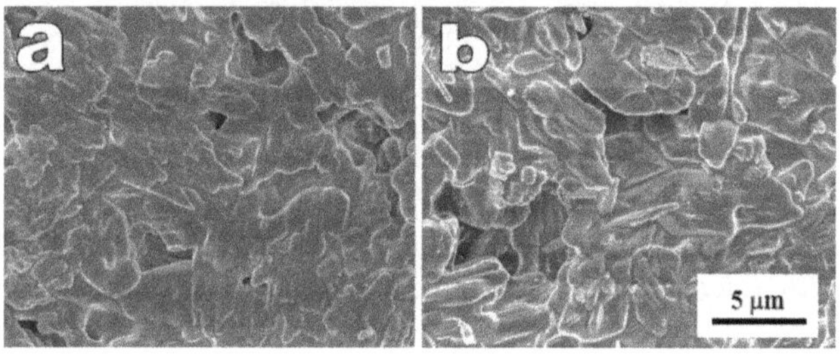

Figure 9.50. SEM images of the 1500 °C-sintered samples with Sb_2O_3 of (a) 10 mol% and (b) 20 mol%. Reproduced with permission from [17]. Copyright 2004 Elsevier.

shown in figure 9.51(a). Therefore, it could be concluded that the effect of Bi_2O_3 on the sintering behavior of the mullite ceramics was much stronger than that of Sb_2O_3, even though Bi_2O_3 has a higher melting temperature. In other words, Bi_2O_3 acted as a more effective liquid phase sintering aid than Sb_2O_3.

Mechanochemical activation has also been employed to treat the mixtures doped with three oxides. Selected results are presented in this section. For example, figure 9.52 shows XRD patterns of the samples doped with different contents of SnO_2 from the mixtures mechanochemically activated with WC milling media after sintering at 1100 °C for 4 h. After mechanochemical activation, the effect of SnO_2 on the mullite phase formation was altered. The mullite phase formation temperature was decreased by about 200 °C. Moreover, without the application of mechanochemical activation, the phase formation of mullite was retarded as the concentration of SnO_2 was increased. In contrast, after mechanochemical activation, the formation tendency of mullite was enhanced with increasing concentration of SnO_2.

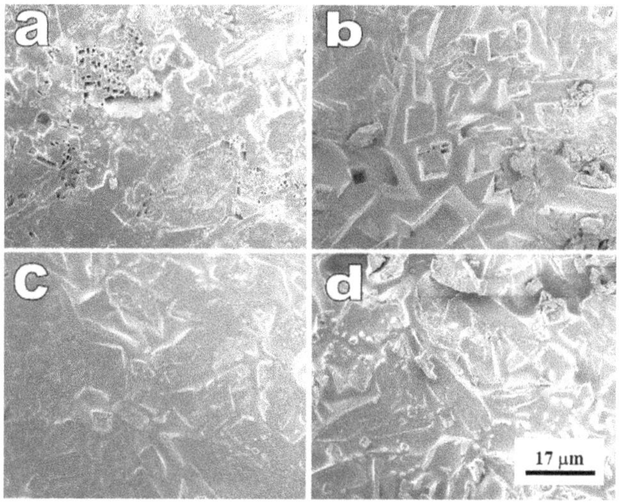

Figure 9.51. SEM images of 20 mol%-Bi_2O_3-doped samples sintered at: (a) 1100 °C, (b) 1200 °C, (c) 1300 °C and (d) 1400 °C for 4 h. Reproduced with permission from [17]. Copyright 2004 Elsevier.

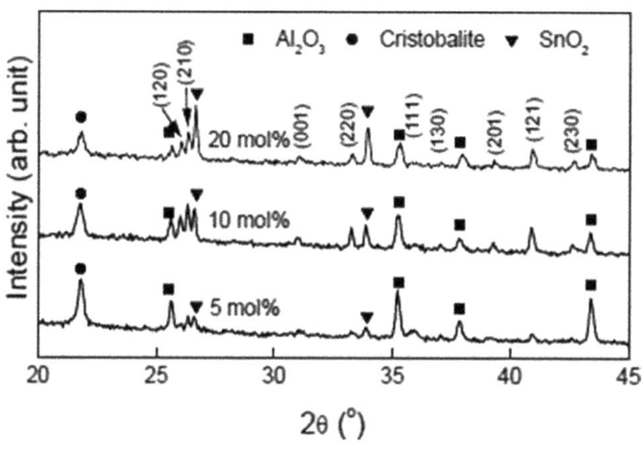

Figure 9.52. XRD patterns of the samples doped with different contents of SnO_2 from the mixtures mechanochemically activated with WC milling media after sintering at 1100 °C for 4 h.

This could be qualitatively attributed to the particle refinement of the starting materials. Similarly, SnO_2 was detected in the XRD patterns.

Figure 9.53 shows SEM images of the samples doped with 20 mol% SnO_2 from the mixture mechanochemically activated with WC milling media, after sintering at different temperatures. The mullite phase was formed at 1100 °C, while anisotropic grain growth occurred at the same time. Mullite whiskers with relatively regular morphology were obtained after sintering at 1200 °C. A further increase in sintering temperature had almost no effect on the microstructure of the samples and the morphology of the mullite whiskers, up to 1500 °C. The results of the samples doped

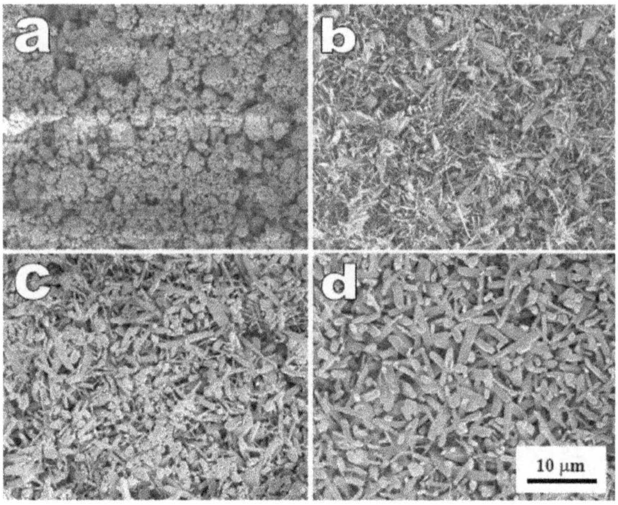

Figure 9.53. SEM images of the samples doped with 20 mol% SnO_2 from the mixture mechanochemically activated with WC milling media, after sintering at different temperatures for 4 h: (a) 1000 °C, (b) 1100 °C, (c) 1200 °C and (d) 1300 °C.

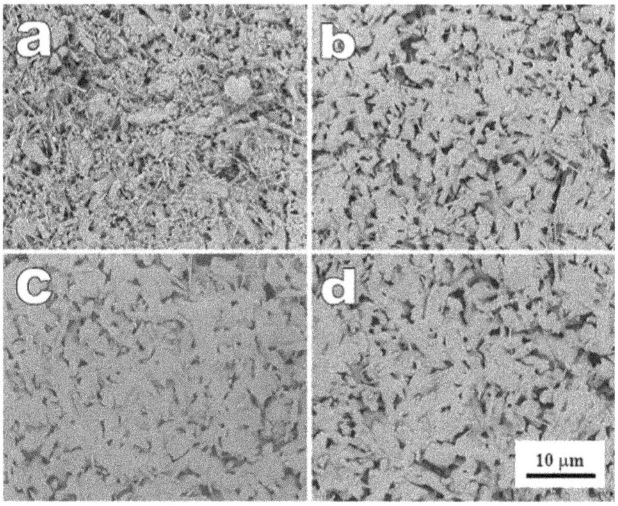

Figure 9.54. SEM images of the samples doped with 5 mol% Sb_2O_3 from the mixture mechanochemically activated with WC milling media, after sintering at different temperatures for 4 h: (a) 1200 °C, (b) 1300 °C, (c) 1400 °C and (d) 1500 °C.

with 5 mol% and 10 mol% SnO_2 were nearly the same, but the morphology of the mullite whiskers was more regular.

SEM images of the samples doped with Sb_2O_3 from the mixtures mechanochemically activated with WC milling media after sintering at different temperatures are shown in figures 9.54–9.56. Although the three groups of samples all had anisotropic grain growth, the morphology of the mullite grains were different for different

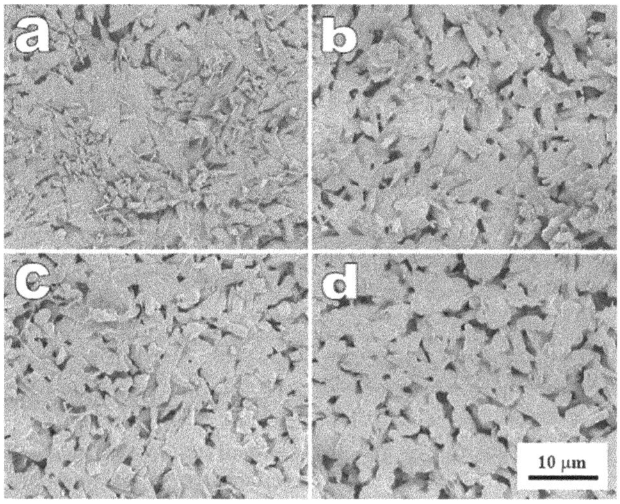

Figure 9.55. SEM images of the samples doped with 10 mol% Sb_2O_3 from the mixture mechanochemically activated with WC milling media, after sintering at different temperatures for 4 h: (a) 1200 °C, (b) 1300 °C, (c) 1400 °C and (d) 1500 °C.

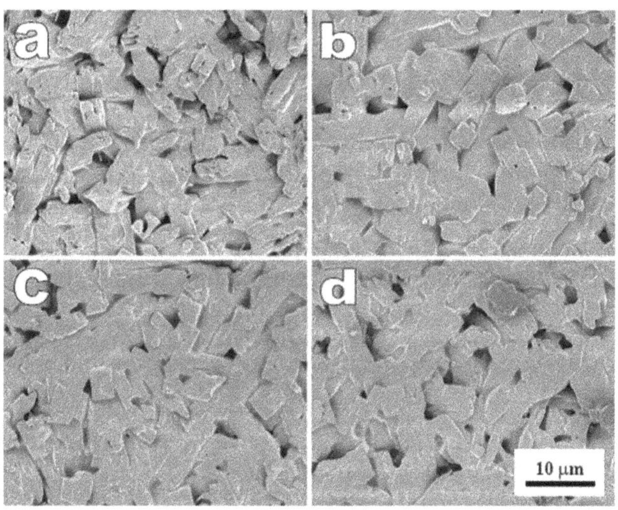

Figure 9.56. SEM images of the samples doped with 20 mol% Sb_2O_3 from the mixture mechanochemically activated with WC milling media, after sintering at different temperatures for 4 h: (a) 1200 °C, (b) 1300 °C, (c) 1400 °C and (d) 1500 °C.

concentrations of Sb_2O_3. For the samples with 5 mol% Sb_2O_3, after sintering at 1100 °C for 4 h, thin and thick whiskers coexisted, as shown in figure 9.54(a). As the sintering temperature was gradually increased, the number of thin wishers was reduced, while the size distribution of the thick whiskers tended to be uniform. The dimensions of the mullite whiskers in the samples doped with 10 mol% Sb_2O_3 were larger than that of the

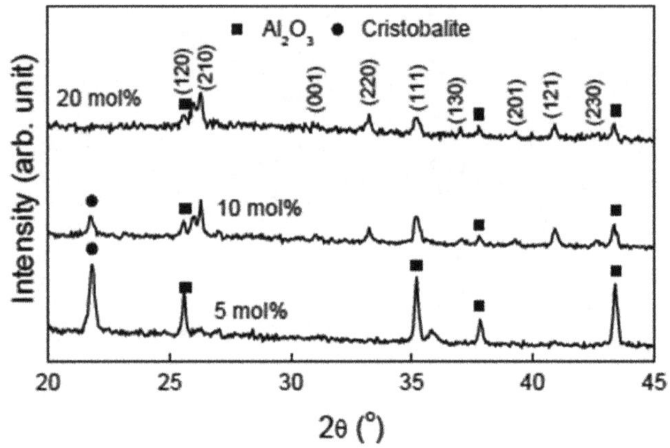

Figure 9.57. XRD patterns of the samples doped with different concentrations of Bi_2O_3 from the mixtures mechanochemically activated with stainless steel milling media after sintering at 1100 °C for 4 h.

samples with 5 mol% Sb_2O_3. Interestingly, the samples doped with 20 mol% Sb_2O_3 contained mullite grains with a cuboid morphology.

Figure 9.57 shows the XRD patterns of the samples doped with Bi_2O_3 from the mixtures mechanochemically activated with stainless steel milling media after sintering at 1100 °C. Obviously, the mullite phase formation temperature was reduced due to the mechanochemical activation. The sample doped with 5 mol% Bi_2O_3 had no mullite phase. In comparison, the mullite phase was dominant in the sample with 10 mol% Bi_2O_3, while the mullitization process of the sample with 20 mol% Bi_2O_3 was almost complete. In addition, the samples from the mixtures mechanochemically activated with WC milling media exhibited completed mullitization.

Figure 9.58 shows SEM images of the samples doped with 20 mol% Bi_2O_3 from the mixture mechanochemically activated with stainless steel milling media, after sintering at different temperatures. The sample sintered at 1000 °C was porous with spherical particles. The sample sintered at 1100 °C–1500 °C exhibited mullite whiskers with cuboid morphology, which was closer to that synthesized using a wet-chemical method [14], however, the cuboid mullite whiskers were densely packed together. In addition, the samples doped with 5 mol% and 10 mol% also displayed very dense microstructures.

SEM images of the samples doped with Bi_2O_3 from the mixtures mechanochemically activated with WC milling media, after sintering at different temperatures, are depicted in figures 9.59–9.61. For the samples doped with 5 mol% Bi_2O_3, mullite whiskers with relatively regular shape were formed after sintering at 1100 °C. The morphology of the mullite whiskers was not changed with increasing sintering temperature thereafter. With increasing concentration of Bi_2O_3, the dimensions and morphology of the whiskers varied significantly. The size was reduced, while the morphology became more irregular. This observation was different from that in the mixtures without the presence of any dopant. In that case, anisotropic grain growth occurred if the mullite phase was formed before densification. Once the anisotropic

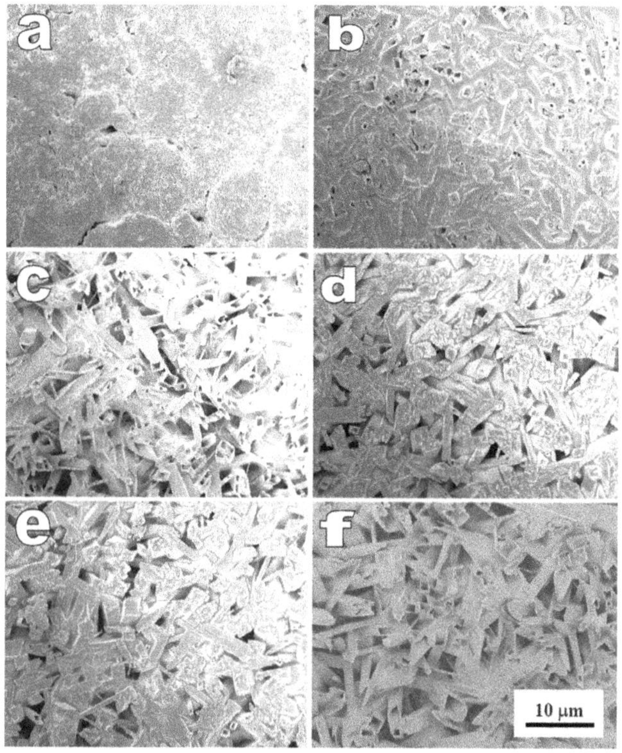

Figure 9.58. SEM images of the samples doped with 20 mol% Bi_2O_3 from the mixture mechanochemically activated with stainless steel milling media, after sintering at different temperatures for 4 h: (a) 1000 °C, (b) 1100 °C, (c) 1200 °C, (d) 1300 °C, (e) 1400 °C and (f) 15 000 °C. Reproduced with permission from [22]. Copyright 2018 Elsevier.

grain growth began, the sample would not be densified, while mullite whiskers with regular shape were obtained. However, if dopants were used, the scenario was different. For the samples doped with Bi_2O_3, the Bi_2O_3 would promote the formation of the liquid phase, which in turn accelerated the mullite phase formation. At the same time, Bi_2O_3 was also favorable for the densification process. As a result, with high concentrations of Bi_2O_3 the anisotropic grain growth was restricted to a certain degree, thus leading to less regular mullite whiskers.

9.3 Anisotropic grain growth of pure mullite

Mechanochemical activation has been applied to a commercial mullite powder in order to demonstrate the anisotropic grain growth [23]. The powder had particle sizes in the range 2–10 μm, which were refined to be less than 1 μm after milling for 10 h with WC milling media. Anisotropic grain growth of mullite was detected in the activated mullite powder, starting at a relatively low temperature of 1200 °C. It was confirmed that anisotropic grain growth of mullite was present if the sample was not fully densified. In this case, 20 g mullite powder (99.5+% purity) was milled using a Retsch PM400 type planetary ball milling machine. A 250 ml tungsten carbide vial

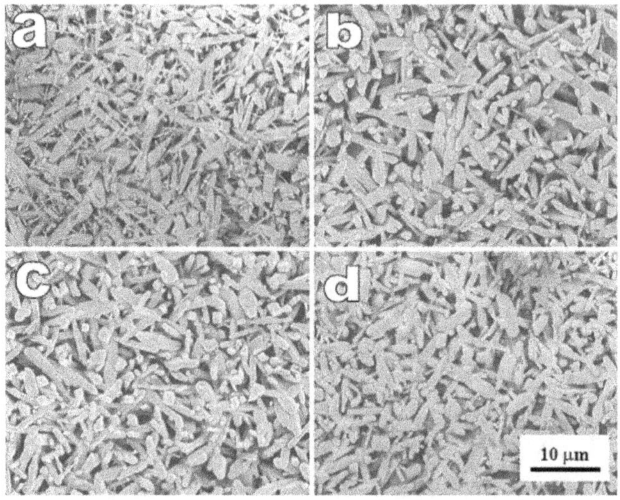

Figure 9.59. SEM images of the samples doped with 5 mol% Bi_2O_3 from the mixture mechanochemically activated with WC milling media, after sintering at different temperatures for 4 h: (a) 1100 °C, (b) 1200 °C, (c) 1300 °C and (d) 1400 °C.

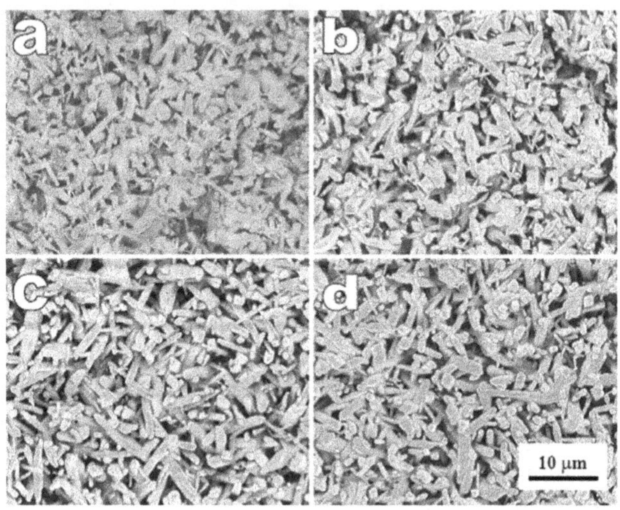

Figure 9.60. SEM images of the samples doped with 10 mol% Bi_2O_3 from the mixture mechanochemically activated with WC milling media, after sintering at different temperatures for 4 h: (a) 1100 °C, (b) 1200 °C, (c) 1300 °C and (d) 1400 °C.

together with 100 tungsten carbide balls of 10 mm were utilized as the milling media. The weight ratio of ball-to-powder was about 40:1 and the milling speed was set to be 200 rpm.

Figure 9.62 shows XRD patterns of the mullite powders without and with activation for 10 h. The commercial mullite powder (JCPDS No. 15–776) contained a small quantity of Al_2O_3, with diffraction peaks at 25.5° and 43.5°. After activation

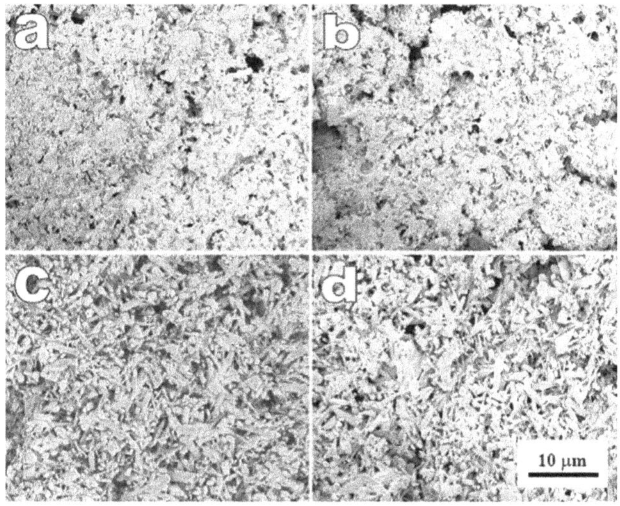

Figure 9.61. SEM images of the samples doped with 20 mol% Bi_2O_3 from the mixture mechanochemically activated with WC milling media, after sintering at different temperatures for 4 h: (a) 1100 °C, (b) 1200 °C, (c) 1300 °C and (d) 1400 °C.

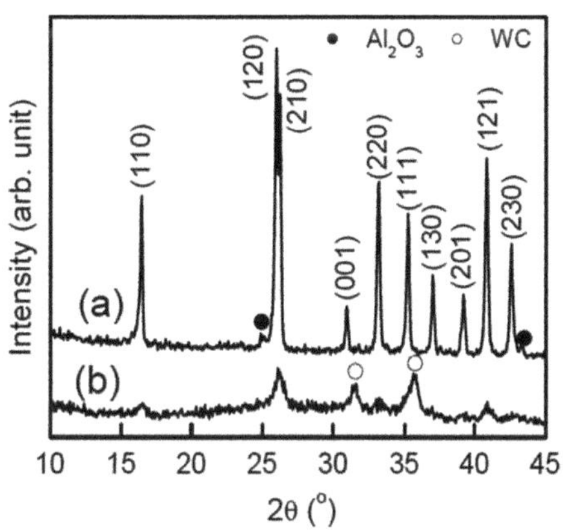

Figure 9.62. XRD patterns of the mullite powder before (a) and after (b) high-energy ball milling for 10 h. Reproduced with permission from [23]. Copyright 2007 Wiley.

for 10 h, the diffraction peaks of mullite phase were tremendously reduced, suggesting the great refinement of the mullite particles. A similar effect was also reported when Si_3N_4 media were used, where the mullite was amorphized after milling for 10 days [24]. In addition, WC was detected in the XRD pattern of the milled mullite powder. The refinement of the mullite due to the high-energy activation was also confirmed by SME images, as shown in figure 9.63.

Figure 9.64 shows the XRD patterns of the samples from the as-received and activated mullite powders after sintering at 1500 °C for 4 h. The patterns were consistent with each other, confirming that the refined mullite powder was highly recrystallized after sintering at high temperatures. However, there was also a small

Figure 9.63. SEM images of the mullite powder before (a) and after (b) high-energy ball milling for 10 h. Reproduced with permission from [23]. Copyright 2007 Wiley.

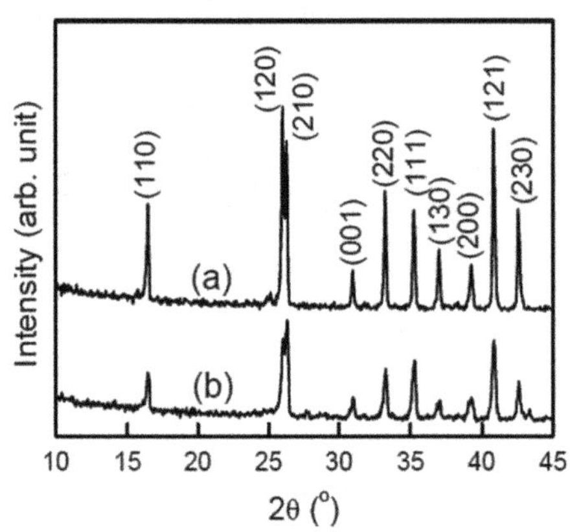

Figure 9.64. XRD patterns of the 1500 °C-sintered samples derived from the mullite powder (a) before and (b) after milling for 10 h. Reproduced with permission from [23]. Copyright 2007 Wiley.

difference in the XRD patterns between the two samples. On one hand, the diffraction peaks of the activated sample were lower than those of the sample taken directly from the commercial powder. On the other hand, the relative intensities of the diffraction peaks were different, such as the peaks of (120) and (210). The intensity of (120) was higher than that of (210) in the unactivated sample, while this was reversed in the activated sample. This suggested that they were different in grain morphology.

SEM images of the samples from the mullite powders before and after activation, sintered at different temperatures for 4 h, are shown in figures 9.65 and 9.66, respectively. The samples taken directly from the commercial mullite powder had grains with irregular shapes, which were similar to the original powder. In

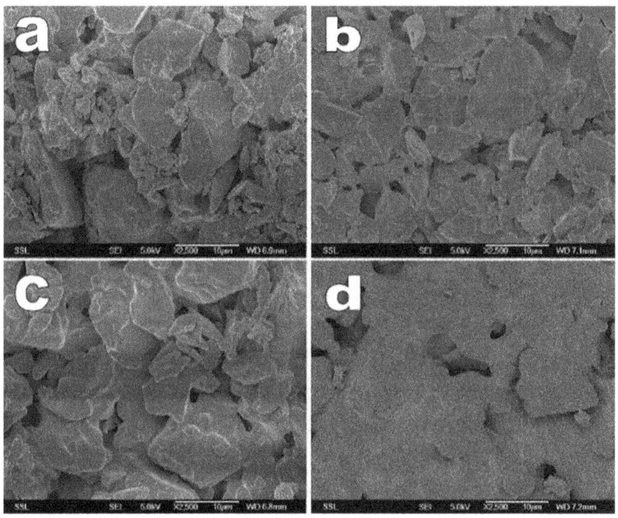

Figure 9.65. SEM images of the samples derived from the unmilled mullite powder, sintered for 4 h at: (a) 1200 °C, (b) 1300 °C, (c) 1400 °C and (d) 1500 °C. Reproduced with permission from [23]. Copyright 2007 Wiley.

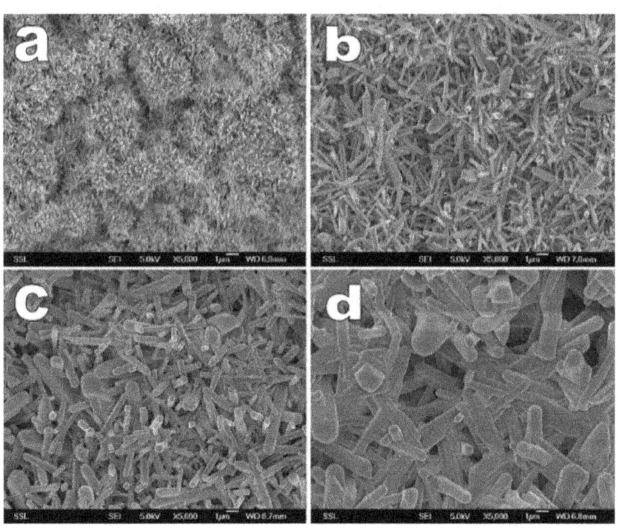

Figure 9.66. SEM images of the samples derived from the milled mullite powder, sintered for 4 h at: (a) 1200 °C, (b) 1300 °C, (c) 1400 °C and (d) 1500 °C. Reproduced with permission from [23]. Copyright 2007 Wiley.

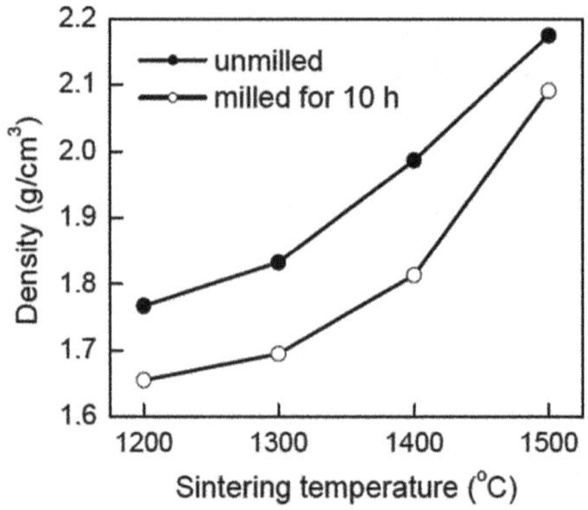

Figure 9.67. Measured densities of the sintered mullite ceramics as a function of sintering temperature. Reproduced with permission from [23]. Copyright 2007 Wiley.

comparison, the samples from the activated powder demonstrated a different grain growth behavior. Mullite whiskers were present in the sample sintered at 1200 °C, as illustrated in figure 9.66(a). With increasing sintering temperature, the dimensions of the mullite whiskers gradually increased.

The occurrence of anisotropic grain growth in the activated mullite powder was also indirectly confirmed by the difference in the measured densities between the two groups of samples. As seen in figure 9.67, the measured densities of the samples from the activated powder were always lower than those of samples from the commercial powder without activation. Due to the formation of the mullite whiskers, densification of the powder was prevented. The commercial powder followed the normal densification process. This result further confirmed our statement on the anisotropic grain growth of mullite ceramics from the mixtures of oxides with high-energy mechanochemical activation and doping with various dopants.

Acknowledgments

Shenzhen Technology University (SZTU) is acknowledged for the financial support of a start-up grant (2018) and also the Natural Science Foundation of Top Talent of SZTU (grant no. 2019010801002).

References

[1] Kong L B, Zhang T S, Ma J, Boey F and Zhang R F 2004 Mullite phase formation in oxide mixtures in the presence of Y_2O_3, La_2O_3 and CeO_2 *J. Alloys Compd.* **372** 290–9

[2] Huling J C and Messing G L 1989 Hybrid gels for homoepitactic nucleation of mullite *J. Am. Ceram. Soc.* **72** 1725–29

[3] Huling J C and Messing G L 1991 Epitactic nucleation of spinel in aluminonsilicate gels and its effect on mullite crystallization *J. Am. Ceram. Soc.* **74** 2374–81

[4] Huling J C and Messing G L 1992 Chemistry crystallization relations in molecular mullite gels *J. Non-Cryst. Solids* **147** 213–21

[5] Kolitsch U, Seifert H J, Ludwig T and Aldinger F 1999 Phase equilibria and crystal chemistry in the Y_2O_3–Al_2O_3–SiO_2 system *J. Mater. Res.* **14** 447–55

[6] She J H, Mechnich P, Schmucker M and Schneider H 2001 Low-temperature reaction-sintering of mullite ceramics with an Y_2O_3 addition *Ceram. Int.* **27** 847–52
[7] She J H, Mechnich P, Schmucker M and Schneider H 2002 Reaction-bonding behavior of mullite ceramics with Y_2O_3 addition *J. Eur. Ceram. Soc.* **22** 323–28
[8] She J H, Ohji T, Mechnich P, Schmucker M and Schneider H 2001 Phase development of Y_2O_3-doped reaction-bonded mullite ceramics *J. Mater. Sci. Lett.* **20** 2141–43
[9] She J H, Ohji T, Mechnich P, Schmucker M and Schneider H 2002 Mullitization and densification of Y_2O_3-doped reaction-bonded mullite ceramics with different compositions *Mater. Chem. Phys.* **76** 88–91
[10] Mechnich P, Schmucker M and Schneider H 1999 Reaction sequence and microstructural development of CeO_2-doped reaction-bonded mullite *J. Am. Ceram. Soc.* **82** 2517–22
[11] Mechnich P, Schneider H, Schmucker M and Saruhan B 1998 Accelerated reaction bonding of mullite *J. Am. Ceram. Soc.* **81** 1931–7
[12] Shi Z M, Liang K M and Gu S R 2001 Effects of CeO_2 on phase transformation towards cordierite in $MgO–Al_2O_3–SiO_2$ system *Mater. Lett.* **51** 68–72
[13] Sacks M D, Wang K, Scheiffele G W and Bozkurt N 1997 Effect of composition on mullitization behavior of α-Al_2O_3/silica microcomposite powders *J. Am. Ceram. Soc.* **80** 663–72
[14] Kong X Y, Wang Z L and Wu J S 2003 Rectangular single-crystal mullite microtubes *Adv. Mater.* **15** 1445–9
[15] Kong L B, Chen Y Z, Zhang T S, Ma J, Boey F and Huang H 2004 Effect of alkaline-earth oxides on phase formation and morphology development of mullite ceramics *Ceram. Int.* **30** 1319–23
[16] Bartsch M, Saruhan B, Schmucker M and Schneider H 1999 Novel low-temperature processing route of dense mullite ceramics by reaction sintering of amorphous SiO_2-coated γ-Al_2O_3 particle nanocomposites *J. Am. Ceram. Soc.* **82** 1388–92
[17] Kong L B, Zhang T S, Ma J and Boey F 2003 Some main group oxides on mullite phase formation and microstructure evolution *J. Alloys Compd.* **359** 292–9
[18] Caballero A and Ocana M 2002 Synthesis and structural characterization by x-ray absorption spectroscopy of tin-doped mullite solid solutions *J. Am. Ceram. Soc.* **85** 1910–4
[19] Senda T and Bradt R C 1991 Grain growth of zinc oxide during the sintering of zinc oxide–antimony oxide ceramics *J. Am. Ceram. Soc.* **74** 1296–302
[20] Dey D and Bradt R C 1992 Grain growth of ZnO during Bi_2O_3 liquid-phase sintering *J. Am. Ceram. Soc.* **75** 2925–34
[21] Nalin M, Poulain M, Poulain M, Ribeiro S J L and Messaddeq Y 2001 Antimony oxide based glasses *J. Non-Cryst. Solids* **284** 110–16
[22] Xiao Z H *et al* 2018 Effect of Bi_2O_3 on phase formation and microstructure evolution of mullite ceramics from mechanochemically activated oxide mixtures *Ceram. Int.* **44** 13841–47
[23] Kong L B, Zhang T S, Ma J and Boey F Y C 2007 Anisotropic grain growth in mullite powders as a result of high-energy ball milling *J. Am. Ceram. Soc.* **90** 4055–58
[24] Schmucker M, Schneider H and MacKenzie K J D 1998 Mechanical amorphization of mullite and thermal recrystallization *J. Non-Cryst. Solids* **226** 99–104

IOP Publishing

Functional Ceramics Through Mechanochemical Activation

Ling Bing Kong

Chapter 10

Other oxides

Ling Bing Kong, Zhuohao Xiao, Xiuying Li, Shijin Yu, Wenxiu Que, Yin Liu, Tianshu Zhang, Kun Zhou and Hongfang Zhang

10.1 Introduction

Various other oxide-based powders or ceramics have been synthesized or fabricated using high-energy mechanochemical activation from commercial oxides and/or carbonates. By using these activated mixtures, ceramics could be obtained using a one-step sintering process, without involving the calcination step, thus making the whole process even simpler than the conventional ceramic methods. Due to limitations of space, only those results reported by the current authors will be included in this chapter, such as magnesium aluminate ($MgAl_2O_4$), yttrium aluminum garnet ($Y_3Al_5O_{12}$ or YAG for short), yttrium iron garnet ($Y_3Fe_5O_{12}$, or YIG for short), magnesium niobate ($MgNb_2O_6$), zinc niobate ($ZnNb_2O_6$) and (Sn, Ti, Zr)O_2 solid solutions.

10.2 Selected samples

Selected samples prepared using high-energy activation will be discussed in this section. The high-energy mechanochemical activation process has been applied to the system of MgO and Al_2O_3 [1]. Although the phase formation of $MgAl_2O_4$ was not triggered by mechanochemical activation, the reaction between MgO and Al_2O_3 was largely promoted. For example, the $MgAl_2O_4$ spinel phase with an average grain size of about 100 nm was obtained after sintering at 900 °C from a mixture of MgO and Al_2O_3 activated for 12 h. The activated mixture could be densified directly by skipping the calcination step, leading to $MgAl_2O_4$ ceramics with a relative density of 98% and grain sizes in the range 2–5 μm, after sintering at 1550 °C for 2 h.

$MgAl_2O_4$ spinel ceramics are of important technological interest for refractory and structural applications at elevated temperatures, because of the high melting point of the compound, its promising mechanical strength and outstanding chemical durability [2–4]. $MgAl_2O_4$ ceramics can be made to be transparent, thus having

doi:10.1088/978-0-7503-2191-4ch10

potential applications as transparent armor in defense, host media in solid-state lasers and biomaterials in medicine [5–10].

$MgAl_2O_4$ spinel powders were conventionally synthesized using the solid-state reaction method, with oxide $MgO/MgCO_3$ and Al_2O_3 as the precursors [11]. In this method, the reaction of MgO and Al_2O_3 to form the compound $MgAl_2O_4$ required very high calcination temperatures of 1400 °C–1600 °C. Accordingly, the sintering temperature of the powders prepared in this way was even higher (e.g. >1700 °C). Therefore, the fabrication of $MgAl_2O_4$ ceramics using the solid-state process is not cost-effective. To reduce the sintering temperature, various methods have been developed to synthesize $MgAl_2O_4$ powders with fine particles, thus having a high densification rate and low calcination/sintering temperatures [12–16]. However, wet-chemical processes have encountered several problems, such as complicated processing, expensive precursors and special environmental requirements. In this regard, mechanochemical activation could be an alternative method to develop $MgAl_2O_4$ ceramics.

In the experiment, commercial powders of MgO (99.99+% purity, Aldrich Chemical Company Inc., USA) and Al_2O_3 (99.9+% purity, Aldrich Chemical Company Inc., USA) were used as the starting materials, which were mixed according to the composition of $MgAl_2O_4$. The mixture was activated with high-energy ball milling, which was conducted using a Retsch PM400 type planetary ball milling system operating in air at room temperature for 12 h. A 250 ml WC vial and 100 WC balls with a diameter of 1 cm were used as the milling media, while the milling speed was 200 rpm and the ball-to-powder weight ratio was about 40:1.

Figure 10.1 shows the XRD pattern of the mixture of MgO and Al_2O_3 after mechanochemical activation for 12 h. All the peaks belonged to MgO and Al_2O_3, suggesting that the chemical reaction between the two components was not triggered in the activation process. Nevertheless, the powders were highly refined, as evidenced by the broadened and weakened diffraction peaks of the two starting oxides. Figure 10.2 shows an SEM image of the as-activated mixture. The particle sizes were mainly distributed in the range 100–300 nm, among which there were also

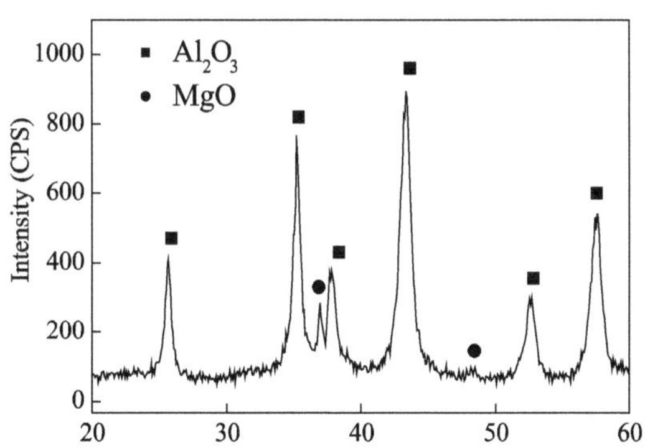

Figure 10.1. XRD pattern of the mixture of MgO and Al_2O_3 mechanochemically activated for 12 h. Reproduced with permission from [1]. Copyright 2002 Elsevier.

Figure 10.2. SEM image of the mixture of MgO and Al_2O_3 mechanochemically activated for 12 h. Reproduced with permission from [1]. Copyright 2002 Elsevier.

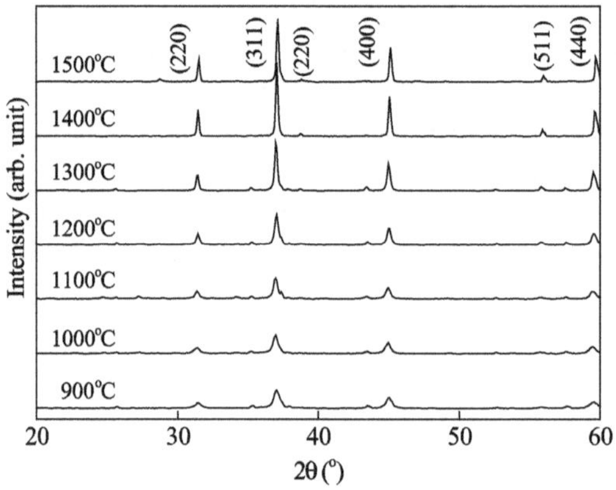

Figure 10.3. XRD patterns of the samples from the activated mixture sintered at different temperatures for 2 h. Reproduced with permission from [1]. Copyright 2002 Elsevier.

small particles with sizes <100 nm. Due to the significant refinement, the formation temperature of the $MgAl_2O_4$ spinel phase was greatly reduced. In addition, the activated mixed powder could be used to fabricate $MgAl_2O_4$ ceramics, without the requirement of a calcination step. As a result, the production cycle was effectively shortened compared to conventional ceramic processing.

Figure 10.3 shows the XRD patterns of the as-activated mixture and the samples after sintering at different temperatures for 2 h. It was found that the $MgAl_2O_4$ spinel phase was obtained after sintering at 900 °C, implying that the MgO and Al_2O_3 in the activated mixture possessed high reactivity due their refined particles. This temperature is lower than that required by the conventional solid-state reaction process by about 400 °C. Although the temperature of 900 °C was higher than some of those for the wet-chemical processes, the mechanochemical activation method was obviously much simpler. Also, the milling process could be further optimized to enhance the reactivity of the precursors.

The XRD peaks of the sample sintered at 900 °C were relatively broad, which implied that the powder had a small particle size and low crystallinity. The intensity of the diffraction peaks of the $MgAl_2O_4$ spinel phase gradually increased as the sintering temperature was increased to 1200 °C. After sintering at temperatures ⩾1300 °C, the diffraction peaks were greatly strengthened and sharpened. This observation was consistent with the SEM results of the sintered samples, as illustrated in figures 10.4 (high magnification) and 10.5 (low magnification).

As observed in figure 10.4(a), the average grain size of the sample after sintering at 900 °C was about 100 nm, which was smaller than that of the as-activated powder. This was probably because the particles were further refined during the reaction between MgO and Al_2O_3. As the sintering temperature was raised to 1000 °C, the grain growth was negligible, as revealed in figures 10.4(a) and (b). Even after sintering at 1100 °C and 1200 °C, the grain growth was still not very significant, as seen in figures 10.4(c) and (d). However, these two samples contained large grains with sizes in the range 0.5–1.5 μm and clear facets, suggesting that they had higher crystallinity.

The number of large grains quickly increased as the sintering temperature was increased to 1300 °C, whereas the grain sizes increased to 1–2 μm. However, small particles without an increase in size were still visible in the samples sintered at high temperatures. A further increase in sintering temperature to 1400 °C led to

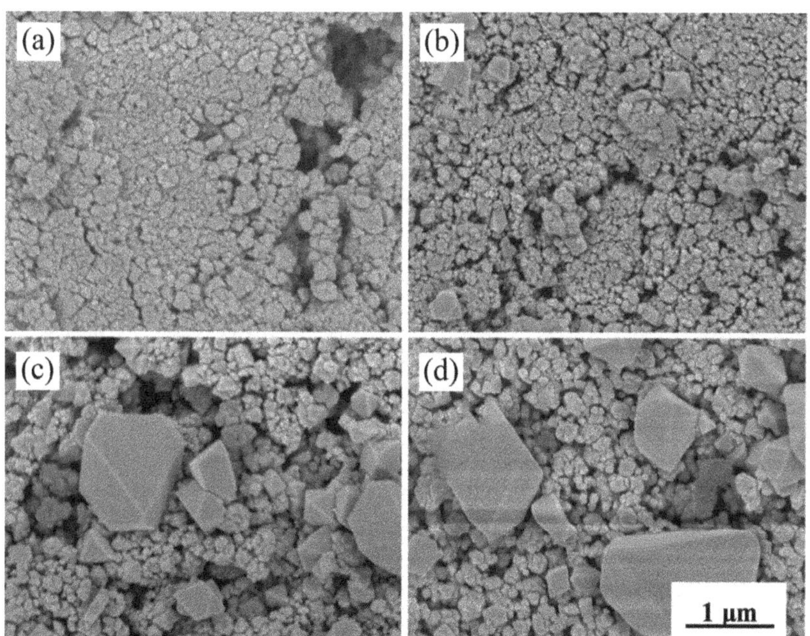

Figure 10.4. SEM images of samples sintered for 2 h at different temperatures: (a) 900 °C, (b) 1000 °C, (c) 1100 °C and (d) 1200 °C. Reproduced with permission from [1]. Copyright 2002 Elsevier.

Figure 10.5. SEM images of the samples sintered for 2 h at different temperatures: (a) 1300 °C, (b) 1400 °C, (c) 1500 °C and (d) 1550 °C. Reproduced with permission from [1]. Copyright 2002 Elsevier.

further grain growth. In addition, the small particles were completely absent, suggesting that they had been consumed during the growth of the large grains, as demonstrated in figures 10.5(a) and (b). It was found that the average grain size only needed to be increased by a relatively low amount for the sintering temperature to be raised from 1300 °C to 1400 °C. Continuing grain growth was observed as the sintering temperature was increased to 1500 °C and 1550 °C. At the same time, the samples became much denser in microstructure, as illustrated in figures 10.5(c) and (d).

Figure 10.6 shows the relative densities of the samples as a function of sintering temperature. The relative density of the samples was less than 80% when the sintering temperature was $\leqslant$1300 °C. A significant increase in relative density occurred thereafter. The sample sintered at 1500 °C had a relative density of about 96%, while the sample sintered at 1550 °C was nearly fully densified, with a relative density of about 98%. The variation trend in the relative density of the $MgAl_2O_4$ ceramics with the sintering temperature was in good agreement with the evolution of their microstructure. In other words, both the grain growth and densification were facilitated when the sintering temperature was above 1300 °C, i.e. 1300 °C was a critical temperature for the grain growth and densification of the $MgAl_2O_4$ ceramics.

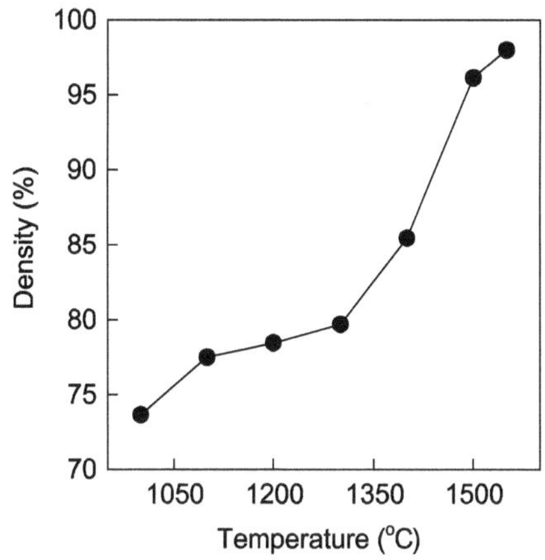

Figure 10.6. Relative densities of the $MgAl_2O_4$ samples derived from the activated mixture as a function of sintering temperature. Reproduced with permission from [1]. Copyright 2002 Elsevier.

$Y_3Al_5O_{12}$ ceramics have important properties for potential applications in solid-state lasers, rare-earth based phosphors and refractory coating of electronic devices. YAG is also a promising candidate for high-temperature structural materials, because of its low creep rate at high temperatures, high thermal stability and excellent chemical durability [17–23]. YAG powders were conventionally prepared through a solid-state reaction of Y_2O_3 and Al_2O_3 powders, which required a phase formation temperature $\geqslant$1600 °C. In this case, the YAG powder would have very poor sinterability due to the large particle size. Accordingly, the processing of YAG ceramics needed a very high sintering temperature, in particular when developing YAG transparent ceramics.

Mechanochemical activation was employed to treat the mixture of Y_2O_3 and Al_2O_3 in order to increase the reactivity between the two oxides [24]. The $Y_3Al_5O_{12}$ garnet phase was obtained at 1000 °C from the activated mixture, which was much lower than that required by the conventional solid-state reaction process and comparable to those for most of the wet-chemical synthetic routes. The reduction in the YAG phase formation temperature was readily ascribed to the refined Y_2O_3 and Al_2O_3. Commercial powders of Y_2O_3 (99.9+% purity, Aldrich Chemical Company Inc., USA) and Al_2O_2 (99.9+% purity, Aldrich Chemical Company Inc., USA) powders were mixed according to the nominal composition of $Y_3Al_5O_{12}$, followed by mechanochemical activation for 12 h with WC milling media.

Figure 10.7 shows the XRD patterns of the as-activated samples and the samples after sintering at different temperatures. It was observed that the as-activated mixture contained both Y_2O_3 and Al_2O_3, suggesting that they did not react in the high-energy activation process. Similarly, the diffraction peaks of the two oxides are largely weakened, owing to the reduction in particle size and crystallinity of Y_2O_3 and Al_2O_3. A similar result was observed when synthesizing $Y_{1-x}Ce_xAlO_3$ using a mechanochemical process with Y_2O_3 and $Al(OH)_3$ as the starting materials [25]. Without the presence of the designed compound, $Y_{1-x}Ce_xAlO_3$, instead disordered

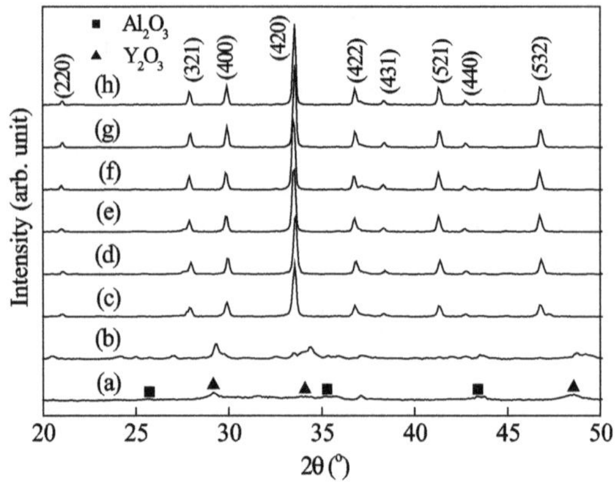

Figure 10.7. XRD patterns of the as-activated mixture of Y_2O_3 and Al_2O_3 and the samples sintered for 2 h at different temperatures: (a) as-milled, (b) 900 °C, (c) 1000 °C, (d) 1100 °C, (e) 1200 °C, (f) 1300 °C, (g) 1400 °C and (h) 1500 °C. Reproduced with permission from [24]. Copyright 2002 Elsevier.

non-equilibrium phases were present after the mechanochemical activation. However, the phase formation of $Y_{1-x}Ce_xAlO_3$ was achieved from the amorphous precursors at a relatively low temperature of 1150 °C. In comparison, if the solid-state reaction process was used, the phase formation temperature was as high as 1600 °C.

Although the energy applied to the mixture of Y_2O_3 and Al_2O_3 was not sufficiently high to facilitate the reaction to form $Y_3Al_5O_{12}$ during the activation process, the phase formation temperature was greatly reduced. After sintering at 900 °C for 2 h, $Y_3Al_5O_{12}$ was still not visible in the XRD pattern, but the intensity of the diffraction peaks was increased compared to that of the as-activated powder, implying that grain growth of the milled Y_2O_3 and Al_2O_3 occurred during the sintering process. After sintering at 1000 °C the $Y_3Al_5O_{12}$ phase was obtained, as illustrated in figure 10.7(c). This phase formation temperature of YAG was much lower than that required when using the solid-state reaction process [26]. It was very close to the calcination temperature used to form YAG in most of the wet-chemical synthesis routes, e.g. chemical coprecipitation [27] and the sol–gel process [28].

The highly enhanced phase formation of YAG was simply because of the increased reactivity of Y_2O_3 and Al_2O_3, after the high-energy mechanochemical activation. In addition, the reaction sequence to the final YAG phase in the activated mixture of Y_2O_3 and Al_2O_3 was different from that encountered by the conventional solid-state reaction method and most of the chemical processes, where the yttrium aluminum monoclinic ($Y_4Al_2O_9$ or YAM) and yttrium aluminum perovskite ($YAlO_3$ or YAP) are usually formed as an intermediate phase, prior to the presence of the YAG phase. This difference could also be correlated with the refined precursors, although further studies are needed to obtain more convincing clarification.

Figure 10.8 shows SEM images of the activated mixture after sintering at different temperatures for 2 h. Variation in the average grain size of the YAG

Figure 10.8. SEM images of the activated mixture after sintering for 2 h at different temperatures: (a) 1000 °C, (b) 1100 °C, (c) 1200 °C, (d) 1300 °C, (e) 1400 °C and (f) 1500 °C. Reproduced with permission from [24]. Copyright 2002 Elsevier.

samples as a function of sintering temperature is illustrated in figure 10.9. It was found that the particles of the sample sintered at 1000 °C had sizes in the range 0.1–0.2 μm, very close to those of the as-activated mixture powder, which suggested that no significant grain growth took place at this temperature. In addition, both the grain size and grain morphology of the YAG sample were similar to those of the samples synthesized using the wet-chemical synthesis methods [27]. The average grain size of the YAG samples monotonically increased from about 0.3 μm to 1.5 μm, as the sintering temperature was increased from 1100 °C to 1500 °C. In addition, the sample sintered at 1300 °C exhibited the highest density, according to the SEM images, as revealed in figure 10.8(d). Further increase in sintering temperature led to samples with more pores, which could be ascribed to the overgrowth of the YAG grains.

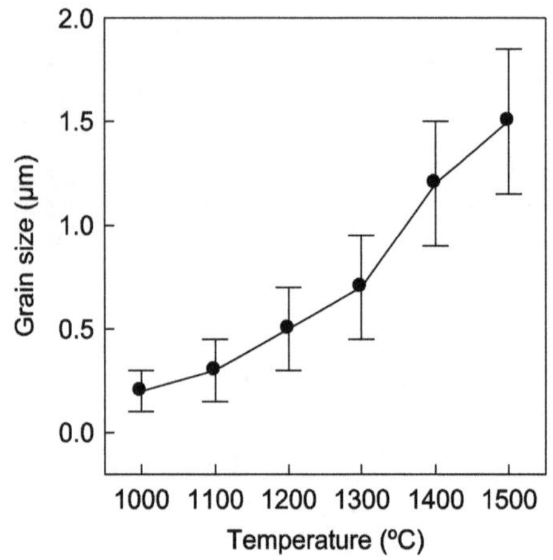

Figure 10.9. Grain size of the sintered samples as a function of sintering temperature. Reproduced with permission from [24]. Copyright 2002 Elsevier.

Figure 10.10. SEM image of the as-activated mixture of Y_2O_3 and Fe_2O_3.

$Y_3Fe_5O_{12}$ (YIG) is an important ferromagnetic material among the three groups of ferrites. YIG ceramics could be used as filters in microwave circuitry, electronic resonators and laser devices [29–32]. The mechanochemical process was applied to a mixture of commercial Y_2O_3 and Fe_2O_3 oxides in order to increase the reactivity of the precursors to form $Y_3Fe_5O_{12}$. The YIG phase was formed from the activated mixture after sintering at 900 °C. Figure 10.10 shows an SEM image of the as-activated mixture, indicating that the particle size was about 100 nm.

Figure 10.11 depicts the XRD patterns of the as-activated mixture and the samples sintered at different temperatures. The as-milled sample contained only the two precursor oxides, with diffraction peaks that were greatly broadened, implying that the particles were significantly refined. In other words, the reaction between the two components was not induced by the high-energy activation. After sintering at

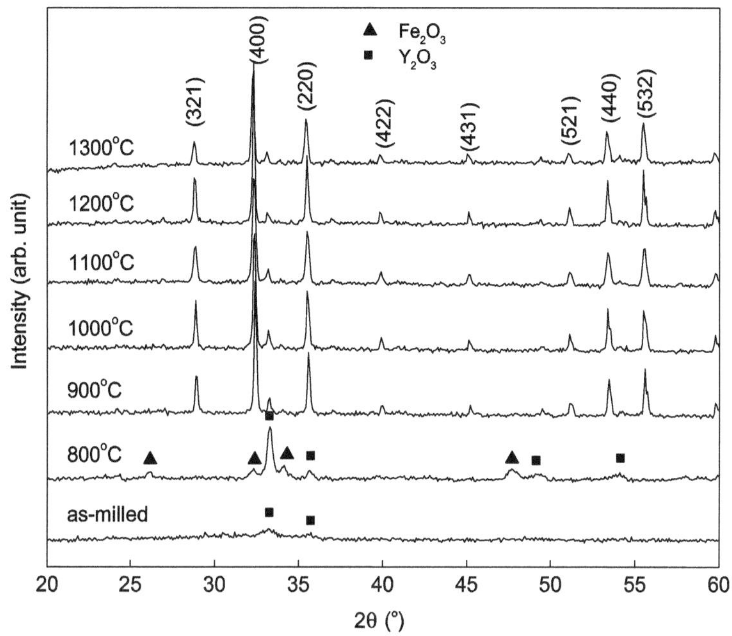

Figure 10.11. XRD patterns of the as-activated mixture and the samples sintered at different temperatures.

800 °C, the major phases were still from the oxides of Y_2O_3 and Fe_2O_3, suggesting that this temperature was not high enough for the phase formation of YIG. However, the sample sintered at 900 °C displayed single-phase YIG, which meant that 900 °C was sufficiently high to trigger the YIG phase formation. This temperature was much lower than those observed in the conventional solid-state reaction, due to the increased reactivity of the precursors. For the samples sintered at temperatures in the range 1000 °C–1300 °C, the phase composition was not changed.

Figure 10.12 shows SEM images of the samples after sintering at different temperatures for 2 h. After sintering at 900 °C, the sample exhibited a porous microstructure, together with agglomerated particles of irregular morphology. The samples sintered at 1000 °C and 1100 °C shared a similar microstructure, while the grain size increased and the number of smaller particles decreased with increasing sintering temperature. At the same time, the grains tended to be more and more regular. As the sintering temperature was increased to 1200 °C, the smaller particles almost all disappeared, while the microstructure consisted of regular grains, although there were still pores. Finally, in the sample sintered 1300 °C for 2 h, typical grains with facets of ceramics were present, while the sample was nearly fully densified.

In a separate study, $Y_3Fe_5O_{12}$ powder was obtained from a mixture of Y_2O_3 and Fe_2O_3 by mechanochemical activation [33]. According to the XRD patterns, Y_2O_3 and Fe_2O_3 reacted to form $FeYO_3$ as an intermediate phase after activation for 2 h, while the YIG phase was detected in the sample milled for 4 h. Complete phase

Figure 10.12. SEM images of the samples sintered for 2 h at different temperatures: (a) 900 °C, (b) 1000 °C, (c) 1100 °C, (d) 1200 °C and (e) 1300 °C.

formation of YIG was achieved after activation for 8 h. The as-obtained YIG powder had an average particle size of 150 nm. Dense YIG ceramics with a relative density of about 95% were derived from the mechanochemically activated powder after sintering at 1425 °C for 10 h without involving the calcination step.

This result was somehow surprising, because the milling media were stainless steel. The mechanochemical activation was conducted with a Fritsch Pulverisette 5 planetary ball mill machine. A stainless steel vial with a volume of 500 cm^3 and stainless steel balls with a diameter of 10 mm were used for the experiments. The ball-to-powder weight ratio was 40:1, while the rotation speed of the disks with vials was 320 rpm. The activation experiment was carried out in air. One of the possible reasons for the YIG phase formation could be attributed to the fact that the

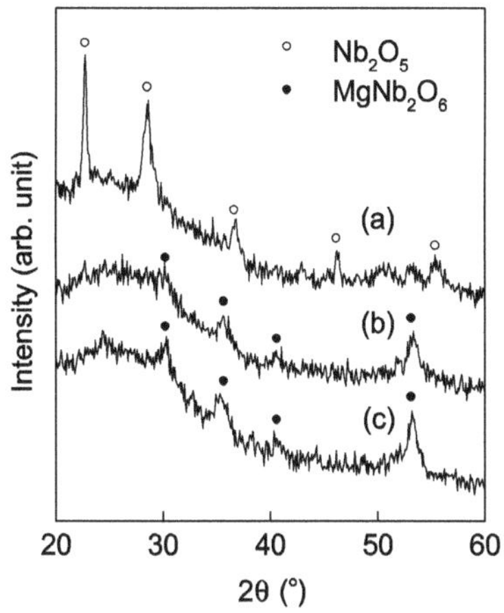

Figure 10.13. XRD patterns of the mixture of MgO and Nb_2O_5 activated for different times: (a) 5 h, (b) 10 h and (c) 15 h. Reproduced with permission from [44]. Copyright 2002 Elsevier.

precursor oxide powders had higher reactivity. However, it is necessary to clarify this as a research topic in future studies.

$MgNb_2O_6$ ceramics have promising microwave dielectric properties, with potential applications as dielectric resonators and microwave device substrates [34, 35]. $MgNb_2O_6$ with high reactivity has been extensively used to synthesize $Pb(Mg_{1/3}Nb_{2/3})O_3$ ceramics, which is the so-called Columbite process [36, 37]. The conventional solid-state reaction method to prepare $MgNb_2O_6$ requires a calcination temperature ⩾1100 °C [38, 39]. Various chemical processes have been reported to synthesize $MgNb_2O_6$ at relatively low temperatures [40–43].

It was demonstrated that $MgNb_2O_6$ could be crystallized from the amorphous mixture of MgO and Nb_2O_5 at a low temperature of 500 °C, while complete crystallization was achieved at 700 °C [44]. The mixture was mechanochemically activated through high-energy ball milling. The activation was carried out with WC milling media, with a duration of 5 h. In addition, no significant grain growth occurred after the activated powders were sintered at temperatures ⩽900 °C. Full densification to $MgNb_2O_6$ ceramics was realized at 1100 °C directly from the activated powders without any requirement for the calcination step.

Commercial powders of MgO (99.9+% purity, Aldrich Chemical Company Inc., USA) and Nb_2O_5 (99.9+% purity, Aldrich Chemical Company Inc., USA) were mixed according to the composition of $MgNb_2O_6$. The mixture was milled with a Retsch PM400 type planetary ball milling system in air at room temperature for times of 5 h, 10 h and 15 h, at a milling speed of 200 rpm. The milling was stopped for 5 min every 30 min to avoid overheating. A 250 ml tungsten carbide vial and 100 tungsten carbide balls with a diameter of 10 mm were used as the milling media, corresponding to a ball-to-powder ratio of about 40:1.

Figure 10.13 shows the XRD patterns of the mixture activated for different times. The sample milled for 5 h contained the main diffraction peaks of Nb_2O_5, while MgO was detected in the XRD pattern. This could be attributed to the fact that the weight percentage of MgO in the mixture was relatively low, while the diffraction of Nb_2O_5 was much stronger. In addition, no diffraction peaks of $MgNb_2O_6$ were present. Also, a broad hump was visible over 20°–40° in the XRD pattern, as seen in figure 10.13(a), implying that an amorphous phase was produced, as a result of the high-energy activation for 5 h.

Further increasing the milling time to 10 h, the diffraction peaks of Nb_2O_5 were completely absent in the XRD pattern, whereas weak peaks of $MgNb_2O_6$ were present, as illustrated in figure 10.13(b). In addition, the amorphous hump was stronger than that of the sample activated for 5 h, as observed in figure 10.13(a). As the milling time was increased to 15 h, the $MgNb_2O_6$ phase exhibited a higher degree of crystallization, which was well reflected by the increased diffraction peaks in the XRD pattern, as revealed in figure 10.13(c). It could thus be expected that complete crystallization would be achievable with prolonged activation.

This result suggested that the phase formation of $MgNb_2O_6$ during the high-energy activation proceeded according to the sequence of (i) refinement of the precursor oxides, (ii) amorphorization of the oxides and (iii) crystallization of $MgNb_2O_6$ from the amorphous state. Such a phase formation process was distinctively different from those of other materials, such as lead titanate ($PbTiO_3$ or PT), lead zirconate titanate ($PbZr_{0.48}Ti_{0.52}O_3$ or PZT), lead magnesium niobate ($PbMg_{1/3}Nb_{2/3}O_3$ and PMN) and bismuth titanate ($Bi_4Ti_3O_{12}$), as discussed in chapter 3. In those cases, all the desired phases were formed directly from the precursors without the production of the intermediate amorphous state.

The as-activated mixtures were directly used to make $MgNb_2O_6$ ceramics, while the phase evolution and microstructure development as a function of sintering temperature were studied. Figure 10.14 depicts the XRD patterns of the samples from the mixtures activated for different times, after sintering at different temperatures for 8 h. After sintering at 500 °C, $MgNb_2O_6$ was detected in all three samples. However, the diffraction peaks of Nb_2O_5 were still present in the XRD pattern of the sample from the mixture activated for 5 h. Nb_2O_5 was even detectable in the sample from the mixture activated for 10 h. However, no Nb_2O_5 was visible in the as-activated mixture (10 h), as observed in figure 10.13(b). This could be ascribed to the fact that the diffraction peaks of Nb_2O_5 were too weak to be detected in the as-activated sample, while it was crystallized after sintering at 500 °C for 8 h.

Therefore, it was concluded that crystallization of $MgNb_2O_6$ from the amorphous state was triggered at 500 °C. However, such a temperature was not sufficiently high to facilitate a reaction between the remaining Nb_2O_5 and MgO. In other words, the crystallization from the amorphous state and the reaction between the two components had different activation energies, which deserves further investigation. In addition, the broad diffraction peaks of the samples implied that the $MgNb_2O_6$ phase crystallized from an amorphous state consisting of small grains, while the crystallinity was still not very high. This observation further confirmed the formation of the amorphous state as a result of high-energy mechanochemical activation.

As the sintering temperature was increased to 700 °C, all the samples exhibited a single phase of $MgNb_2O_6$, while the remaining oxides and the amorphous phase

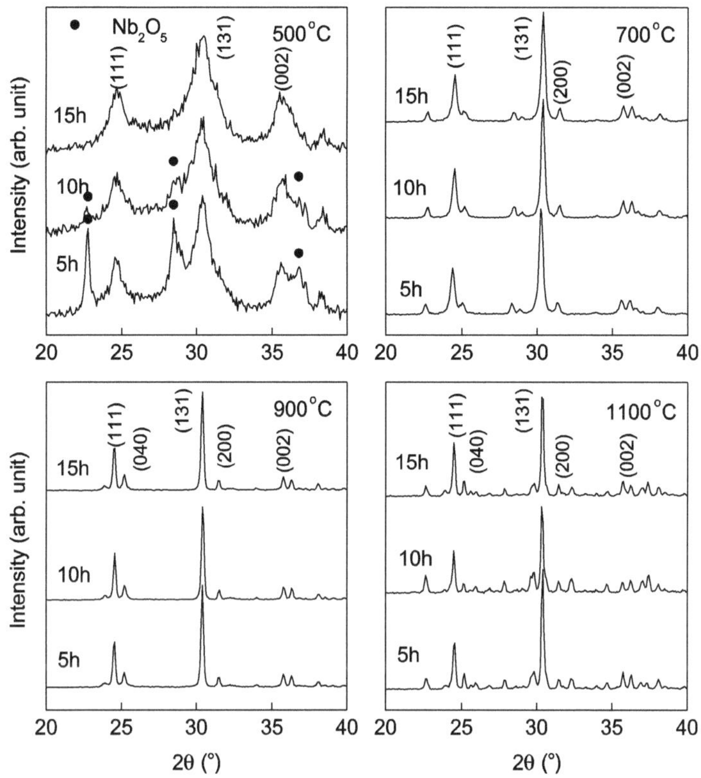

Figure 10.14. XRD patterns of the samples from the mixtures activated for different times, after sintering at different temperatures. Reproduced with permission from [44]. Copyright 2002 Elsevier.

disappeared. This meant that this temperature was sufficiently high to trigger both the reaction and the amorphous–crystal phase transition. According to the XRD patterns, the $MgNb_2O_6$ phase had an orthorhombic structure with lattice constants of $a = 5.699$, $b = 14.192$ and $c = 5.033$, which were consistent with the literature values (JCPDS No. 33–875). This temperature for the phase formation of $MgNb_2O_6$ was lower than that required when using the conventional solid-state reaction method. A further increase in sintering temperature just resulted in an increase in the diffraction peaks of $MgNb_2O_6$. In other words, there was no grain growth of $MgNb_2O_3$ with increasing sintering temperature.

SEM images of the samples from the mixtures activated for different times, after sintering for 8 h at 900 °C and 1100 °C, are depicted in figures 10.15 and 10.16, respectively. The samples sintered at 900 °C were highly porous with an average grain size of less than 1 μm. In addition, there were only spherical grains. In comparison, after sintering at 1100 °C all the samples were fully densified, without the presence of pores. At the same time, rod-like grains with a length of 10 μm and thicknesses of 1–2 μm were formed in all three samples. For the samples sintered at the low temperatures of 500 °C and 700 °C, they had grains with sizes almost the same as those of the as-activated mixtures, implying that no grain growth occurred if

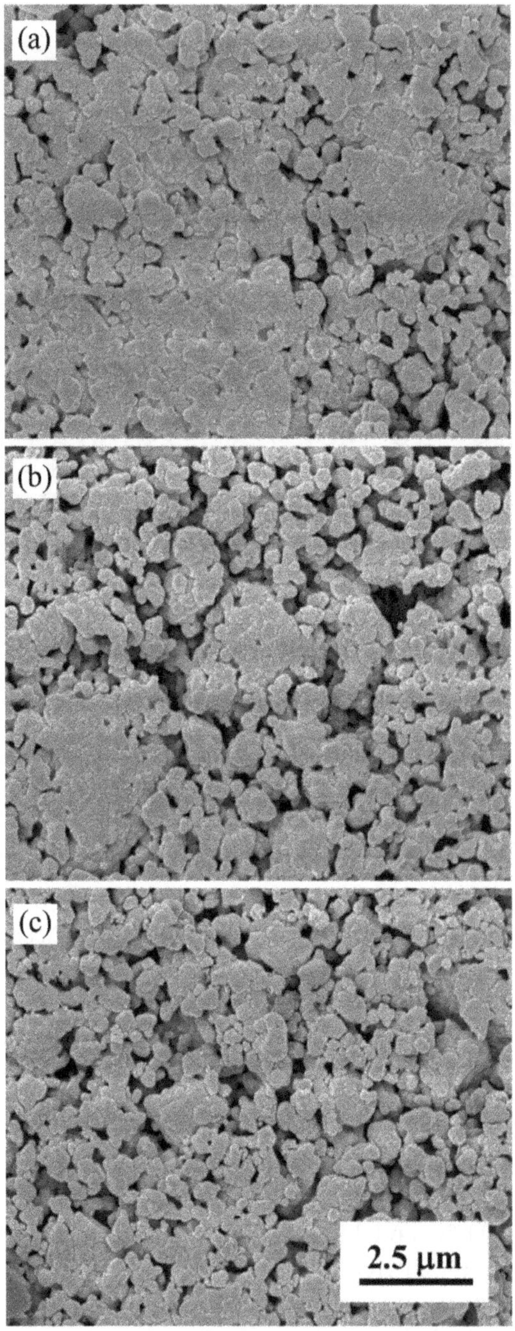

Figure 10.15. SEM images of the samples from the mixtures activated for different times, after sintering at 900 °C for 8 h: (a) 5 h, (b) 10 h and (c) 15 h. Reproduced with permission from [44]. Copyright 2002 Elsevier.

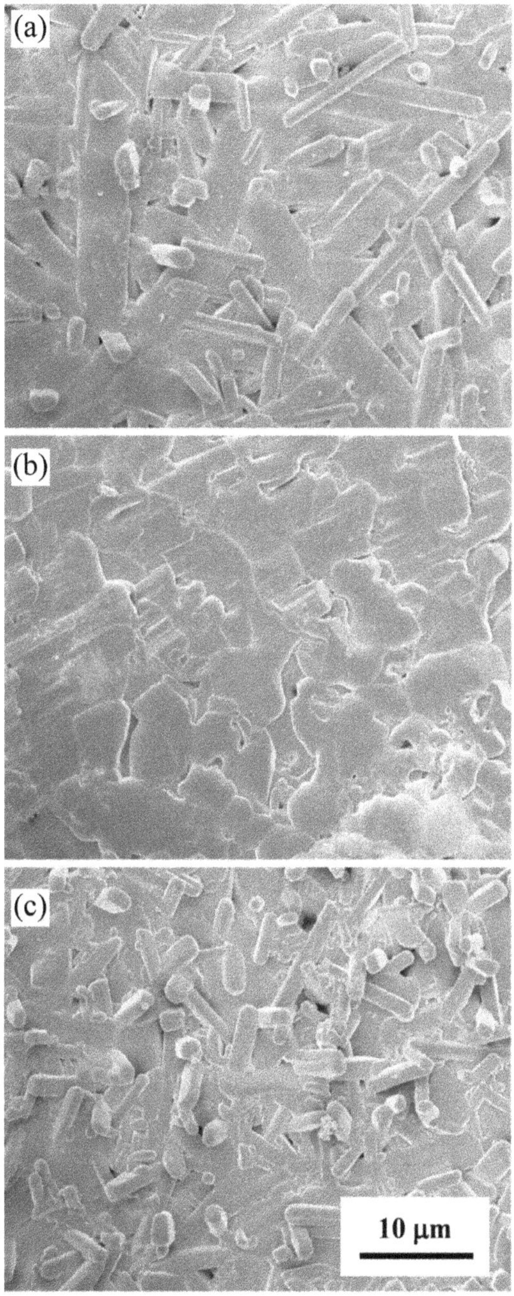

Figure 10.16. SEM images of the samples from the mixtures activated for different times, after sintering at 1100 °C for 8 h: (a) 5 h, (b) 10 h and (c) 15 h. Reproduced with permission from [44]. Copyright 2002 Elsevier.

the sintering temperature was <900 °C. Obviously, the densification temperature was much lower than those observed for other methods, demonstrating the advantages of mechanochemical activation in the fabrication of ceramics.

$ZnNb_2O_6$ ceramics are also suitable candidates for applications as microwave dielectric materials [45–49]. The columbite-structured $ZnNb_2O_6$ is also widely employed as a precursor to fabricate lead zinc niobate ($PbZn_{1/3}Nb_{2/3}O_3$ or PZN) based ferroelectric ceramics, because PZN-based ceramics could not be obtained directly from the oxide precursors using the conventional solid-state reaction approach [50–52]. Conventionally, $ZnNb_2O_6$ powder was synthesized using the solid-state reaction process, where ZnO and Nb_2O_5 reacted during the calcination at temperatures in the range 800 °C–1100 °C. Similarly to $MgNb_2O_6$, $ZnNb_2O_6$ powder has also been prepared by various wet-chemical methods in order to achieve high sinterability [53–55].

ZnO and Nb_2O_5 mixtures with partial reaction to $ZnNb_2O_6$ were treated with high-energy activation for times of 5–15 h [56]. The activated powders were of very high reactivity, so that almost single-phase $ZnNb_2O_6$ was obtained, after the mixtures were sintered at a low temperature of 500 °C. The grain growth and microstructure evolution of the samples from the mixtures activated for different times were studied as a function of sintering temperature. Fully dense $ZnNb_2O_6$ ceramics, with average grain sizes of 5 μm, 3.5 μm and 2.4 μm, were fabricated after sintering at 1100 °C from the mixtures activated for 5 h, 10 h and 15 h, respectively. Commercial powders of ZnO (99.9+% purity, Aldrich Chemical Company Inc., USA) and Nb_2O_5 (99.9+% purity, Aldrich Chemical Company Inc., USA) were used as the starting materials.

Figure 10.17 shows the XRD patterns of the mixture of ZnO and Nb_2O_5 activated for different times. The diffraction peaks belonged to ZnO and Nb_2O_5, as well as the $ZnNb_2O_6$ phase. Therefore, it was concluded that the reaction between ZnO and Nb_2O_5 to $ZnNb_2O_6$ was partially induced as a result of the high-energy

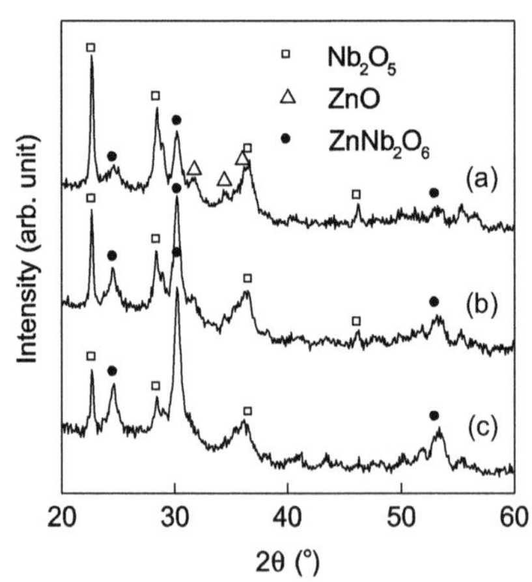

Figure 10.17. XRD patterns of the mixtures of ZnO and Nb_2O_5 after mechanochemical activation for different times: (a) 5 h, (b) 10 h, and (c) 15 h. Reproduced with permission from [56]. Copyright 2002 Elsevier.

mechanochemical activation for 5 h. The weakened peaks of ZnO and Nb_2O_5 implied that they were highly refined. In addition, the newly formed $ZnNb_2O_6$ phase also displayed a small particle size.

According to the XRD patterns, the crystal sizes of the ZnO, Nb_2O_5 and $ZnNb_2O_6$ phases were 32 nm, 56 nm and 28 nm, respectively. These values were slightly lower than those observed in the SEM images, because the SEM images reflect the particle sizes. As the milling time was increased the quantity of $ZnNb_2O_6$ gradually increased, while the amounts of ZnO and Nb_2O_5 were reduced. In addition, the reaction between ZnO and Nb_2O_5 was still incomplete after the mixture was activated for 15 h.

Figure 10.18 shows the XRD patterns of the samples from the activated mixtures after sintering at different temperatures for 8 h. As demonstrated in figure 10.18(a), $ZnNb_2O_6$ was the major phase in all the samples after sintering at 500 °C. This observation indicated that the residual ZnO and Nb_2O_5 reacted to form $ZnNb_2O_6$. The phase formation temperature of $ZnNb_2O_6$ from the activated mixture of ZnO and Nb_2O_5 was relatively low, compared to that used in the solid-state reaction, which is usually >800 °C. The crystal size of the $ZnNb_2O_6$ was about 30 nm according to the XRD, which was comparable to that of the as-activated mixtures.

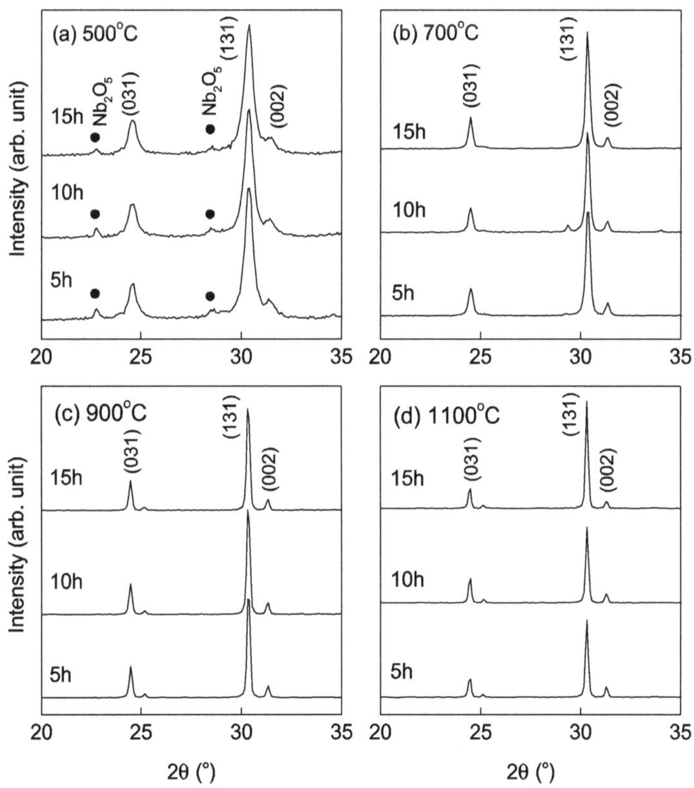

Figure 10.18. XRD patterns of the samples from the activated mixtures after sintering at different temperatures. Reproduced with permission from [56]. Copyright 2002 Elsevier.

In other words, the samples exhibited no grain growth although the reaction occurred.

As the sintering temperature was increased to 700 °C, all the samples possessed $ZnNb_2O_6$ as the single phase. At the same time, the diffraction peaks were greatly increased and sharpened, suggesting that grain growth of the $ZnNb_2O_6$ phase took place during the sintering process. However, SEM images revealed that the grain size of the samples was not >100 nm, as presented later. In this regard, the enhanced diffraction peaks of the samples were attributed to the improvement in crystallinity of the $ZnNb_2O_6$ phase. The XRD patterns of the samples sintered at 900 °C and 1100 °C were almost unchanged, with the diffraction peaks being enhanced gradually, reflecting the grain growth of the samples.

SEM images of the samples from the mixtures with different activation times after sintering at temperatures of 700 °C–1100 °C are illustrated in figures 10.19–10.21. The grain size of the samples gradually increased with increasing sintering temperature. Specifically, the average grain sizes of the samples from the mixture activated for 5 h, after sintering at 700 °C, 900 °C and 1100 °C, were 0.1 μm, 1.5 μm and 4 μm, respectively. At a fixed sintering temperature, the grain size decreased with increasing activation time. This observation has not been clearly explained, with a possible reason being contamination from the milling media during high speed milling. It was found that, after sintering at 900 °C, the samples possessed a fairly dense microstructure, in particular for the samples from the mixtures activated for 10 h and 15 h. In addition, the $ZnNb_2O_6$ ceramics reached a relative density of 97% after the activated mixtures were sintered at 1100 °C for 8 h.

Figure 10.22 shows the densification curves of the mixtures activated for different times. In the shrinkage curves, an expansion began at about 600 °C, which was readily ascribed to the reaction of the residual ZnO and Nb_2O_5 in the activated mixtures, because the magnitude of the expansion was gradually decreased with increasing activation time. A similar result was observed when preparing lead zirconate titanate ($PbZr_{0.52}Ti_{0.48}O_3$, or PZT) ceramics from the oxide mixtures, in which PZT was partially formed after mechanochemical activation for relatively short times [57].

The densification characteristics of the samples were consistent with the XRD results, as illustrated in figures 10.17 and 10.18(a). The expansion could be attributed to the fact that the volume of the newly formed $ZnNb_2O_6$ was larger than the summation of those of the two oxides. However, the temperature of the reaction of ZnO and Nb_2O_5 was higher than that according to the XRD patterns. This was not surprising because the densification behavior was monitored in a dynamic way at a heating rate of 10 °C min^{-1} without dwelling, while the sintering occurred in a static state with a long dwelling time.

There is increasingly demand for communication devices at microwave frequencies, in particular for 5G systems [58]. Microwave applications require dielectric materials with a high dielectric constant, low dielectric loss tangent and minimized temperature coefficient of resonant frequency [59]. Zirconium titanate based ceramics, with a formula of $(Zr_{1-x}Ti_x)O_2$, have been considered as suitable candidates for such applications [60].

Elsevier.

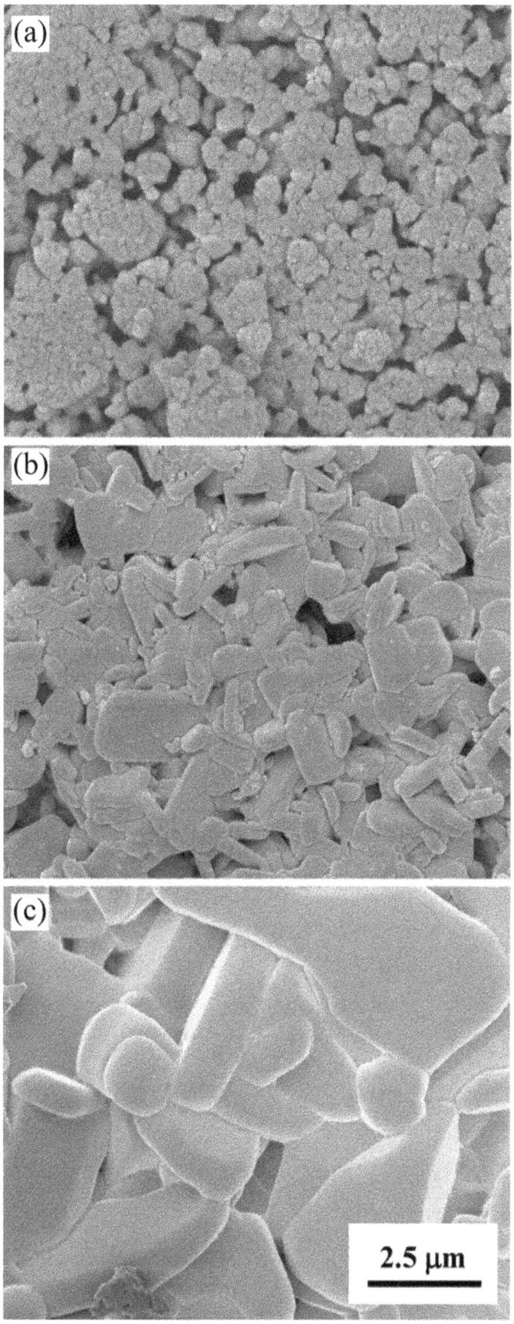

Figure 10.19. SEM images of the samples from the mixture activated for 5 h, after sintering for 8 h at different temperatures: (a) 700 °C, (b) 900 °C and (c) 1100 °C. Reproduced with permission from [56]. Copyright 2002 Elsevier.

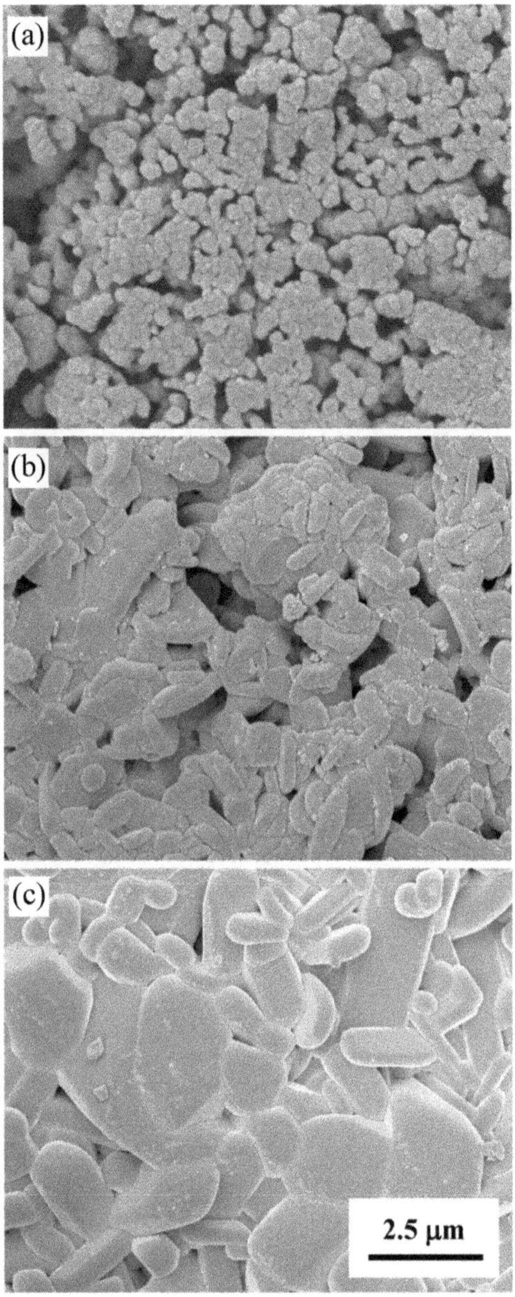

Figure 10.20. SEM images of the samples from the mixture activated for 10 h, after sintering for 8 h at different temperatures: (a) 700 °C, (b) 900 °C and (c) 1100 °C. Reproduced with permission from [56]. Copyright 2002 Elsevier.

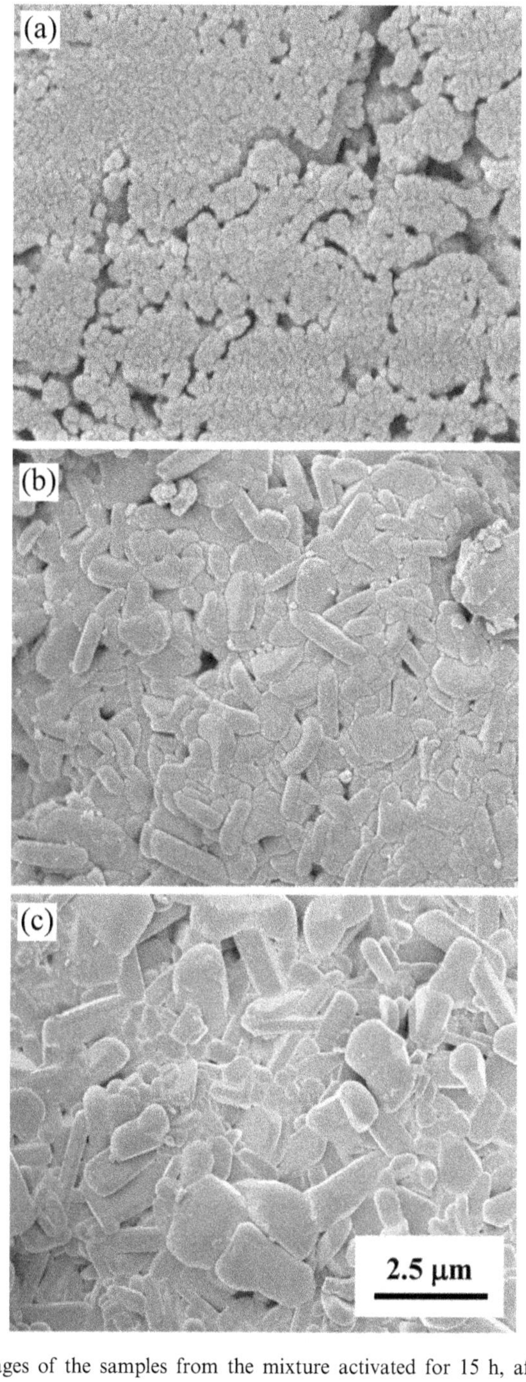

Figure 10.21. SEM images of the samples from the mixture activated for 15 h, after sintering for 8 h at different temperatures: (a) 700 °C, (b) 900 °C and (c) 1100 °C. Reproduced with permission from [56]. Copyright 2002 Elsevier.

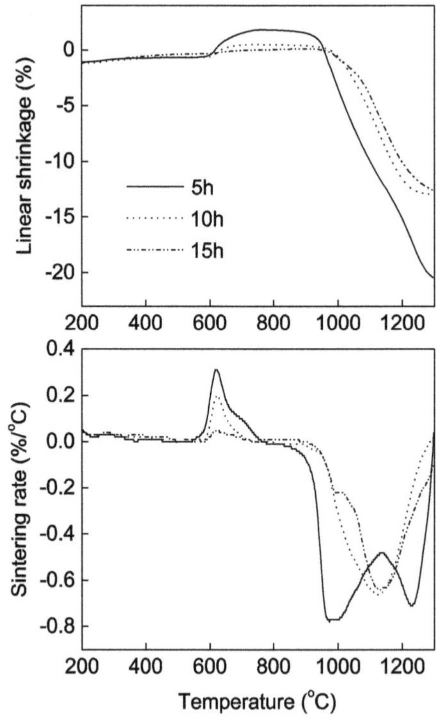

Figure 10.22. Densification curves of the mixtures activated for different times. Reproduced with permission from [56]. Copyright 2002 Elsevier.

Zirconium titanate ceramics could be fabricated using the solid-state reaction method, with oxides of ZrO_2 and TiO_2 as the precursors [61]. In this case, the mixture of ZrO_2 and TiO_2 powders was calcined at a temperature of 1100 °C, while the sintering temperatures were in the range 1400 °C–1600 °C. To reduce the sintering temperature of the $(Zr_{1-x}Ti_x)O_2$ ceramics, it was necessary to use sintering aids [62]. The use of sintering aids could degrade the dielectric properties of the materials. As a result, the synthesis of zirconium titanate powders and the fabrication of zirconium titanate ceramics without the presence of sintering aids are of technological significance.

Solid-state solutions, $(Zr_{1-x}Ti_x)O_2$ (x = 0.44, 0.48, 0.52 and 0.60), with a srilankite structure were prepared with ZrO_2 and TiO_2 oxides as the starting materials using high-energy mechanochemical activation [63]. The mixtures were activated for times of 3 h, 8 h, 20 h and 30 h. The activated mixtures were used to obtain the corresponding ceramics by sintering at different temperatures. $(Zr_{1-x}Ti_x)O_2$ ceramics with nearly full densification could be achieved at relatively low temperatures.

Figure 10.23 shows the XRD patterns for the mixtures of ZrO_2 and TiO_2, with different compositions, which were mechanochemically activated for different times. The four groups of samples, with different ratios of ZrO_2 and TiO_2, exhibited similar phase evolution with activation time. Figure 10.24 depicts the XRD patterns of the commercial ZrO_2 and TiO_2 powders. As demonstrated in figure 10.23(a), the peak

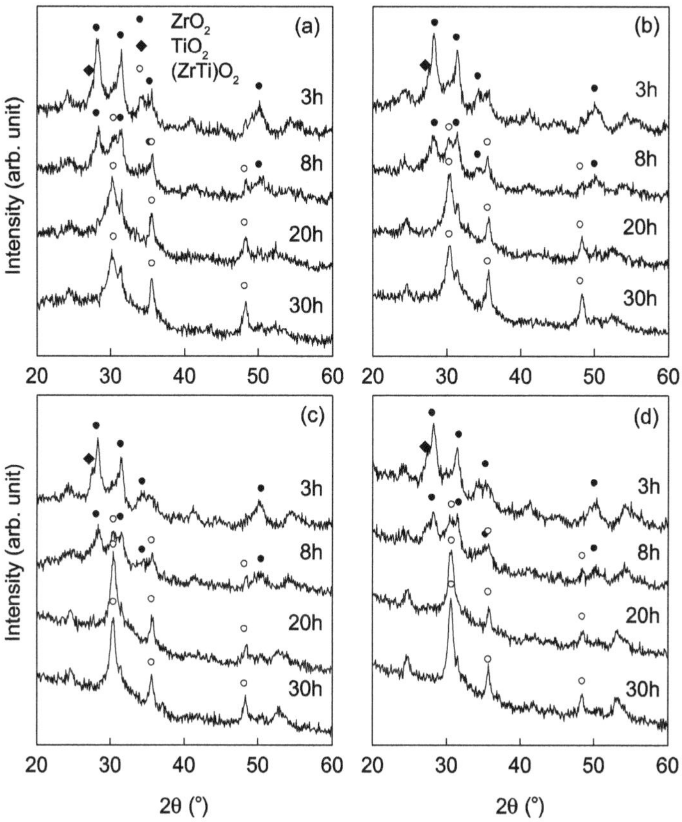

Figure 10.23. XRD patterns of the as-activated mixtures of ZrO_2 and TiO_2 oxides with a nominal composition of $(Zr_{1-x}Ti_x)O_2$ having different values of x: (a) 0.44, (b) 0.48, (c) 0.52 and (d) 0.60. Reproduced with permission from [63]. Copyright 2002 Elsevier.

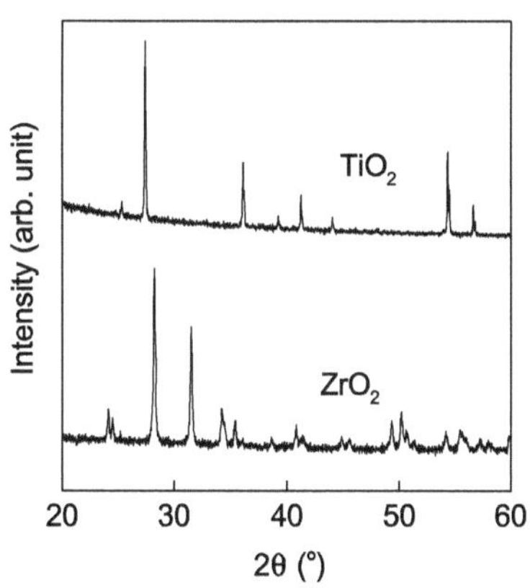

Figure 10.24. XRD patterns of the commercial ZrO_2 and TiO_2 oxides. Reproduced with permission from [63]. Copyright 2002 Elsevier.

intensity of the newly formed zirconium titanate phase gradually increased as the activation time was increased. For example, after activation for 3 h the diffraction peaks of the two oxides were largely weakened, compared to those of the samples without activation, as revealed in figure 10.24. The reduction in peak intensity was attributed to the refinement of the particles of the oxide powders. As the activation time was increased to 8 h, the diffraction peaks were further broadened, suggesting further refinement of the powders.

In addition, a new diffraction peak at $2\theta \approx 30°$ was observed, corresponding to the newly formed zirconium titanate phase with an orthorhombic crystal structure (srilankite). In other words, ZrO_2 and TiO_2 reacted as the activation proceeded to 8 h. However, the diffraction peaks of the starting oxides were still superior to those of the new phase. As the activation time was increased to 20 h, single-phase srilankite was achieved, indicating that the reaction was complete. The crystal size of the srilankite phase was about 30 nm, according to the XRD patterns with the Scherrer formula. A further increase of the activation time to 30 h led to a weak enhancement of the diffraction peaks of the $(Zr_{1-x}Ti_x)O_2$ phase.

In comparison, an amorphous state was observed as a result of mechanochemical activation in a separate study [64]. This different observation could be ascribed to the differences in the properties of the starting materials or equipment. For example, a Pulverisette type 7 planetary micro-mill was employed in that study, while a type 5 planetary mill was adopted in this case. Also, the difference in the parameters of the activation could be responsible for the observation as an additional factor. WC vials and balls were used as the milling media. In a typical experiment, 3 g of powder was placed in a cylindrical vial with a volume of 45 ml, together with four balls 12 mm in diameter and ten balls 10 mm in diameter. The amorphous state crystallized at relatively low temperatures.

Figure 10.25 shows the lattice constants of the zirconium titanate phase as a function of composition. The lattice parameters were obtained with the XRD patterns of the samples mechanochemically activated for 30 h. The three values were monotonically decreased with increasing content of titanium in the samples, simply because the Ti^{4+} ion has a smaller radius than the Zr^{4+} ion. In other words, the solid-state solutions of $(Zr_{1-x}Ti_x)O_2$ were achieved during the high-energy mechanochemical activation process.

XRD patterns of the samples from the activated mixtures with different compositions, after sintering at 700 °C, 900 °C and 1100 °C, are plotted in figures 10.26–10.28. After sintering at 700 °C, the diffraction patterns of the samples almost did not vary, compared to the as-activated mixtures. For example, the samples from the mixture activated for 3 h contained just ZrO_2 and TiO_2, suggesting that the reaction between the two oxides did not take place. After sintering at 700 °C, the peak intensity of TiO_2 was increased. In the samples from the mixtures activated for 8 h, the peaks of the newly produced zirconium titanate phase strengthened. A single phase of zirconium titanate with high crystallinity was achieved in the samples from the mixtures activated for 20 h and 30 h.

In the samples sintered at 900 °C, the phase formation of zirconium titanate was further boosted. This was clearly demonstrated by the XRD patterns of the samples

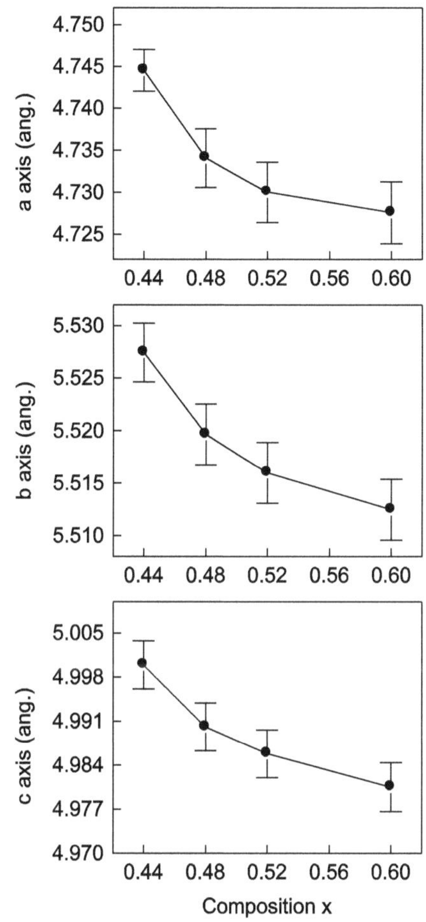

Figure 10.25. Lattice constants of the $(Zr_{1-x}Ti_x)O_2$ samples with an orthorhombic structure as a function of composition, according to the XRD patterns of the samples milled for 30 h. Reproduced with permission from [63]. Copyright 2002 Elsevier.

from the mixtures activated for 3 h and 8 h, after sintering at 900 °C and 700 °C, as revealed in figures 10.27 and 10.26, respectively. The zirconium titanate phase was visible in the samples from the mixtures activated for 3 h, while nearly single-phase was present in the samples from the mixtures activated for 8 h at these sintering temperatures. The phase formation temperature of the zirconium titanate compounds was much lower than those needed when the conventional solid-state reaction method was employed. This observation was attributed to the significant refinement of the starting oxide particles. The average grain size of the zirconium titanate samples from the mixtures activated for 20 h and 30 h was obviously increased, as reflected by the enhanced diffraction peaks.

Careful inspection of figures 10.26 and 10.27 indicated that the diffraction peaks of ZrO_2 and TiO_2 were present in the samples of $(Zr_{0.56}Ti_{0.44})O_2$ and $(Zr_{0.40}Ti_{0.60})O_2$, as demonstrated in figures 10.27(a) and (d), respectively. This observation

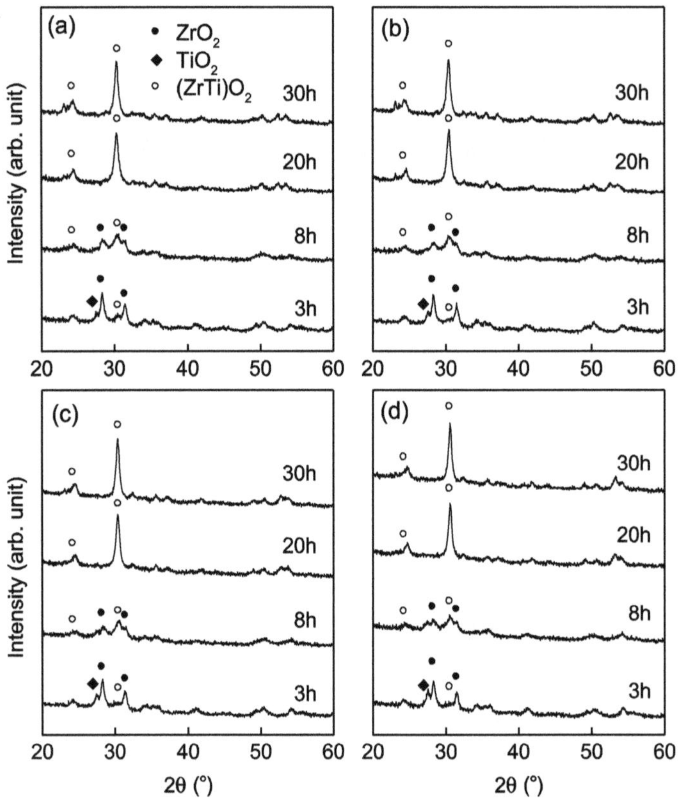

Figure 10.26. XRD patterns of the samples from the mixtures of $(Zr_{1-x}Ti_x)O_2$ with different compositions, after sintering at 700 °C: (a) 0.44, (b) 0.48, (c) 0.52 and (d) 0.60. Reproduced with permission from [63]. Copyright 2002 Elsevier.

suggested that the solid solution of $(Zr_{1-x}Ti_x)O_2$ was stable in the compositions between $(Zr_{0.52}Ti_{0.48})O_2$ and $(Zr_{0.48}Ti_{0.52})O_2$ at this sintering temperature. In other words, the stability range of the $(Zr_{1-x}Ti_x)O_2$ composition was dependent on the sintering temperature.

As the sintering temperature was increased to 1100 °C, the reaction between ZrO_2 and TiO_2 to form srilankite zirconium titanate in the samples from the mixtures activated for 3 h and 8 h was completely finished, as observed in figures 10.26(a) and 10.27(d). However, the peaks of ZrO_2 in $(Zr_{0.56}Ti_{0.44})O_2$ and TiO_2 in $(Zr_{0.40}Ti_{0.60})O_2$ were increased after sintering at 900 °C. In other words, the average grain size of the zirconium titanate phase was largely enhanced due to the sintering at higher temperatures. In addition, the samples of $(Zr_{0.52}Ti_{0.48})O_2$ and $(Zr_{0.48}Ti_{0.52})O_2$ were of single-phase, further indicating the stable solid solution range of $(Zr_{1-x}Ti_x)O_2$.

Figure 10.29 shows the SEM results of the as-activated mixture of $(Zr_{0.52}Ti_{0.48})O_2$ and the samples from the mixture activated for 3 h after sintering at 700 °C, 900 °C and 1100 °C. Figure 10.30 shows SEM images of the samples from the mixtures of $(Zr_{0.52}Ti_{0.48})O_2$ activated for 8 h, 20 h and 30 h at 1100 °C. The average grain sizes of the $(Zr_{0.52}Ti_{0.48})O_2$ samples before and after sintering at different temperatures, as

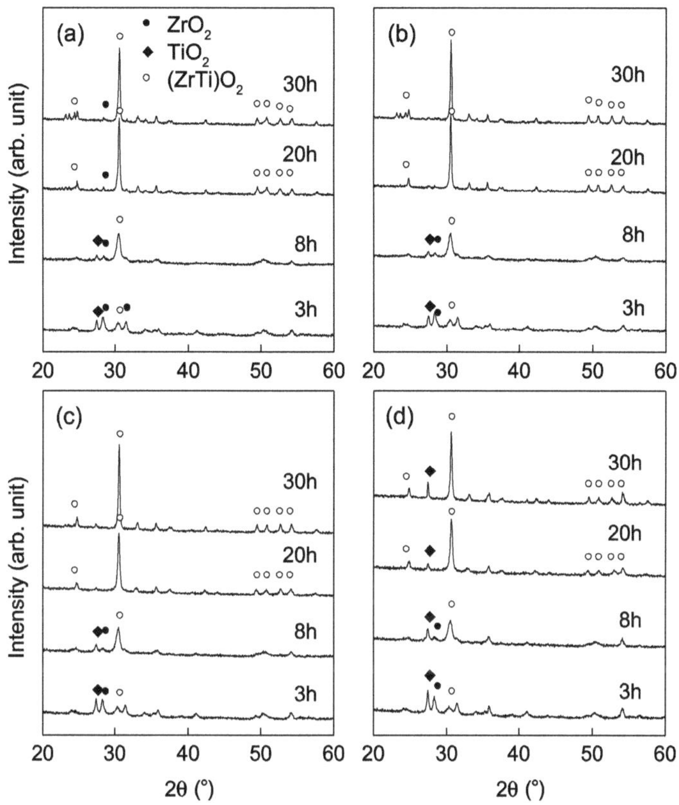

Figure 10.27. XRD patterns of the $(Zr_{1-x}Ti_x)O_2$ mixtures with different compositions, after sintering at 900 °C: (a) 0.44, (b) 0.48, (c) 0.52 and (d) 0.60. Reproduced with permission from [63]. Copyright 2002 Elsevier.

a function of activation times, are plotted in figure 10.31. The samples with other compositions exhibited a similar variation trend in microstructure and grain growth.

The average particle size of the as-activated mixtures was within 100 nm after milling for just 3 h, although the reaction between the constituent oxides was still not induced. The particle size values of the samples were very close to those derived from the XRD patterns with the Scherrer equation. As observed in figure 10.29(b), both the grain size and grain morphology of the samples were not altered after they were sintered at 700 °C for 8 h. The grain size increased just slightly as the sintering temperature was increased to 900 °C. However, the grain size quickly increased to 0.5–0.7 μm after the samples were sintered at 1100 °C.

In addition, for the samples sintered at 1100 °C for 8 h, the average grain size of the samples increased with increasing milling time. This observation could be ascribed to the fact that the mixtures activated for different times contained different compositions in terms of the phase and properties of morphology. The mixtures activated for short times contained residual ZrO_2 and TiO_2, so that the grain growth of the zirconium titanate phase took place along with the occurrence of densification. In contrast, the mixtures activated for longer times only experienced grain

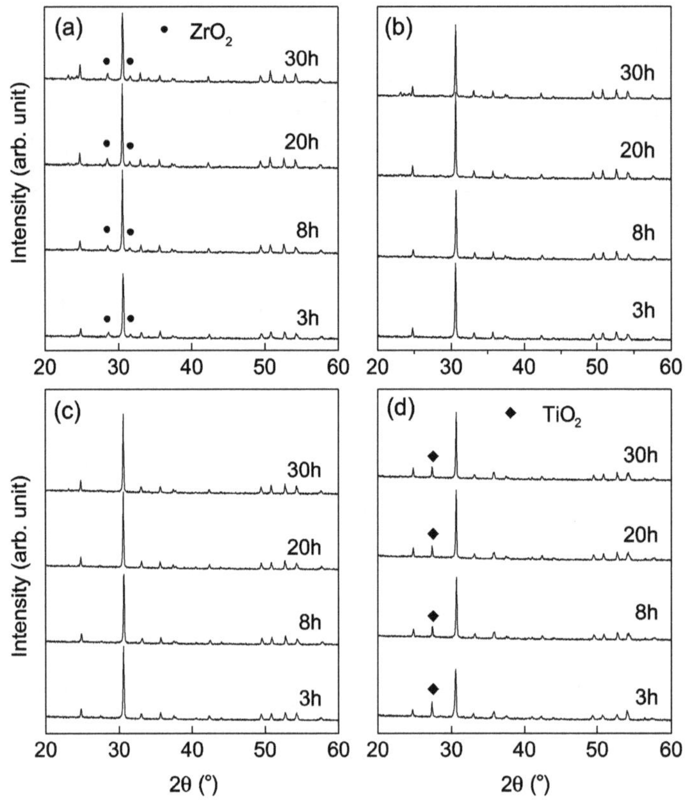

Figure 10.28. XRD patterns of the samples from the $(Zr_{1-x}Ti_x)O_2$ mixtures with different compositions, after sintering at 1100 °C: (a) 0.44, (b) 0.48, (c) 0.52 and (d) 0.60. Reproduced with permission from [63]. Copyright 2002 Elsevier.

growth. Nearly fully densified $(Zr_{1-x}Ti_x)O_2$ ceramics could be obtained from the mixtures after sintering at 1100 °C, as long as the mechanochemical activation time was sufficiently long. In fact, it is not very important whether the reaction is completed or not during the activation process. Instead, an incomplete reaction actually facilitated the reactive sintering process, thus enhancing the densification of the samples. A short activation time meant a short production cycle, thus leading to cost-effective products.

SnO_2 is known as cassiterite and has a tetragonal rutile structure with lattice parameters of a = 0.4738 nm and c = 0.3188 nm. Rutile TiO_2 has the lattice constants of a = 0.4593 nm and c = 0.2959 nm. Because they have a similar crystal structure, solid solutions of $(Sn_xTi_{1-x})O_2$ could be obtained [65–67]. $(Sn_xTi_{1-x})O_2$ nanoparticles exhibit a strong catalytic performance, while $(Sn_xTi_{1-x})O_2$ ceramics could be used as microwave dielectric materials.

As an example, solid solution $Sn_{0.5}Ti_{0.5}O_2$ was derived from a mixture of SnO_2 and TiO_2 that was mechanochemically activated for 5 h [68]. Although the reaction between TiO_2 and SnO_2 was not triggered during the activation process, the particle size was significantly affected. After activation for 10 h, single-phase $Sn_{0.5}Ti_{0.5}O_2$

Figure 10.29. SEM images of the $(Zr_{0.52}Ti_{0.48})O_2$ samples from the mixture mechanochemically activated for 3 h: (a) as-milled and annealed for 8 h, (b) 700 °C, (c) 900 °C and (d) 1100 °C. Reproduced with permission from [63]. Copyright 2002 Elsevier.

Figure 10.30. SEM images of the $(Zr_{0.52}Ti_{0.48})O_2$ samples from the mixtures activated for different times after sintering at 1100 °C: (a) 8 h, (c) 20 h and (d) 30 h. Reproduced with permission from [63]. Copyright 2002 Elsevier.

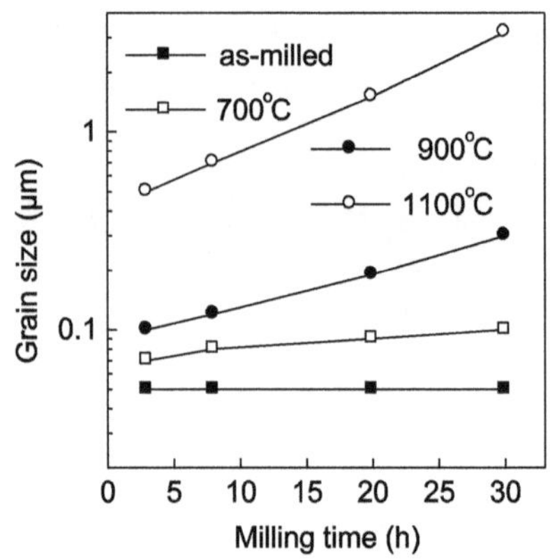

Figure 10.31. Grain sizes of the $(Zr_{0.52}Ti_{0.48})O_2$ samples before and after sintering as a function of activation time. Reproduced with permission from [63]. Copyright 2002 Elsevier.

was formed. The as-activated mixtures were used to develop $Sn_{0.5}Ti_{0.5}O_2$ ceramics by sintering at different temperatures. However, it was found that the $Sn_{0.5}Ti_{0.5}O_2$ phase was decomposed after sintering at temperatures >900 °C. For example, sintering at 1000 °C resulted in phase separation, forming a Sn-rich phase with a composition of $Sn_{0.77}Ti_{0.23}O_2$ and a Ti-rich phase with a formula of $Sn_{0.17}Ti_{0.83}O_2$. However, when the solid solution of $Sn_xTi_{1-x}O_2$ forms cassiterite SnO_2 and rutile TiO_2 using the conventional solid-state reaction process, the phase formation temperature was >1200 °C [69].

The XRD patterns of the mixtures of SnO_2 and TiO_2 mechanochemically activated for 5 h, 10 h and 15 h are depicted in figure 10.32(a). Figure 10.32(b) shows the XRD patterns of the commercial SnO_2 and TiO_2 powders without activation. The sample activated for 5 h contained the component oxides without the formation of the solid solution, but the powders were greatly refined because the diffraction peaks were tremendously broadened and reduced. After activation for 10 h, the solid solution $Sn_{0.5}Sn_{0.5}O_2$ was produced as a new phase in the XRD pattern. This new phase was not changed as the activation time was prolonged.

Figure 10.33 depicts XRD patterns of the samples from the activated mixtures after sintering at temperatures in the range 500 °C–1100 °C for 8 h. After sintering at 500 °C, the XRD patterns of the samples were no different from those of the as-activated mixtures. This result further indicated that the solid solution was not formed in the mixture activated for 5 h. As the sintering temperature was increased to 700 °C, the single phase of the solid solution $Sn_{0.5}Ti_{0.5}O_2$ was observed in all samples, as revealed in figure 10.33(b). In other words, the two oxides reacted to form the solid solution from the mixture activated for 5 h at this temperature. This phase formation temperature was much lower than those needed when using the conventional solid-state reaction and wet-chemical methods [70].

The XRD patterns of the samples sintered at 700 °C and 900 °C did not vary in terms of peak position, while the peak intensity gradually increased. This observation suggested that the sintering only resulted in grain growth and enhanced crystallization, while the solid solution was stable up to 900 °C, as demonstrated

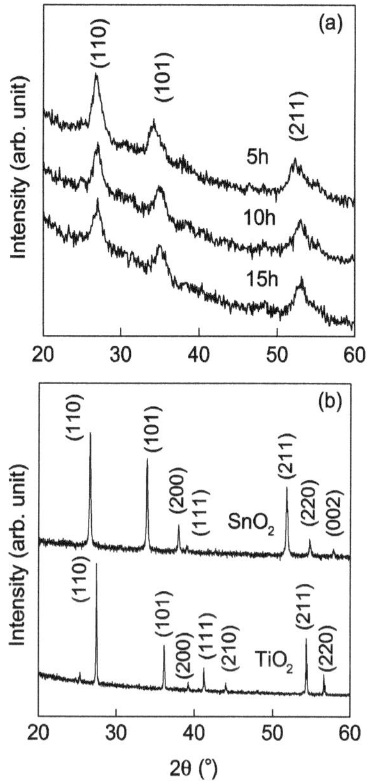

Figure 10.32. (a) XRD patterns of the mixtures activated for different times. (b) XRD patterns of TiO_2 and SnO_2. Reproduced with permission from [68]. Copyright 2002 Elsevier.

in figures 10.33(c) and (d). The samples sintered at 900 °C had lattice constants of $a = 0.4673$ nm and $c = 0.3067$ nm, both of which were between those of SnO_2 and TiO_2, as stated previously.

However, after sintering at 1000 °C, the diffraction peaks of the XRD patterns of the samples diverged, as seen in figure 10.33(d). In other words, phase separation occurred in the samples at this temperature. The two sets of patterns could be ascribed to two phases, both with rutile structure. The major phase had lattice parameters of $a = 0.4687$ nm and $c = 0.3144$ nm, while the minor phase possessed lattice constants of $a = 0.4619$ nm and $c = 0.2997$ nm. By considering the lattice parameters of SnO_2 and TiO_2, it was estimated that the major phase was rich in Sn, while the minor phase was rich in Ti. Accordingly, the two phase were $Sn_{0.77}Ti_{0.23}O_2$ and $Sn_{0.17}Ti_{0.83}O_2$. Therefore, the solubility of Ti^{4+} in SnO_2 was higher than that of Sn^{4+} in TiO_2. This observation could be attributed to the fact that the ionic radius of Ti^{4+} (0.68 Å) was smaller than that of Sn^{4+} (0.71 Å) [70]. In addition, these two phases were not altered as the sintering temperature was further increased to 1100 °C, as illustrated in figure 10.33(e).

Figure 10.34 shows SEM images of the samples from the mixture activated for 10 h after sintering at temperatures of 900 °C–1100 °C. The SEM images of the samples

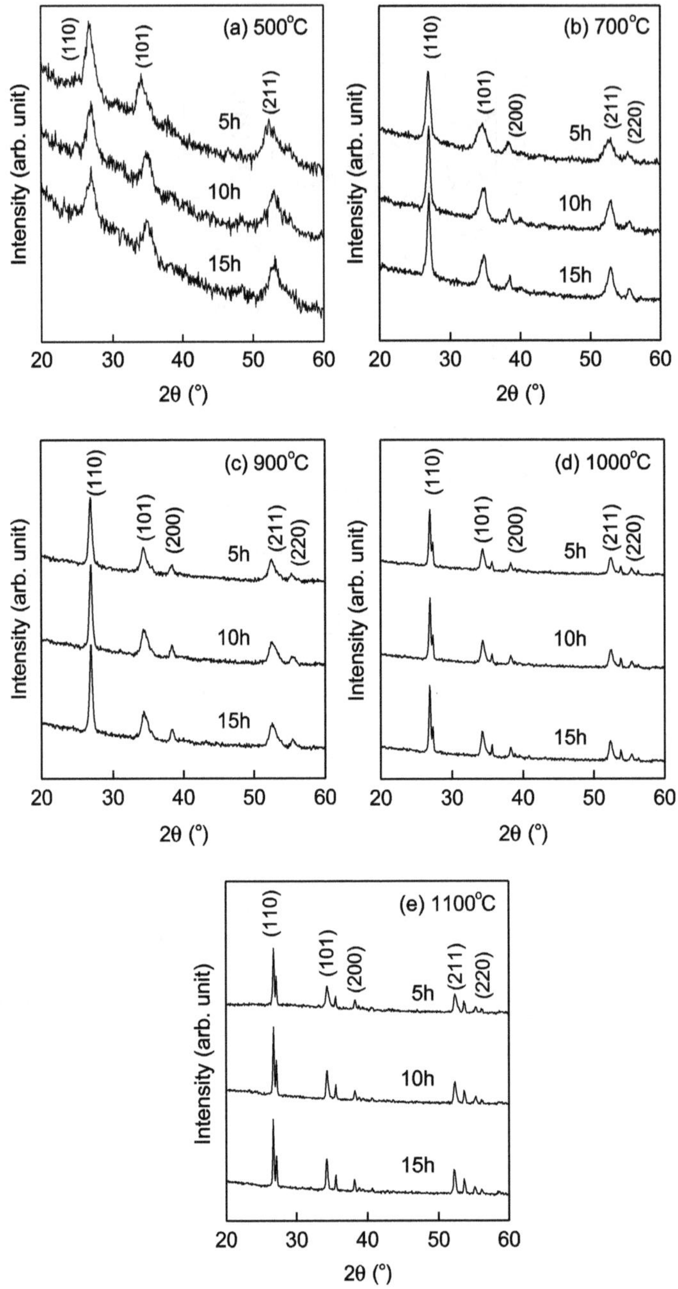

Figure 10.33. XRD patterns of the samples from the mixtures activated for different times, after sintering at different temperatures for 8 h. Reproduced with permission from [68]. Copyright 2002 Elsevier.

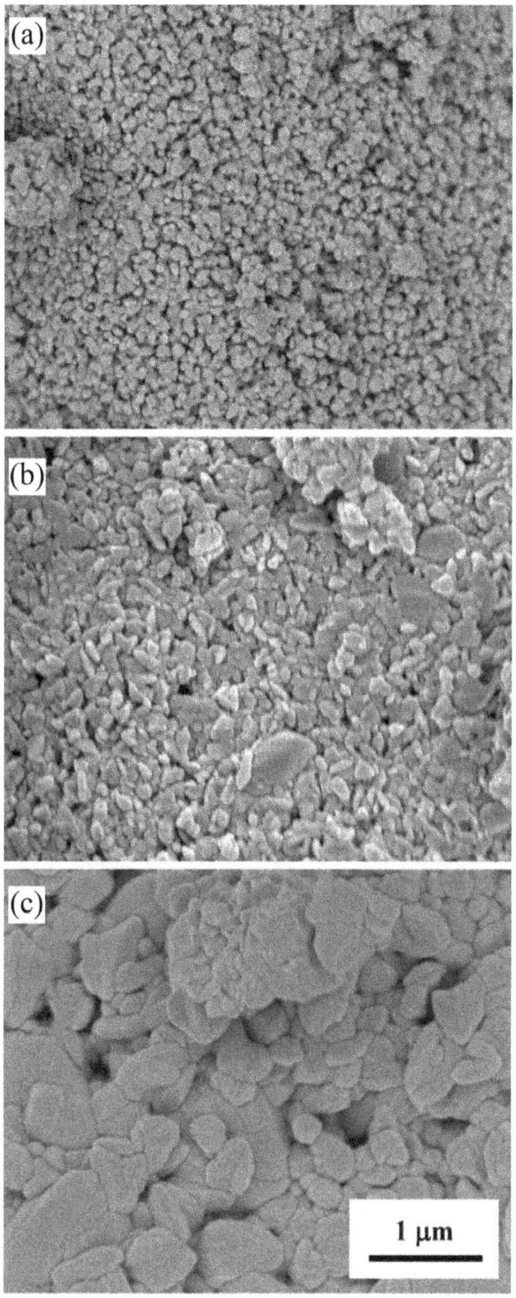

Figure 10.34. SEM images of the samples from the mixture activated for 10 h after sintering at different temperatures: (a) 900 °C, (b) 1000 °C and (c) 1100 °C. Reproduced with permission from [68]. Copyright 2002 Elsevier.

from the mixtures activated for 5 h and 15 h, after they were sintered at 1100 °C for 8 h, are shown in figure 10.35. The sample sintered at 900 °C had grain sizes similar to those of the as-activated mixtures and samples sintered at low temperature, suggesting that no grain growth took place at this temperature, as revealed in figure 10.34(a). However, the XRD patterns indicated that the diffraction peaks were narrowed and enhanced after sintering 700 °C. This difference implied that the samples sintered at temperatures $\leqslant 700$ °C experienced enhanced crystallization instead of grain growth.

Sintering at 1000 °C resulted in grain growth, with grain sizes of 0.2–0.3 μm. In addition, there were also large grains with sizes of 0.5–0.6 μm. After sintering at 1100 °C, the average size was increased by about 1 μm. The samples from the mixtures activated for 5 h and 15 h encountered a similar grain growth behavior. At a fixed sintering temperature, the grain size of the samples was increased with increasing activation time, as observed in figures 10.34(c) and 10.35. Therefore, the grain sizes of the samples sintered at a given temperature could simply be controlled by using

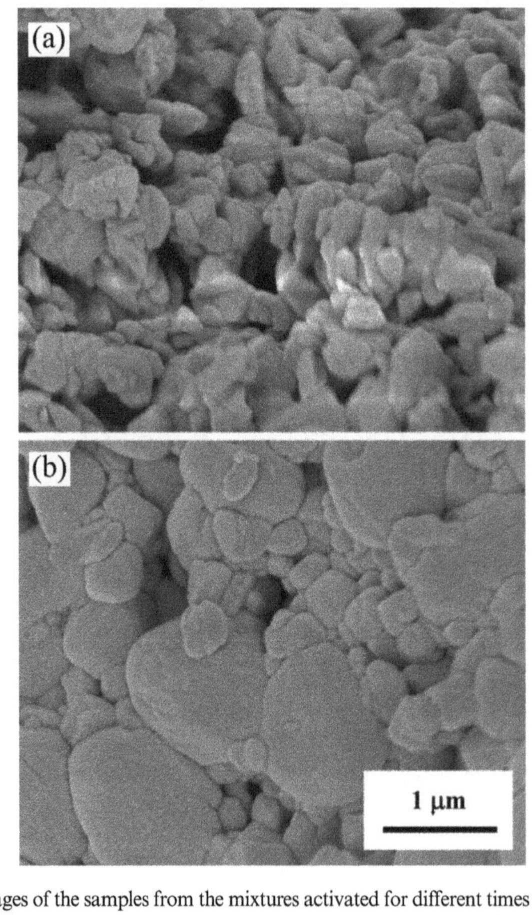

Figure 10.35. SEM images of the samples from the mixtures activated for different times after sintering at 1100 °C: (a) 5 h and (b) 15 h. Reproduced with permission from [68]. Copyright 2002 Elsevier.

mixtures activated for different time durations. However, this observation was different from that observed for other materials, where the grain size decreased with prolonged activation. This could be because the effect of mechanochemical activation on the grain growth behaviors of different materials is different.

10.3 Summary

The effects of mechanochemical activation (with WC milling media) on the phase formation, grain growth and microstructures of selected oxide ceramics have been presented and discussed in this chapter. It was demonstrated that, depending on the properties of the materials, mechanochemical activation could either induce direction phase formation of the desired compounds or reduce their phase formation temperature. Although the underlying mechanisms have not been well clarified, it is suggested that the energy requirement for a given reaction could play an important role in determining whether it could be induced by mechanochemical activation or not, because the level of the energy applied to the milled systems was almost the same, given that the milling conditions used for all the experiments were nearly fixed. In addition, properties such as electrical, dielectric and magnetic properties should be studied to further demonstrate the advantages of the mechanochemical activation process.

Acknowledgments

Shenzhen Technology University (SZTU) is acknowledged for the financial support of a start-up grant (2018) and also the Natural Science Foundation of Top Talent of SZTU (grant no. 2019010801002).

References

[1] Kong L B, Ma J and Huang H 2002 $MgAl_2O_4$ spinel phase derived from oxide mixture activated by a high-energy ball milling process *Mater. Lett.* **56** 238–43

[2] Bakkar W and Lindsay J G 1967 Reactive magnesia spinel, preparation and properties *Am. Ceram. Soc. Bull.* **46** 1094–97

[3] White G S, Jones R V and Crawford J H 1982 Optical spectra of $MgAl_2O_4$ crystals exposed to ionizing radiation *J. Appl. Phys.* **53** 265–70

[4] Yoshida H, Biswas P, Johnson R and Mohan M K Flash-sintering of magnesium aluminate spinel ($MgAl_2O_4$) ceramics *J. Am. Ceram. Soc.* **100** 554–62

[5] Balabanov S S, Belyaev A V, Novikova A V, Permin D A, Rostokina E Y and Yavetskiy R P 2018 Densification peculiarities of transparent $MgAl_2O_4$ ceramics—effect of LiF sintering additive *Inorg. Mater.* **54** 1045–50

[6] Fu P, Lu W Z, Lei W, Xu Y, Wang X H and Wu J M 2013 Transparent polycrystalline $MgAl_2O_4$ ceramic fabricated by spark plasma sintering: microwave dielectric and optical properties *Ceram. Int.* **39** 2481–87

[7] Bonnefont G, Fantozzi G, Trombert S and Bonneau L 2012 Fine-grained transparent $MgAl_2O_4$ spinel obtained by spark plasma sintering of commercially available nanopowders *Ceram. Int.* **38** 131–40

[8] Esposito L, Piancastelli A and Martelli S 2013 Production and characterization of transparent $MgAl_2O_4$ prepared by hot pressing *J. Eur. Ceram. Soc.* **33** 737–47

[9] Morita K, Kim B N, Yoshida H and Hiraga K 2010 Densification behavior of a fine-grained $MgAl_2O_4$ spinel during spark plasma sintering (SPS) *Scr. Mater.* **63** 565–68

[10] Morita K, Kim B N, Yoshida H, Zhang H B, Hiraga K and Sakka Y 2012 Effect of loading schedule on densification of $MgAl_2O_4$ spinel during spark plasma sintering (SPS) processing *J. Eur. Ceram. Soc.* **32** 2303–9

[11] Bratton R J 1971 Sintering and grain-growth kinetics of $MgAl_2O_4$ *J. Am. Ceram. Soc.* **52** 141–43

[12] Gusmano G, Nunziante P, Traversa E and Chiozzini G 1991 The mechanism of $MgAl_2O_4$ spinel formation from the thermal decomposition of coprecipitated hydroxides *J. Eur. Ceram. Soc.* **7** 31–9

[13] Bratton R J 1969 Coprecipitates yielding $MgAl_2O_4$ spinel powders *Am. Ceram. Soc. Bull.* **48** 759–62

[14] Varnier O, Hovnanian N, Larbot A, Bergez P, Cot L and Charpin J 1994 Sol–gel synthesis of magnesium aluminum spinel from a heterometallic alkoxide *Mater. Res. Bull.* **29** 479–88

[15] Yang N and Chang L 1992 Structural inhomogeneity and crystallization behavior of aerosol-reacted $MgAl_2O_4$ powders *Mater. Lett.* **15** 84–8

[16] Pati R K and Pramanik P 2000 Low-temperature chemical synthesis of nanocrystalline $MgAl_2O_4$ spinel powder *J. Am. Ceram. Soc.* **83** 1822–24

[17] Ikesue A, Kinoshita T, Kamata K and Yoshida K 1995 Fabrication and optical properties of high-performance polycrystalline Nd:YAG ceramics for solid-state lasers *J. Am. Ceram. Soc.* **78** 1033–40

[18] Chaim R, Kalina M and Shen J Z 2007 Transparent yttrium aluminum garnet (YAG) ceramics by spark plasma sintering *J. Eur. Ceram. Soc.* **27** 3331–37

[19] Chen Z H, Li J T, Hu Z G and Xu J J 2008 Fabrication of YAG transparent ceramics by two step sintering process *J. Inorg. Mater.* **23** 130–34

[20] Gong H, Zhang J, Tang D Y, Xie G Q, Huang H and Ma J 2011 Fabrication and laser performance of highly transparent Nd:YAG ceramics from well-dispersed Nd:Y_2O_3 nano-powders by freeze-drying *J. Nanopart. Res.* **13** 3853–60

[21] Ba X W *et al* 2013 Optimization of dispersing agents for preparing YAG transparent ceramics *J. Rare Earths* **31** 507–11

[22] Esposito L, Piancastelli A, Bykov Y, Egorov S and Eremeev A 2013 Microwave sintering of Yb:YAG transparent laser ceramics *Opt. Mater.* **35** 761–65

[23] Hostasa J, Esposito L, Alderighi D and Pirri A 2013 Preparation and characterization of Yb-doped YAG ceramics *Opt. Mater.* **35** 798–803

[24] Kong L B, Ma J and Huang H 2002 Low temperature formation of yttrium aluminum garnet from oxides via a high-energy ball milling process *Mater. Lett.* **56** 344–48

[25] Sakurai K and Guo X 2001 Mechanical solid-state formation of $Y_{1-x}Ce_xAlO_3$ and its application as an x-ray scintillator *Mater. Sci. Eng.* A **304–306** 403–7

[26] Lee S H, Kochawattana S, Messing G L, Dumm J Q, Quarles G and Castillo V 2006 Solid-state reactive sintering of transparent polycrystalline Nd:YAG ceramics *J. Am. Ceram. Soc.* **89** 1945–50

[27] Li J, Ikegami T, Lee J and Mori T 2000 Well-sinterable $Y_3Al_5O_{12}$ powders from carbonate precursor *J. Mater. Res.* **15** 1514–23

[28] Manalert R and Rahaman M N 1996 Sol–gel processing and sintering of yttrium aluminum garnet (YAG) powders *J. Mater. Sci.* **31** 3453–58

[29] Yang Q H, Zhang H W, Liu Y L, Wen Q Y and Jia L J 2008 The magnetic and dielectric properties of microwave sintered yttrium iron garnet (YIG) *Mater. Lett.* **62** 2647–50

[30] Sadhana K, Murthy S R and Praveena K 2015 Structural and magnetic properties of Dy^{3+} doped $Y_3Fe_5O_{12}$ for microwave devices *Mater. Sci. Semicond. Process.* **34** 305–11

[31] Grosseau P, Bachiorrini A and Guilhot B 1997 Preparation of polycrystalline yttrium iron garnet ceramics *Powd. Technol.* **93** 247–51

[32] Cruickshank D 2003 1–2 GHz dielectrics and ferrites: overview and perspectives *J. Eur. Ceram. Soc.* **23** 2721–26

[33] Mergen A and Qureshi A 2009 Characterization of YIG nanopowders by mechanochemical synthesis *J. Alloys Compd.* **478** 741–44

[34] Hsu C S, Huang C L, Tseng J F and Huang C Y 2003 Improved high-Q microwave dielectric resonator using CuO-doped $MgNb_2O_6$ ceramics *Mater. Res. Bull.* **38** 1091–99

[35] Huang C L and Chiang K H 2008 Improved high-Q microwave dielectric material using B_2O_3-doped $MgNb_2O_6$ ceramics *Mater. Sci. Eng.* A **474** 243–46

[36] Chaipanich A and Tunkasiri T 2007 Effect of milling time on the properties of $Pb(Mg_{1/3}Nb_{2/3})O_3$ ceramics using the starting precursors PbO and $MgNb_2O_6$ *Curr. Appl Phys.* **7** 281–84

[37] Hong Y K, Park H B and Kim S J 1998 Preparation of powder using a citrate–gel derived columbite $MgNb_2O_6$ precursor and its dielectric properties *J. Eur. Ceram. Soc.* **18** 613–19

[38] Singh K N and Bajpai P K 2010 Synthesis, characterization and dielectric relaxation of phase pure columbite $MgNb_2O_6$: optimization of calcination and sintering *Physica* B **405** 303–12

[39] Ananta S, Brydson R and Thomas N W 1999 Synthesis, formation and characterisation of $MgNb_2O_6$ powder in a Columbite-like phase *J. Eur. Ceram. Soc.* **19** 355–62

[40] Camargo E R, Longo E and Leite E R 2000 Synthesis of ultra-fine Columbite powder $MgNb_2O_6$ by the polymerized complex method *J. Sol–Gel Sci. Technol.* **17** 111–21

[41] Shanker V and Ganguli A K 2003 Comparative study of dielectric properties of $MgNb_2O_6$ prepared by molten salt and ceramic method *Bull. Mater. Sci.* **26** 741–44

[42] Pasrich R and Ravi V 2005 Preparation of nanocrystalline $MgNb_2O_6$ by citrate gel method *Mater. Lett.* **59** 2146–48

[43] Navale S C, Gaikwad A B and Ravi V 2006 Synthesis of $MgNb_2O_6$ by coprecipitation *Mater. Res. Bull.* **41** 1353–56

[44] Kong L B, Ma J, Huang H and Zhang R F 2002 Crystallization of magnesium niobate from mechanochemically derived amorphous *J. Alloys Compd.* **340** L1–4

[45] Lee H J, Hong K S, Kim S J and Kim I T 1997 Dielectric properties of MNb_2O_6 compounds (where M = Ca, Mn, Co, Ni, or Zn) *Mater. Res. Bull.* **32** 847–55

[46] Kim D W, Ko K H and Hong K S Influence of copper (II) oxide additions to zinc niobate microwave ceramics on sintering temperature and dielectric properties *J. Am. Ceram. Soc.* **84** 1286–90

[47] Wee S H, Kim D W and Yoo S I 2004 Microwave dielectric properties of low-fired $ZnNb_2O_6$ ceramics with $BiVO_4$ addition *J. Am. Ceram. Soc.* **87** 871–74

[48] Zhang Y C, Li L T, Yue Z X and Gui Z L 2003 Effects of additives on microstructures and microwave dielectric properties of $ZnNb_2O_6$ ceramics *Mater. Sci. Eng.* B **99** 282–85

[49] Gao F, Liu J J, Hong R Z, Li Z and Tian C S 2009 Microstructure and dielectric properties of low temperature sintered $ZnNb_2O_6$ microwave ceramics *Ceram. Int.* **35** 2687–92

[50] Gururaja T R, Safari A and Halliyal A 1986 Preparation of perovskite PZN–PT ceramic powder near the morphotropic phase boundary *Am. Ceram. Soc. Bull.* **65** 1601–03

[51] Shrout T T and Halliyal A 1987 Preparation of lead-based ferroelectric relaxors for capacitors *Am. Ceram. Soc. Bull.* **66** 704–11

[52] Villegas M, Caballero A C, Moure C, Durán P, Fernández J F and Newnham R E 2000 Influence of processing parameters on the sintering and electrical properties of $Pb(Zn_{1/3}Nb_{2/3})O_3$-based ceramics *J. Am. Ceram. Soc.* **83** 141–46

[53] Dai J H, Zhang C, Shi L Y, Song W W, Wu P W and Huang X 2012 Low-temperature synthesis of $ZnNb_2O_6$ powders via hydrothermal method *Ceram. Int.* **38** 1211–14

[54] Guo L Z, Dai J H, Tian J T, Zhu Z B and He T 2007 Molten salt synthesis of $ZnNb_2O_6$ powder *Mater. Res. Bull.* **42** 2013–16

[55] Wu S Y, Liu X Q and Chen X M 2009 Low temperature synthesis of $ZnNb_2O_6$ fine powders by wet-chemical processes *Ferroelectrics* **388** 114–19

[56] Kong L B, Ma J, Huang H, Zhang R F and Zhang T S 2002 Zinc niobate derived from mechanochemically activated oxides *J. Alloys Compd.* **347** 308–13

[57] Kong L B, Ma J, Zhu W and Tan O K 2001 Reaction sintering of partially reacted system for PZT ceramics via a high-energy ball milling *Scr. Mater.* **44** 345–50

[58] Cava R J 2001 Dielectric materials for applications in microwave communications *J. Mater. Chem.* **11** 54–62

[59] Reaney I M and Iddles D 2006 Microwave dielectric ceramics for resonators and filters in mobile phone networks *J. Am. Ceram. Soc.* **89** 2063–72

[60] McHale A E and Roth R S 1986 Low-temperature phase relationships in the system ZrO_2–TiO_2 *J. Am. Ceram. Soc.* **69** 827–32

[61] Azough F, Freer R and Petzelt J 1993 A Raman spectral characterization of ceramics in the system ZrO_2–TiO_2 *J. Mater. Sci.* **28** 2273–76

[62] Wakino K, Minai K and Tamura H 1983 Microwave characteristics of $(Zr, Sn)TiO_4$ and BaO–PbO–Nd_2O_3–TiO_2 dielectric resonators *J. Am. Ceram. Soc.* **67** 278–81

[63] Kong L B, Ma J, Zhu W and Tan O K 2002 Phase formation and thermal stability of $(Zr_{1-x}Ti_x)O_2$ solid-state solution via a high-energy ball milling process *J. Alloys Compd.* **335** 290–96

[64] Stubičar M, Bermanec V, Stubičar N, Dudrnovski D and Drumes D 2001 Microstructure evolution of an equimolar powder mixture of ZrO_2–TiO_2 during high-energy ball milling and post-annealing *J. Alloys Compd.* **316** 316–20

[65] Park M, Mitchell T E and Heuer A H 1975 Subsolidus equilibria in the TiO_2–SnO_2 system *J. Am. Ceram. Soc.* **58** 43–7

[66] Wolfram G and Gobel H E 1981 Existence range, structural and dielectric properties of $Zr_xTi_ySn_zO_4$ ceramics ($x + y + z = 2$) *Mater. Res. Bull.* **16** 1455–63

[67] Zhao Y *et al* 2010 Surfactant-free synthesis uniform $Ti_{1-x}Sn_xO_2$ nanocrystal colloids and their photocatalytic performance *Appl. Catalysis* B **100** 68–76

[68] Kong L B, Ma J and Huang H 2002 Preparation of the solid solution $Sn_{0.5}Ti_{0.5}O_2$ from an oxide mixture via mechanochemical process *J. Alloys Compd.* **336** 315–19

[69] Naidu H P and Virkar A B 1998 Low-temperature TiO_2–SnO_2 phase diagram using the molten-salt method *J. Am. Ceram. Soc.* **81** 2176–80

[70] Kulshreshtha S K, Sasikala R and Sudarsan V 2001 Non-random distribution of cations in $Sn_{1-x}Ti_xO_2$ ($0.0 \leqslant x \leqslant 1.0$): a ^{119}Sn MAS NMR study *J. Mater. Chem.* **11** 930–35

Lightning Source UK Ltd.
Milton Keynes UK
UKHW050017130220
358597UK00003B/46

9 780750 321891